AF446860

CARDIOVASCULAR BIOLOGY OF PURINES

Developments in Cardiovascular Medicine

198. Antoine Lafont, Eric Topol (eds.): *Arterial Remodeling: A Critical Factor in Restenosis.* 1997   ISBN 0-7923-8008-8
199. Michele Mercuri, David D. McPherson, Hisham Bassiouny, Seymour Glagov (eds.):*Non-Invasive Imaging of Atherosclerosis*   ISBN 0-7923-8036-3
200. Walmor C. DeMello, Michiel J. Janse(eds.): *Heart Cell Communication in Health and Disease*   ISBN 0-7923-8052-5
201. P.E. Vardas (ed.): *Cardiac Arrhythmias Pacing and Electrophysiology.* The Expert View. 1998   ISBN 0-7923-4908-3
202. E.E. van der Wall, P.K. Blanksma, M.G. Niemeyer, W. Vaalburg and H.J.G.M. Crijns (eds.) *Advanced Imaging in Coronary Artery Disease, PET, SPECT, MRI, I VUS, EBCT. 1998*   ISBN 0-7923-5083-9
203. R.L. Wilensky (ed.) *Unstable Coronary Artery Syndromes, Pathophysiology, Diagnosis and Treatment. 1998.*   ISBN 0-7923-8201-3
204. J.H.C. Reiber, E.E. van der Wall (eds.): *What's New in Cardiovascular Imaging?* 1998   ISBN 0-7923-5121-5
205. Juan Carlos Kaski, David W. Holt (eds.): *Myocardial Damage Early Detection by Novel Biochemical Markers. 1998.*   ISBN 0-7923-5140-1
207. Gary F. Baxter, Derek M. Yellon, *Delayed Preconditioning and Adaptive Cardioprotection. 1998.*   ISBN 0-7923-5259-9
208. Bernard Swynghedauw, *Molecular Cardiology for the Cardiologist, Second Edition* 1998.   ISBN 0-7923-8323-0
209. Geoffrey Burnstock, James G.Dobson, Jr., Bruce T.Liang, Joel Linden (eds): *Cardiovascular Biology of Purines*. 1998.   ISBN: 0-7923-8334-6

Previous volumes are still available

KLUWER ACADEMIC PUBLISHERS - DORDRECHT/BOSTON/LONDON

CARDIOVASCULAR BIOLOGY OF PURINES

edited by

GEOFFREY BURNSTOCK, D. Sc.
Autonomic Neuroscience Institute
Royal Free and University College Medical School, London, UK

JAMES G. DOBSON, JR., Ph.D.
University of Massachusetts, USA

BRUCE T. LIANG, M.D.
University of Pennsylvania, USA

JOEL LINDEN, Ph.D.
University of Virginia, USA

Distributors for North, Central and South America:
Kluwer Academic Publishers
101 Philip Drive
Assinippi Park
Norwell, Massachusetts 02061 USA
Telephone (781) 871-6600
Fax (781) 871-6528
E-Mail <kluwer@wkap.com>

Distributors for all other countries:
Kluwer Academic Publishers Group
Distribution Centre
Post Office Box 322
3300 AH Dordrecht, THE NETHERLANDS
Telephone 31 78 6392 392
Fax 31 78 6546 474
E-Mail <services@wkap.nl>
Electronic Services <http://www.wkap.nl>

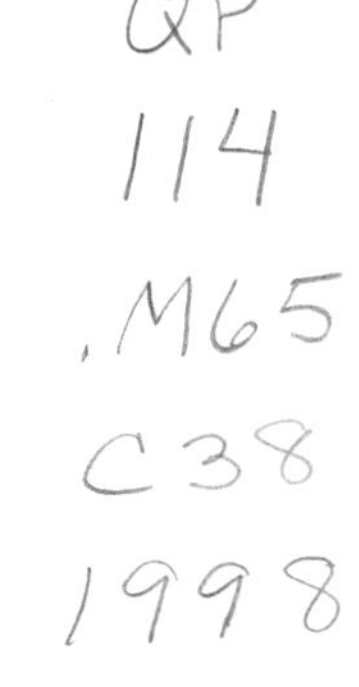

 **Library of Congress Cataloging-in-Publication Data**

A C.I.P. Catalogue record for this book is available from the
Library of Congress.

Cover Figure:
This is a high magnification (x280) of intrarenal arteries which demonstrate the $P2X_1$ immunoreactivity on the vascular smooth muscle cells on IA (interlobular artery) and AA (afferent arteriole).
Reprinted with permission from Chan, C.M., Unwin, R.J., Bardini, M., Oglesby, I.B., Ford, A.P.D.W., Townsend-Nicholson, A. & Burnstock, G. (1998). Localization of the $P2X_1$ purinoceptors by autoradiography and immunohistochemistry in rat kidneys. *Am. J. Physiol.* 274, F799-F804 American Physiological Society.

LIST OF CONTRIBUTORS

ALLGAIER, Clemens, Institut für Pharmakologie und Toxikologie, Universitat Leipzig, Universtatsklinikum, Hartelstrasse 16-18, 04107 Leipzig, Germany

APASOV, Sergey, Laboratory of Immunology, NIAID, NIH Building 10, Room 11N-311, National Institutes of Health, Bethesda, MD 20892-0001, USA

ARMOUR, J. Andrew, Department of Physiology and Biophysics, Dalhousie University, Sir Charles Tupper Bldg., Halifax, Nova Scotia, Canada B3H 4H7

ARMSTRONG, John, Laboratory of Immunology, NIAID, NIH Building 10, Room 11N-311, National Institutes of Health, Bethesda, MD 20892-0001, USA

DECKING, Ulrich, Department of Cardiovascular Physiology, Heinrich-Heine Institut für Herz- und Kteislaufphysiologie, Universität Dusseldorf, Universitatsstraße 1, D-40225 Dusseldorf, Germany

DI VIRGILIO, Francesco, Department of Experimental and Diagnostic Medicine, Instituto Medicine, Instituto di Patoligia Generale, Universita degli Studi di Ferrara, Via L. Borsari, 46, I-44100 Ferrara, Italy

DOBSON, James G., Jr., Department of Physiology, University of Massachusetts Medical Center, 55 Lake Avenue North, Worcester, MA 01655-0127, USA

FENTON, Richard A., Department of Physiology, University of Massachusetts Medical Center, 55 Lake Avenue North, Worcester, MA 01655-0127, USA

FALZONI, Simonetta, Department of Experimental and Diagnostic Medicine Instituto di Patoligia Generale, Universita degli Studi di Ferrara, Via L. Borsari, 46, I-44100 Ferrara, Italy

GRANGER, Harris J., Department of Medical Physiology, Texas A&M University, College of Medicine, 336 Reynolds Medical Building, College Station, TX 77843-1114, USA

GREEN, David W., Cardiovascular Drug Discovery, Bristol-Myers Squibb Pharmaceutical Institute, Route 206 and Providence Road, Princeton, NJ 08543-4000, USA

GROVER, Gary J., Cardiovascular Drug Discovery, Bristol-Myers Squibb Pharmaceutical Institute, Route 206 and Providence Road, Princeton, NJ 08543-4000, USA

HARDEN, T. Kendall, Department of Pharmacology, University of North Carolina, CB#7365, Chapel Hill, NC 27599, USA

HEADRICK, John P., Rotary Center for Cardiovascular Research, Griffith University Gold Coast Campus, Southport, QLD, Australia 4217

HOOD, John, Department of Medical Physiology, Texas A&M University, College of Medicine, 336 Reynolds Medical Building, College Station, TX 77843-1114, USA

ILLES, Peter, Institut für Pharmakologie und Toxikologie, Universität Leipzig, Universtatsklinikum, Hartelstrasse 16-18, 04107 Leipzig, Germany

JACOBSON, Kenneth A., NIDDK, Room B1A-17, Building 8A, National Institutes of Health, Bethesda, MD 20892-1008, USA

KENNEDY, Charles, Department of Physiology and Pharmacology, University of Strathclyde, Royal College, 204 George Street, Glasgow, Scotland G1XW

KIEHN, Jasper, Pharmakologisches Institut, Universitats-Krankenhaus Eppendorf, Martinistraße 52, D-20246 Hamburg, Germany

KING, Brian F., Autonomic Neuroscience Institute, Royal Free Hospital School of Medicine, Rowland Hill Street, Hampstead, London NW3 2PF, United Kingdom

KOSHIBA, Masahiro, Laboratory of Immunology, NIAID, NIH Building 10, Room 11N-311, National Institutes of Health, Bethesda, MD 20892-0001, USA

LERMAN, Bruce B., Division of Cardiology, Department of Medicine, The New York Hospital, Cornell University Medical Center, Starr-4, 525 East 68th Street, New York, NY 10021, USA

LIANG, Bruce T., Department of Medicine, Cardiovascular Division, 504 Johnson Pavilion, University of Pennsylvania Medical Center, 3610 Hamilton Walk, Philadelphia, PA 19104, USA

LINDEN, Joel, Department of Medicine and Molecular Physiology, University of Virginia, Box 158, Charlottesville, VA 22908-0001, USA

LORBAR, Mojca, Department of Physiology, University of Massachusetts Medical Center, 55 Lake Avenue North, Worcester, MA 01655-0127, USA

MARTIN, Pauline L., Discovery Therapeutics, Inc., Richmond, VA 23230, USA

MATHERNE, G. Paul, Department of Pediatrics, University of Virginia, MR-4, Box 14, Charlottesville, VA 22908-0001, USA

MEININGER, Cynthia, Department of Medical Physiology, Texas A&M University, University, College of Medicine, 345 Reynolds Medical Building, College Station, TX 77843-1114, USA

NEUMANN, Joachim, Institut für Pharmakologie und Toxikologie, Westfälische Westfälische Wilhelms-Universität Munster, Domagkstraße 12, D-48129 Münster, Germany

OLSSON, Ray A., Department of Medicine, MDC Box 19, University of South Frida, Florida College of Medicine, 12901 Bruce B. Downs Blvd., Tampa, FL 33612-4742, USA

PEARSON, Jeremy D., King's College, University of London, United Kingdom

RALEVIC, Vera, School of Biomedical Sciences, Queen's Medical Centre, University Nottingham, Nottingham, NG7, 2UH, United Kingdom

RUBINO, Annalisa, Autonomic Neuroscience Institute, Royal Free Hospital School of Medicine, Rowland Hill Street, Hampstead, London NW3 2PF, United Kingdom

SCHRADER, Jürgen, Department of Cardiovascular Physiology, Heinrich-Heine Institut für Herz- und Kteislaufphysiologie, Universität Dusseldorf, Universitatsstraße 1, D-40225 Dusseldorf, Germany

SITKOVSKY, Michail, Laboratory of Immunology, NIAID, NIH Building 10, Room 11N-311, National Institutes of Health, Bethesda, MD 20892-0001, USA

SLOTWINER, David J., Division of Cardiology, Department of Medicine, The New York Hospital - Cornell Medical Center, Starr-4, 525 East 68th Street, New York, NY 10021, USA

STEIN, Birgitt, Pharmakologisches Institut, Universitats-Krankenhaus Eppendorf Matinistraße 52, D-20246 Hamburg, Germany

STUMPE, Thomas, Department of Cardiovascular Physiology, Heinrich-Heine Institut für Herz-und Kteislaufphysiologie, Universität Dusseldorf, Universitatsstraße 1, D-40225 Dusseldorf, Germany

SULLIVAN, Gail, Department of Internal Medicine, Box 6012, MR-4 Annex 2126, Health Science Center, University of Virginia, Charlottesville, VA 22098, USA

ZICHE, Marina, Department of Pharmacology, University of Florence, Florence, Italy

ZIMMERMANN, Herbert, Biozentrum der Universität, Frankfurt/Main, Germany

Preface

Following many years when a great deal of attention was directed towards the intracellular roles of purines, there is exploding interest in the field of extracellular purinergic signalling. In this book we focus on the actions of purines in cardiovascular biology where it is clear that they play major roles in both normal and pathophysiological conditions. Activation of different purinoceptor subtypes by purines can regulate cardiac contractility and electrical activity, modulate catecholamine-mediated responses both pre- and post-junctionally, trigger and mediate ischaemic preconditioning, cause vasodilatation and vasoconstriction and enhance endothelial proliferation and apoptosis as well as inhibit platelet and neurotrophil function. The book covers the cardiovascular actions mediated by the major P1 and P2 subclasses of purinoceptors and emphasizes the interactions between these two signalling systems. The book covers topics ranging from molecular and cellular to systemic and clinical. It also aims to highlight how basic advances have led to the identification of novel targets for cardiovascular therapeutic developments. We hope that our book will prove to be timely and helpful.

Geoffrey Burnstock

James G. Dobson, Jr

Bruce T. Liang

Joel Linden

MOLECULAR BIOLOGY AND PHARMACOLOGY OF RECOMBINANT ADENOSINE RECEPTORS

Joel Linden, University of Virginia, USA; Kenneth A. Jacobson, NIDDK, National Institutes of Health, USA

I. SUMMARY

The physiological effects of adenosine are mediated by four homologous G protein coupled receptors, A_1, A_{2A}, A_{2B} and A_3, that have been cloned from multiple species. There is no difference in the structure of receptor subtypes cloned from different tissues of the same species but there are substantial differences in the pharmacological properties of receptors between species. There have been recent significant advances in the development of receptor subtype-selective compounds, particularly A_{2A} and A_3 antagonists. The anti-asthma drug, enprofylline, is a weak but somewhat selective antagonist of A_{2B} receptors, and inosine is a weak, but selective agonist of rodent A_3 receptors. The A_1 and A_3 receptors both couple to inhibitory Gi/o G proteins, but can activate phospholipase C via G protein $\beta\gamma$ subunits, but the A_1 receptor desensitizes much more slowly. The A_{2A} and A_{2B} receptors both couple to Gs, but the A_{2B} receptor also can couple to Gq to mobilize calcium.

II. BACKGROUND

Tissues responses to adenosine and adenine nucleotides have been noted for decades (Drury and Szent-Gyorgyi, 1929). Based on differences in the responses of various tissues the purines, Burnstock proposed that there are distinct receptors that bind adenosine or ATP designated P1 and P2 receptors, respectively (Burnstock, 1978). P1 or adenosine receptors consist of four G protein coupled receptor subtypes, A_1, A_{2A}, A_{2B} and A_3. The P2 receptors bind adenine nucleotides, uracil nucleotides, and/or diadenosine polyphosphates (Miras-Portugal *et al.*, 1996) and are composed of at least 11 subtypes that comprise both ligand-gated channels (P2X) and G protein coupled receptors (P2Y). All four of the known adenosine receptor subtypes have been detected multiple times and deposited in the Genbank-National Center for Biotechnology Information (NCBI) expressed sequence tag database. This most probably indicates that all of the adenosine receptor subtypes have now been cloned, although it is possible that there are additional receptor transcripts that are either very rare or only distantly related to the known receptors.

Despite improved pharmacological tools, determining which adenosine receptor subtype or subtypes mediates a particular tissue responses is still difficult in some circumstances. There are several reasons for this. Many, if not most tissues express multiple subtypes of adenosine receptors. In addition, there are substantial differences in the pharmacological properties of the various receptor subtypes between species (Ukena *et al.*, 1986; Tucker *et al.*, 1994; Linden, 1994). Species differences are especially large in the case of the A_3 adenosine receptor antagonists (Linden *et al.*, 1993). Compounds that are touted as selective agonists or antagonists may be highly selective for one subtype of receptor over another, e.g. A_1 over A_{2A}, but not as selective if the other receptor subtypes are included in the comparison. As an example, 8-cyclopentyl-1,3-dipropylxanthine (CPX) is very widely used as an "A_1 selective" antagonist. This is based on the high selectivity of this compound for A_1 over A_{2A} receptors. However, if A_{2B} receptors are included in a comparison of human receptors, CPX has only moderate selectivity (Linden *et al.*, 1993). Another factor that can contribute to difficulty in determining which receptor subtype mediates a particular physiologic response stems from the fact that the potency of agonists in functional responses depends in part on receptor density. This may explain why coronary arteries, which apparently have a large receptor reserve, appear to be exquisitely sensitive to A_{2A} agonists (Martin *et al.*, 1997). Hence a selective agonist of one receptor subtype may be a potent agonist in an assay of functional responses mediated by another. Hence, there continues to be a pressing need to develop improved selective ligands that distinguish between adenosine receptor subtypes and to better characterize the pharmacological properties of various compounds in multiple species.

The ability to screen and characterize new compounds has been greatly facilitated by the cloning and expression of recombinant adenosine receptors from various species, including human. This chapter summarizes the properties of adenosine receptor subtypes cloned from multiple species and recent developments in the production of selective ligands. Cell lines that express recombinant human or rat adenosine receptors are now widely available. However, physiologic studies of cardiovascular responses to adenosine continue to be carried out in a number of other animal species, particularly rabbit, mouse, guinea pig and dog. Hence, the most sure way to pharmacologically characterize adenosine receptor mediated responses in a given species is to clone and express and characterize individual receptors from that species. This point is illustrated by the recent demonstration that the A_3 adenosine receptor mediates ischemic preconditioning in the rabbit (Tracey *et al.*, 1997), a conclusion that was made possible by cloning and characterization of recombinant rabbit A_1 and A_3 adenosine receptors (Wang *et al.*, 1997; Hill *et al.*, 1997). Recent advances in the molecular biology and pharmacology of adenosine receptors are of great utility to the cardiovascular investigator. For additional information the reader is referred to several recent review articles (Van Rhee and Jacobson, 1996; Shryock and Belardinelli, 1997; Feoktistov and Biaggioni, 1997; Jacobson, 1998).

III. TEXT

A. Adenosine receptor clones

The table below is a summary of adenosine receptor subtypes and Genbank accession numbers of the adenosine receptor that have been cloned to date from various species. Within a given species there are no convincing differences in the coding sequences of cDNA's derived from different tissues.

Species	Tissue	Genbank #	Reference
A_1			
Chick	adipose	U28380, S78192	(Aguilar *et al.*, 1995)
Cow	brain	X63592	(Tucker *et al.*, 1992)
Cow	brain	M86261	(Olah *et al.*, 1992)
Dog	thyroid	X14051	(Libert *et al.*, 1989)
G. pig	brain	G507154	(Meng *et al.*, 1994)
Human	hippocampus	S45235	(Libert *et al.*, 1992)
Human	hippocampus	S56143	(Townsend-Nicholson and Shine, 1992)
Human	hippocampus	L22214	(Ren and Stiles, 1994)
Human	heart, brain, kidney	X68485	-
Human	heart	AB004662	-
Mouse	mast cell	U05671	(Marquardt *et al.*, 1994)
Rabbit	kidney	L01700	(Bhattacharya *et al.*, 1993)
Rat	brain	M64299	(Mahan *et al.*, 1991)
Rat	brain	M69045	(Reppert *et al.*, 1991)
Rat	neonatal	Y12519	(Pfaff and Karschin, 1997)
Rat	heart	AB001089	-
Rat	gut	AF042079	-
A_{2A}			
Dog	thyroid	X14052	(Libert *et al.*, 1989)
G. pig	brain	U04201	(Meng *et al.*, 1994)
G. pig	leukocyte	D63674	(Hirano *et al.*, 1996)
Human	hippocampus	S46950	(Furlong *et al.*, 1992)
Human	heart, brain	X68486	-
Human	HL60	M97370	-
Human	genomic	U40770, U40771	(Le *et al.*, 1996)
Mouse	mast cell	U05672	(Marquardt *et al.*, 1994)
Mouse	genomic	Y13346	(Ledent *et al.*, 1997)
Rat	hypothalamus	L08102	(Fink *et al.*, 1992)
Rat	brain	M91214	(Chern *et al.*, 1992)
A_{2B}			
Chicken	embryonic	Y12601	(Worpenberg *et al.*, 1997)
Human	hippocampus	M97759	(Pierce *et al.*, 1992)
Human	brain	X68487	(Jacobson *et al.*, 1995)
Mouse	mast cell	U05673	(Marquardt *et al.*, 1994)
Mouse	embryonic	AA438195, AA104569	Expressed Sequence Tags
Rat	hypothalamus	M91466	(Stehle *et al.*, 1992)
A_3			
Dog	mast cell	U54792	(Auchampach *et al.*, 1997)
Human	striatum	L22607	(Salvatore *et al.*, 1993)
Human	heart	L20463	(Sajjadi and Firestein, 1993)
Human	testis	X76981	(Meyerhof *et al.*, 1994)
Human	genomic	L77729, L77730	-

Mouse	testis	L20331	(Wilkie *et al.*, 1993)
Mouse	mast cell	L05673	(Marquardt *et al.*, 1994)
Rabbit	lung	U90718	(Hill *et al.*, 1997)
Rat	testis	X59249	(Meyerhof *et al.*, 1991)
Rat	brain	M94152	(Zhou *et al.*, 1992)
Sheep	pars tuberalis	S65334	(Linden *et al.*, 1993)

B. Adenosine receptor structure

All adenosine receptors belong to the superfamily of receptors that span the plasma membrane seven times and signal via trimeric G proteins. Below is shown an amino acid alignment of all adenosine receptor subtypes from all species that have been cloned to date. The transmembrane segments are relatively highly conserved and consist primarily of hydrophobic amino acids, as indicated on the alignment. Ligands are thought to bind within a cleft formed by a circular array of the transmembrane binding segments and interact with amino acids in TM 3, 5, 6 and 7(Van Rhee and Jacobson, 1996).

C. Adenosine receptor coupling and regulation

A_1 and A_3 adenosine receptors both are coupled to inhibitory G proteins, Gi/o. Coupling of A_1 receptors to Gi proteins results not only in inhibition of adenylyl cyclase, but also in the modulation of ion channel conductance's. In fact, the A_1 adenosine receptor has recently been expression cloned from rat cerebellar cells by coupling to Kir channels (Pfaff and Karschin, 1997). A_1 and A_3 adenosine receptors also both can couple to stimulation of phospholipase C, probably due to the release of $\beta\gamma$ subunits from activated G proteins, inasmuch as these responses are inhibited by pertussis toxin (White *et al.*, 1992; Ramkumar *et al.*, 1993). A significant difference between these two receptors is that the A_3 receptor desensitizes within minutes, whereas the A_1 receptors desensitize slowly over the course of hours (Palmer and Stiles, 1997). The A_{2A} and A_{2B} adenosine receptors undergo desensitization with an intermediate $t_{1/2}$ of 15-20 minutes. These A_2 desensitization responses appears to be mediated in part by phosphorylation of the receptors with a G protein-coupled receptor kinase since the rate of desensitization is attenuated in NG108 cells transfected with a dominant negative mutant of GRK2 (Mundell *et al.*, 1997).

It has long been know that A_{2A} and A_{2B} receptors are coupled to the stimulatory G protein, Gs. Recent evidence indicates that A_{2B} receptors are dually coupled to Gs and Gq/11since activation of A_{2B} receptors activates phospholipase C and mobilizes calcium via a pertussis toxin-insensitive mechanism. The degranulation of canine mast cells appears to be mediated by the calcium-mobilizing action of the A_{2B} receptor (Auchampach *et al.*, 1997). Another recently discovered interesting feature of A_{2B} receptors is the finding that their transcription is increased in myelomonocytic cells that have been stimulated to undergo transformation (Worpenberg *et al.*, 1997).

```
                    1                                                     50
       Rat_A2B      ~~~~~~~MQL ETQDALYVAL ELVIAALAVA GNVLVCAAVG ASSALQTPTN
     Mouse_A2B      ~~~~~~~MQL ETQDALYVAL ELVIAALAVA GNVLVCAAVG ASSALQTPTN
     Human_A2B      ~~~~~~~MLL ETQDALYVAL ELVIAALSVA GNVLVCAAVG TANTLQTPTN
       Rat_A2A      ~~~~~~~~~~ ~MGSSVYITV ELAIAVLAIL GNVLVCWAVW INSNLQNVTN
     Mouse_A2A      ~~~~~~~~~~ ~MGSSVYIMV ELAIAVLAIL GNVLVCWAVW INSNLQNVTN
     Human_A2A      ~~~~~~~~MP IMGSSVYITV ELAIAVLAIL GNVLVCWAVW LNSNLQNVTN
       Dog_A2A      ~~~~~~~~MS TMGSWVYITV ELAIAVLAIL GNVLVCWAVW LNSNLQNVTN
  Guinea_pig_A2A    ~~~~~~~~~~ ~MSSSVYITV ELVIAVLAIL GNVLVCWAVW INSNLQNVTN
       Human_A1     ~~~~~MPPSI SAFQAAYIGI EVLIALVSVP GNVLVIWAVK VNQALRDATF
   Guinea_pig_A1    ~~~~~MPHSV SAFQAAYIGI EVLIALVSVP GNVLVIWAVK VNQALRDATF
         Rat_A1     ~~~~~MPPYI SAFQAAYIGI EVLIALVSVP GNVLVIWAVK VNQALRDATF
         Cow_A1     ~~~~~MPPSI SAFQAAYIGI EVLIALVSVP GNVLVIWAVK VNQALRDATF
         Dog_A1     ~~~~~MPPAI SAFQAAYIGI EVLIALVSVP GNVLVIWAVK VNQALRDATF
      Rabbit_A1     ~~~~~MPPSI SAFQAAYIGI EVLIALVSVP GNVLVIWAVK VNQALRDATF
       Chick_A1     ~~~~~MAQSV TAFQAAYISI EVLIALVSVP GNILVIWAVK MNQALRDATF
       Human_A3     ~MPNNSTALS LA.NVTYITM EIFIGLCAIV GNVLVICVVK LNPSLQTTTF
         Dog_A3     ~MAVNGTALL LA.NVTYITV EILIGLCAIV GNVLVIWVVK LNPSLQTTTF
       Sheep_A3     ~MPVNSTAVS WT.SVTYITV EILIGLCAIV GNVLVIWVVK LNPSLQTTTF
      Rabbit_A3     ~MPDNSTTLF LAIRASYIVF EIVIGVCAVV GNVLVIWVIK LNPSLKTTTF
         Rat_A3     MKANNTTTSA LWLQITYVTM EAAIGLCAVV GNMLVIWVVK LNRTLRTTTF
       Consensus    ~~~~~~~~~~ ~~~~~~Y~~~ E~~I~~~~~~ GN~LV~~~~~
                    ~~~~L~~~T~
                    51                                                    100

       Rat_A2B      YFLVSLATAD VAVGLFAIPF AITISLGFCT DFHSCLFLAC FVLVLTQSSI
     Mouse_A2B      YFLVSLATAD VAVGLFAIPF AITISLGFCT DFHGCLFLAC FVLVLTQSSI
     Human_A2B      YFLVSLAAAD VAVGLFAIPF AITISLGFCT DFYGCLFLAC FVLVLTQSSI
       Rat_A2A      FFVVSLAAAD IAVGVLAIPF AITISTGFCA ACHGCLFFAC FVLVLTQSSI
     Mouse_A2A      FFVVSLAAAD IAVAVLAIPF AITISTGFCA ACHGCLFFAC FVLVLTQSSI
     Human_A2A      YFVVSLAAAD IAVGVLAIPF AITISTGFCA ACHGCLFIAC FVLVLTQSSI
       Dog_A2A      YFVVSLAAAD IAVGVLAIPF AITISTGFCA ACHNCLFFAC FVLVLTQSSI
  Guinea_pig_A2A    YFVVSLAAAD IAVGVLAIPF AITISTGFCA ACHGCLFFAC FVLVLTQSSI
       Human_A1     CFIVSLAVAD VAVGALVIPL AILINIGPQT YFHTCLMVAC PVLILTQSSI
   Guinea_pig_A1    CFIASLAVAD VAVGALVIPL AILINIGPQT YFHTCLMVAC PVLILTQSSI
         Rat_A1     CFIVSLAVAD VAVGALVIPL AILINIGPQT YFHTCLMVAC PVLILTQSSI
         Cow_A1     CFIVSLAVAD VAVGALVIPL AILINIGPRT YFHTCLKVAC PVLILTQSSI
         Dog_A1     CFIVSLAVAD VAVGALVIPL AILINIGPRT YFHTCLMVAC PVLILTQSSI
      Rabbit_A1     CFIVSLAVAD VAVGALVIPL AILINIGPET YFHTCLMVAC PVLILTQSSI
       Chick_A1     CFIVSLAVAD VAVGALVIPL AIIINIGPQT EFYSCLMMAC PVLILTESSI
       Human_A3     YFIVSLALAD IAVGVLVMPL AIVVSLGITI HFYSCLFMTC LLLIFTHASI
         Dog_A3     YFIVSLALAD IAVGVLVMPL AIVISLGITI QFYNCLFMTC LLLIFTHASI
       Sheep_A3     YFIVSLALAD IAVGVLVMPL AIVISLGVTI HFYSCLFMTC LMLIFTHASI
      Rabbit_A3     YFIFSLALAD TAVGFLVMPL AIVISLGITI GFYSCLVMSC LLLVFTHASI
         Rat_A3     YFIVSLALAD IAVGVLVIPL AIAVSLEVQM HFYACLFMSC VLLVFTHASI
       Consensus    ~F~~SLA~AD ~AV~~~~~P~ AI~~~~~~~~ ~~~~CL~~~C
                    ~~L~~T~~SI
                    101                                                   150

       Rat_A2B      FSLLAVAVDR YLAIRVPLRY KGLVTGTRAR GIIAVLWVLA FGIGLTPFLG
     Mouse_A2B      FSLLAVAVDR YLAIRVPLRY KGLVTGTRAR GIIAVLWVLA FGIGLTPFLG
     Human_A2B      FSLLAVAVDR YLAICVPLRY KSLVTGTRAR GVIAVLWVLA FGIGLTPFLG
       Rat_A2A      FSLLAIAIDR YIAIRIPLRY NGLVTGVRAK GIIAICWVLS FAIGLTPMLG
     Mouse_A2A      FSLLAIAIDR YIAIRIPLRY NGLVTGMRAK GIIAICWVLS FAIGLTPTAG
     Human_A2A      FSLLAIAIDR YIAIRIPLRY NGLVTGTRAK GIIAICWVLS FAIGLTPMLG
       Dog_A2A      FSLLAIAIDR YIAIRIPLRY NGLVTGTRAK GIIAVCWVLS FAIGLTPMLG
  Guinea_pig_A2A    FSLLTITIDR YIAIRIPLRY NGLVTCTRAK GIIAICWVLS FAIGLTPMLG
       Human_A1     LALLAIAVDR YLRVKIPLRY KMVVTPRRAA VAIAGCWILS FVVGLTPMFG
   Guinea_pig_A1    LALLAIAVDR YLRVKIPLRY KTVVTPRRAA VAIAGCWILS LVVGLTPMFG
         Rat_A1     LALLAIAVDR YLRVKIPLRY KTVVTQRRAA VAIAGCWILS LVVGLTPMFG
         Cow_A1     LALLAIAVDR YLRVKIPLRY KTVVTPRRAV VAITGCWILS FVVGLTPMFG
         Dog_A1     LALLAIAVDR YLRVKIPLRY KTVVTPRRAA VAIAGCWILS FVVGLTPLFG
      Rabbit_A1     LALLAIAVDR YLRVKIPLRY KAVVTPRRAA VAIAGCWILS LVVGLTPMFG
       Chick_A1     LALLAIAVDR YLRVKIPVRY KSVVTPRRAA VAIACCWIVS FLVGLTPMFG
```

5

```
   Human_A3   MSLLAIAVDR YLRVKLTVRY KRVTTHRRIW LALGLCWLVS FLVGLTPMFG
     Dog_A3   MSLLAIAVDR YLRVKLTVRY RRVTTQRRIW LALGLCWLVS FLVGLTPMFG
   Sheep_A3   MSLLAIAVDR YLRVKLTVRY RRVTTQRRIW LALGLCWLVS FLVGLTPMFG
  Rabbit_A3   MSLLAIAVDR YLRVKLTVRY RRVTTQRRIW LALGLCWVVS LLVGFTPMFG
     Rat_A3   MSLLAIAVDR YLRVKLTVRY RTVTTQRRIW LFLGLCWLVS FLVGLTPMFG
  Consensus   --LL----DR Y-------RY ----T--R-- ------W---
             ---G-TP--G


           151                                                200

    Rat_A2B   WNSKDRATSN CTEPGDGITN KSCC...PVK CLFENVVPMS YMVYFNFFGC
  Mouse_A2B   WNSKDSATSN CTELGDGIAN KSCC...PLT CLFENVVPMS YMVYFNFFGC
  Human_A2B   WNSKDSATNN CTEPWDGTTN ESCC...LVK CLFENVVPMS YMVYFNFFGC
    Rat_A2A   WN.......N CSQ..KDGNS TKTCGEGRVT CLFEDVVPMN YMVYYNFFAF
  Mouse_A2A   WN.......N CSQ..KDENS TKTCGEGRVT CLFEDVVPMN YMVYYNFFAF
  Human_A2A   WN.......N CGQPKEGKNH SQGCGEGQVA CLFEDVVPMN YMVYFNFFAC
    Dog_A2A   WN.......N CSQPKEGRNY SQGCGEGQVA CLFEDVVPMN YMVYYNFFAF
Guinea_pig_A2A WN.......N CSQPKGDKNH SESCDEGQVT CLFEDVVPMN YMVYYNFFAF
   Human_A1   WNNLSAVER. ......AWAA NGSMGEPVIK CEFEKVISME YMVYFNFFVW
Guinea_pig_A1  WNNLSKIEM. ......AWAA NGSVGEPVIK CEFEKVISME YMVYFNFFVW
     Rat_A1   WNNLSVVEQ. ......DWRA NGSVGEPVIK CEFEKVISME YMVYFNFFVW
     Cow_A1   WNNLSAVER. ......DWLA NGSVGEPVIE CQFEKVISME YMVYFNFFVW
     Dog_A1   WNRLGEAQR. ......AWAA NGSGGEPVIK CEFEKVISME YMVYFNFFVW
  Rabbit_A1   WNNLREVQR. ......AWAA NGSVGEPVIK CEFEKVISME YMVYFNFFVW
   Chick_A1   WNNLNKVLG. ......TRDL NVSHSEFVIK CQFETVISME YMVYFNFFVW
   Human_A3   WNMKLTSEY. ......HRNV T......FLS CQFVSVMRMD YMVYFSFLTW
     Dog_A3   WNMKLTSEH. ......QRNV T......FLS CQFSSVMRMD YMVYFSFFTW
   Sheep_A3   WNMKLSSA.. ......DENL T......FLP CRFRSVMRMD YMVYFSFFLW
  Rabbit_A3   WNMKPTLES. ......ARNY S......DFQ CKFDSVIPME YMVFFSFFTW
     Rat_A3   WNRKVTLEL. ......SQNS S......TLS CHFRFVVGLD YMVFFSFITW
  Consensus   WN-------- ---------- ---------- C-F--V----
             YMV---F---


           201                                                250
    Rat_A2B   VLPPLLIMMV IYIKIFMVAC KQLQHMELM. ...EHSRTTL QREIHAAKSL
  Mouse_A2B   VLPPLLIMLV IYIKIFMVAC KQLQSMELM. ...DHSRTTL QREIHAAKSL
  Human_A2B   VLPPLLIMLV IYIKIFLVAC RQLQRTELM. ...DHSRTTL QREIHAAKSL
    Rat_A2A   VLLPLLLMLA IYLRIFLAAR RQLKQMESQP LPGERTRSTL QKEVHAAKSL
  Mouse_A2A   VLLPLLLMLA IYLRIFLAAR RQLPQMESQP LPGERTRSTL QKEVHAAKSL
  Human_A2A   VLVPLLLMLG VYLRIFLAAR RQLKQMESQP LPGERARSTL QKEVHAAKSL
    Dog_A2A   VLVPLLLMLG VYLRIFLAAR RQLKQMESQP LPGERARSTL QKEVHAAKSL
Guinea_pig_A2A VLVPLLLMLG IYLRIFLAAR RQLKQMESQP LPGERTRSTL QKEVHPAKSL
   Human_A1   VLPPLLLMVL IYLEVFYLIR KQLNKKVS.. ASSGDPQKYY GKELKIAKSL
Guinea_pig_A1  VLPPLLLMVL IYLEVFYLIR KQLSKKVS.. ASSGDPQKYY GKELKIAKSL
     Rat_A1   VLPPLLLMVL IYLEVFYLIR KQLNKKVS.. ASSGDPQKYY GKELKIAKSL
     Cow_A1   VLPPLLLMVL IYMEVFYLIR KQLSKKVS.. ASSGDPQKYY GKELKIAKSL
     Dog_A1   VLPPLLLMVL IYLEVFYLIR RQLGKKVS.. ASSGDPQKYY GKELKIAKSL
  Rabbit_A1   VLPPLLLMVL IYLEVFYLIR RQLSKKAS.. ASSGDPHKYY GKELKIAKSL
   Chick_A1   VLPPLLLMLL IYLEVFNLIR TQLNKKVS.. SSSNDPQKYY GKELKIAKSL
   Human_A3   IFIPLVVMCA IYLDIFYIIR NKLSLNLS.. .NSKETGAFY GREFKTAKSL
     Dog_A3   ILIPLVVMCA IYLDIFYVIR NKLNQNFS.. .SSKETGAFY GREFKTAKSL
   Sheep_A3   ILVPLVVMCA IYFDIFYIIR NRLSQSFS.. .GSRETGAFY GREFKTAKSL
  Rabbit_A3   ILIPLLLMCA LYVYIFYIIR NKLVQSFS.. .SFKETGAFY RREFKTAKSL
     Rat_A3   ILIPLVVMCI IYLDIFYIIR NKLSQNLT.. .GFRETRAFY GREFKTAKSL
  Consensus   ---PL-M-- -Y---F---- --L------- ----------
             --E---AKSL


           251                                                300
    Rat_A2B   AMIVGIFALC WLPVHAINCI TLFHPALAKD KPKWVMNVAI LLSHANSVVN
  Mouse_A2B   AMIVGIFALC WLPVHAINCI TLFHPALAKD KPKWVMNVAI LLSHANSVVN
  Human_A2B   AMIVGIFALC WLPVHAVNCV TLFQPAQGKN KPKWAMNMAI LLSHANSVVN
    Rat_A2A   AIIVGLFALC WLPLHIINCF TFFCST.CRH APPWLMYLAI ILSHSNSVVN
  Mouse_A2A   AIIVGLFALC WLPLHIINCF TFFCST.CQH APPWLMYLAI ILSHSNSVVN
  Human_A2A   AIIVGLFALC WLPLHIINCF TFFCPD.CSH APLWLMYLAI VLSHTNSVVN
    Dog_A2A   AIIVGLFALC WLPLHIINCF TFFCPE.CSH APLWLMYLTI VLSHTNSVVN
```

6

```
Guinea_pig_A2A  AIIVGLFALC CLPLNIINCF TFFCPE.CDH APPWLMYLTI ILSHGNSVVN
      Human_A1  ALILFLFALS WLPLHILNCI TLFCPSC..H KPSILTYIAI FLTHGNSAMN
Guinea_pig_A1   ALILFLFALS WLPLHILNCI TLFCPTC..H KPTILTYIAI FLTHGNSAMN
        Rat_A1  ALILFLFALS WLPLHILNCI TLFCPTC..Q KPSILIYIAI FLTHGNSAMN
        Cow_A1  ALILFLFALS WLPLHILNCI TLFCPSC..H MPRILIYIAI FLSHGNSAMN
        Dog_A1  ALILFLFALS WLPLHILNCI TLFCPSC..R KPSILMYIAI FLTHGNSAMN
     Rabbit_A1  ALILFLFALS WLPLHILNCV TLFCPSC..Q KPSILVYTAI FLTHGNSAMN
      Chick_A1  ALVLFLFALS WLPLHILNCI TLFCPSC..K TPHILTYIAI FLTHGNSAMN
      Human_A3  FLVLFLFALS WLPLSIINCI IYF..NG..E VPQLVLYMGI LLSHANSMMN
        Dog_A3  FLVLFLFAFS WLPLSIINCI TYF..HG..E VPQIILYLGI LLSHANSMMN
      Sheep_A3  LLVLFLFALC WLPLSIINCI LYF..DC..Q VPQTVLYLGI LLSHANSMMN
     Rabbit_A3  FLVLALFAGC WLPLSIINCV TYF..KC..K VPDVVLLVGI LLSHANSMMN
        Rat_A3  FLVLFLFALC WLPLSIINFV SYF..NV..K IPEIAMCLGI LLSHANSMMN
     Consensus  ------FA-- -LP----N-- --F------- -P-------I
                -L-H-NS--N


                301                                              350

        Rat_A2B  PIVYAYRNRD FRYSFHRIIS RYVLCQTDTK GGSGQAGGQS TFSLSL*~~~
      Mouse_A2B  PIVYAYRNRD FRYSFHKIIS RYVLCQAETK GGSGQAGAQS TLSLGL~~~~
      Human_A2B  PIVYAYRNRD FRYTFHKIIS RYLLCQADVK SGNGQAGVQP ALGVGL*~~~
        Rat_A2A  PFIYAYRIRE FRQTFRKIIR THVLRRQEPF QAGGSSAWAL AAHSTEGEQV
      Mouse_A2A  PFIYAYRIRE FRQTFRKIIR THVLRRQEPF QAGGSSAWVA GAHSTEGEQV
      Human_A2A  PFIYAYRIRE FRQTFRKIIR SHVLRQQEPF KAAGTSARVL AAHGSDGEQV
        Dog_A2A  PFIYAYRIRE FRQTFRKIIR SHVLRRREPF KAGGTSARAL AAHGSDGEQI
Guinea_pig_A2A   PLIYAYRIRE FRQTFRKIIR SHILRRRELF KAGGTSARAS AAHSPEGEQV
      Human_A1   PIVYAFRIQK FRVTFLKIWN DHFRCQPAP. .PIDEDLPEE RPDD*~~~~~
Guinea_pig_A1    PIVYAFRIQK FRVTFLKIWN DHFRCQPEP. .PIDEDLPEE KVDD~~~~~~
        Rat_A1   PIVYAFRIHK FRVTFLKIWN DHFRCQPKP. .PIDEDLPEE KAED*~~~~~
        Cow_A1   PIVYAFRIQK FRVTFLKIWN DHFRCQPAP. .PIDEDAPAE RPDD*~~~~~
        Dog_A1   PIVYAFRIQK FRVTFLKIWN DHFRCQPTP. .PVDEDPPEE APHD*~~~~~
     Rabbit_A1   PIVYAFRIHK FRVTFLKIWN DHFRCRPAPA GDGDEDLPEE KPND*~~~~~
      Chick_A1   PIVYAFRIKK FRTAFLQIWN QYFCCKTNKS SSSSTAETVN ~~~~~~~~~~
      Human_A3   PIVYAYKIKK FKETYLLILK ACVVCHPSDS LDTSIEKNSE *~~~~~~~~~
        Dog_A3   PIVYAYKIKK FKETYLLIFK TYMICQSSDS LDSSTE*~~~ ~~~~~~~~~~
      Sheep_A3   PIVYAYKIKK FKETYLLILK ACVMCQPSKS MDPSTEQTSE *~~~~~~~~~
     Rabbit_A3   PIVYACKIQK FKETYLLIFK ARVTCQPSDS LDPSSEQNSE *~~~~~~~~~
        Rat_A3   PIVYACKNKK VQRNHFVILR ACRLCQTSDS LDSNLEQTTE *~~~~~~~~~
     Consensus   P--YA----- -------I-- ---------- ----------
                 ---------~~
                351                                              400
        Rat_A2B  ~~~~~~~~~~ ~~~~~~~~~~ ~~~~~~~~~~ ~~~~~~~~~~ ~~~~~~~~~~
      Mouse_A2B  ~~~~~~~~~~ ~~~~~~~~~~ ~~~~~~~~~~ ~~~~~~~~~~ ~~~~~~~~~~
      Human_A2B  ~~~~~~~~~~ ~~~~~~~~~~ .......... .......... ..........
        Rat_A2A  SLRLNGHPLG VWANGSATHS GRRPNGYTLG LGGGGSAQGS PR.....DVE
      Mouse_A2A  SLRLNGHPLG VWANGSAPHS GRRPNGYTLG PGGGGSTQGS PG.....DVE
      Human_A2A  SLRLNGHPPG VWANGSAPHP ERRPNGYALG LVSGGSAQES QGNTGLPDVE
        Dog_A2A  SLRLNGHPPG VWANGSAPHP ERRPNGYTLG LVSGGIAPES HGDMGLPDVE
Guinea_pig_A2A   SLRLNGHPPG VWANGSALRP EQRPNGYVLG LVSGRSAQRS HGDASLSDVE
      Human_A1   ~~~~~~~~~~ ~~~~~~~~~~ ~~~~~~~~~~ ~~~~~~~~~~ ~~~~~~~~~~
Guinea_pig_A1    ~~~~~~~~~~ ~~~~~~~~~~ ~~~~~~~~~~ ~~~~~~~~~~ ~~~~~~~~~~
        Rat_A1   ~~~~~~~~~~ ~~~~~~~~~~ ~~~~~~~~~~ ~~~~~~~~~~ ~~~~~~~~~~
        Cow_A1   ~~~~~~~~~~ ~~~~~~~~~~ ~~~~~~~~~~ ~~~~~~~~~~ ~~~~~~~~~~
        Dog_A1   ~~~~~~~~~~ ~~~~~~~~~~ ~~~~~~~~~~ ~~~~~~~~~~ ~~~~~~~~~~
     Rabbit_A1   ~~~~~~~~~~ ~~~~~~~~~~ ~~~~~~~~~~ ~~~~~~~~~~ ~~~~~~~~~~
      Chick_A1   ~~~~~~~~~~ ~~~~~~~~~~ ~~~~~~~~~~ ~~~~~~~~~~ ~~~~~~~~~~
      Human_A3   ~~~~~~~~~~ ~~~~~~~~~~ ~~~~~~~~~~ ~~~~~~~~~~ ~~~~~~~~~~
        Dog_A3   ~~~~~~~~~~ ~~~~~~~~~~ ~~~~~~~~~~ ~~~~~~~~~~ ~~~~~~~~~~
      Sheep_A3   ~~~~~~~~~~ ~~~~~~~~~~ ~~~~~~~~~~ ~~~~~~~~~~ ~~~~~~~~~~
     Rabbit_A3   ~~~~~~~~~~ ~~~~~~~~~~ ~~~~~~~~~~ ~~~~~~~~~~ ~~~~~~~~~~
        Rat_A3   ~~~~~~~~~~ ~~~~~~~~~~ ~~~~~~~~~~ ~~~~~~~~~~ ~~~~~~~~~~
     Consensus   ~~~~~~~~~~ ~~~~~~~~~~ ~~~~~~~~~~ ~~~~~~~~~~ ~~~~~~~~~~


                401                              438
        Rat_A2B  ~~~~~~~~~~ ~~~~~~~~~~ ~~~~~~~~~~ ~~~~~~~~
```

```
    Mouse_A2B   ~~~~~~~~~~  ~~~~~~~~~~  ~~~~~~~~~~  ~~~~~~~~
    Human_A2B   ~~~~~~~~~~  ~~~~~~~~~~  ~~~~~~~~~~  ~~~~~~~~
      Rat_A2A   LPTQERQEG.  QEHPGLRGHL  VQARVGASSW  SSEFAPS*
    Mouse_A2A   LLTQEHQEG.  QEHPGLGDHL  AQGRVGTASW  SSEFSPS*
    Human_A2A   LLSHELKGVC  PEPPGLDDPL  AQDGAGVS*~  ~~~~~~~~
      Dog_A2A   LLSHELKGAC  PESPGLEGPL  AQDGAGVS*~  ~~~~~~~~
Guinea_pig_A2A  LLSHEHKGTC  PESPSLEDPP  AHGGAGVS*~  ~~~~~~~~
      Human_A1  ~~~~~~~~~~  ~~~~~~~~~~  ~~~~~~~~~~  ~~~~~~~~
 Guinea_pig_A1  ~~~~~~~~~~  ~~~~~~~~~~  ~~~~~~~~~~  ~~~~~~~~
        Rat_A1  ~~~~~~~~~~  ~~~~~~~~~~  ~~~~~~~~~~  ~~~~~~~~
        Cow_A1  ~~~~~~~~~~  ~~~~~~~~~~  ~~~~~~~~~~  ~~~~~~~~
        Dog_A1  ~~~~~~~~~~  ~~~~~~~~~~  ~~~~~~~~~~  ~~~~~~~~
     Rabbit_A1  ~~~~~~~~~~  ~~~~~~~~~~  ~~~~~~~~~~  ~~~~~~~~
      Chick_A1  ~~~~~~~~~~  ~~~~~~~~~~  ~~~~~~~~~~  ~~~~~~~~
      Human_A3  ~~~~~~~~~~  ~~~~~~~~~~  ~~~~~~~~~~  ~~~~~~~~
        Dog_A3  ~~~~~~~~~~  ~~~~~~~~~~  ~~~~~~~~~~  ~~~~~~~~
      Sheep_A3  ~~~~~~~~~~  ~~~~~~~~~~  ~~~~~~~~~~  ~~~~~~~~
     Rabbit_A3  ~~~~~~~~~~  ~~~~~~~~~~  ~~~~~~~~~~  ~~~~~~~~
        Rat_A3  ~~~~~~~~~~  ~~~~~~~~~~  ~~~~~~~~~~  ~~~~~~~~
     Consensus  ~~~~~~~~~~  ~~~~~~~~~~  ~~~~~~~~~~  ~~~~~~~~
```

<u>D. Adenosine receptor pharmacology</u>
Ligands and radioligands that have been developed recently that have improved affinity and selectivity for the various adenosine receptor subtypes. Some of the most selective compounds are summarized in the Table and Figures 1 and 2 (showing K_i values measured in binding assays at rat A_1 / A_{2A} / A_3 receptors) below. The choice of the best radioligand for a given assay system depends on several factors, including species, which of the four adenosine receptors are expressed in a given tissue or cell, and in solubility and biological distribution.

	Compound	Selective ligand	Radioligand
A_1	Agonists	CCPA, CPA	[³H]CPA
	Antagonists	CVT-124, WRC-0571, CPX	[³H]CPX
A_{2A}	Agonists	CGS21680, WRC0470, HENECA	[³H]CGS21680, ¹²⁵I-APE
	Antagonists	ZM241385, SCH58216	[³H]ZM, ¹²⁵I-ZM, [³H]SCH
A_{2B}	Agonists	-	-
	Antagonists	enprofylline	-
A_3	Agonists	Cl-IB-MECA, inosine	¹²⁵I-AB-MECA, ¹²⁵I-ABA
	Antagonists	MRS1220, MRS1191, L249313	-

A_1 Receptors: In general, for adenosine agonists (Figure 1), numerous modifications of the N^6-position with hydrophobic moieties (such as CPA, N^6-cyclopentyladenosine; CHA, N^6-cyclohexyladenosine; the primary amine functionalized agonist ADAC; and R-PIA, (R)-N^6-phenylisopropyladenosine) provide selectivity for A_1 receptors. The 2-chloro analogue CCPA (2-chloro-N^6-cyclopentyladenosine) (Klotz *et al.*, 1989) has slightly greater A_1 receptor selectivity than the parent CPA. The affinities of these N^6-substituted adenosine derivatives at A_3 receptors are often intermediate between their respective A_1 and A_{2A} affinities.

Selective antagonists for A_1 receptors (Figure 2) include principally 8-cycloalkyl xanthine derivatives, such as CPX (1,3-dipropyl-8-cyclopentylxanthine). CPX (also known as DPCPX) is ~500-fold selective for A_1 vs A_{2A} receptors generally across species (Jacobson and van Rhee, 1997). CVT-124 (S-1,3-dipropyl-8[2-(5,6-epoxynorbornyl)]xanthine) is a conformational restricted enantiomer that is the most A_1 selective xanthine antagonist yet characterized (Pfister et al., 1997). However, the greates selectivity of antagonism of A_1 over A_3 receptors has been achieved with the non-xanthine WRC-0571 (Martin et al., 1996).

A_{2A} Receptors: The 5'-derivatized analogue NECA is a highly potent but non-selective agonist. A_{2A} -selective adenosine agonists derived from NECA have also been developed. Evaluation of 2-position modifications of NECA led to the identification of CGS 21680 (2-[4-[(2-carboxyethyl)phenyl]ethylamino]-5'-N-ethylcarboxamidoadenosine), which is 140- fold selective for A_{2A} vs A_1 receptors (Hutchison *et al.*, 1989), with a K_i value of 15 nM at A_{2A} receptors. The related *p*-hydroxyphenylpropionylaminoethylamino conjugate, PAPA-APEC, can be radioiodinated to obtain a high affinity radioigand (Jacobson and van Rhee, 1997). CGS 21680 and its derivatives are inactive at A_{2B} receptors (Hide *et al.*, 1992). The SAR (structure activity relationship) at the 2-position of such adenosine-5'-uronamide derivatives has also been extensively probed, leading to such derivatives as the 2-alkynyl-5'-uronamide derivative, HENECA (Cristalli *et al.*, 1995) which is a very potent A_{2A} agonist (K_i = 2.2 nM).

Selectivity for A_{2A} receptors in xanthines is observed for 7-methyl-8-styrylxanthines, such as KF17837 (Shimada et al., 1992), and CSC (8-(3-

Fig. 1. ADENOSINE AGONISTS

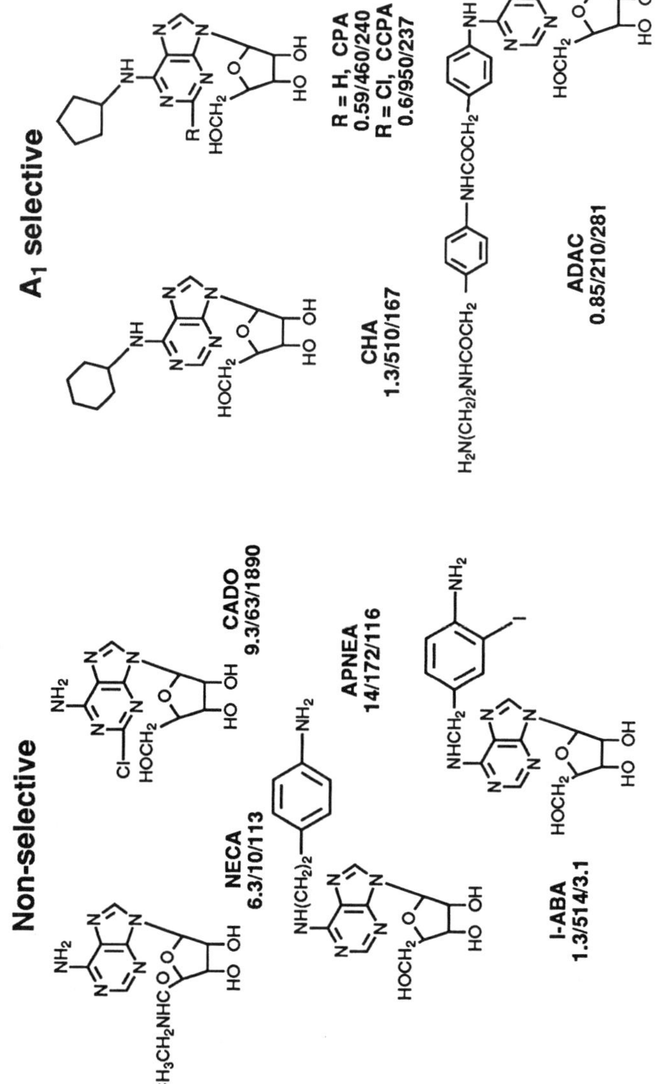

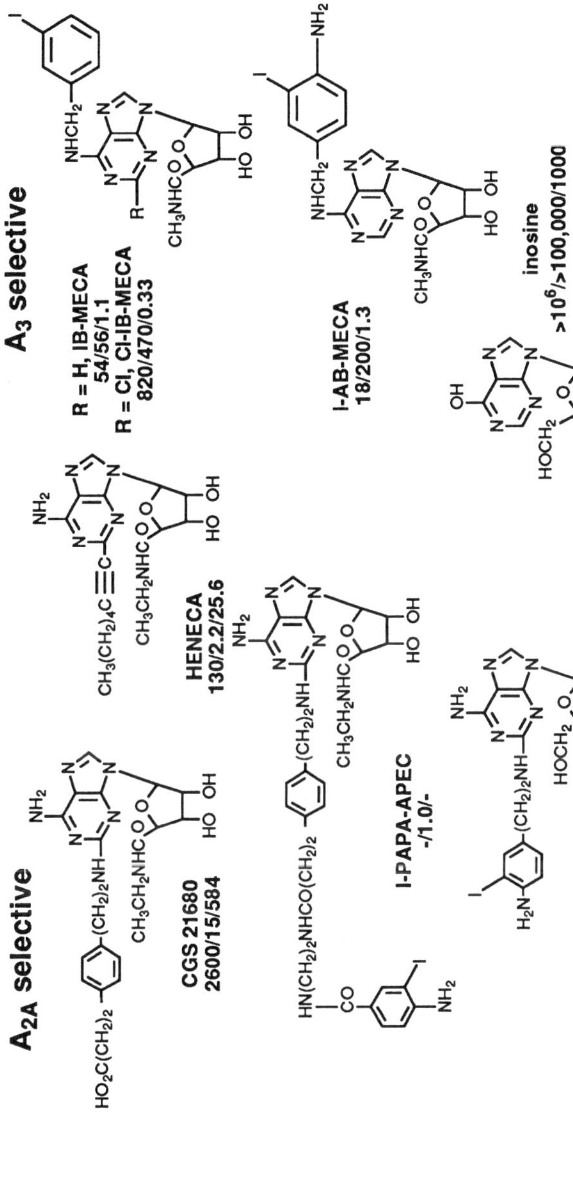

A_3 selective

R = H, IB-MECA
54/56/1.1
R = Cl, Cl-IB-MECA
820/470/0.33

I-AB-MECA
18/200/1.3

inosine
>10^6/>100,000/1000

HENECA
130/2.2/25.6

A_2A selective

CGS 21680
2600/15/584

I-PAPA-APEC
-/1.0/-

I-APE
>150/1.3/-

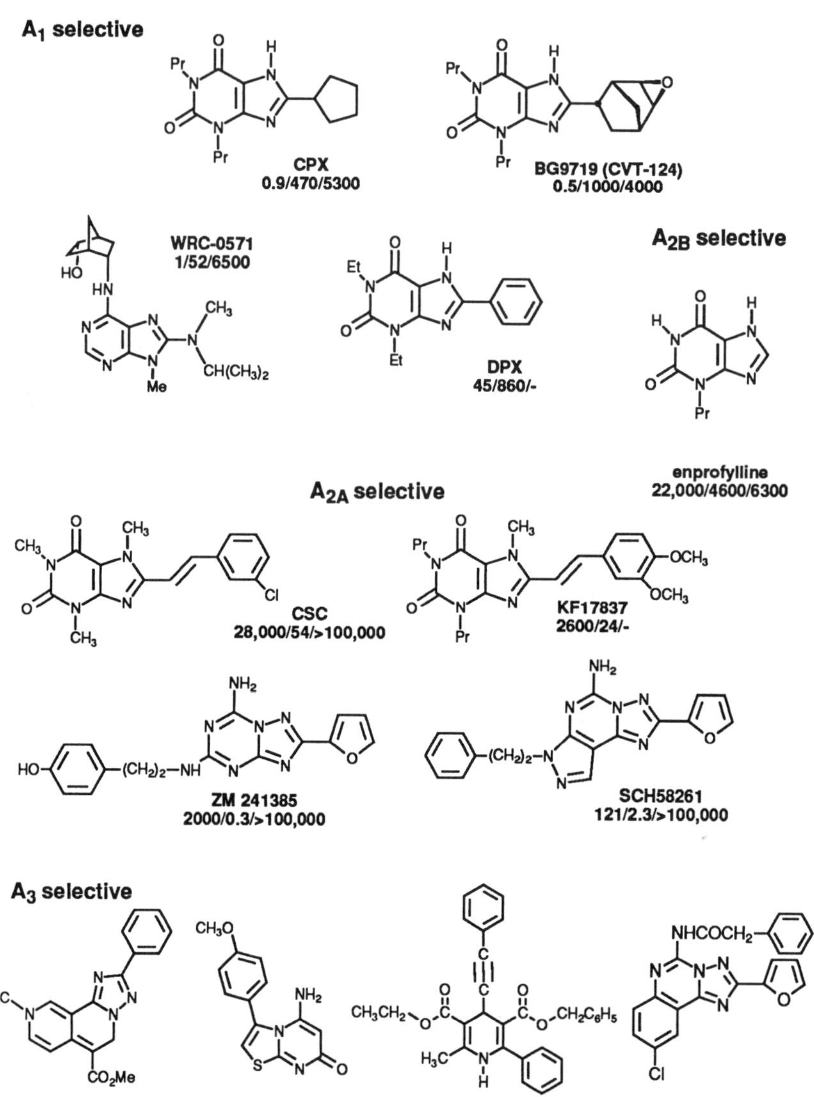

12

chlorostyryl)caffeine) (Jacobson et al., 1993), which displayed selectivity for A_{2A} vs A_1 receptors of 62-fold and 520-fold, respectively. Styrylxanthines in dilute solution however suffer from sensitivity to photoisomerization. Fortunately, the new A_{2A} selective non-xanthine antagonists, ZM241385 (Keddie et al., 1996), iodo-ZM241385 (Palmer et al., 1995) and SCH58261 (Dionisotti et al., 1996; Zocchi et al., 1996)) are highly selective for A_{2A} over A_3 receptors and should be useful for clearly distinguishing between A_{2A}- and A_3-mediated physiological responses.

A_{2B} Receptors: There is still a need for selective agonists and antagonists of A_{2B} adenosine receptors. Enprofylline has been found to be a weak, but partially selective A_{2B} antagonist (Robeva et al., 1996). Radiolabeled xanthines such as $[^3H]DPX$ ($[^3H]$1,3-diethyl-8-phenylxanthine) and ^{125}I-ABOPX (^{125}I-3-(4-amino-3-iodobenzyl)-8-oxyacetate-1-propyl-xanthine) can be used to label recombinant A_{2B} receptors, but are not useful for detecting the low levels of A_{2B} receptors in tissues. Since NECA is a nonselective agonist of all adenosine receptors, but CGS21680 is highly selective for A_{2A} over A_{2B} receptors, a functional response trigger by NECA, but not CGS21680, and insensitive to blockade by pertussis toxin or antagonists of A_1 and A_3 receptors, can be attributed to A_{2B} receptors.

A_3 Receptors: Previously, N^6-p-aminophenylethyladenosine (APNEA) had been used as an agent to activate A_3 receptors in the presence of non-A_3 antagonists, although APNEA is actually 8-fold selective for A_1 receptors (Kim et al., 1994). In combination with WRC-0571, the A_1/A_3 radioligand, ^{125}I-ABA was used to specifically label A_3 receptors on guinea pig lung, and to identify inosine as a low affinity ($K_I = 10$ mM) but selective agonist of rodent A_3 receptors (Jin et al., 1997).

The first A_3 receptor selective agonists, IB-MECA and Cl-IB-MECA have been reported recently (Kim et al., 1994). Another related derivative, $[^{125}I]$I-AB-MECA, although less selective for this subtype, bound to cloned human A_3 receptors expressed in HEK-293 cells with a K_d value of 0.59 nM (Ji et al., 1996). IB-MECA and Cl-IB-MECA (Jacobson, 1998), although highly selective agonists of A_3 adenosine receptors in binding assays, when applied to tissues may activate A_{2A} receptors, particularly when the receptors are present in high density. Moreover, A_{2A}-selective agonists such as CGS21680 (Johansson and Fredholm, 1995), HENECA (Monopoli et al., 1994) and WRC-0470 (Martin et al., 1997) will sometimes activate A_3 receptors.

The dihydropyridine 3-ethyl 5-benzyl-2-methyl-6-phenyl-4-phenylethynyl-1,4-($\pm$)-dihydropyridine-3,5-dicarboxylate (MRS1191) (Jiang $et\ al.$, 1997) was relatively potent at both rat and human A_3 receptors, with K_i values of 1.4 μM and 31 nM, respectively, and highly selective (>1300-fold) for human A_3 vs. human A_1 receptors. MRS1220 (a triazoloquinazoline, Kim $et\ al.$, 1996) is even more potent at human A_3 receptors, with a K_i value of 0.65 nM, however it is not selective for rat A_3 receptors. These competitive antagonists effectively antagonized the inhibitory effects of an agonist on adenylate cyclase in cells expressing the recombinant human A_3 receptor (Jacobson $et\ al.$, 1997).

The Merck group (Jacobson $et\ al.$, 1996) has identified two antagonists which are highly selective for human A_3 receptors: the triazolonaphthyridine derivative L-249313, which behaves as a non-competitive antagonist, and the thiazolopyrimidine derivative L-268605, which is competitive in binding.

REFERENCES

Aguilar JS, Tan F, Durand I and Green RD (1995) Isolation and characterization of an avian A1 adenosine receptor gene and a related cDNA clone. *Biochemical Journal* 307:729-734.

Auchampach JA, Jin J, Wan TC, Caughey GH and Linden J (1997) Canine mast cell adenosine receptors: cloning and expression of the A3 receptors and evidence that degranulation is mediated by the A2B receptor. *Mol Pharmacol* 52:846-860.

Bhattacharya S, Dewitt DL, Burnatowska-Hledin M, Smith WL and Spielman WS (1993) Cloning of an adenosine A1 receptor-encoding gene from rabbit. *Gene* 128:285-288.

Burnstock, G. A basis for distinguishing two types of purinergic receptor. In: *Cell membrane receptors for drugs and hormones: a multidisciplinary approach*, edited by L. Bolis and R.W. Straub. New York: Raven Press, 1978, p. 107-118.

Chern Y, King K and Lai HT (1992) Molecular cloning of a novel adenosine receptor gene from rat brain. *Biochem Biophys Res Commun* 185:304-309.

Cristalli G, Camaioni E, Vittori S, Volpini R, Borea PA, Conti A, Dionisotti S, Ongini E, and Monopoli A (1995) 2-Aralkynyl and 2-heteroalkynyl derivatives of adenosine-5'-N-ethyluronamide as selective A_{2a} adenosine receptor agonists. *J Med Chem* 38:1462-1472.

Dionisotti S, Ferrara S, Molta C, Zocchi C and Ongini E (1996) Labeling of A2A adenosine receptors in human platelets by use of the new nonxanthine antagonist radioligand [3H]SCH 58261. *J Pharmacol Exp Ther* 278:1209-1214.

Drury AN and Szent-Gyorgyi A (1929) The physiological activity of adenine compounds with special reference to their action upon the mammalian heart. *J Physiol (London)* 68:213-237.

Feoktistov I and Biaggioni I (1997) Adenosine A(2B) Receptors [Review]. *Pharmacol Rev* 49:381-402.

Fink JS, Weaver DR, Rivkees SA, Peterfreund RA, Pollack AE, Adler EM and Reppert SM (1992) Molecular cloning of the rat A_2 adenosine receptor: selective co-expression with D_2 dopamine receptors in rat striatum. *Mol Brain Res* 14:186-195.

Furlong TJ, Pierce KD, Selbie LA and Shine J (1992) Molecular characterization of a human brain adenosine A_2 receptor. *Mol Brain Res* 15:62-66.

Hide I, Padgett WL, Jacobson KA, and Daly JW (1992) A_{2A} adenosine receptors from rat striatum and rat pheochromocytoma PC12 Cells - Characterization with radioligand binding and by activation of adenylate cyclase. *Mol Pharmacol* 41: 352-359.

Hill RJ, Oleynek JJ, Hoth CF, Kiron MAR, Weng WF, Wester RT, Tracey WR, Knight DR, Buchholz RA and Kennedy SP (1997) Cloning, expression and pharmacological characterization of rabbit adenosine A(1) and A(3) receptors. *J Pharmacol Exp Ther* **280**:122-128.

Hirano D, Aoki Y, Ogasawara H, Kodama H, Waga I, Sakanaka C, Shimizu T and Nakamura M (1996) Functional coupling of adenosine A2a receptor to inhibition of the mitogen-activated protein kinase cascade in Chinese hamster ovary cells. *Biochemical Journal* **316**: 81-86.

Hutchison AJ, Webb RL, Oei HH, Ghai GR, Zimmerman MB, and Williams M (1989) CGS 21680C, an A_2 selective adenosine receptor agonist with preferential hypotensive activity. *J Pharmacol Exp Ther* **251**: 47-55.

Jacobson KA, Gallo-Rodriguez C, Melman N, Fischer B, Maillard M, van Bergen A, van Galen PJM, and Karton Y (1993) Structure-activity relationships of 8-styrylxanthines as A_2-selective adenosine antagonists. *J Med Chem* **36**:1333-1342.

Jacobson M, Chakravarty PK, Johnson RG, and Norton R (1996) Novel selective non-xanthine A3 adenosine receptor antagonists. *Drug Devel Res* **37**:131.

Jacobson KA and van Rhee AM (1997) Development of selective purinoceptor agonists and antagonists, Chap. 6, in Purinergic Approaches in Experimental Therapeutics, K.A. Jacobson and M.F. Jarvis, eds., Wiley, New York, pp. 101-128.

Jacobson KA, Park KS, Jiang J-l., Kim Y-C, Olah ME, Stiles GL, and Ji, Xd (1997) Pharmacological characterization of novel A_3 adenosine receptor-selective antagonists. *Neuropharmacology*, **36**:1157-1165.

Jacobson KA (1998) Adenosine A3 receptors: novel ligands and paradoxical effects. *TIPS* **19**:184-191.

Jacobson MA, Johnson RG, Luneau CJ and Salvatore CA (1995) Cloning and chromosomal localization of the human A_{2b} adenosine receptor gene (ADORA2B) and its pseudogene. *Genom* **27**:374-376.

Ji XD, Melman N, and Jacobson KA (1996) Interactions of flavonoids and other phytochemicals with adenosine receptors. *J Med Chem* **39**: 781-788.

Jiang J-l, van Rhee AM, Chang L, Patchornik A, Evans P, Melman N, and Jacobson KA (1997) Structure activity relationships of 4-phenylethynyl-6-phenyl-1,4-dihydropyridines as highly selective A3 adenosine receptor antagonists. *J Med Chem* **40**: 2596-2608.

Jin X, Shepherd RK, Duling BR and Linden J (1997) Inosine binds to A3 adenosine receptors and stimulates mast cell degranulation. *J Clin Invest* **100**:2849-2857.

Johansson B and Fredholm BB (1995) Further characterization of the binding of the adenosine receptor agonist [^{3}H]CGS 21680 to rat brain using autoradiography. *Neuropharmacology* **34**:393-403.

Keddie JR, Poucher SM, Shaw GR, Brooks R and Collis MG (1996) In vivo characterisation of ZM 241385, a selective adenosine A2A receptor antagonist. *Eur J Pharmacol* **301**:107-113.

Kim HO, Ji XD, Siddiqi SM, Olah ME, Stiles GL, and Jacobson KA (1994) 2-Substitution of N^6-benzyladenosine-5'-uronamides enhances selectivity for A_3 adenosine receptors. *J Med Chem* **37**: 3614-3621.

Kim, Y-C, Ji X-d, and Jacobson KA (1996) Derivatives of the triazoloquinazoline adenosine antagonist (CGS15943) are selective for the human A_3 receptor subtype. *J Med Chem* **39**: 4142-4148.

Klotz, KN, Lohse MJ, Schwabe U, Cristalli G, Vittori S, and Grifantini M (1989) 2-Chloro-N^6-[^{3}H]cyclopentyladenosine ([^{3}H]CCPA)--a high affinity agonist radioligand for A1 adenosine receptors. *Naunyn Schmiedebergs Arch Pharmacol* **340**: 679-683.

Le F, Townsend-Nicholson A, Baker E, Sutherland GR and Schofield PR (1996) Characterization and chromosomal localization of the human A2a adenosine receptor gene: ADORA2A. *Biochemical & Biophysical Research Communications* **223**:461-467.

Ledent C, Vaugeois JM, Schiffmann SN, Pedrazzini T, Elyacoubi M, Vanderhaeghen JJ, Costentin J, Heath JK, Vassart G and Parmentier M (1997) Aggressiveness, hypoalgesia and high blood pressure in mice lacking the adenosine a(2a) receptor. *Nature* **388**:674-678.

Libert F, Parmentier M, Lefort A, Dinsart C, Van Sande J, Maenhaut C, Simons M-J, Dumont JE and Vassart G (1989) Selective amplification and cloning of four new members of the G protein-coupled receptor family. *Science* **244**:569-572.

Libert F, Van Sande J, Lefort A, Czernilofsky A, Dumont JE, Vassart G, Ensinger HA and Mendla KD (1992) Cloning and functional characterization of a human A1 adenosine receptor. *Biochemical & Biophysical Research Communications* **187**:919-926.

Linden J (1994) Cloned adenosine A_3 receptors: Pharmacological properties, species differences and receptor functions. *Trends Pharmacol Sci* **15**:298-306.

Linden J, Taylor HE, Robeva AS, Tucker AL, Stehle JH, Rivkees SA, Fink JS and Reppert SM (1993) Molecular cloning and functional expression of a sheep A_3 adenosine receptor with widespread tissue distribution. *Mol Pharmacol* **44**:524-532.

Linden J, Tucker AL, Robeva AS, Graber SG and Munshi R (1993) Properties of recombinant adenosine receptors. *Drug Devel Res* **28**:232-236.

Mahan LC, McVittie LD, Smyk-Randall EM, Nakata H, Monsma FJ, Jr., Gerfen CR and Sibley DR (1991) Cloning and expression of an A_1 adenosine receptor from rat brain. *Mol Pharmacol* **40**:1-7.

Marquardt DL, Walker LL and Heinemann S (1994) Cloning of two adenosine receptor subtypes from mouse bone marrow-derived mast cells. *J Immunol* **152**:4508-4515.

Martin PL, Barrett RJ, Linden J and Abraham WM (1997) Pharmacology of 2-cyclohexylmethylidenehydrazinoadenosine (WRC-0470), A novel, short-acting adenosine A(2A) Receptor agonist that produces selective coronary vasodilation. *Drug Dev Res* **40**:313-324.

Martin PL, Wysocki RJ, Barrett RJ, May JM and Linden J (1996) Characterization of 8-(N-methylisopropyl)amino-N^6-(5'- endohydroxy-endonorbornyl)-9-methyladenine (WRC-0571), a highly potent and selective, non-xanthine antagonist of A_1 adenosine receptors. *J Pharmacol Exp Ther* **276**: 490-499.

Meng F, Xie G, Chalmers D, Morgan C, Watson SJ, Jr. and Akil H (1994) Cloning and expression of the A_{2a} adenosine receptor from guinea pig brain. *Neurochem Res* **19**:613-621.

Meng F, Xie GX, Chalmers D, Morgan C, Watson SJJ and Akil H (1994) Cloning and characterization of a pharmacologically distinct A1 adenosine receptor from guinea pig brain. *Brain Research Molecular Brain Research* **26**:143-155.

Meyerhof, W., B. Hamm, C. Schoenrock, S. Fehr, I. Wulfsen, and D. Richter. Functional characterization and differential expression of a human adenosine A3 receptor and rat somatostatin receptor subtypes. In: *VIP, PACAP and related peptides*, edited by G. Rosselin. Singapore: World Scientific, 1994, p. 203-218.

Meyerhof W, Müller-Brechlin R and Richter D (1991) Molecular Cloning of a novel putative G-protein coupled receptor expressed during rat spermiogenesis. *Febs Lett* **284**:155-160.

Miras-Portugal MT, Castro E, Mateo J and Pintor J (1996) The diadenosine polyphosphate receptors: P2D purinoceptors. *Ciba Foundation Symposium* **198**:35-47;:discussion 48-52.

Monopoli A, Conti A, Zocchi C, Casati C, Volpini R, Cristalli G and Ongini E (1994) Pharmacology of the new selective A2a adenosine receptor agonist 2-hexynyl-5'-N-ethylcarboxamidoadenosine. *Arzneimittel-Forschung* **44**:1296-1304.

Mundell SJ, Benovic JL and Kelly E (1997) A dominant negative mutant of the G protein-coupled receptor kinase 2 selectively attenuates adenosine A2 receptor desensitization. *Mol Pharmacol* **51**:991-998.

Olah ME, Ren H, Ostrowski J, Jacobson KA and Stiles GL (1992) Cloning, expression, and characterization of the unique bovine A1 adenosine receptor. Studies on the ligand binding site by site-directed mutagenesis. *J Biol Chem* **267**:10764-10770.

Palmer TM, Poucher SM, Jacobson KA and Stiles GL (1995) ^{125}I-4-(2-[7-amino-2-{2-furyl}{1,2,4}triazolo{2,3-a}{1, 3,5}triazin-5-yl-amino]ethyl)phenol, a high affinity antagonist radioligand selective for the A_{2a} adenosine receptor. *Mol Pharmacol* **48**:970-974.

Palmer TM and Stiles GL (1997) Structure-function analysis of inhibitory adenosine receptor regulation. *Neuropharmacology* **36**:1141-1147.

Pfaff T and Karschin A (1997) Expression cloning of rat cerebellar adenosine A1 receptor by coupling to Kir channels. *Neuroreport* **8**:2455-2460.

Pfister JR, Belardinelli L, Lee G, Lum RT, Milner P, Stanley WC, Linden J, Baker SP and Schreiner G (1997) Synthesis and biological evaluation of the enantiomers of the potent and selective A(1)-adenosine antagonist 1,3-dipropyl-8-[2- (5,6-epoxynorbonyl)]xanthine. *J Med Chem* **40**:1773-1778.

Pierce KD, Furlong TJ, Selbie LA and Shine J (1992) Molecular cloning and expression of an adenosine A2b receptor from human brain. *Biochem Biophys Res Commun* **187**:86-93.

Ramkumar V, Stiles GL, Beaven MA and Ali H (1993) The A_3 adenosine receptor is the unique adenosine receptor which facilitates release of allergic mediators in mast cells. *J Biol Chem* **268**:16887-16890.

Ren H and Stiles GL (1994) Characterization of the human A_1 adenosine receptor gene. Evidence for alternative splicing. *J Biol Chem* **269**:3104-3110.

Reppert SM, Weaver DR, Stehle JH and Rivkees SA (1991) Molecular cloning and characterization of a rat A1-adenosine receptor that is widely expressed in brain and spinal cord. *Molecular Endocrinology* **5**:1037-1048.

Robeva AS, Woodard R, Jin X, Gao Z, Bhattacharya S, Taylor HE, Rosin DL and Linden J (1996) Molecular characterization of recombinant human adenosine receptors. *Drug Dev Res* **39**: 243-252.

Sajjadi FG and Firestein GS (1993) cDNA cloning and sequence analysis of the human A3 adenosine receptor. *Biochim Biophys Acta Mol Cell Res* **1179**:105-107.

Salvatore CA, Jacobson MA, Taylor HE, Linden J and Johnson RG (1993) Molecular cloning and characterization of the human A_3 adenosine receptor. *Proc Natl Acad Sci USA* **90**:10365-10369.

Shimada J, Suzuki F, Nonaka H, Ishii A, and Ichikawa, S (1992) (E)-1,3-Dialkyl-7-methyl-8-(3,4,5-Trimethoxystyryl)xanthines - Potent and Selective Adenosine-A2 Antagonists. *J Med Chem* **35**: 2342-2345.

Shryock JC and Belardinelli L (1997) Adenosine and adenosine receptors in the cardiovascular system: biochemistry, physiology, and pharmacology. [Review] [123 refs]. *American Journal of Cardiology* #19;79:2-10.

Stehle JH, Rivkees SA, Lee JJ, Weaver DR, Deeds JD and Reppert SM (1992) Molecular cloning and expression of the cDNA for a novel A$_2$-adenosine receptor subtype. *Mol Endocrinol* **6**:384-393.

Townsend-Nicholson A and Shine J (1992) Molecular cloning and characterisation of a human A1 adenosine receptor cDNA. *Mol Brain Res* **16**:365-370.

Tracey WR, Magee W, Masamune H, Kennedy SP, Knight DR, Buchholz RA and Hill RJ (1997) Selective adenosine a(3) receptor stimulation reduces ischemic myocardial injury in the rabbit heart. *Cardiovascular Research* **33**:410-415.

Tucker AL, Linden J, Robeva AS, D'Angelo DD and Lynch KR (1992) Cloning and expression of a bovine adenosine A$_1$ receptor cDNA. *Febs Lett* **297**:107-111.

Tucker AL, Robeva AS, Taylor HE, Holeton D, Bockner M, Lynch KR and Linden J (1994) A$_1$ adenosine receptors: two amino acids are responsible for species differences in ligand recognition. *J Biol Chem* **269**:27900-27906.

Ukena D, Jacobson KA, Padgett WL, Ayala C, Shamim MT, Kirk KL, Olsson RA and Daly JW (1986) Species differences in structure-activity relationships of adenosine agonists and xanthine antagonists at brain A$_1$ adenosine receptors. *Febs Lett* **209**:122-128.

Van Rhee AM and Jacobson KA (1996) Molecular architecture of G protein-coupled receptors [review]. *Drug Dev Res* **37**:1-38.

White TE, Dickenson JM, Alexander SPH and Hill SJ (1992) Adenosine A$_1$-receptor stimulation of inositol phospholipid hydrolysis and calcium mobilisation in DDT$_1$ MF-2 cells. *Br J Pharmacol* **106**:215-221.

Wilkie TM, Chen Y, Gilbert DJ, Moore KJ, Yu L, Simon MI, Copeland NG and Jenkins NA (1993) Identification, chromosomal location, and genome organization of mammalian G-protein-coupled receptors. *Genom* **18**:175-184.

Worpenberg S, Burk O and Klempnauer KH (1997) The chicken adenosine receptor 2B gene is regulated by v-myb. *Oncogene* **15**:213-221.

Zhou QY, Li C, Olah ME, Stiles GL and Civelli O (1992) Molecular cloning and characterization of an adenosine receptor: the A3 adenosine receptor. *Proc Natl Acad Sci USA* **89**:7432-7436.

Zocchi C, Ongini E, Conti A, Monopoli A, Negretti A, Baraldi PG and Dionisotti S (1996) The non-xanthine heterocyclic compound SCH 58261 is a new potent and selective A_{2a} adenosine receptor antagonist. *J Pharmacol Exp Ther* **276**:398-404.

CARDIAC PHYSIOLOGY OF ADENOSINE

James G. Dobson, Jr. and Richard A. Fenton, Department of Physiology, University of Massachusetts Medical School and Graduate School of Biomedical Sciences, Worcester, Massachusetts U.S.A.

I. SUMMARY

Adenosine present at physiological levels in the interstitial fluid of the heart interacts with adenosine A_1 and A_2 receptors. The lability of adenosine allows the antiadrenergic action resulting from A_1 stimulation to be modulated by the concentration of interstitial adenosine. The higher the interstitial adenosine level the greater the attenuation of ß-adrenergic-elicited myocardial contractile and metabolic responses. The adenosine A_1 receptor interacts with an inhibitory G-protein that reduces the ß-adrenoceptor-mediated increase in cyclic AMP at a point in the transmembrane coupling process between the ß receptor and adenylyl cyclase. The adenosine A_2 receptor elicits a positive inotropic effect in ventricular myocytes via cyclic AMP dependent and independent mechanisms. However, the physiological significance of the A_2 adenosinergic-mediated responses is far from fully understood.

II. INTRODUCTION

Although adenosine was recognized to have important and potent cardiovascular actions for as many as 70 years (Drury and Szent-Gyorgi, 1929), study of these actions has accelerated greatly over the last several decades. In the heart adenosine dilates the coronary vasculature (Berne, 1963; Gerlach, et al., 1963), depresses sinoatrial (SA) and atrioventricular (AV) nodal activity (West and Belardinelli, 1985; Schutz, et al., 1986; Schutz and Tuisl, 1981; Balwierczak, et al., 1991; Stoggall and Shaw, 1990) and depresses ventricular (Conti, et al., 1995) and isolated ventricular myocyte (Fenton, et al., 1991) automaticity. Adenosine has also been implicated in myocardial preconditioning (Mullane and Bullough, 1995; Parratt. 1995) and angiogenesis (Ethier, et al., 1993; Adair, et al., 1990). The nucleoside is important in myocardial adenine nucleotide metabolism (Belardinelli, et al., 1989) and reduces the responsiveness of the aging adult heart to ß-adrenergic stimulation (Dobson, et al., 1990; Headrick, 1996). Many of these interesting topics will be covered elsewhere is this text.

In this chapter the following topics will be considered. Upon interacting with cardiac presynaptic nerve endings adenosine attenuates the release of norepinephrine caused by adrenergic nerve stimulation (Hedqvist and Fredholm, 1979; Wakade and Wakade, 1978). The nucleoside also attenuates the stimulatory action of the released catecholamines that interact with ß-adrenoceptors in both the ventricular and atrial myocardium. The latter action is well known as the antiadrenergic action of adenosine and occasionally has been referred to as the retaliatory (Newby, et al., 1990) or indirect (Dobson and Fenton, 1983; Dobson, et al., 1987b; Olsson and Pearson, 1990) effect of the nucleoside. Finally, adenosine exerts direct actions on atrial function and possibly on ventricular function independent of ß-adrenergic stimulation.

The focus of this chapter is to provide a brief review and present the current state of knowledge regarding the actions of adenosine. Emphasis will be placed on the effects that physiological levels of adenosine exert in the heart and the mechanisms by which the adenosine actions are expressed. It is the intent herein to examine the role of adenosine as an important mediator of cardiac function in the normal and diseased heart.

III. TEXT

Interstitial Levels of Adenosine in the Heart

Interstitial or extracellular adenosine is believed to be important in exerting the myocardial actions of the nucleoside (Dobson, et al., 1986; Dobson and Schrader, 1984). The primary rationale for this concept is that the adenosine receptors are thought to exist in the myocyte sarcolemma, a position where they can recognize external ligands (Linden, 1991; Olsson and Pearson, 1990; Olah and Stiles, 1995; Wang and Belardinelli, 1994). Thus, it is important to know the actual concentration of adenosine in the interstitium. Numerous estimates of interstitial adenosine have been made using the collection of samples from epicardial surface transudates (Fenton and Dobson, 1992; Fenton and Dobson, 1987; Fenton, et al., 1990; DeDeckere and Ten Hoor, 1977), ventricular dams (Heller and Mohrmann, 1988), ventricular surface absorbent disks (Gidday, et al., 1992), ventricular wells (Hanley, et al., 1983; Gidday, et al., 1988), and microdialysis probes (Van Wylen, et al., 1990). Perhaps the most advantageous of the above techniques is the epicardial surface transudate method of Fenton (Fenton and Dobson, 1992; Fenton and Dobson, 1987), because very small samples (2-5 µl) can be collected from the surface of the heart every 15 seconds allowing temporal evaluation of adenosine levels. Transudate adenosine content is thought to reflect interstitial levels because the mesothelial membrane on the surface of the heart insignificantly affects the concentration of adenosine in the epicardial transudate fluid (Fenton, et al., 1990). The overall consensus of all techniques considered is that the interstitial concentration of adenosine greatly exceeds those levels of the nucleoside measured in the coronary effluent (Dobson, et al., 1987b; Dobson and Schrader, 1984). Depending on the animal species studied and the techniques employed interstitial levels of adenosine have been found to range from 50 to 500 nM in the oxygenated myocardium (Fenton and Dobson, 1987; Van Wylen, et al., 1990), and to increase to significantly higher levels with ß-adrenergic stimulation (Fenton, et al., 1990), hypoxia (Fenton and Dobson, 1987; Fenton and Dobson, 1993) or ischemia (Fenton, et al., 1990).

Myocardial Adenosine A_1 Receptor Mediated Responses

Myocardial Response to ß-Adrenergic Stimulation

It is well known that neural adrenergic stimulation of the heart elicits a profound increase in the functional aspects of the myocardium. Upon nerve stimulation, norepinephrine is released from presynaptic nerve endings. The neurotransmitter molecules diffuse through the synaptic cleft and interact with adrenergic receptors, primarily $ß_1$, on the postsynaptic membrane of myocardial cells. When these receptors are activated a sequence of events involving the stimulatory guanine nucleotide binding protein, G_S, results in an increase in adenylyl cyclase activity (Epstein, et al., 1971; Romano, et al., 1989). The intermediary events leading to ß-adrenergic stimulation of adenylyl cyclase are thought to involve a complex transduction process (Birnbaumer, et al., 1985; Clapham and Neer, 1993). Agonist binding to the receptor initiates the formation of a ternary complex composed of the agonist, receptor, and G_s. G_s is a heterotrimeric protein composed of α_S, ß and γ subunits. In the ternary complex state GDP is bound to the α_s subunit of G_s and the receptor has high affinity for agonist. The formation of the agonist-receptor-G_s complex promotes the binding of GTP to the α_s subunit with concomitant release of GDP. The ternary complex is thought to be destabilized by GTP binding as G_s dissociates into α_s-GTP and ßγ subunits and the receptor reverts to a lower affinity for agonists. The GTP-liganded α_s subunit then activates the catalytic unit of adenylyl cyclase. The activation of the catalytic moiety is reversed by the hydrolysis of GTP by a GTPase on the α_s subunit. The initial formation of the ternary complex appears to be important for agonist stimulation of adenylyl cyclase activity because it facilitates the GTP activation of α_s.

The cyclase catalyzes the formation of cyclic AMP (Dobson and Mayer, 1973; Dobson, et al., 1976) which in turn activates cyclic AMP-dependent protein kinase (PKA) activity (Brown, et al., 1978; Dobson, 1981a; Dobson, 1978a). PKA is responsible for the phosphorylation of a number of proteins, including myosin C-protein, troponin I, phospholamban, calcium channel protein, and myosin light chains that are involved in contraction and phosphorylase kinase activity (Dobson, 1981b; Barany and Barany, 1981; Lindemann, et al., 1983; Stull, 1980). The latter enzyme is involved in the phosphorylation of glycogen phosphorylase and glycogen breakdown (Dobson and Mayer, 1973; Dobson, 1981a; Dobson, et al., 1976). Thus, the ß-adrenergic catecholamines induce a marked augmentation of myocardial contractile and metabolic function.

Inhibition by Adenosine of Neuronal Norepinephrine Release

Adenosine decreases the release of norepinephrine from sympathetic nerve terminals (Wakade and Wakade, 1978). This presynaptic action has been observed to occur in rabbit (Hedqvist and Fredholm, 1979), dog (Lokhandivala, 1979) and rat (Richardt, et al., 1987; Richardt, et al., 1989) hearts. The effect of adenosine is mimicked by two adenosine A_1 receptor agonists, N^6-cyclohexyladenosine (CHA) and N^6-(2-phenylisopropyl) adenosine (PIA), at 0.1 to 1 μM indicating that A_1 receptors are involved (Londos and Wolff, 1977). Moreover, the release of norepinephrine is increased in the presence of the adenosine antagonists such as theophylline or 8-phenyltheophylline (Londos and Wolff, 1977). The effects of adenosine are not influenced by desipramine indicating a mechanism independent from norepinephrine reuptake. Thus, endogenous myocardial levels of adenosine which range from 50 to 500 nM in the oxygenated myocardium are believed to be sufficient to act as a negative feedback regulator that inhibits norepinephrine release in the intact heart (Richardt, et al., 1989). It is thought that the accumulation of endogenous myocardial adenosine which is known to occur with ischemia (Fenton and Dobson, 1987) is sufficient to inhibit the exocytotic release of norepinephrine (Richardt, et al., 1987). It is possible this effect of adenosine provides cardioprotection to the ischemic heart, in part, by reducing the norepinephrine available for stimulating the flow-deprived myocardium.

Cardiac Antiadrenergic Action of Adenosine

While ß-adrenergic stimulation is well known to be a potent stimulus to markedly augment myocardial metabolism and contractility, the full responsiveness of the catecholamines postsynaptically is prevented by the antiadrenergic actions of adenosine (Schrader, et al., 1977; Dobson, 1978b; Endoh and Yamashita, 1980; Dobson, 1983b). At about the time adenosine was reported to reduce the inotropic response to catecholamine stimulation in the heart (Schrader, et al., 1977), the nucleoside was found to reduce the catecholamine induced elevation of myocardial glycogen phosphorylase activity (Dobson, 1978b). Adenosine has been found to attenuate catecholamine-induced increases in the inotropic state of rat (Rockoff and Dobson, 1980; Dobson, 1983a; Dobson and Fenton, 1983) and guinea pig (Dobson, 1983c) atria, rabbit papillary muscle (Endoh and Yamashita, 1980) and isolated perfused hearts (Schrader, et al., 1977; Dobson, 1983b; Dobson and Schrader, 1984; Dobson, et al., 1986; Brown, et al., 1990). The underlying mechanism appears to involve, in part, a reduction in ß-adrenergic-elicited increases in Ca^{2+} transient magnitude (Fenton, et al., 1991). The effect of adenosine can not be mimicked by adenine, inosine, adenosine 5'-monophosphate (AMP), or guanosine (Rockoff and Dobson, 1980). It is important that the antiadrenergic effect can be demonstrated in the *in situ* rat (Romano, et al., 1991) and dog (Sato, et al., 1992) heart, for this suggests that the antiadrenergic actions of adenosine are important in the intact working heart. Thus, adenosine acts as a negative feedback modulator of ß-adrenoceptor-mediated responses in the heart (Dobson, 1983b; Dobson and Fenton, 1983). The greater the interstitial level of adenosine the more pronounced is the attenuation of catecholamine induced responses (Dobson, 1983b; Dobson, et al., 1986; Dobson, 1983a; Dobson, 1996).

The antiadrenergic effect has been demonstrated in single ventricular myocytes of the rat (Belardinelli and Isenberg, 1983). However, a more complete description of the antiadrenergic effect in individual rat ventricular myocytes has been presented (Dobson and Fenton, 1997) and is illustrated in Figure 1. A selective A_1 receptor agonist, PIA, reduced the isoproterenol-elicited increases in peak shorting, the rate of shortening and relengthening (relaxation) and the catecholamine induced decrease in the duration of shortening. The specific A_1 receptor antagonist, 1,3-dipropyl-8-cyclopentylxanthine (DPCPX), prevented the antiadrenergic actions of PIA.

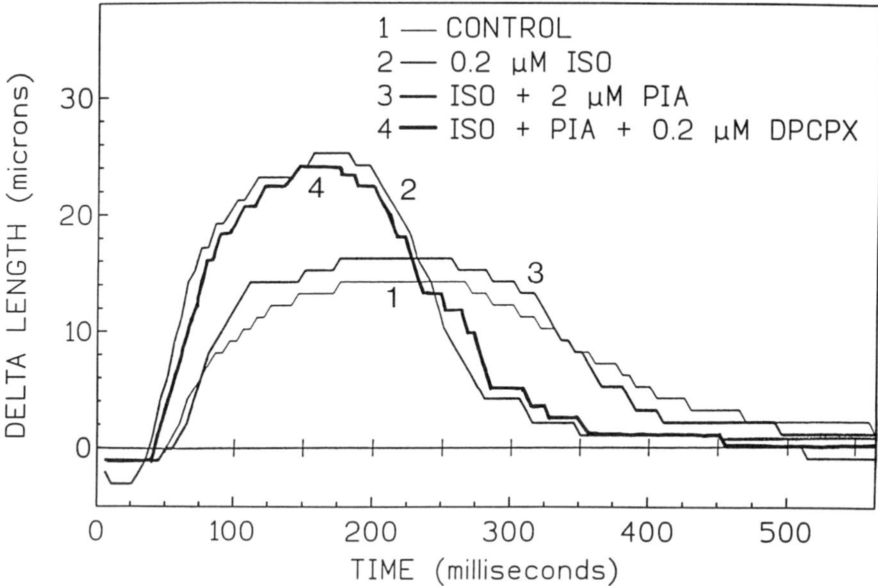

Figure 1. The antiadrenergic effect of an adenosine A_1 receptor agonist on the contraction of an individual rat ventricular myocyte. The myocyte was exposed to sequential addition every 8-10 min of isoproterenol (ISO), ISO + PIA, and ISO + PIA + DPCPX as indicated. The shortening (Delta Length) of a myocyte contracting at 0.2 Hz was recorded with a line scan camera operating at 200 Hz as described previously (Dobson and Fenton, 1997).

Pharmacological and biochemical studies indicate that the antiadrenergic action of adenosine and its analogs is mediated by the extracellular adenosine A_1 receptor (Romano, et al., 1989; Romano and Dobson, 1990). In cardiac tissues the antiadrenergic effects are prevented by the alkylxanthines such as theophylline, 8-phenyltheophylline or DPCPX (Romano, et al., 1989; Romano and Dobson, 1990; Romano, et al., 1991; Dobson, et al., 1996). DPCPX is a specific A_1 receptor antagonist (Burns, et al., 1987). Adenosine deaminase which promotes the conversion of adenosine to inosine prevents the antiadrenergic actions of adenosine (Dobson, et al., 1987a; Dobson, et al., 1987b; Dobson, et al., 1986; Dobson and Fenton, 1993). Selective A_1 receptor agonists such as PIA or the highly selective A_1 receptor agonist, 2-chloro-N^6-cyclopentyladenosine (CCPA), exert a more potent antiadrenergic action compared to adenosine because such analogs are generally less susceptible to tissue metabolism (Dobson, et al., 1986). The use of inhibitors of adenosine deaminase, such as erythro-9-(2-hydroxy-3-nonyl)adenine (EHNA), and transport, such as S-(4-nitrobenzyl)-6-thioinosine (NBMPR or NBTI) or dipyridamole, potentiate the antiadrenergic action of adenosine (Dobson, 1983a).

Muscarinic agonists cause an antiadrenergic action in the heart (Blukoo-Allotey, et al., 1969; Ingebretsen, et al., 1980; Dobson, 1983a). While atropine prevents the muscarinic-induced antiadrenergic effects, the antagonist is without effect on antiadrenergic effects caused by adenosine (Dobson, 1983a). It is well known that the acetylcholine released from parasympathetic nerves in the myocardium attenuates adrenoceptor-mediated contractile and metabolic responses in the heart. While both muscarinic and adenosinergic stimulation elicits antiadrenergic effects in the heart, muscarinic stimulation attenuates the maximal antiadrenergic actions elicited by PIA (Shusterman, et al., 1994). The maximal inhibition of the contractile response caused by 25 nM isoproterenol was reduced by 76% with 10 µM PIA in perfused hearts as illustrated in Figure 2. However, in the presence of carbachol, a muscarinic agonist, at 10 nM, the maximal inhibition of the isoproterenol-elicited contractile response caused by PIA was reduced to only 41%. The IC_{50} for PIA in the absence or presence of carbachol was 0.107 and 0.158 µM, respectively. Because the values for IC_{50} were not different and maximal PIA inhibition was reduced by carbachol, it can be concluded that there is competition between adenosinergic and muscarinic transduction pathways involving the G_i proteins.

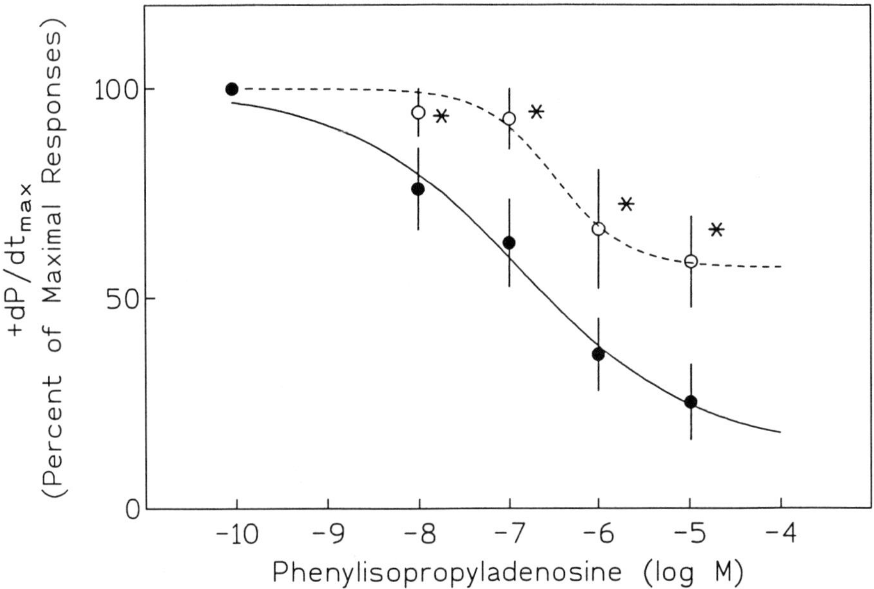

Figure 2. Modulation of the antiadrenergic effect of the adenosine A_1 receptor agonist PIA in perfused rat hearts by a muscarinic agonist carbachol. Hearts were constant flow perfused and stimulated to contract at 375/min. Isoproterenol at 25 nM was infused for 10 sec at each dose of PIA either in the absence (closed circles) or presence (open circles) of 10 nM carbachol and the percent of the maximal isoproterenol contractile response (+dP/dt_max) recorded. Values are the mean ±SE for 6 hearts. Asterisks denote a statistically significant difference from values obtained in the absence of carbachol.

Since the myocardial interstitial levels of adenosine are in the nanomolar range for the oxygenated heart and both adenosine A_1 and A_2 receptor agonists exert their effects at these concentrations, it is likely that adenosine can elicit A_1- or A_2-adenosinergic effects in the myocardium. Upon administration of an A_1 receptor antagonist such as DPCPX at 0.1 to 1 µM to perfused rat hearts the contractile response to ß-adrenergic stimulation was increased due to relief of the antiadrenergic action of endogenous adenosine (Fenton, et al., 1995; Dobson, et al., 1996; Romano, et al., 1991). Theophylline, also an adenosine receptor antagonist, when used at a concentration that does not increase myocardial cyclic AMP by inhibiting phosphodiesterase activity causes an enhancement of ß-adrenergic elicited responses in the heart (Dobson and Fenton, 1993; Dobson, 1983b; Dobson, et al., 1991). Treatment of the oxygenated heart with adenosine deaminase, to degrade endogenous adenosine to inosine , which does not display antiadrenergic properties, results in an increase in ß-adrenergic-induced contractile responses (Dobson, et al., 1986; Dobson and Fenton, 1993; Dobson, et al., 1996). These findings reveal that a greater contractile response is observed with ß-adrenergic stimulation of the oxygenated myocardium indicating that the endogenous adenosine is exerting an antiadrenergic effect. Therefore, adenosine in the well oxygenated heart serves as an important negative feedback modulator of ß-adrenoceptor-mediated responses (Dobson, et al., 1986; Dobson, et al., 1987a; Dobson, et al., 1996).

Because adenosine serves an important role as an antiadrenergic agent in the oxygenated heart where the endogenous interstitial levels of adenosine are in low nanomolar range, a logical extension would be that interventions causing a further increase in interstitial adenosine, such as with ß-adrenergic stimulation, hypoxia or ischemia, would result in a greater attenuation of catecholamine-induced responses (Fenton and Dobson, 1987). This notion is supported by studies in isolated hearts that were perfused under hypoxic conditions and stimulated with a low concentration of isoproterenol (Fenton and Dobson, 1993). The isoproterenol-induced increase in interstitial adenosine levels was greater with hypoxia than with normoxia. In this preparation the contractile responsiveness of the hypoxic heart to isoproterenol stimulation was reduced with hypoxia. This reduction was reversed with adenosine A_1 receptor blockade using DPCPX. It can be concluded from these studies that the contractile responsiveness of the hypoxic heart to ß-adrenergic stimulation is limited by adenosine, thus fostering reduced energy consumption by the hypoxic myocardium. Therefore, adenosine appears to play an important role as an antiadrenergic agent in the survival of the hypoxic and, perhaps, the ischemic (Fenton and Dobson, 1987; Fenton, et al., 1995) myocardium subjected to ß-adrenergic stimulation.

Direct Effects of Adenosine (non-Antiadrenergic)

In atrial tissue, as in the ventricular myocardium, adenosine reduces the ß-adrenergic elicited increase in contractility via A_1 receptors (Dobson, 1983a; Dobson, 1983c; Rockoff and Dobson, 1980). However, unlike in the ventricular myocardium, adenosine also elicits a negative inotropic effect independent of catecholamine stimulation (Dobson, 1983c; Dobson, 1983a; Rockoff and Dobson, 1980; Grossman and Furchgott, 1964; DeGubareff and Sleator, 1965). In atrial tissue adenosine reduces the calcium current (I_{Ca}) and causes hyperpolarization via a inhibitory guanine nucleotide binding protein, G_i, coupled to potassium channel protein which increases ionic conductance (Honey, et al., 1930; Visentin, et al., 1990). Dissociation of the antiadrenergic and direct effects has not been possible because the concentrations of adenosine required to produce these effects overlap. In addition the two effects when elicited by PIA could not be dissociated based on the dosages employed (Linden, et al., 1985; Bohm, et al., 1985).

Mechanisms Involved in the Adenosine A_1 Receptor Responses

The antiadrenergic action of adenosine is mediated by A_1 receptors (Romano, et al., 1989; Romano and Dobson, 1990) that inhibit coupling between the ß-adrenoceptor and the mediated responses (Schrader, et al., 1977; Dobson, 1978b; Rockoff and Dobson, 1980; Dobson, 1983b). In rat ventricular membranes adenosine decreases the formation of the high affinity complex that is composed of the ß-adrenergic agonist, ß-adrenoceptor and stimulatory guanine nucleotide binding protein, G_s (Romano, et al., 1988). This effect of adenosine reduces the catecholamine-elicited activation of adenylyl cyclase (LaMonica, et al., 1985; Romano, et al., 1989; Romano and Dobson, 1990) and formation of cyclic AMP (Dobson, 1978b; Dobson, 1983b; Schrader, et al., 1977). In turn, there is a reduction of the catecholamine-induced increase in PKA (Dobson, 1983b; Dobson, et al., 1987a) and glycogen phosphorylase (Dobson, 1978b; Dobson, 1983b) activities, phosphorylation of myocardial proteins (Fenton and Dobson, 1984) and enhancement of ventricular myocyte membrane inward Ca^{2+} current (Isenberg and Belardinelli, 1984), intracellular Ca^{2+} transient (Fenton, et al., 1991) and positive inotropy (Schrader, et al., 1977; Dobson, 1983b; Fenton, et al., 1991; Rockoff and Dobson, 1980; Romano, et al., 1991). In the ventricular myocardium adenosine does not attenuate the positive inotropic effect caused by increasing the extracellular concentration of calcium, α-adrenoceptor agonists and dibutyryl cAMP (Belardinelli and Isenberg, 1983; Endoh and Yamashita, 1980).

The antiadrenergic action of adenosine is mediated by A_1 receptors that interact with an inhibitory guanine nucleotide binding protein, G_i (Romano, et al., 1989; Romano and Dobson, 1990). The inhibitory G_i protein is a heterotrimeric protein composed of α_i, ß and γ subunits. The inhibition of adenylyl cyclase activity may occur by several mechanisms. Upon activation by a coupled A_1 receptor G_i dissociates into α_i and ßγ subunits which results in a reduced adenylyl cyclase activity (LaMonica, et al., 1985; Romano, et al., 1989). This occurs either by a direct inhibition of the cyclase catalytic subunit by $G_{i\alpha}$, by the action of the ßγ from G_i to sequester $G_{s\alpha}$ subunits and prevent them from activating the catalytic subunit or perhaps by a direct action of $G_{i\beta\gamma}$ on the catalytic unit (Hazeki and Ui, 1981; Clapham and Neer, 1993). Treatment of cardiac preparations with pertussis toxin, which inactivates the inhibitory G proteins by ADP-ribosylation, prevents the actions of adenosine (Bohm, et al., 1989; Brown, et al., 1990). The antiadrenergic action of adenosine appears to be proximal to the point of adenylyl cyclase/cyclic AMP formation (Romano and Dobson, 1990). Such information has been garnered by using forskolin or guanylylimido-diphosphate (GPP(NH)P) which directly increases the catalytic activity of adenylyl cyclase. The increase in cyclase caused by these agents is not significantly reduced by PIA indicating that the adenosine analog does not directly inhibit the catalytic unit of adenylyl cyclase. Furthermore, it is thought that $G_{i\alpha}$-GPP(NH)P can directly inhibit the catalytic unit that is under forskolin stimulation in dually regulated adenylyl cyclase systems (Seamon and Daly, 1982). The inhibitory action of GPP(NH)P is only minimally potentiated by PIA (Romano and Dobson, 1990). These findings suggest that adenosine is able to significantly attenuate ß-adrenergic-enhanced adenylyl cyclase activity, but is unable to significantly inhibit non-receptor activated adenylyl cyclase.

Recently, it has been found that protein kinase C (PKC) may play a role in a prolonged antiadrenergic action of adenosine (Perlini, et al., 1997). In these studies 33 µM adenosine was infused for 30 min into the perfusion fluid of isolated rat hearts. An antiadrenergic action of adenosine was observed during the adenosine infusion period and for 30 min after cessation of the infusion. Only the antiadrenergic effect observed 30 min after exposure to adenosine was prevented by administration of the PKC inhibitor, chelerythrine, suggesting a role for PKC in the persistent antiadrenergic action of adenosine.

Myocardial Adenosine A_2 Receptor Mediated Responses

Adenosine A_2 receptor stimulation elicits increases in contractile function in neonatal avian (Xu, et al., 1992; Liang and Haltiwanger, 1995; Liang and Morley, 1996) and adult rat (Xu, et al., 1996; Dobson and Fenton, 1997) ventricular myocytes and hearts of anesthetized dogs (Gerencer, et al., 1992). It is interesting that adenosine has been reported to have a positive contractile effect with an EC_{50} of 10 nM in rat ventricular myocytes (Gruver, et al.,, 1994). Some studies utilizing mammalian ventricular myocyte preparations indicate that A_2 receptor stimulation causes an increase in myocardial adenylyl cyclase activity and cyclic AMP (Dobson and Fenton, 1997; Romano, et al., 1989; Stein, et al., 1993; Stein, et al., 1994) while other studies do not (Wilken, et al., 1992; Shryock, et al., 1993). Despite an increase in cyclic AMP, A_2 receptor agonists have been reported to be without effect on the mechanical performance of the mammalian cardiac myocyte (Stein, et al., 1994).

The adenosine A_2 agonist, 2-p-(-2-carboxyethyl)phenethyl-amino-5'-N-ethyl-carboxamido-adenosine (CGS-21680), has been found to increase rat ventricular myo-cyte extent of shortening, duration of shortening, time-to-peak shortening, time-to-75% relaxation and maximal rate of shortening with an average EC_{50} of 95 nM. The typical contractile responses of a ventricular myocyte to 0.1 to 10 µM CGS-21680 are illustrated in Figure 3. The adenosine A_2 receptor mediated effects presented herein results mainly from A_{2a} receptors, because A_{2a} and A_{2b} receptor subtypes are coupled in a high (EC_{50}, ≈ 0.1 µM)- and low (EC_{50}, ≈ 10 µM)- affinity manner, respectively, through the G_S protein to activate adenylyl cyclase (Bruns, et al., 1986; Bruns, et al., 1986; Brackett and Daly, 1994; Liang and Haltiwanger, 1995). The A_2 antagonist, 9-chloro-2-(2-furanyl-5,6-di-hydro-1,2,4-triazolo-(1,5-C)quinazolin-5-imine (CGS-15943), prevented the increase in contractile force caused by CGS-21680 as illustrated in Figure 4. A more selective A_2 antagonist, 8-(3-chloro-styryl)caffeine (CSC), also prevented the CGS-21680-induced contractile response (Dobson and Fenton, 1997). It is interesting that fetal avian atrial myocytes do not appear to be functionally responsive to A_2 agonists (Xu, et al., 1992).

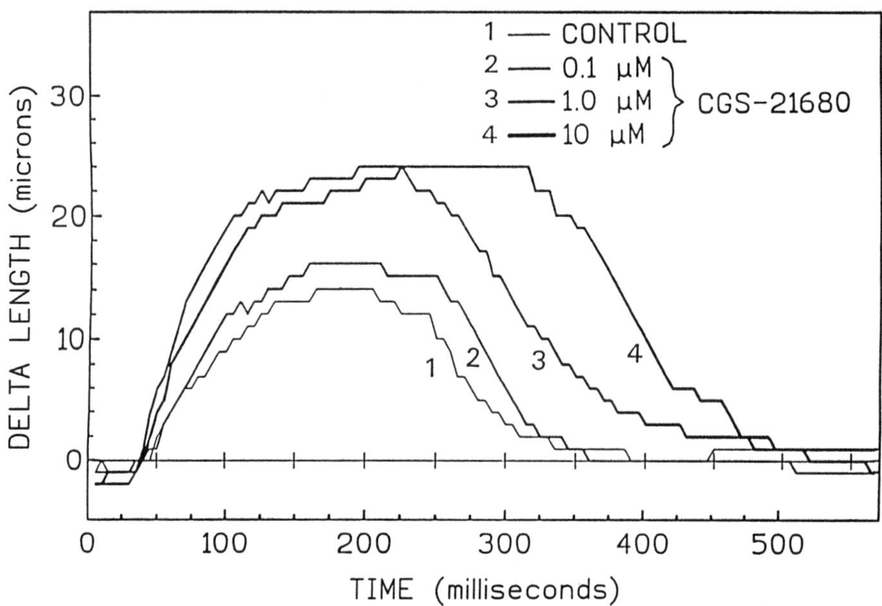

Figure 3. Effect of an adenosine A_2 agonist on an individual contracting rat ventricular myocyte. The myocyte was exposed to increasing concentrations of CGS-21680 as indicated. See legend of Figure 1 for further details.

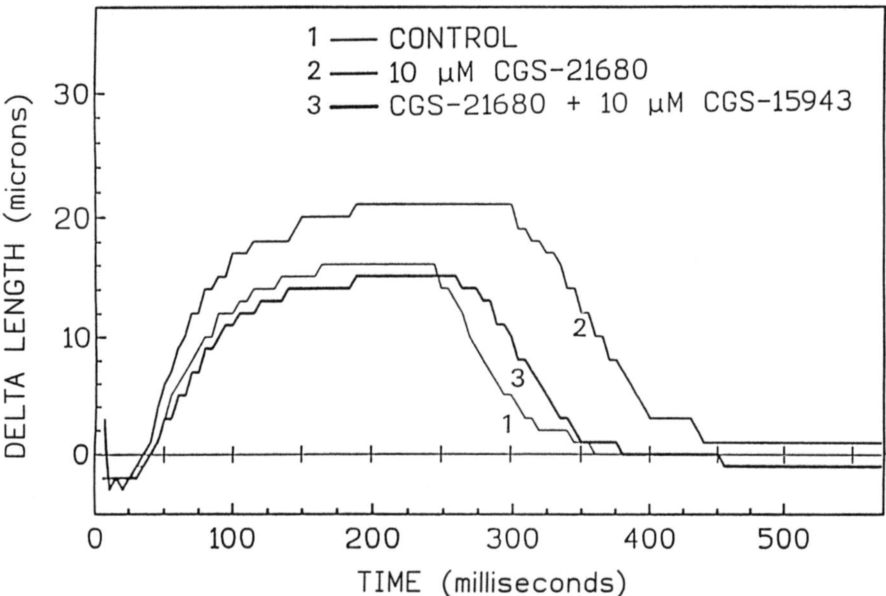

Figure 4. Effect of an adenosine A_2 receptor antagonist, CGS-15943, on the action of A_2 receptor agonist CGS-21680 in an individual contracting rat ventricular myocyte. The myocyte was exposed to CGS-21680 and CGS-15943 as indicated. See legend of Figure 1 for further details.

The adenosine A_2 receptor elicited increase in myocyte contractile performance does not fully resemble the ß-adrenoceptor mediated increase in contractility (Dobson and Fenton, 1997) as illustrated in Figure 5. While ß and A_2 receptor stimulation each increase the extent of shortening and the maximal rate of shortening, the duration of shortening, time-to-peak shortening and time-to-75% relaxation are decreased by ß-adrenergic stimulation, but increased by A_2 adenosinergic stimulation. ß-adrenergic stimulation increases the maximum rate of myocyte relaxation, whereas A_2-adenosinergic stimulation does not effect this contractile variable. The differences between adrenergic- and adenosinergic-induced contractile responses suggest that the inotropic responses are mediated by different mechanisms. The A_2-adenosinergic-induced inotropic response is not an abbreviated contraction as is observed with ß-adrenergic stimulation. In fact it has been suggested (Dobson and Fenton, 1997) that the positive inotropic response caused by A_2 adenosinergic stimulation is more similar to the inotropic response observed with an increase in extracellular Ca^{2+} concentration (Rockoff and Dobson, 1980; Murray, et al., 1985; Sonnenblick, 1962). Ca^{2+} influx has been reported to be increased with adenosine A_2 receptor agonists in neonatal avian ventricular myocytes (Liang and Morley, 1996). At low concentrations of adenosine agonists there does not appear to be an activation of adenylyl cyclase and an increase in the formation of cyclic AMP (Dobson and Fenton, (1997). Thus, the mechanism by which A_2-adenosinergic stimulation causes an increase in contractile performance is not understood. At this point, there appears to be both cyclic AMP dependent and independent pathways involved (Dobson and Fenton, 1997; Liang and Morley, 1996). In perfused rat hearts treated with the adenosine A_2 antagonist, CGS-15943, there was a reduced contractile response to ß-adrenergic stimulation with isoproterenol as shown in Figure 6. Although it is recognized that this agent exhibits A_1 antagonism as well, the results suggest that there may be synergism between ß-adrenergic and A_2-adenosinergic-elicited inotropy due to the presence of endogenous adenosine.

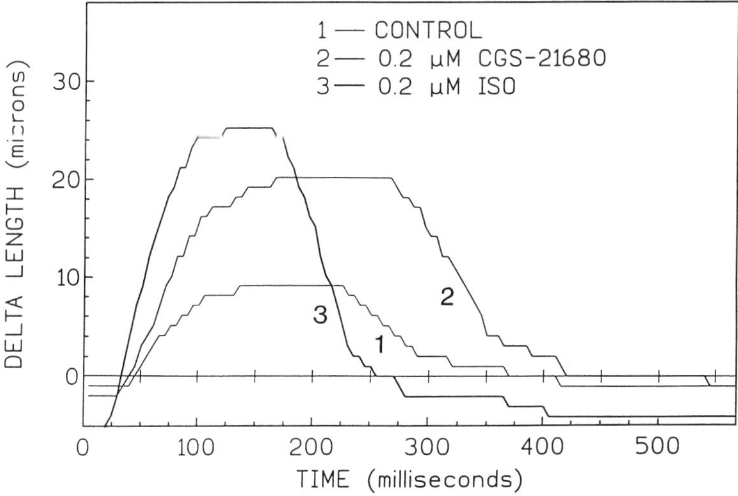

Figure 5. Comparison of the effect of A_2-adenosinergic and ß-adrenergic receptor stimulation on a rat ventricular myocyte. A contracting ventricular myocyte was exposed to CGS-21680 for 5 min, washed for 15 min and then exposed to isoproterenol for 3 min as indicated. See legend of Figure 1 for further details.

Other Nucleoside and Nucleotide Effects in Cardiac Muscle

Over the years there have been numerous reports regarding the effects of higher concertrations of adenosine, inosine and ATP on ventricular contractile function. Rather than present a review, it is the intent to reveal the current status of the knowledge on the subject. Administration of adenosine at high concentrations of 100-1000 µM to ventricular muscle preparations has been reported to cause either an increase (Chiba and Himori. 1975; Bruckner, et al., 1985; Legssyer, et al., 1988) or no change (Burnstock and Meghji, 1983) in contractile force development. ATP at 10-100 µM increases force development in rat papillary muscles by 12 to 15% (Legssyer, et al., 1988). Even inosine at doses of 300 µg to 3 mg increased contractile force development in blood-perfused canine atria (Furukawa and Chiba, 1980). Adenosine at relatively high concentrations of 100 µM have been reported to decrease contractility in perfused rat hearts by 20-30% (Wannenburg, et al. 1994). The latter effects were not antiadrenergic because ß-adrenergic antagonists were employed. Thus, the effects of high concentrations of adenosine that are of unknown physiological importance await further study under highly defined conditions such as might be possible with isolated ventricular myocyte preparations.

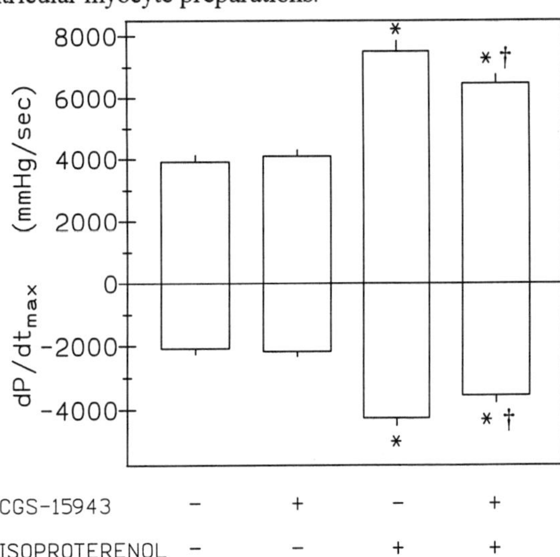

Figure 6. Effect of adenosine A_2 receptor antagonist on ß-adrenergic elicited contractile response ($+dP/dt_{max}$) in isolated rat hearts. Hearts were constant flow perfused and exposed to 1 µM CGS-15943 and/or 10 µM isoproterenol as indicated. Values are the mean ±SE for 4 hearts.

Importance of Adenosine in the *In Vivo* Heart

While the antiadrenergic effect of adenosine has been convincingly documented in perfused hearts, papillary muscles, atrial muscles, isolated ventricular myocytes and ventricular myocytes membranes, less has been done to verify that adenosine is antiadrenergic in the blood perfused heart. The antiadrenergic action of adenosine has been demonstrated in vivo (Romano, et al., 1991; Sato, et al., 1992; Lai, et al., 1992; Gerencer, et al., 1992) but these effects are not universally observed (Seitelberger, et al., 1984; Schipke, et al., 1987). Both in open- and closed-chest preparations of anesthetized rats in which the hearts were paced, PIA reduced the positive inotropic responses caused by intravenous isoproterenol or epinephrine (Romano, et al., 1991). The specific A_1- receptor antagonist, DPCPX, inhibited the antiadrenergic actions of PIA in the preparations. Such findings indicate that A_1 adenosinergic-mediated mechanisms are functional in the blood perfused heart and provide strong support for an important physiological role for adenosine in modulating cardiac contractile responses elicited by ß-adrenergic stimulation. While relatively low concentrations of adenosine and A_2 receptor agonists are known to elicit an increase in contractile performance in various cardiac preparations, much more information needs to be obtained regarding any possible *in vivo* effects. The physiological importance of A_{2a} adenosinergic-mediated contractile effects is not yet fully appreciated let alone understood. However, it has been suggested that the A_2 receptor may enhance the contractile state in the compromised myocardium when interstitial levels of adenosine are significantly elevated (Dobson, 1996).

IV. CONCLUSIONS AND FUTURE DIRECTIONS

The antiadrenergic actions of adenosine are of importance in curtailing the expression of ß-adrenoceptor-mediated responses in the normal oxygenated working heart, the heart subjected to excessive catecholamine stimulation, hypoxia, ischemia or any combination of these situations as schematically presented in Figure 7. Adenosine exerts this action via A_1 receptors which attenuates the signal transduction of ß-adrenergic agonist receptor interaction ultimately decreasing activation of adenylyl cyclase, formation of cyclic AMP, PKA activation, protein phosphorylation and metabolic and mechanical responses. Adenosine A_2 receptors, on the other hand, elicit a positive inotropic effect that is mediated via cyclic AMP dependent and independent mechanisms. The physiological importance of the latter adenosinergic-induced responses is far from fully appreciated.

Figure 7. Schematic representation of the effects elicited by adenosine A_1 and A_2 receptor stimulation on myocardial function.

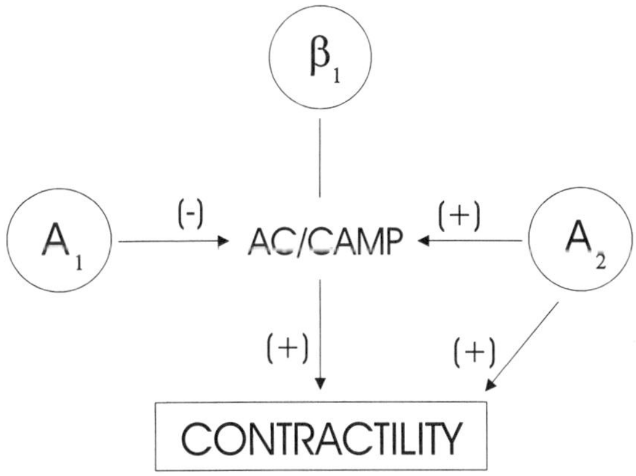

The role of the antiadrenergic action of adenosine in the diseased myocardium should be explored further. While there is some understanding of how the adenosine A_1 receptors mediate their effects a good deal remains to be revealed regarding the molecular mechanisms involved. The importance of adenosine A_2 receptor stimulation in the functioning of the heart is just beginning to become apparent and many burning questions remain. With respect to the molecular mechanisms involved, very little is known, thus this would appear to be a fertile area of investigation.

V. REFERENCES

Adair TH, Gay WJ, Montani JP. Growth regulation of the vascular system: evidence for a metabolic hypothesis. *Am J Physiol, 259:* R393-R404, 1990.

Balwierczak JL, Sharif R, Krulan CM, Field FP, Weiss GB, Miller MJ. Comparative effects of a selective adenosine A_2 receptor agonist, CGS21680, and nitroprusside in vascular smooth muscle. *Eur J Pharmacol, 196:* 117-123, 1991.

Barany M, Barany K. Protein phosphorylation in cardiac and vascular smooth muscle. *Am J Physiol, 241:* H117-H128, 1981.

Belardinelli L, Isenberg G. Actions of adenosine and isoproterenol on isolated mammalian ventricular myocytes. *Circ Res, 53:* 287-297, 1983.

Belardinelli L, Linden J, Berne RM. The cardiac effects of adenosine. *Prog Cardiovasc Dis, 32:* 73-97, 1989.

Berne RM. Cardiac nucleotides in hypoxia: A possible role in regulation of coronary blood flow. *Am J Physiol, 204:* 317-322, 1963.

Birnbaumer L, Codina J, Mattera R, Cerione RA, Hildebrandt JD, Sunyer T, Rojas FJ, Caron MG, Lefkowitz RJ, Iyengar R. Regulation of hormone receptors and adenylyl cyclases by guanine nucleotide binding N proteins. *Rec Prog Hormone Res, 41:* 41-99, 1985.

Blukoo-Allotey JA, Vincent NH, Ellis S. Interactions of acetylcholine and epinephrine on contractility, glycogen and phosphorylase activity of isolated mammalian hearts. *J Pharmacol Exper Ther, 170:* 27-36, 1969.

Bohm M, Bruckner R, Meyer W, Nose M, Schmitz W, Scholz H, Starbatty J. Evidence for adenosine receptor-mediated isoprenaline antagonistic effects of the adenosine analogs PIA and NECA on force of contraction in guinea-pig atrial and ventricular cardiac preparations. *Naunyn-Schmiedeberg's Arch Pharmacol, 331:* 131-139, 1985.

Bohm M, Schmitz W, Scholz H, Wilken A. Pertussis toxin prevents adenosine receptor- and m-cholinoceptor-mediated sinus rate slowing and AV conduction block in the guinea-pig heart. *Naunyn-Schmiedeberg's Arch Pharmacol, 339:* 152-158, 1989.

Brackett LE, Daly JW. Functional characterization of the A_{2b} adenosine receptor in NIH 373 fibroblasts. *Biochem Pharmacol, 47:* 801-814, 1994.

Brown DF, Honeyman TW, Dobson JG, Jr. Properties of epinephrine-induced activation of cardiac adenosine 3',5'-monophosphate dependent protein kinase. *Biochim Biophys Acta, 544:* 462-473, 1978.

Brown LA, Humphrey SM, Harding SE. The anti-adrenergic effect of adenosine and its blockade by pertussus toxin: a comparative study in myocytes isolated from guinea-pig, rat and failing human hearts. *Brit J Pharmacol, 101:* 484-488, 1990.

Bruckner R, Fenner A, Meyer W, Nobis T-M, Schmitz W, Scholz H. Cardiac effects of adenosine and adenosine analogs in guinea-pig atrial and ventricular preparations: evidence against a role of cyclic AMP and cyclic GMP. *J Pharmacol Exper Ther, 234:* 766-774, 1985.

Bruns RF, Lu GA, Pugsley TA. Adenosine receptor subtypes: binding studies. In: Gerlach E and Becker BF (eds.), *Topics and Perspectives in Adenosine Research.* New York: Springer-Verlag, 1986, pp. 59-73.

Bruns RF, Lu GH, Pugsley TA. Characterization of the A_2 adenosine receptor labeled by [^{3}H]NECA in rat striatal membrane. *Mol Pharmacol, 29:* 331-346, 1986.

Burns RF, Lu GH, Pugsley TA. Adenosine receptor subtypes: binding studies. In: Gerlach E and Becker BF (eds.), *Topics and Perspectives in Adenosine Research.* Berlin: Springer-Verlag, 1987, pp. 59-73.

Burnstock G, Meghji P. The effect of adenyl compounds in the rat heart. *Brit J Pharmacol, 79:* 211-218, 1983.

Chiba S, Himori N. Different inotropic responses to adenosine on the atrial and ventricular muscle of the dog heart. *Japanese J Pharmacol, 25:* 489-491, 1975.

Clapham DE, Neer EJ. New roles for G-protein ßγ-dimers in transmembrane signalling. *Nature, 365:* 403-406, 1993.

Conti JB, Belardinelli L, Utterback DB, Curtis AB. Endogenous adenosine is an antiarrhythmic agent. *Circulation, 91:* 1761-1767, 1995.

DeDeckere EAM, Ten Hoor P. A modified Langendorff technique for metabolic investigations. *Pflugers Arch - Eur J Physiol, 370:* 103-105, 1977.

DeGubareff T, Sleator WJ. Effects of caffeine on mammalian atrial muscle and its interaction with adenosine and calcium. *J Pharmacol Exper Ther, 148:* 202-214, 1965.

Dobson JG, Jr. Protein kinase regulation of cardiac phosphorylase activity and contractility. *Am J Physiol, 234:* H638-H645, 1978a.

Dobson JG, Jr. Reduction by adenosine of the isoproterenol-induced increase in cyclic adenosine 3',5'-monophosphate formation and glycogen phosphorylase activity in rat heart muscle. *Circ Res, 43:* 785-792, 1978b.

Dobson JG, Jr. Cyclic AMP-dependent activation of protein kinases in the myocardium. In: Delius W, Gerlach E, Grobecker H, and Kubler W (eds.), *Catecholamines and the Heart.* New York: Springer-Verlag, 1981a, pp. 128-141.

Dobson JG, Jr. Catecholamine-induced phosphorylation of cardiac muscle proteins. *Biochim Biophys Acta, 675:* 123-131, 1981b.

Dobson JG, Jr. Adenosine reduces catecholamine contractile responses in oxygenated and hypoxic atria. *Am J Physiol, 245:* H468-H474, 1983a.

Dobson JG, Jr. Mechanism of adenosine inhibition of catecholamine-induced elicited responses in heart. *Circ Res, 52:* 151-160, 1983b.

Dobson JG, Jr. Interaction between adenosine and inotropic interventions in guinea pig atria. *Am J Physiol, 245:* H475-H480, 1983c.

Dobson JG, Jr. Adenosine and adrenergic mediated effects in the heart. In: Abd-Elfattah AS and Wechsler AS (eds.), *Purines and Myocardial Protection.* Norwell,MA: Kluwer, 1996, pp. 359-372.

Dobson JG, Jr., Fenton RA. Antiadrenergic effects of adenosine in the heart. In: Berne RM, Rall TW, and Rubio R (eds.), *Regulatory Function of Adenosine.* Boston: Nijhoff, 1983, pp. 363-376.

Dobson JG, Jr., Fenton RA. Adenosine inhibition of β-adrenergic induced responses in aged hearts. *Am J Physiol, 265:* H494-H503, 1993.

Dobson JG, Jr., Fenton RA. Adenosine A_{2a} receptor function in rat ventricular myocytes. *Cardiovasc Res, 34:* 337-347, 1997.

Dobson JG, Jr., Fenton RA, Romano FD. The cardiac antiadrenergic effect of adenosine. In: Pelleg A, Michelson EL, and Dreifus LS (eds.), *Cardiac Electrophysiology and Pharmacology of Adenosine and ATP; Basic and Clinical Aspects.* New York: Alan R. Liss, 1987a, pp. 331-343.

Dobson JG, Jr., Fenton RA, Romano FD. The antiadrenergic actions of adenosine in the heart. In: Gerlach E and Becker BF (eds.), *Topics and Perspectives in Adenosine Research.* Berlin: Springer-Verlag, 1987b, pp. 356-368.

Dobson JG, Jr., Fenton RA, Romano FD. Increased myocardial adenosine production and reduction of β-adrenergic contractile response in aged hearts. *Circ Res, 66:* 1381-1390, 1990.

Dobson JG, Jr., Fenton RA, Romano FD. Adenosine and the reduced responsiveness of the aged heart to adrenergic stimulation. In: Imai S and Nakazawa M (eds.), *Role of Adenosine and Adenine Nucleotides in Biological Systems*. Amsterdam: Elsevier, 1991, pp. 377-386.

Dobson JG, Jr., Fenton RA, Sawmiller DR. The contractile response of the ventricular myocardium to adenosine A_1 and A_2 receptor stimulation. In: Das DK (ed.), *Myocardial Reperfusion*. Annals N.Y. Acad. Sci. 1996, pp. 64-73.

Dobson JG, Jr., Mayer SE. Mechanisms of activation of cardiac glycogen phosphorylase in ischemia and anoxia. *Circ Res, 33:* 412-420, 1973.

Dobson JG, Jr., Ordway RW, Fenton RA. Endogenous adenosine inhibits catecholamine contractile responses in normoxic hearts. *Am J Physiol, 251:* H455-H462, 1986.

Dobson JG, Jr., Ross J, Jr., Mayer SE. The role of cyclic adenosine 3',5'-monophosphate and calcium in the regulation of contractility and glycogen phosphorylase activity in guinea pig papillary muscle. *Circ Res, 39:* 388-395, 1976.

Dobson JG, Jr., Schrader J. Role of extracellular and intracellular adenosine in the attenuation of catecholamine evoked responses in guinea pig heart. *J Mol Cell Cardiol, 16:* 813-822, 1984.

Drury AN, Szent-Gyorgi A. The physiological activity of adenine compounds with especial reference to their action upon the mammalian heart. *J Physiol (London), 68:* 213-237, 1929.

Endoh M, Yamashita S. Adenosine antagonizes the positive inotropic action mediated via β-, but not α-adrenoceptors in the rabbit papillary muscle. *Eur J Pharmacol, 65:* 445-448, 1980.

Epstein SE, Levey GS, Skelton CL. Adenylate cyclase and cyclic AMP. Biochemical links in the regulation of myocardial contractility. *Circulation, 43:* 437-448, 1971.

Ethier MF, Chander V, Dobson JG, Jr. Adenosine stimulates proliferation of human endothelial cells in culture. *Am J Physiol, 265:* H131-H138, 1993.

Fenton RA, Dobson JG, Jr. Adenosine and calcium alter adrenergic-induced intact heart protein phosphorylation. *Am J Physiol, 246:* H559-H565, 1984.

Fenton RA, Dobson JG, Jr. Measurement by fluorescence of interstitial adenosine levels in normoxic, hypoxic and ischemic perfused rat hearts. *Circ Res, 60:* 177-184, 1987.

Fenton RA, Dobson JG, Jr. Fluorometric quantitation of adenosine concentration in small samples of extracellular fluid. *Anal Biochem, 207:* 134-141, 1992.

Fenton RA, Dobson JG, Jr. Hypoxia enhances isoproterenol-induced increase in heart interstitial adenosine depressing β-adrenergic contractile responses. *Circ Res, 72:* 571-578, 1993.

Fenton RA, Galeckas KJ, Dobson JG, Jr. Endogenous adenosine reduces depression of cardiac function induced by β-adrenergic stimulation during low flow perfusion. *J Mol Cell Cardiol, 27:* 2373-2383, 1995.

Fenton RA, Moore EDW, Fay FS, Dobson JG, Jr. Adenosine reduces the Ca^{2+} transients of isoproterenol-stimulated rat ventricular myocytes. *Am J Physiol, 261:* C1107-C1114, 1991.

Fenton RA, Tsimikas S, Dobson JG, Jr. Influence of β-adrenergic stimulation and contraction frequency on heart interstitial adenosine. *Circ Res, 66:* 457-468, 1990.

Furukawa Y, Chiba S. Inotropic and chronotropic responses to inosine in isolated and blood-perfused dog atria. *Eur J Pharmacol, 67:* 339-345, 1980.

Gerencer RZ, Finegan BA, Clanachan AS. Cardiovascular selectivity of adenosine receptor agonists in anaesthetized dogs. *Brit J Pharmacol, 107:* 1048-1056, 1992.

Gerlach E, Deuticke B, Dreisbach RH. Der nucleotid-abbau im herzmuskel bei sauerstoffmangel und seine mogliche bedeutung fur die coronardurchblutung. *Naturwissenschaften, 50:* 228-229, 1963.

Gidday JM, Hill HE, Rubio R, Berne RM. Estimates of left ventricular interstitial fluid adenosine during catecholamine stimulation. *Am J Physiol, 254:* H207-H216, 1988.

Gidday JM, Kaiser DM, Rubio R, Berne RM. Heterogeneity and sampling volume dependence of epicardial adenosine concentrations. *J Mol Cell Cardiol, 24:* 351-364, 1992.

Grossman A, Furchgott RF. The effects of various drugs on calcium exchange in the isolated guinea-pig left auricle. *J Pharmacol Exper Ther, 145:* 162-172, 1964.

Gruver EJ, Toupin D, Smith TW, Marsh JD. Acadesine improves tolerance to ischemic injury in rat cardiac myocytes. *J Mol Cell Cardiol, 26:* 1187-1195, 1994.

Hanley F, Messina LM, Baer RW, Uhlig PN, Hoffman JIE. Direct measurement of left ventricular interstitial adenosine. *Am J Physiol, 245:* H327-H335, 1983.

Hazeki O, Ui M. Modification by islet-activating protein of receptor-mediated regulation of cyclic AMP accumulation in isolated rat heart cells. *J Biol Chem, 256:* 2856-2862, 1981.

Headrick JP. Impact of aging on adenosine levels, A_1/A_2 responses, arrhythmogenesis, and energy metabolism in rat heart. *Am J Physiol, 270:* H897-H906, 1996.

Hedqvist P, Fredholm BB. Inhibitory effect of adenosine on adrenergic neuroeffector transmission in the rabbit heart. *Acta Physiol Scand, 105:* 120-122, 1979.

Heller LJ, Mohrmann DE. Estimates of interstitial adenosine from surface exudates of isolated rat hearts. *J Mol Cell Cardiol, 20:* 509-523, 1988.

Honey RM, Ritchie WT, Thomson WAR. The action of adenosine upon the human heart. *J Medicine, 23:* 485-490, 1930.

Ingebretsen CG, Rabin RA, Allen DO. Acetylcholine modulation of phosphorylase and contractility in rat hearts exposed to anoxia or isoproterenol. *Biochem Pharmacol, 29:* 1681-1686, 1980.

Isenberg G, Belardinelli L. Ionic basis for the antagonism between adenosine and isoproterenol and isolated mammalian ventricular myocytes. *Circ Res, 55:* 309-325, 1984.

Lai WT, Wu SN, Sung RJ. Negative dromotropism of adenosine under beta-adrenergic stimulation with isoproterenol. *Amer J Cardiol, 70:* 1427-1431, 1992.

LaMonica DA, Frohloff N, Dobson JG, Jr. Adenosine inhibiton of catecholamine-stimulated cardiac membrane adenylate cyclase. *Am J Physiol, 248:* H737-H744, 1985.

Legssyer A, Poggioli J, Renard D, Vassort G. ATP and other adenine compounds increase mechanical activity and inositol trisphosphate production in rat heart. *J Physiol, 401:* 185-199, 1988.

Liang BT, Haltiwanger B. Adenosine A_{2a} and A_{2b} receptors in cultured fetal chick heart cells. High- and low-affinity coupling to stimulation of myocyte contractility and cAMP accumulation. *Circ Res, 76:* 242-251, 1995.

Liang BT, Morley JF. A new cyclic AMP-independent, G_s-mediated stimulatory mechanism via the adenosine A_{2a} receptor in the intact cardiac cell. *J Biol Chem, 271:* 18678-18685, 1996.

Lindemann JP, Jones LR, Hathaway DR, Henry BG, Watanabe AM. β-Adrenergic stimulation of phospholamban phosphorylation and Ca^{2+}-ATPase activity in guinea pig ventricles. *J Biol Chem, 258:* 464-471, 1983.

Linden J. Structure and function of A_1 adenosine receptors. *FASEB J, 5:* 2668-2676, 1991.

Linden J, Hollen CE, Patel A. The mechanism by which adenosine and cholinergic agents reduce contractility in rat myocardium. *Circ Res, 56:* 728-735, 1985.

Lokhandivala MF. Inhibition of cardiac sympathetic neurotransmission by adenosine. *Eur J Pharmacol, 60:* 353-357, 1979.

Londos C, Wolff J. Two distinct adenosine-sensitive sites on adenylate cyclase. *Proc Nat Acad Sci, U S A, 74:* 5482-5486, 1977.

Mullane K, Bullough D. Harnessing an endogenous cardioprotective mechanism: Cellular sources and sites of action of adenosine. *J Mol Cell Cardiol, 27:* 1041-1054, 1995.

Murray JJ, Dobson JG, Jr., Reed PW. Effects of divalent cation ionophore A23187 on cardiac contractile parameters. *Am J Physiol, 249:* H1195-H1203, 1985.

Newby AC, Worku Y, Meghji P, Nakazawa M, Skladanowski AC. Adenosine: A retaliatory metabolite or not? *NIPS, 5:* 67-70, 1990.

Olah ME, Stiles GL. Adenosine receptor subtypes: Characterization and therapeutic regulation. *Ann Rev Pharmacol Toxicol, 35:* 581-606, 1995.

Olsson RA, Pearson JD. Cardiovascular Purinoceptors. *Physiol Rev, 70:* 761-845, 1990.

Parratt JR. Possibilities for the pharmacological exploitation of ischaemic preconditioning. *J Mol Cell Cardiol, 27:* 991-1000, 1995.

Perlini S, Khoury E, Chung ES, Fenton RA, Dobson JG, Jr., Meyer TE. Adenosine mediates sustained antiadrenergic depression via activation of protein kinase C in the rat heart. *Circulation, 96:* I-449, 1997 (Abstract).

Richardt G, Waas W, Kranzhofer R, Mayer E, Schomig A. Adenosine inhibits exocytotic release of endogenous noradrenaline in rat heart: a protective mechanism in early myocardial ischemia. *Circ Res. 61:* 117-123, 1987.

Richardt G, Waas W, Kranzhofer R, Cheng B, Lohse MJ, Schomig A. Interaction between the release of adenosine and noradrenaline during sympathetic stimulation. A feed-back mechanism in rat heart. *J Mol Cell Cardiol, 21:* 269-277, 1989.

Rockoff JB, Dobson JG, Jr. Inhibition by adenosine of catecholamine-induced increase in rat atrial contractility. *Am J Physiol, 239:* H365-H370, 1980.

Romano FD, Dobson JG, Jr. Adenosine modulates β-adrenergic signal transduction in guinea pig heart ventricular membranes. *J Mol Cell Cardiol, 22:* 1359-1370, 1990.

Romano FD, Fenton RA, Dobson JG, Jr. The adenosine R_i agonist, phenylisopropyladenosine, reduces high affinity isoproterenol binding to the β-adrenergic receptor of rat myocardial membranes. *Sec Mess Phosphoproteins, 12:* 29-43, 1988.

Romano FD, Macdonald SG, Dobson JG, Jr. Adenosine receptor coupling to adenylate cyclase of rat ventricular myocyte membranes. *Am J Physiol, 257:* H1088-H1095, 1989.

Romano FD, Naimi TS, Dobson JG, Jr. Adenosine attenuation of catecholamine-enhanced contractility of rat heart in vivo. *Am J Physiol, 260:* H1635-H1639, 1991.

Sato H, Hori M, Kitakaze M, Takashima S, Inoue M, Kitabatake A, Kamada T. Endogenous adenosine blunts β-adrenoceptor-mediated inotropic response in hypoperfused canine myocardium. *Circulation, 85:* 1594-1603, 1992.

Schipke J, Heusch G, Thamer V. Evidence against the adenosine-catecholamine antagonism in the canine heart in situ. *Arzneim -Forsch, 37:* 1345-1347, 1987.

Schrader J, Baumann G, Gerlack E. Adenosine as inhibitor of myocardial effects of catecholamines. *Pflugers Arch, 372:* 29-35, 1977.

Schutz W, Freissmuth M, Hausleithner V, Tuisl E. Cardiac sarcolemmal purity is essential for the verification of adenylate cyclase inhibition via A_1-adenosine receptors. *Naunyn-Schmiedeberg's Arch Pharmacol, 333:* 156-162, 1986.

Schutz W, Tuisl E. Evidence against adenylate cyclase-coupled adenosine receptors in the guinea pig heart. *Eur J Pharmacol, 76:* 285-288, 1981.

Seamon KB, Daly JW. Guanosine 5'-(β,γ-imido) triphosphate inhibition of forskolin-activated adenylate cyclase is mediated by the putative inhibitory guanine nucleotide regulatory protein. *J Biol Chem, 257:* 11591-11596, 1982.

Seitelberger R, Schutz W, Schlappack O, Raberger G. Evidence against the adenosine-catecholamine antagonism under in vivo conditions. *Naunyn-Schmiedeberg's Arch Pharmacol, 325:* 234-239, 1984.

Shryock J, Song Y, Wang D, Baker SP, Olsson RA, Belardinelli L. Selective A_2-adenosine receptor agonists do not alter action potential duration, twitch shortening, or cyclic AMP accumulation in guinea pig, rat, or rabbit isolated ventricular myocytes. *Circ Res, 72:* 194-205, 1993.

Shusterman MS, Fenton RA, Dobson JG, Jr. Carbachol influences the antiadrenergic action of phenylisopropyladenosine on heart contractility. *FASEB J, 8:* A844, 1994 (Abstract).

Sonnenblick EH. Force-velocity relations in mammalian heart muscle. *Am J Physiol, 202:* 931-939, 1962.

Stein B, Mende U, Neumann J, Schmitz W, Scholz H. Pertussis toxin unmasks stimulatory myocardial A_2-adenosine receptors on ventricular cardiomyocytes. *J Mol Cell Cardiol, 25:* 655-659, 1993.

Stein B, Schmitz W, Scholz H, Seeland C. Pharmacological characterization of A_2-adenosine receptors in guinea pig ventricular myocytes. *J Mol Cell Cardiol, 26:* 403-414, 1994.

Stoggall SM, Shaw JS. The coexistence of adenosine A_1 and A_2 receptors in guinea-pig aorta. *Eur J Pharmacol, 190:* 329-335, 1990.

Stull JT. Phosphorylation of contractile proteins in relation to muscle function. *Adv Cyclic Nucleotide Res, 13:* 39-93, 1980.

Van Wylen DGL, Willis J, Sodhi J, Weiss RJ, Lasley RD, Mentzer RM. Cardiac microdialysis to estimate interstitial adenosine and coronary blood flow. *Am J Physiol, 258:* H1642-H1649, 1990.

Visentin S, Wu S-N, Belardinelli L. Adenosine-induced changes in atrial action potential: contribution of Ca and K currents. *Am J Physiol, 258:* H1070-H1078, 1990.

Wakade AR, Wakade TD. Inhibition of noradrenaline release by adenosine. *J Physiol (London), 282:* 35-49, 1978.

Wang D, Belardinelli L. Mechanism of the negative inotropic effect of adenosine in guinea-pig atrial myocytes. *Am J Physiol, 267:* H2420-H2429, 1994.

Wannenburg T, DeTombe PP, Little WC. Effect of adenosine on contractile state and oxygen consumption in isolated rat hearts. *Am J Physiol, 267:* H1429-H1436, 1994.

West GA, Belardinelli L. Sinus slowing and pacemaker shift caused by adenosine in rabbit SA node. *Pflugers Arch, 403:* 66-74, 1985.

Wilken A, Tawfik-Schlieper H, Schwabe U. Evidence against the presence of A_2 adenosine receptors on guinea pig ventricular myocytes. *Eur J Pharmacol, 192:* 161-163, 1992.

Xu D, Kong H, Liang BT. Expression and pharmacological characterization of a stimulatory subtype of adenosine receptor in fetal chick ventricular myocytes. *Circ Res, 70:* 56-65, 1992.

Xu H, Stein B, Liang BT. Characterization of a stimulatory adenosine A_{2a} receptor in adult rat ventricular myocyte. *Am J Physiol, 270:* H1655-H1661, 1996.

VASCULAR BIOLOGY AND PHARMACOLOGY OF ADENOSINE RECEPTORS

Pauline L. Martin, Ph.D, Discovery Therapeutics, Inc., USA; Ray A. Olsson, University of South Florida, Tampa, FL 33612

I. BACKGROUND

Nearly seventy years have passed since Drury and Szent-Gyorgyi (1929) described some of the cardiovascular effects of adenosine. The "Modern Era" of adenosine research began 35 years ago, when Berne (1963) and Gerlach et al. (1963) independently proposed that adenosine controls hypoxic coronary vasodilation. The eve of those anniversaries is a fitting time to review the actions of adenosine on blood vessels. Some older reviews (Belardinelli et al., 1989; Olsson and Pearson, 1990; Collis, 1991) are still useful. Subsequent updates of the pharmacology of adenosine receptors and the regulatory roles of adenosine (Tucker and Linden, 1992; Muller et al., 1996; Olsson, 1996; Ongini and Fredholm, 1996; Feoktistov and Biaggioni, 1997) allow the authors to focus this review on important new developments that have occurred over the past two or three years. Readers wanting a broader understanding should consult those earlier articles.

II. TEXT

A. Vascular adenosine receptors and effectors

Of the four known adenosine receptors, the A_1AR, $A_{2A}AR$ and the $A_{2B}AR$, appear to directly control vasomotion. Adenosine is a renal vasoconstrictor (Drury and Szent- Gyorgyi, 1929). The A_1AR agonists R-PIA and CPA dilate arterioles of 25-40 μm diameter in rat diaphragm muscle, and CPX blocks that effect (Danialou et al., 1997). Evidence from in vitro experiments on epicardial coronary arteries suggests that the A_1AR might participate in coronary vasomotor control (Merkel et al., 1992; Hussain and Mustafa, 1995), but as yet there is no support for that possibility from in vivo experiments. The A_3AR is highly expressed in smooth muscle cells of the aorta of neonatal and young rats, but not of adult rats (Zhao et al., 1997). Reverse transcribed polymerase chain reaction and Northern blots identified the receptor mRNA. The A_3AR agonists IB-MECA and R-PIA but not CPA exerted inhibition of forskolin stimulation of cAMP accumulation that was insensitive to CPX. Antisense oligonucleotides against the A_3AR reduced receptor mRNA levels, potentiated forskolin-promoted accumulation of cAMP and reduced the inhibitory effect of A_3AR agonists on cAMP accumulation. However, those observations do not speak to the physiological importance of the A_3AR in resistance vessels or in adult animals. Other experiments in rats show that A_3AR agonists cause hypotension (Fozard and Carruthers, 1993), presumably a reflection of the dilation of resistance vessels. Subsequent work showed the agonists act indirectly, stimulating the release of histamine from perivascular mast cells (Hannon et al., 1995; Fozard et al., 1996; Shepherd et al., 1996).

Growing evidence implicates both the $A_{2A}AR$ and the $A_{2B}AR$ in the regulation of tissue perfusion, but at different levels in the circulation. That evidence suggests the high-affinity $A_{2A}AR$ acts at the level of resistance vessels and the low-affinity $A_{2B}AR$ at the level of conductance vessels. Direct microscopy of the coronary microcirculation shows that adenosine acts on arterioles having diameters ~40 μm (Kuo et al., 1995) and appears to act in concert with flow-induced vasodilation in larger vessels upstream (Kuo and Chancellor, 1995; Liao and Kuo, 1997). A wealth of pharmacological evidence supports the notion that adenosine initiates vasodilation at the $A_{2A}AR$ (Hamilton et al., 1987; Ueeda et al., 1991; Zocchi et al., 1996; Keddie et al., 1996; Belardinelli et al., 1998). In addition to its location on resistance vessels, two pharmacological properties of the $A_{2A}AR$ may also contribute to its pre-eminence over the $A_{2B}AR$ in flow regulation. First, the $A_{2A}AR$ is a high affinity receptor that binds ligands at the low-nanomolar level whereas the $A_{2B}AR$ is a low affinity receptor that requires micromolar concentrations of ligand for activation. Second, at least in the coronary circulation, there is a large reserve of "spare" $A_{2A}ARs$. Depending on the agonist, occupancy of as few as 1.5 per cent of the coronary $A_{2A}ARs$ is sufficient to achieve maximum vasodilation (Shryock et al., 1998).

The $A_{2B}AR$ also mediates vasodilation, though the evidence is less extensive and largely indirect. The lack of agonists and antagonists that are selective for the $A_{2B}AR$ has greatly hindered assessment of the physiological significance of that receptor. As a coronary vasodilator in the guinea pig, the $A_{2A}AR$ agonist CGS 21680 is approximately 4 times more potent than the unselective agonist NECA (Ueeda et al., 1991), but is 800 times less potent as a relaxant of the aorta (Martin, 1992, 1993). Such a result suggests that the $A_{2A}AR$ occurs in the coronary arteries and the $A_{2B}AR$ in the aorta. Further, the highly selective $A_{2A}AR$ antagonist ZM241385 was 50-100 times more potent in blocking relaxation of the coronary arteries than of the aorta (Poucher et al., 1996). Another $A_{2A}AR$ antagonist, SCH58261, failed to block relaxation of guinea pig aorta by NECA (Zocchi et al., 1996). Micromolar concentrations of adenosine, 2-chloroadenosine, NECA and CPA hyperpolarized the smooth muscle cells of coronary conductance vessels (diameter 200-300 μm), but the $A_{2A}AR$ agonists CGS21680 and DPMA did not, further evidence of action at an $A_{2B}AR$ (Mutafova-Jambolieva and Keef, 1997). That study also provided evidence that downstream elements in the signal pathway included adenylate cyclase and the sulfonylurea-sensitive K_{ATP} channel.

<u>B. Edothelium in adenosine vasodilation</u>
Endothelial cells contain adenosine receptors that could be important in the regulation of blood flow. Human aorta endothelial cells contain the mRNA of both the $A_{2A}AR$ and the $A_{2B}AR$ (Iwamoto et al., 1994), but whether those cells express both receptors equally is unknown. Adenosine analogues stimulate the accumulation of cAMP in coronary microvascular endothelial cells from the guinea pig with an agonist pharmacological profile of an $A_{2A}AR$ (Schiele and Schwabe, 1994).

There are at least two ways that adenosine could initiate endothelium-dependent vasomotion. The model of Daut et al. (1994) proposes: activation of endothelial $A_{2A}AR$, $\longrightarrow$ activation of protein kinase A $\longrightarrow$ phosphorylation of K_{ATP} channels, which increases the probability of channel opening $\longrightarrow$ endothelial cell hyperpolarization $\longrightarrow$ calcium influx (Lückhoff and B.sse, 1990a,b) activation of eNOS $\longrightarrow$ NO production $\longrightarrow$ smooth muscle relaxation. The two unique features of that model are (a) the involvement of the K_{ATP} channel and (b) the participation of eNOS. There is abundant evidence that the K_{ATP} channel is an important effector of adenosine vasodilation (Daut et al., 1990; Dart and Standen, 1993; Kleppisch and Nelson, 1995; Kuo and Chancellor, 1995; Mutafova-Yambolieva and Keef, 1997). However, evidence for the possible involvement of eNOS weakens in the transition from in vitro to in vivo experiments. Endothelial cells in culture generate NO when exposed to adenosine (Li et al, 1995). Adenosine and $A_{2A}AR$ agonists hyperpolarize human umbilical vein endothelial cell membranes and stimulate the uptake of L-arginine via system y+, NO synthesis and the accumulation of cGMP (Sobrevia et al., (1997). The $A_{2A}AR$ antagonist ZM241385 antagonized those effects. It is unlikely that the increase in L-arginine influx was the stimulus to NO production because arginine concentration rarely limits the catalytic rate of eNOS and, in any event, the culture medium contained 0.1 mM L-arginine. Further, measurements of cytoplasmic calcium concentration excluded the possibility that the stimulus to NO production was calcium influx promoted by hyperpolarization of the cell membrane. Rather, inhibition of tyrosine kinase activity by genistein attenuated both L-arginine transport and NO synthesis.

Intravital microscopy of the beating, blood-perfused hearts of anesthetized dogs do not support the idea that adenosine acts through NO (Jones et al., 1995). Those experiments confirmed that adenosine preferentially dilates small arterioles (diameter < 100 µm). The eNOS inhibitor NG-nitro-L-arginine (L-NAME) constricted larger coronary arteries (diameter > 100 µm) and simultaneously dilated the smaller arterioles. In that situation the control of vascular resistance had shifted from the arterioles to larger, more proximal vessels. The subsequent administration of adenosine or stimulation of metabolism by cardiac pacing did not further dilate the small arterioles. Because L-NAME and adenosine acted downstream of the site of NO production, the authors concluded that adenosine cannot act through NO. Of course, the dilation of downstream microvessels by adenosine could increase blood flow and thereby indirectly stimulate NO production by increasing the endothelial shear stress on proximal vessels.

The experiments of Jones et al. refute one line of evidence supporting the idea that NO mediates the vasoactivity of adenosine. In isolated epicardial coronary arteries NOS inhibitors cause a rightward shift of the adenosine concentration-coronary relaxation curve, a result interpreted as evidence that NO mediates the coronary vasoactivity of adenosine (Vials and Burnstock, 1993, Abebe et al., 1995; Skinner and Marshall, 1996). However, those experiments did not take into account that the inhibitors also increase basal vascular tone and thus act as functional antagonists of adenosine (Olsson, 1996). Because in beating heart preparations eNOS inhibitors dilate the microvessels containing the $A_{2A}AR$ maximally (Jones et al, 1995), the rightward shift of the adenosine dose-response curve caused by NOS inhibitors may also reflect the action of adenosine at larger, less sensitive vessels containing the $A_{2B}AR$.

Experiments in eNOS knockout mice show that NO is not an obligatory mediator of the vasoactivity of adenosine. Perfusion of isolated hearts of wild type and eNOS-/- mice with either 250 nM or 1 µM adenosine elicited similar increases in coronary flow (G°decke et al., 1998). Although the knockout mice were hypertensive, basal coronary flow was similar to that of wild type mice. Even though coronary resistance was elevated in the $eNOS^{-/-}$ mice, the basal rate of adenosine release was similar in the two groups, evidence that adenosine was not compensating for the loss of the contribution of NO to basal tone.

Electrotonic coupling of endothelial cells to underlying smooth muscle cells via myoendothelial gap junctions (Little et al., 1995a,b; Xia and Duling, 1995) provides a second mechanism for endothelium-dependent relaxation. That model suggests that the propagation of hyperpolarization from the endothelial to the smooth muscle cell shuts the voltage-operated channels that provide the calcium that supports contraction. Conversely, raising intracellular calcium concentration in vascular smooth muscle cells leads to a rise in the calcium concentration in endothelial cells and the production of NO. The authors interpret those observations as evidence that calcium diffuses down a concentration gradient through myoendothelial gap junctions into endothelial cells, where it activates NOS. Adenosine might act in still other ways, for example, by promoting the phosphorylation of small heat-shock proteins (Baell et al., 1997) or through small G proteins that alter the calcium sensitivity of actomyosin (Uehata et al., 1997).

<u>C. Adenosine receptor agonists and antagonist ligands as investigational tools</u>

Long and intense searches for selective ligands have identified agonists and antagonists that are both potent and selective for the A_1AR and the $A_{2A}AR$ in vitro as well as in vivo. At this time there are neither agonists nor antagonists that are selective for the $A_{2B}AR$. Similar agonist profiles and important species differences in antagonist profiles are obstacles to the development of ligands that can discriminate between the A_1AR and the A_3AR. For example, binding to A_1ARs accounts for >95 % of the binding of the "selective" A_3AR agonist [125I]AB-MECA to rat brain membranes (Shearman and Weaver, 1997). Newer A_3AR antagonists such as MRS-1097 and MRS-1191 (Jiang et al, 1996) or L-249313 and L-268605 (Jacobson et al., 1997) appear to be selective in vitro, but have not yet undergone rigorous evaluation in vivo.

The ability of pharmacologists to study all four human adenosine receptors at last permits evaluations of potency and selectivity in the primary species of interest, man (Klotz et al., 1997). Even though access to human adenosine receptors puts drug development on a more realistic basis, selectivity ratios based on ligand binding assays often do not predict selectivity in vivo. As pointed out above, factors such as the existence of "spare" receptors are can importantly influence the biological activity of a drug and may vary from organ to organ. Indeed, when one receptor couples to two effectors in the same organ, for example, in the case of the cardiac A_1AR that inhibits adenylate cyclase and also opens K+ channels, the degree of receptor reserve of the two pathways can differ markedly (Srinavas et al., 1997).

<u>D. Transgenic mice lacking the $A_{2A}AR$ gene</u>

The successful breeding of mice lacking the $A_{2A}AR$ gene (Ledent et al., 1997) affords new insights into the role of that receptor in the cardiovascular system. $A_{2A}AR^{-/-}$ mice had elevated basal blood pressure and heart rate and were refractory to the hypotension and reflex tachycardia usually caused by the $A_{2A}AR$ agonist CGS21680. The hypertension suggests a role for the $A_{2A}AR$ in setting basal systemic vascular resistance, but whether those receptors are in the central nervous system or in peripheral sites is unclear. Likewise, the mechanism of the tachycardia is uncertain.

43

<u>E. Adenosine receptors in human disease</u>

Koglin and von Scheidt (1997) describe three patients with syndrome X and two patients with transplanted hearts who had normal coronary angiograms and vasodilator responses to intracoronary papaverine and acetylcholine but no responses to intracoronary adenosine infused at rates of 80 or 160 µg/min. One subject re-examined 12 months later was still refractiory to adenosine but not to acetylcholine. Whether the defect was at the level of adenosine receptors or more distal in the transduction chain is uncertain.

<u>F. Role of adenosine in angiogenesis</u>

Hypoxia or the addition of adenosine to the culture medium stimulates the proliferation of endothelial cells in vitro, an effect that is blocked by 8-phenyltheophylline (Meininger et al., 1988). Culture medium from cells grown under hypoxia also stimulated endothelial cell proliferation, but the mitogen was not identified. A subsequent study (Sexl et al., 1997) showed that activation of the $A_{2A}AR$ initiated proliferation through tyrosine phosphorylation of the p42 and p44 isoforms of mitogen activated protein kinase (MAP kinase). Neither inactivation of $Gi\alpha$ by pertussis toxin, down-regulation of $Gs\alpha$ by pretreating with cholera toxin, or inhibitors of protein kinase C affected the phosphorylation of MAP kinase. However, an inhibitor of MAP kinase kinase, MEK1, abolished the effect of $A_{2A}AR$ activation. Activation of the $A_{2A}AR$ also increased levels of p21 (ras). Thus, the signaling pathway operates through p21 (ras) and MEK1, but is independent of Gs, Gi and PKC. This appears to be among the first examples of coupling of the $A_{2A}AR$ to an effector other than adenylate cyclase.

Since the formation of tubes of endothelial cells is an early event in angiogenesis (Risau, 1997), adenosine may play a role in that process. However, adenosine is clearly not the primary stimulus in hypoxia or ischemic angiogenesis (Carmeliet and Collen, 1997; Dor and Keshet, 1997). Growth factors such as the peptide vascular endothelial growth factor (VEGF-A) are much more important. In addition to its receptor-mediated actions on endothelial cell proliferation, adenosine may stimulate DNA synthesis by a mechanism independent of receptors (Ethier and Dobson, 1997).

ACKNOWLEDGMENTS

This work was supported in part by the American Heart Association-Ed C. Wright Chair in Cardiovascular Research, University of South Florida. The authors are grateful to Prof. Luiz Belardinelli and Dr. John C. Shryock, University of Florida, for generously furnishing us preprints of their work.

REFERENCES

Abebe W, Hussain T, Olanrewaju H, Mustafa SJ. Role of nitric oxide in adenosine receptor-mediated relaxation of porcine coronary artery, Am J Physiol 1995;269:H1672-H1678.

Beall AC, Kat K, Goldenring JR, Rasmussen H, Brophy CM. Cyclic nucleotide-dependent vasorelaxation is associated with the phosphorylation of a small heat shock-related peptide. J Biol Chem 1997;272:11283-7.

Belardinelli L, Shryock JC, Snowdy S, Zhang Y, Monopoli A, Lozza G, Ongini E, Olsson RA, Dennis DM. The A2A adenosine receptor mediates coronary vasodilation. J Pharmacol Exp Ther In Press 1998.

Berne RM. Cardiac nucleotides in hypoxia: Possible role in regulation of coronary blood flow. Am J Physiol 1963;204:317-22.

Carmeliet E, Collen D. Genetic analysis of blood vessel formation. Role of endothelial versus smooth muscle cells. Trends Cardiovasc Med 1997;7:271-81.

Collis MG. Adenosine receptors in isolated tissue preparations. Nucleosides and Nucleotides 1991;10:1057-66.

Danialou G, Vicaut E, Sambe A, Aubier M, Boczskowski J. Predominant role of A1 adenosine receptors in mediating adenosine induced vasodilation of rat diaphragmatic arterioles: involvement of nitric oxide and ATP-dependent K+ channels. Br J Pharmacol 1997;121:1355-63.

Dart C, Standen NB. Adenosine-activated potassium current in smooth muscle cells isolated from the pig coronary artery. J Physiol 1993;471:767-86.

Daut J, Maier-Rudolph W, von Beckerath N, Mehrke G, G¸nther K, Goedel-Meinen L. Hypoxic dilation of coronary arteries is mediated by ATP-sensitive potassium channels. Science 1990;247:1341-4.

Daut J, Standen NB, Nelson MT. The role of the membrane potential of endothelial and smooth muscle cells in the regulation of coronary blood flow. J Cardiovasc Electrophysiol 1994;5:154-81.

Dor Y, Keshet E. Ischemia-driven angiogenesis. Trends Cardiovasc Med 1997;7:289-294.

Dora KA, Doyle MP, Duling BR. Elevation of intracellular calcium in smooth muscle causes endothelial cell generation of NO in arterioles. Proc Natl Acad Sci USA. 1997;94:6529-34.

Drury AN, Szent-Gyˆrgyi A. The physiological activity of adenine compounds with especial reference to their actions on the mammalian heart. J Physiol 1929;68:213-36.

Ethier M, Dobson JG, Jr. Adenosine stimulus of DNA synthesis in human endothelial cells. Am J Physiol 1997;272:H1470-H1479.

Feoktistov I, Biaggioni I. Adenosine A2B receptors. Pharmacol Revs 1997;49:381-402.

Fozard JR, Carruthers AM. Adenosine A3 receptors mediate hypotension in the angiotensin II-supported circulation of the rat. Br J Pharmacol 1993;109:3-5.

Fozard JR, Pfannkuche H-J, Schurrman H-J. Mast cell degranulation following adenosine A3-receptor activation in rats. Eur J Pharmacol 1996;298:293-7.

Gerlach E, Deuticke B, Dreisbach RH. Der Nucleotid-Abbau im Herzmuskel bei Sauerstoffmangel und seine mˆgliche Bedeutung f¸r die Coronardurchblutung. Naturwissenschaften 1963;50:228-9,.

Gˆdecke A, Decking UKM, Ding Z, Hirchenhain J, Bidmon H-J, Gˆdecke S, Schrader J. Coronary hemodynamics in endothelial NO synthase knockout mice. Circ Res 1998;82:186-94.

Hamilton HW, Taylor MD, Steffen RP, Haleen SJ, Bruns RF. Correlation of adenosine receptor affinities and cardiovascular activity. Life Sci 1987;41:2295-302.

Hannon JP, Fozard JR, Pfannkuche H-J. A role for mast cells in adenosine A3 receptor-mediated hypotension in the rat. Br J Pharmacol 1995;115:945-52.

Hussain T, Mustafa SJ. Binding of A1 adenosine receptor ligand [3H]8-cyclopentyl-1,3-dipropylxanthine in coronary smooth muscle. Circ Res 1995;77:194-8.

Iwamoto T, Umemura S, Uchibori T, Kogi T, Takagi N, Ishii M. Identification of adenosine A2 receptor-cAMP system in human aortic endothelial cells. Biochem Biophys Res Commun 1994;199:905-10.

Jacobson MA, Chakravarty PK, Johnson RG, Norton R. Novel selective non-xanthine A3 adenosine receptor antagonists. Drug Devel Res 1996;37:131.

Jiang J, van Rhee AM, Chang L, Patchornik A, Ji XD, Evans P, Melman N, Jacobson KA. Structure-activity relationships of 4-(phenyethynyl)-6-phenyl-1,4-dihydropyridines as highly selective A3 adenosine receptor antagonists. J Med Chem 1997;40:2596-608.

Jones CJH, Kuo L, Davis MJ, De Fily DV, Chilian WM. Role of nitric oxide in the coronary microvascular responses to adenosine and increased metabolic demand. Circulation 1995;91:1807-13.

Keddie JR, Poucher SM, Shaw GR, Brooks R, Collis MG. In vivo characterization of ZM 241385, a selective adenosine A2A receptor antagonist. Eur J Pharmacol 1996;301:107-13.

Kleppisch T, Nelson MT. Adenosine activates ATP-sensitive potassium channels in arterial myocytes via A2 receptors and cAMP-dependent protein kinase. Proc Natl Acad Sci USA 1995;92:12441-5.

Klotz K-N, Hessling J, Hegler J, Owman C, Kull B, Fredholm BB, Lohse MJ. Comparative pharmacology of human adenosine receptor subtypes ñ characterization of stably transfected receptors in CHO cells. Naunyn-Schmiedebergs Arch Pharmacol 1998;357:1-9.

Koglin G, von Scheidt W. Isolated defect of adenosine-mediated coronary vasodilation: Functional evidence for a new microangiopathic entity. J Am Coll Cardiol 1997;30:103-7.

Kuo L, Davis MJ, Chilian WM. Longitudinal gradients for endothelium-dependent and independent vascular responses in the coronary microcirculation. Circulation 1995;92:518-25.

Kuo l, Chancellor JD. Adenosine potentiates flow-induced dilation of coronary arterioles by activating K_{ATP} channels in endothelium. Am J Physiol 1995;269:H541-H549.

Ledent C, Vaugeols J-M, Schiffmann SN, Pedrazzini T, El Yacoubi M, Vanderhaeghen J-J, Costentin J, Heath JK, Vassart G, Parmentier M. Aggressiveness, hypoalgesia and high blood pressure in mice lacking the A2a adenosine receptor. Nature 1997;388:674-8.

Li JM, Fenton RA, Cutler BS, Dobson JG Jr. Adenosine enhances nitric oxide production by vascular endothelial cells. Am J Physiol 1995;269:C519-C523.

Liao JC, Kuo l. Interaction between adenosine and flow-induced dilation in coronary microvascular network. Am J Physiol 1997;272:H1571-H1581.

Little TL, Beyer EB, Duling BR. Connexin 43 and connexin 40 gap junctional proteins are present in arteriolar smooth muscle and endothelium in vivo. Am J Physiol 1995a;268:H729-H739.

Little TL, Xia J, Duling BR. Dye tracers define different endothelial and smooth muscle coupling patterns within the arteriolar wall. Circ Res 1995b;76:498-504.

Lückhoff A, Busse R. Calcium influx into endothelial cells and formation of endothelium-derived relaxing factor is controlled by the membrane potential. Pfl‚gers Arch 1990a;416:305-11.

Lückhoff A, Busse R. Activators of potassium channels enhance calcium influx into endothelial cells as a consequence of potassium currents. Naunyn-Schmiedebergs Arch Pharmacol 1990b;342:94-9.

Martin PL. Relative agonist potencies of C2-substituted analogues of adenosine: evidence for A2B receptors in the guinea pig aorta. Eur J Pharmacol 1992;216:235-42.

Martin PL, Ueeda M, Olsson RA. 2-Phenylethoxy-9-methyladenine, an adenosine receptor antagonist that discriminates between A2 adenosine receptors in the aorta and coronary vessels of the guinea pig. J Pharmacol Exp Ther 1993;265:248-53.

Meininger CJ, Schelling ME, Granger HJ. Adenosine and hypoxia stimulate proliferation and migration of endothelial cells. Am J Physiol 1988;255:H544-H562.

Merkel LA, Lappe RW, Rivera LM, Cox BF, Perrone MH. Demonstration of vasorelaxant activity with an A1-selective adenosine agonist in porcine coronary artery. J Pharmacol Exp Ther 1992;260:437-43.

Muller JM, Davis MJ, Chilian WM. Integrated regulation of pressure and flow in the coronary microcirculation. Cardisovasc Res 1992;32:668-78.

Mutafova-Yambolieva V, Keef KD. Adenosine-induced hyperpolarization in guinea pig coronary artery involves A2b receptors and K_{ATP} channels. Am J Physiol 1997;273:H2687-H2695.

Olsson RA, Pearson JD. Cardiovascular purinoceptors. Pharmacol Rev 1990;70:761-845.

Olsson RA. Adenosine receptors in the cardiovascular system. Drug Devel Res 1996;39:301-7.

Ongini E, Fredholm BB. Pharmacology of adenosine A2A receptors. Trends Pharmacol Sci 1996;17:364-72.

Poucher SM, Keddie JR, Singh P, Stoggall SM, Caulkett PW, Jones G, Collis MG. The in vitro pharmacology of ZM241385, a potent, non-xanthine A2a selective adenosine receptor antagonist. Br J Pharmacol 1995;115:1096-1102.

Risau E. Mechanisms of angiogenesis. Nature 1997;386:671-4.

Schiele JO, Schwabe U. Characterization of the adenosine receptor in microvascular endothelial cells. Eur J Pharmacol 1994;269:51-8.

Sexl V, Mancusi G, Holler C, Gloria-Maercker E, Schutz W, Freissmuth M. Stimulation of the mitogen-activated protein kinase via the A2A-adenosine receptor in primary human endothelial cells. J Biol Chem 1997;272:5792-9.

Shearman LP, Weaver DR. [125I]4-aminobenzyl-5í-N-methylcarboxamidoadenosine ([125I]AB-MECA) labels multiple adenosine receptor subtypes in rat brain. Brain Res 1997;745:10-20.

Shepherd RK, Linden J, Duling BR. Adenosine-induced vasoconstriction in vivo. Role of the mast cell and A3 receptor. Circ Res 1996;78:627-34.

Shryock JC, Snowdy S, Baraldi PG, Cacciari B, Spalluto G, Monopoli A, Ongini E, Baker SP, Belardinelli L. A2A-adenosine receptor reserve for coronary vasodilation. Circulation In Press, 1998.

Skinner MR, Marshall JM. Studies on the roles of ATP, adenosine and nitric oxide in mediating muscle vasodilation induced in the rat by acute systemic hypoxia. J Physiol 1996;495:553-60.

Sobrevia L, Yudilevich DL, Mann GE. Activation of A2-purinoceptors by adenosine stimulates L-arginine transport (system y+) and nitric oxide synthesis in human fetal endothelial cells. J Physiol 1997;499:135-40.

Srinavas M, Shryock JC, Dennis DM, Baker SP, Belardinelli L. Differential A1 adenosine receptor reserve for two actions of adenosine on guinea pig atrial myocytes. Mol Pharmacol 1997;52:683-91.

Tucker A, Linden J. Cloned adenosine receptors and cardiovascular responses to adenosine. Cardiovasc Res 1992;27:62-7.

Ueeda M, Thompson RD, Padgett WL, Secunda S, Daly JW, Olsson RA. Cardiovascular actions of adenosines, but not adenosine receptors, differ in rat and guinea pig. Life Sci 1991;49:1351-8.

Uehata M, Ishizaki T, Satoh H, Ono T, Kawahara T, Morishita T, Tamakawa H, Yamagami K, Inui J, Maekawa M, Narumiya S. Calcium sensitization of smooth muscle mediated by Rho-associated protein kinase in hypertension. Nature 1997;389:990-4.

Vials A, Burnstock G. A2-purinoceptor-mediated relaxation in the guinea pig vasculature: a role for nitric oxide. Br J Pharmacol 1993;109:424-9.

Xia J, Duling BR. Electromechanical coupling and the conducted vasomotor response. Am J Physiol. 1995;269:H2022-H2030.

Zhao Z, Francis CE, Ravid K. An A3-subtype adenosine receptor is highly expressed in rat vascular smooth muscle cells: Its role in attenuating adenosine-induced increase in cAMP. Microvasc Res 1997;54:243-52.

Zocchi C, Ongini E, Conti A, Monopoli A, Negretti A, Baraldi PG, Dionisotti S. The non-xanthine heterocyclic compound SCH 58261 is a new potent and selective A2a adenosine receptor antagonist. J Pharmacol Exp Ther 1996;276:398-404.

ROLES OF ADENOSINE IN ANGIOGENESIS

Harris J. Granger, Cynthia Meininger, Marina Ziche and John Hood

Microcirculation Research Institute and Department of Medical Physiology, Texas A&M University System Health Science Center, College Station, TX and Department of Pharmacology, University of Florence, Florence, Italy

I. SUMMARY

Angiogenesis, the formation of new microvessels from a pre-existing vascular bed, is stimulated by tissue hypoxia. Adenosine, a key extracellular signal elicited by hypoxia, plays several roles in initiation and/or modulation of angiogenesis. Adenosine governs the angiogenesis process through its actions on endothelial cell proliferation and migration. In addition, adenosine amplifies the mitogenic signal elicited by other angiogenic mediators such as nitric oxide and vascular endothelial growth factor. Through the interplay of tissue oxygen tension, adenosine, nitric oxide and vascular endothelial growth factor, a complex physiological control system ensures the matching of vascularity and metabolic needs in a variety of organs and circumstances.

II. BACKGROUND

Angiogenesis, the formation of new microvessels from a pre-existing microvascular bed, is part of a key physiological strategy to prevent cellular anoxia following restriction of arterial oxygen supply or acceleration of tissue oxygen demand. Consequently, microvascular proliferation is elicited by hypoxemia and hypoperfusion, and is a hallmark of rapidly growing and/or hypermetabolic tissues. Thus, angiogenesis is observed at high altitude, after partial arterial obstruction, in tumors, in cardiac hypertrophy, in skeletal muscles of trained individuals and during endometrial growth in preparation for implantation of the fertilized ovum. The ability of the angiogenic process to stabilize tissue oxygenation by matching vascularity to metabolic requirements implies the existence of regulatory mechanisms triggered by hypoxia and associated chemical mediators. Adenosine, a recognized signal of cellular hypoxia, plays several roles in matching tissue vascularity to parenchymal metabolic needs.

In this chapter, we first consider the multiple processes responsible for the formation of new microvessels. Next we review the interplay between hypoxia and angiogenesis in vivo and in vitro. After presenting evidence for adenosine modulation of endothelial cell proliferation and migration, we examine the role of adenosine in regulating the formation and action of other angiogenic factors, including nitric oxide and vascular endothelial growth factor. Finally, we present an integrated view of the interactons of adenosine, nitric oxide and vascular endothelial growth factor in overall regulation of angiogenesis in response to parenchymal hypoxia.

III. TEXT

A. The Process Involved in Angiogenesis

Angiogenesis occurs primarily in postcapillary or pericytic venules (Clark ER, et al, 1939; Myrhage R, et al, 1978; Phillips GD, et al, 1991; Folkman J, 1982; Diaz-Flores L, et al, 1992). These microvessels are formed by the convergence of two or more capillaries. Postcapillary venules are composed of a continuous layer of endothelium encircled over a fraction of their ablumenal surface by pericytes (Wagner RC, et al, 1992). These unique microvessels are more permeable than capillaries (Simionescu N, et al, 1978); are the major sites of leukocyte (Lewis RE, et al, 1988) and tumor cell (Ward KA, et al, 1989) adhesion and extravasation; and exhibit the highest density of receptors for inflammatory autacoids (Heltianu C, et al, 1982; Majno G, et al, 1969) and growth factors (Qu-Hong, et al, 1995).

Angiogenesis is a multi-step process (Folkman J, et al, 1992). First, upon detection of the angiogenic factor, endothelial cells of the postcapillary venule must detach from their basement membrane and adjacent cells (Ausprunk DH, et al, 1977), leading to venular hyperpermeability (Flamme I, et al, 1995). Activated endothelial cells secrete enzymes, such as collagenase (Unemori EN, et al, 1992) and tissue plasminogen activator (Gross JL, et al, 1983), that are capable of dissolving the basement membrane. In addition, cell/matrix adhesion mediated by integrins (Cheng YF, et al, 1989) is reduced. Cell/cell attachments via cadherins (Lampugnani MG, et al, 1995) are also downregulated. Detachment of the endothelial cell is followed by migration away from the long-axis of the parent venule and toward the angiogenic stimulus (Ratajska A, et al, 1995). As the cells move by chemotaxis they divide, leading to the formation of a long cord of aligned endothelial cells (Folkman J, et al, 1980; Davis GE, et al, 1995). By a poorly understood process, large vacuoles appear in these cells; the serial registration of these holes in the chain of endothelial cells leads to the formation of a primitive tube (Davis GE, et al, 1996). This tube possesses a lumen but exhibits an abnormally high permeability to fluid and plasma proteins (Mayerhofer A, et al, 1990; D'Amore PA, 1992). After pericytes arrive on the scene and envelop the tubular structure, the exchange microvessel acquires its mature phenotype (Cassin S, et al, 1971).

B. Hypoxia and Angiogenesis

Hypoxia has long been considered a driving force for the sprouting of new microscopic blood vessels (Cassin S, et al, 1971; Kayar SR, et al, 1985). Vascularity is increased in humans moved from sea level to high altitude (Kayser B, et al, 1991). Capillary density is increased in skeletal muscle subjected to chronic motor nerve stimulation by artificial (Mathieu-Costello O, et al, 1996) or natural means (Gute D, et al, 1996). Rapidly growing tumors, known to consume oxygen at high rates and to exhibit low cellular pO_2, present with dense, immature microvascular networks (Warren BA, et al, 1966). Lowering the oxygen content of fluid bathing the chick chorioallantoic membrane leads to neovascularization (Dusseau JW, et al, 1988). These are but a few examples of in vivo studies that point to a key role for hypoxia in triggering angiogenesis in a variety of tissues.

Low pO_2 initiates proliferation of isolated endothelial cells (Burton HW, et al, 1986; Meininger CJ, et al, 1988). Thus, endothelial cells contain all the machinery required to detect hypoxia and to automatically initiate the proliferation process (Figure 1). The ability of these cells to detect and/or respond to hypoxia is blocked by the non-specific adenosine receptor blocker, 8-phenyltheophylline, suggesting a possible role for adenosine as a chemical mediator of angiogenesis (Mcininger CJ, et al, 1988). Endothelial cells are endowed with A_1, A_{2A}, A_{2B} and A_3 receptors (Luty J, et al, 1989; Schiele JO, et al, 1994; Iwamoto T, et al, 1994; Stanimirovic DB, et al, 1994). In in vitro studies, the most likely scenario is that hypoxic endothelial cells release adenosine into the medium and the observed increase in cell division is due to autocrine stimulation of adenosine receptors on the cell surface. In the intact hypoxic tissue, adenosine derives from a number of sources including parenchymal, interstitial and vascular cells.

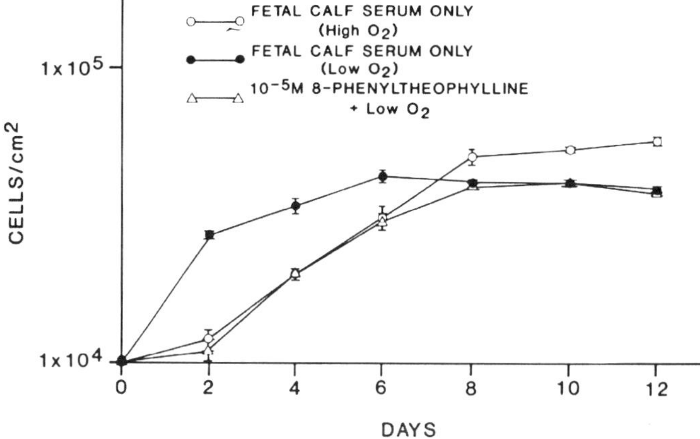

Figure 1. Proliferation of coronary venular endothelial cells (CVEC) in normoxic and hypoxic media. CVEC proliferate faster at the lower oxygen tension. In the presence of 8-phenyltheophylline, an adenosine receptor blocker, hypoxia fails to stimulate proliferation beyond the normoxic rate. Reproduced from Meininger et al, 1988.

C. Adenosine as an endothelial mitogen, chemoattractant and angiogenic factor

The impact of hypoxia on endothelial cells is mimicked by adding adenosine or adenosine analogs to culture medium maintained at normoxic levels (Meininger CJ, et al, 1988; Meininger CJ et al, 1990; Ethier MF, et al, 1993; Rathbone MP, et al, 1992). The proliferative effect of adenosine in bovine coronary venular endothelial cells (CVEC) requires adenosine/receptor interaction since polyadenosine, a large non-transportable polymer, is effective in stimulating cell division and a blocker of adenosine reuptake does not inhibit the response. In human umbilical vein endothelial cells (HUVEC), addition of adenosine deaminase to the medium reduces normal and adenosine-elicited proliferation (Ethier MF, et al, 1993). Inosine and hypoxanthine are not mitogenic for either CVEC (Meininger CJ et al, 1990) or HUVEC (Ethier MF, et al, 1993). Adenosine-induced proliferation of CVEC requires activation of adenylate cyclase via the stimulatory G-protein G_S (Meininger CJ et al, 1990). This latter conclusion is supported by several lines of evidence. First, adenosine and its analogs stimulate a transient increase in cyclic AMP (cAMP) in CVEC. Second, inhibition of adenylate cyclase with 2', 5'-dideoxyadenosine abolishes the adenosine-induced proliferation. Third, pulsed application of cholera toxin, an activator of G_S, stimulates CVEC proliferation. Although we were not successful in identifying the specific adenosine receptor using putative specific receptor blockers and activators, the observed transient rise in cAMP and the similar response achieved by activation of G_S suggest that signaling in CVEC occurs through an A_{2A} receptor/G_S/adenylate cyclase/cAMP/protein kinase A (PKA) cascade (Meininger CJ et al, 1990; Rathbone MP, et al, 1992; Legrand AB, et al, 1989).

In contrast to our findings using CVEC, others have observed a reduction in endothelial cell proliferation with membrane-permeable analogs of cAMP) or stimulation of adenylate cyclase with forskolin (Leitman DC, et al, 1986; Moodie SA, et al, 1991). However, adenosine stimulation and a short pulse of cholera toxin elicit a transient, rather than sustained, elevation of cytosolic cAMP. With membrane-permeable cAMP analogs and continuous exposure to cholera toxin or forskolin, cAMP levels are likely to remain high throughout the experiment. Moreover, chronic exposure to cholera toxin downregulates G_S and tends to throttle signaling processes dependent on this G-protein. In any event, the temporal pattern of second messenger concentration may be a part of the signaling code that directs cellular reactions to different types of activation. Thus, chronic elevation of cAMP may not mimic physiological conditions.

Adenosine activates thymidine incorporation (Sexl V, et al, 1997) and proliferation (Sexl V, et al, 1995) via the ras/MAP kinase cascade in HUVEC . Activation of the A_{2A} receptor leads to conversion of inactive ras-GDP to the active ras-GTP. In addition, MAP kinase activity is increased in conjunction with translocation of the enzyme to the nuclear region. Long-term pretreatment of HUVEC with cholera toxin to downregulate G_S did not inhibit adenosine-elicited MAP kinase stimulation; thus, G_S does not appear to be the link between receptor ligation and ras-GTP formation. Moreover, G_i and G_0 are not involved since pertussis toxin failed to inhibit the adenosine-induced MAP kinase stimulation. Finally, G_q, a pertussis toxin-insensitive isoform responsible for activation of phospholipase $C_{\beta 1}$, is not likely to be recruited since adenosine does not cause increased cytosolic calcium in endothelial cells. Inhibition of MAP kinase kinase, the activator of MAP kinase, prevents phosphorylation of MAP kinase and reduces adenosine-induced thymidine incorporation.

In a recent report, adenosine-stimulated DNA synthesis in HUVEC was blocked by inhibitors of the Na^+/H^+ exchanger or phospholipase A_2 (Either MF, et al, 1997). DNA synthesis was not stimulated by specific receptor agonists or blocked by specific receptor antagonists, suggesting that non-classical, receptor-independent mechanisms may play a role in adenosine stimulation of endothelial proliferation.

Adenosine is a chemoattractant (Meininger CJ, et al, 1988; Teuscher E, et al, 1985; Bull DA, et al, 1993; Montesinos MC, et al, 1997); therefore, endothelial cells migrate toward higher concentrations of adenosine. Adenosine-induced chemotaxis appears to be cAMP-dependent (Montesinos MC, et al, 1997). Thus, at least 2 of the processes (i.e. mitogenesis and migration) involved in angiogensis can be demonstrated in vitro using endothelial cells alone.

Adenosine stimulates neovascularization in vivo. Application of adenosine to the chick chorioallantoic membrane causes increased capillarity (Dusseau JW, et al, 1986). In addition, chronic adenosine administration reduces the minimal resistance of the chich embryo vasculature, implying increased number and size of vascular channels (Adair TH, et al, 1989). Finally, chronic intracoronary infusion of adenosine leads to increased myocardial vascularity (Hudlicka O, et al, 1992). Thus, the participation of adenosine in angiogenesis is supported by studies on cultured cells and intact tissues.

<u>D. Adenosine as a modulator of NO signaling</u>

A number of stimuli, ranging from shear rate (Lamontagne D, et al, 1992; Yuan Y, et al, 1992) to receptor ligation (Moncada S, et al, 1995; Yuan Y, et al, 1993), cause increased production of nitric oxide (NO) by endothelial cells. The predominant form of NO synthase in endothelial cells is appropriately termed ecNOS. This constitutive enzyme is activated by calcium/calmodulin. Thus, receptors linked to phospholipase C stimulate ecNOS via release of internal calcium stores. Although the role of endothelium-derived NO as a paracrine regulator of vascular smooth muscle tone has received the greatest attention, the entire NO/soluble guanylate cyclase (sGC)/cyclic GMP (cGMP)/ cyclic GMP-dependent protein kinase (PKG) cascade operates within endothelial cells (Yuan Y, et al, 1993); thus, NO is also an autocrine regulator of endothelial functions. NO triggers many of the processes involved in angiogenesis including hyperpermeability (i.e. detachment) (Yuan Y, et al, 1993; Yuan Y, et al, 1993), proliferation (Ziche M, et al, 1993; Ziche M, et al, 1993; Ziche M, et al, 1997), migration (Ziche M, et al, 1994) and tube formation (Papapetropoulos A, et al, 1997). In most of these processes, NO acts through the aforementioned cascade leading to activation of PKG and subsequent serine phosphorylation of cellular proteins. A reduced capacity to produce NO has been demonstrated in microvascular endothelium of diabetic rats, perhaps providing a clue to the loss in vascularity asssociated with the disease (Wu G, et al, 1995).

The calcium-independent, inducible form of nitric oxide synthase, iNOS, is expressed in endothelial cells activated by bacterial lipopolysaccharide (LPS) and cytokines such as the interleukins (IL), tumor necrosis factor (TNF), and interferon (IF) (Borgerding RA, et al, 1995; Kanno K, et al, 1994; Cendan JC, et al, 1994;). In addition, iNOS is upregulated in parenchymal and other cells surrounding vascular

5 3

endothelium upon exposure to these agents (Doi K, et al, 1996; Oddis CV, et al, 1996). The inducible NOS is a more active catalyst than ecNOS; therefore, NO production reaches high levels when iNOS is activated in vascular endothelium and surrounding cells. Cytokines capable of inducing iNOS also stimulate angiogenesis (Montrucchio G, et al, 1997), further supporting a central role for NO in neovascularization. Moreover, inhibitors of NOS inhibit angiogenesis elicited by stimulators of iNOS expression (Orucevic A, et al, 1996).

Adenosine is a positive modulator of NO production by ecNOS and iNOS (Li JM, et al, 1995; Iimura O, et al, 1996; Ikeda U, et al, 1997; Ikeda U, et al, 1996; Janigro D, et al, 1996). In porcine carotid artery endothelial cells, a one hour exposure to adenosine increases NO production, presumably through modulation of ecNOS; the response is inhibited by adenosine receptor blockade, supporting a receptor-mediated phenomenon (iNOS (Li JM, et al, 1995). Adenosine also stimulates production of NO by cardiac myocytes previously exposed to IL-1, TNF and/or LPS; this augmentation of NO formation is A_2 receptor-mediated, cyclic AMP-dependent, and requires increased production of iNOS mRNA and protein secondary to PKA activation (Ikeda U, et al, 1997). Similar results are observed after exposure of cytokine-activated cardiac myocytes to dibutryl cAMP (Ikeda U, et al, 1996).

E. Adenosine as amodulator of VEGF signaling

As emphasized above, the intimate relationship between hypoxia and proliferation of capillaries has been a central theme in many investigations of angiogenesis. With the discovery of vascular endothelial growth factor (VEGF), the linkage between hypoxia and vascularity is more clearly defined. VEGF is a secreted mitogen specific for endothelial cells (Neufeld G, et al, 1994). This growth factor increases microvascular permeability (Dvorak HF, et al, 1995), stimulates endothelial migration (Chen Z, et al, 1997; Parenti A, et al, 1996), initiates mitogenesis (Klagsbrun M, et al, 1996), and mediates tube formation (Koolwijk P, et al, 1996). VEGF is produced by a variety of parenchymal (Monacci WT, et al, 1993), tumor (Dvorak HF, et al, 1991), and vascular (Klagsbrun M, et al, 1996; Gu JW, et al, 1997) cells, including endothelium. VEGF production is oxygen-sensitive (Banai S, et al, 1994; Ladoux A, et al, 1993; Shweiki D, et al, 1992; Minchenko A, et al, 1994; Namiki A, et al, 1995); that is, hypoxia stimulates increased formation and secretion of the growth factor. Endothelial cells detect VEGF through two different cell surface receptor kinases, namely, flt1 and KDR/flk1 (Neufeld G, et al, 1994; Klagsbrun M, et al, 1996). The number of VEGF receptors is upregulated by hypoxia (Vaisman N, et al, 1990; Brogi E, et al, 1996). Many of the processes elicited in mature endothelium by VEGF appear to be mediated mainly by the lower affinity KDR receptor. VEGF activates the phosphoinositide (Guo D, et al, 1995; Abedi H, et al, 1997) and MAP kinase (Abedi H, et al, 1997) signaling pathways through the actions of receptor tyrosine kinase. Through its activation of phospholipase $C_{\gamma 1}$, VEGF recruits the ecNOS/NO/sGC/cGMP/PKG signaling cascade (Ku DD, et al, 1993; Ziche M, et al, 1997; Hood JD, et al). Consequently, the major actions of VEGF on endothelium are NO-dependent (Brock TA, et al,

1991; Morbidelli L, et al, 1996; Wu HM, et al, 1996); inhibition of ecNOS, sGC or PKG prevents VEGF stimulation of endothelial cell proliferation (Morbidelli L, et al, 1996) and venular hyperpermeability (Wu HM, et al, 1996). Finally, the VEGF-activated NO signaling pathway stimulates the ras/MAP kinase cascade to enhance phosphorylation of nuclear modulators (Parenti A, et al), thereby triggering cell division.

Adenosine augments VEGF signaling by at least three different mechanisms. First, given adenosine's stimulatory action on ecNOS (Li JM, et al,1995), the rate of NO production elicited by VEGF-induced phospholipase $C_{\gamma 1}$ activation is amplified. Second, adenosine upregulates VEGF mRNA and protein in endothelial and other cells receptor (Takagi H, et al, 1996; Hashimoto E, et al, 1994; Fischer S, et al, 1995). Third, hypoxic upregulation of VEGF receptors on endothelial cells may be mediated via adenosine (Takagi H, et al, 1996).

Figure 2 provides an overview of the potential intracellular signaling pathways involved in adenosine modulation of endothelial cell proliferation, and summarizes the preceeding discussion.

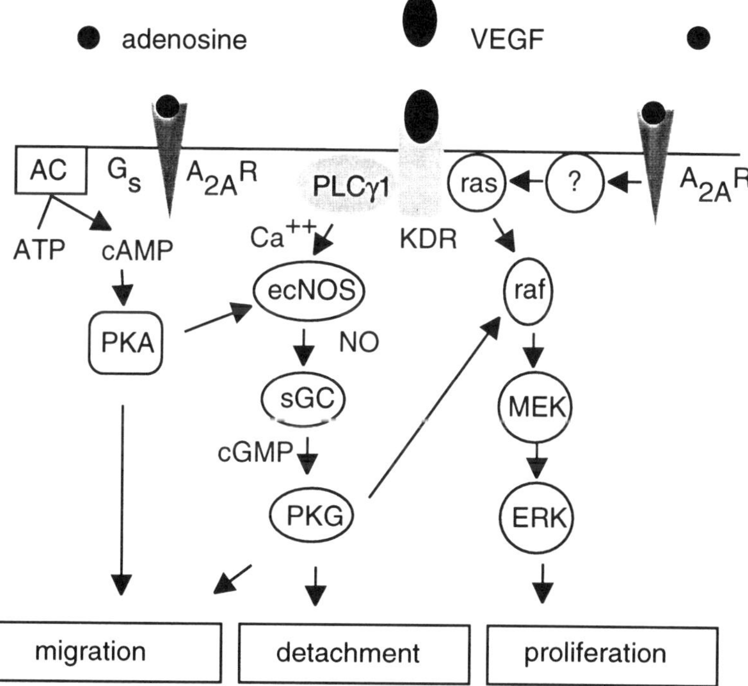

Figure 2.   Potential intracellular signaling pathways involved in adenosine modulation of endothelial cell proliferation. AC, adenylate cyclase. All other abbreviations defined in text.

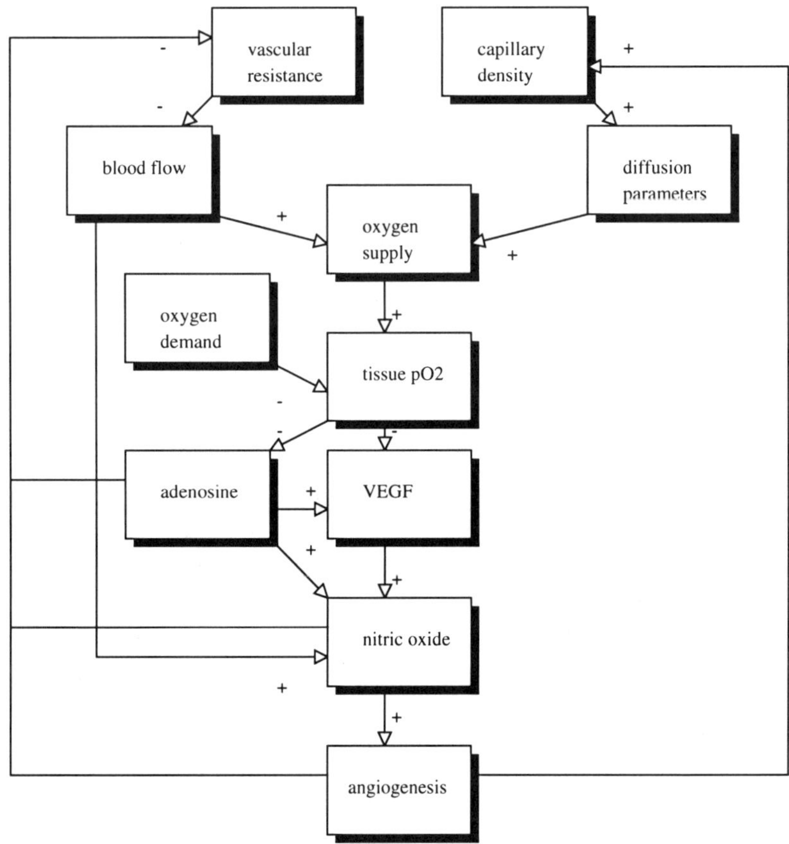

Figure 3.   An integrated view of control of angiogenesis via interactions of tissue pO2, adenosine, VEGF and NO.

<u>F. Integrated control of angiogenesis: interaction of adenosine, NO and VEGF</u>

Figure 3 provides an physiological overview of the linkage between hypoxia and angiogenesis with adenosine, nitric oxide and VEGF as key mediators. Tissue pO_2 at a given moment reflects the balance between oxygen supply and demand, and is stablized by changes in blood flow and number of perfused exchange microvessels (Granger HJ, et al, 1976). Hypoxia lasting more than a few minutes results in the transcription of VEGF mRNA and secretion of the protein into the interstitial fluid.

56

VEGF initiates angiogenesis and stimulates all phases from EC detachment to tube formation via activation of ecNOS and the production of NO. In addition to its angiogenic properties, VEGF elicits endothelium-dependent, NO-mediated vasodilation, thereby increasing blood flow. The increased flow results in elevated shear rate in all vessels including postcapillary venules, thereby augmenting NO production further. Adenosine, a breakdown product of ATP, also accumulates during hypoxia and causes vasodilation, leading to further increases in shear rate and additional stimulation of endothelial NO production. Adenosine may be the chemical linkage between tissue hypoxia and VEGF production. In addition, hypoxia upregulates VEGF receptors on EC, probably through an adenosine-dependent mechanism. Finally, adenosine directly stimulates NO production by EC and parenchymal cells. The net effect of these concerted reactions is to tip the balance strongly in the direction of pro-angiogenic influences. With the formation of new endothelial tubes, oxygen supply returns toward normal; consequently, tissue levels of adenosine, VEGF and NO return toward normal and the stimulus for further neovascularization wanes. This paradigm can explain increased angiogenesis during chronic hypoperfusion, elevated oxygen demand and chronic administration of vasodilators (via the shear-dependent component only). The disappearance of skeletal muscle microvessels in chronic arterial hypertension and diabetes may result from a defect in NO production or an initial phase of excess oxygen supply secondary to hyperperfusion. The schema also applies to tumor circulation as an extreme example of increased oxygen demand.

IV. CONCLUSION AND FUTURE DIRECTIONS

The ability of adenosine to modulate moment-to-moment vascular smooth muscle tone in accordance with tissue metabolic requirements provides an elegant feedback system for locally adjusting blood flow within a time frame of seconds to a few minutes (Berne RM, et al, 1983; Granger HJ, et al, 1980; Morff RJ, et al, 1983). When the physiological range of this local vasomotor feedback is exceeded, another important physiological action of adenosine-its capacity to initiate and/or modulate angiogenesis-becomes evident, and a long-term restructuring of the microvascular bed occurs. Our current knowledge of the angiogenic properties of adenosine point to its potential role in salvaging hypoxic parenchymal cells by effective clinical strategies for augmenting tissue concentrations of this multi-faceted vasoactive agent. Indeed, the reparative process of wound healing, a phenomenon critically dependent on growth of new vessels, is greatly accelerated by application of adenosine or its analogs to damaged tissue in normal and diabetic states (Bull DA, et al, 1993). In spite of the numerous studies described in this review, knowledge of the mechanisms of adenosine-induced proliferation is still very limited. In addition, we know nothing about the possible roles of adenosine in inhibiting anti-angiogenic processes in normal and diseased tissues. Furthermore, the impact of adenosine on growth and proliferation of vascular smooth muscle cells has not been addressed. A significant challenge in vascular biology today is to achieve a clear view of how adenosine and other mediators govern vessel formation and devolution.

ACKNOWLEDGEMENTS

The authors are grateful to Kim Manry for excellent clerical assistance. The authors' research described in this review was supported by MERIT Award HL-21498 from the NHLBI and a grant from the Texas Advanced Technology and Research Program to HJG, grants-in-aid from the American Heart Association to CJM and grants to MZ from Associazione Italiana Ricerca sul Cancro (AIRC) (Special Project "Angiogenesis"), Ministero dell'Universita' e della Ricerca Scientifica e Tecnologica (MURST), and Consiglio Nazionale delle Ricerche (CNR) (Project n. 95.02983.CT14).

V. REFERENCES

Abedi H, Zachary I: Vascular endothelial growth factor stimulates tyrosine phosphorylation and recruitment to new focal adhesions of focal adhesion kinase and paxillin in endothelial cells. J Biol Chem 1997;272:15442-15451.

Adair TH, Montani JP, Strick DM, Guyton AC: Vascular development in chick embryos: a possible role for adenosine. Am J Physiol 1989;256:H240-H246.

Ausprunk DH, Folkman J: Migration and proliferation of endothelial cells in preformed and newly formed blood vessels during tumor angiogenesis. Microvasc Res 1977;14:53-65.

Banai S, Shweiki D, Pinson A, Chandra M, Lazarovici G, Keshet E. Upregulation of vascular endothelial growth factor expression induced by myocardial ischaemia: implications for coronary angiogenesis. Cardiovas Res 1994;28:1176-1179.

Berne RM, Knabb RM, Ely SW, Rubio R: Adenosine in the local regulation of blood flow: a brief overview. Fed Proc 1983;42:3136-3140.

Borgerding RA, Murphy S. Expression of inducible nitric oxide synthase in cerebral endothelial cells is regulated by cytokine-activated astrocytes. J Neurochem 1995;65:1342-1347.

Brock TA, Dvorak HF, Senger DR. Tumor-secreted vascular permeability factor increases cytosolic Ca2+ and von Willebrand factor release in human endothelial cells. Am J Pathol 1991;138:213-218.

Brogi E, Schatteman G, Wu T, Kim EA, Varticovski L, Keyt B, Isner JM. Hypoxia-induced paracrine regulation of vascular endothelial growth factor receptor expression. J Clin Invest 1996;97:469-474.

Bull DA, Seftor EA, Hendrix MJ, Larson DF, Hunter GC, Putnam CW. Putative vascular endothelial cell chemotactic factors: comparison in a standardized migration assay. J Surg Res 1993;55:473-479.

Burton HW, Barclay JK. Metabolic factors from exercising muscle and the proliferation of endothelial cells. Med & Sci Sports & Exer 1986;18:390-395.

Cassin S, Gilbert RD, Bunnell CE, Johnson EM: Capillary development during exposure to chronic hypoxia. Am J Physiol 1971;220:448-451.

Cendan JC, Moldawer LL, Souba WW, Copeland EM III, Lind DS. Endotoxin-induced nitric oxide production in pulmonary artery endothelial cells is regulated by cytokines. Arch Sur 1994;129:1296-1300.

Chen Z, Fisher RJ, Riggs CW, Rhim JS, Lautenberger JA: Inhibition of vascular endothelial growth factor-induced endothelial cell migration by ETS1 antisense oligonucleotides. Cancer Res 1997;57:2013-2019.

Cheng YF, Kramer RH. Human microvascular endothelial cells express integrin-related complexes that mediate adhesion to the extracellular matrix. J Cell Physiol 1989;139:275-279.

Clark ER, Clark, EL: Microscopic observations on the growth of blood capillaries in the living mammal.

Am J Anat 1939; 64:251-265.

D'Amore PA: Capillary growth: a two-cell system. Sem Cancer Biol. 1992;3:49-56.

Davis GE, Camarillo CW: An alpha2 beta1 integrin-dependent pinocytic mechanism involving intracellular vacuole formation and coalescence regulates capillary lumen and tube formation in three-dimensional collagen matrix. Exptl Cell Res 1996;224:39-51,.

Davis GE, Camarillo CW: Regulation of endothelial cell morphogenesis by integrins, mechanical forces, and matrix guidance pathways. Exptl Cell Res 1995;216:113-123.

Diaz-Flores L, Gutierrez R, Varela H: Behavior of postcapillary venule pericytes during postnatal angiogenesis. J Morphol 1992;213:33-39.

Doi K,Akaike T, Horie H, Noguchi Y, Fujii S, Beppu T, Ogawa M, Maeda H: Excessive production of nitric oxide in rat solid tumor and its implication in rapid tumor growth. Cancer. 77(Suppl 8):1598-1604, 1996

Dusseau JW, Hutchins PM, Malbasa DS. Stimulation of angiogenesis by adenosine on the chick chorioallantoic membrane. Circ Res 1986;59:163-170.

Dusseau JW, Hutchins PM. Hypoxia-induced angiogenesis in chick chorioallantoic membranes: a role for adenosine. Resp Physiol 1988;71:33-44.

Dvorak HF, Brown LF, Detmar M, Dvorak AM. Vascular permeability factor/vascular endothelial growth factor, microvascular hyper-permeability, and angiogenesis. Am J Pathol 1995;146:1029-1034.

Dvorak HF, Sioussat TM, Brown LF, Berse B, Nagy JA, Sotrel A, Manseau EJ, Van de Water L, Senger DR: Distribution of vascular permeability factor (vascular endothelial growth factor) in tumors: concentration in tumor blood vessels. J Exp Med 1991;174:1275.

Ethier MF, Chander V, Dobson JG Jr. Adenosine stimulates proliferation of human endothelial cells in culture. Am J Physiol 1993;265:H131-H138.

Ethier MF, Dobson JG. Adenosine stimulation of DNA synthesis in human endothelial cells. Am J Physiol-Heart Circ Physiol 1997;41:H1470-H1479.

Fischer S, Sharma HS, Karliczek GF, Schaper W. Expression of vascular permeability factor/vascular endothelial growth factor in pig cerebral microvascular endothelial cells and its upregulation by adenosine. Brain Res Mol Brain Res 1995;28:141-146.

Flamme I, von Reutern M, Drexler HC, Syed-Ali S, Risau W: Overexpression of vascular endothelial growth factor in the avian embryo induces hypervascularization and increased vascular permeability without alterations of embryonic pattern formation. Developmental Biology. 1995;171:399-414,.

Folkman J, Haudenschild C. Angiogenesis in vitro. Nature 1980;288:551-558.

Folkman J. Shing Y. Angiogenesis. J Biol Chem 1992;267:10931-10934.

Folkman J: Angiogenesis: initiation and control. Ann NY Acad Sci 1982;401:212-215.

Granger HJ, Goodman AH, Granger DN. Role of resistance and exchange vessels in local microvascular control of skeletal muscle oxygenation. Circ Res 1976;38: 379-385.

Granger HJ, Norris CP. Role of adenosine in local control of intestinal circulation in the dog. Circ Res 1980;46: 764-770.

Gross JL, Moscatelli D, Rifkin DB. Increased capillary endothelial cell protease activity in response to angiogenic stimuli in vitro. Proc Natl Acad Sci USA 1983;80:2623-2629.

Gu JW, Adair TH: Hypoxia-induced expression of VEGF is reversible in myocardial vascular smooth muscle cells. Am J Physiol 1997;273:H628-H633.

Guo D, Jia Q, Song HY, Warren RS, Donner DB. Vascular endothelial cell growth factor promotes tyrosine phosphorylation of mediators of signal transduction that contain SH2 domains. Association with endothelial cell proliferation. J Biol Chem 1995;270:6729-6736.

Gute D, Fraga C, Laughlin MH, Amann JF: Regional changes in capillary supply in skeletal muscle of high-intensity endurance-trained rats. J Appl Physiol 1996;81:619-626.

Hashimoto E, Kage K, Ogita T, Nakaoka T, Matsuoka R, Kira Y. Adenosine as an endogenous mediator of hypoxia for induction of vascular endothelial growth factor mRNA in U-937 cells. Biochem Biophys Res Commun 1994;204:318-324.

Heltianu C, Simionescu M, Simionescu N: Histamine receptors of the microvascular endothelium revealed in situ with a histamine-ferritin conjugate: characteristic high-affinity binding sites in venules. J Cell Biol 1982;93:357-364.

Hood J.D, Meininger CJ, Ziche M, Granger HJ. VEGF upregulates ecNOS message, protein and NO production in human endothelial cells. Am J Physiol (in press).

Hudlicka O, Brown M, Egginton S: Angiogenesis in skeletal and cardiac muscle. Physiol Rev 1992;72:369-417.

Iimura O, Kusano E, Amemiya M, Muto S, Ikeda U, Shimada K, Asano Y. Dipydridamole enhances interleukin-1beta-stimulated nitric oxide production by cultured rat vascular smooth muscle cells. Eur J Pharm 1996;296:319-325.

Ikeda U, Kurosaki K, Shimpo M, Okada K, Saito T, Shimada K. Adenosine stimulates nitric oxide synthesis in rat cardiac myocytes. Am J Physiol 1997;273:H59-H65.

Ikeda U, Yamamoto K, Ichida M, Ohkawa F, Murata M, Iimura O, Kusano E, Asano Y, Shimada K. Cyclic AMP augments cytokine-stimulated nitric oxide synthesis in rat cardiac myocytes. J Mol Cell Cardiol 1996;28:789-795.

Iwamoto T, Umemura S, Toya Y, Uchibori T, Kogi K, Takagi N, Ishii M. Identification of adenosine A_2 receptor-cAMP system in human aortic endothelial cells. Biochem Biophys Res Commun 1994;199:905-910.

Janigro D, Wender R, Ransom G, Tinklepaugh DL, Winn HR. Adenosine-induced release of nitric oxide from cortical astrocytes. Neuroreport 1996;7:1640-1644.

Kanno K, Hirata Y, Imai T, Iwashina M, Marumo F. Regulation of inducible nitric oxide synthase gene by interleukin-1 beta in rat vascular endothelial cells. Amer J Physiol: 1994;H2318-H2322.

Kayar SR, Banchero N. Myocardial capillarity in acclimation to hypoxia. Pflügers Arch 1985;404:319-324.

Kayser B, Hoppeler H, Claassen H, Cerretelli P: Muscle structure and performance capacity of Himalayan Sherpas. J Appl Physiol 1991;70:1938-1942.

Klagsbrun M, D'Amore PA: Vascular endothelial growth factor and its receptors. Cytokine Growth Factor Rev 1996; 7(3):259-270.

Koolwijk P, van Erck MG, de Vree WJ, Vermeer MA, Weich HA, Hanemaaijer R, van Hinsbergh VW: Cooperative effect of TNFalpha, bFGF, and VEGF on the formation of tubular structures of human microvascular endothelial cells in a fibrin matrix. Role of urokinase activity. J Cell Biol 1996;132:1177-1188.

Ku DD, Zaleski JK, Liu S, Brock TA. Vascular endothelial growth factor induces EDRF-dependent relaxation in coronary arteries. Am J Physiol 1993;265:H586-H591.

Ladoux A, Frelin C. Hypoxia is a strong inducer of vascular endothelial growth factor mRNA expression in the heart. Biochem Biophys Res Comm. 1993;195:1005-1010.

Lamontagne D, Pohl U, Busse R: Mechanical deformation of vessel wall and shear stress determine the basal release of endothelium-derived relaxing factor in the intact rabbit coronary vascular bed. Circ

Res 1992;70:123-130.

Lampugnani MG, Corada M, Caveda L, Breviario F, Ayalon O, Geiger B, Dejana E. The molecular organization of endothelial cell to cell junctions: differential association of plakoglobin, beta-catenin, and alpha-catenin with vascular endothelial cadherin (VE-cadherin). J Cell Biol 1995;129:203-209.

Legrand AB, Narayanan TK, Ryan US, Aronstam RS, Catravas JD. Modulation of adenylate cyclase activity in cultured bovine pulmonary arterial endothelial cells. Effects of adenosine and derivatiives. Biochem Pharmacol. 1989;38:423-430.

Leitman DC, Fiscus RR, Murad F. Forskolin, phosphodiesterase inhibitors, and cyclic AMP analogs inhibit proliferation of cultured bovine aortic endothelial cells. J Cell Physiol 1986;127:237-243.

Lewis RE, Granger HJ: Diapedesis and the permeability of venous microvessels to protein macro-molecules: the impact of leukotriene B_4 (LTB_4). Microvasc Res 1988;35:27-37.

Li JM, Fenton RA, Cutler BS, Dobson JG Jr. Adenosine enhances nitric oxide production by vascular endothelial cells. Am J Physiol 1995;269:C519-C524.

Luty J, Hunt JA, Nobbs PK, Kelly E, Keen M, MacDermot J. Expression and desensitisation of A_2 purinoceptors on cultured bovine aortic endothelial cells. Cardiovasc Res 1989;23:303-337.

Majno G, Shea SM, Leventhal M: Endothelial contraction induced by histamine-type mediators. J Cell Biol 1969;42:647-654.

Mathieu-Costello O, Agey PJ, Wu L, Hang J, Adair TH: Capillary-to-fiber surface ratio in rat fast-twitch hindlimb muscles after chronicelectrical stimulation. J Appl Physiol 1996;80:904-909.

Mayerhofer A, Bartke, A: Developing testicular microvasculature in the golden hamster, Mesocricetus auratus: a model for angiogenesis under physiological conditions. Acta Anat (Basel) 1990;139:78-85.

Meininger CJ, Granger HJ. Mechanisms leading to adenosine-stimulated proliferation of microvascular endothelial cells. Am J Physiol 1990;258: H198-H206.

Meininger CJ, Schelling ME, Granger HJ: Adenosine and hypoxia stimulate proliferation and migration of endothelial cells. Am J Physiol 1988;255: H554-H562.

Minchenko A, Bauer T, Salceda S, Caro J. Hypoxic stimulation of vascular endothelial growth factor expression in vitro and in vivo. Lab Invest 1994;71:374-379.

Monacci WT, Merrill MJ, Oldfield EH: Expression of vascular permeability factor/vascular endothelial growth factor in normal rat tissues. Am J Physiol 264:C995-C1001, 1993.

Moncada S, Higgs EA: Molecular mechanisms and therapeutic strategies related to nitric oxide. FASEB J 1995;9:1319-1330.

Montesinos MC,Gadangi P, Longaker M, Sung J, Levine J, Nilsen D, Reibman J, Li M, Jiang C-K, Hirschhorn R, Recht PA, Ostad E, Levin RI, Cronstein BN. Wound healing is accelerated by agonists of adenosine A_2 (G_s-linked) receptors. J. Exp. Med. 1997;186:1615-1620.

Montrucchio G, Lupia E, de Martino A, Battaglia E, Arese M, Tizzani A, Bussolino F, Camussi G: Nitric oxide mediates angiogenesis induced in vivo by platelet-activating factor and tumor necrosis factor-alpha. Am J Pathol 1997;151:557-563.

Moodie SA, Martin W: Effects of cyclic nucleotides and phorbol myristate acetate on proliferation of pig aortic endothelial cells. Br J Pharmacol 1991;102:101-106

Morbidelli L, Chang CH, Douglas JG, Granger HJ, Ledda F, Ziche M. Nitric oxide mediates the mitogenic effect of VEGF on coronary venular endothelium. Am J Physiol 1996;270: H411-H415.

Morff RJ, Granger HJ. Contribution of adenosine to arteriolar autoregulation in striated muscle. Am J Physiol 1983;244: H567-H576.

Myrhage R, Hudlicka O: Capillary growth in chronically stimulated adult skeletal muscle as studied by intravital microscopy and histological methods in rabbits and rats. Microvasc Res 1978; 17:73-85.

Namiki A, Brogi E, Kearney M, Kim EA, Wu T, Couffinhal T, Varticovski L, Isner JM. Hypoxia induces vascular endothelial growth factor in cultured human endothelial cells. J Biol Chem 1995;270: 31189-31195.

Neufeld G, Tessler S, Gitay-Goren H, Cohen T, Levi BZ: Vascular endothelial growth factor and its receptors. Prog Growth Fact Res 1994; 5:89-97.

Oddis CV, Simmons RL, Hattler BG, Finkel MS. Protein kinase A activation is required for IL-1-induced nitric oxide production by cardiac myocytes. Am J Physiol 1996;271:C429-C434.

Orucevic A, Lala PK. NG-nitro-L-arginine methyl ester, an inhibitor of nitric oxide synthesis, ameliorates interleukin 2-induced capillary leakage and reduces tumour growth in adenocarcinoma-bearing mice. Br J Cancer 1996;73:189-194.

Papapetropoulos A, Desai KM, Rudic RD, Mayer B, Zhang R, Ruiz-Torres MP, Garcia-Cardena G,. Madri JA, Sessa WC: Nitric oxide synthase inhibitors attenuate transforming-growth-factor-beta 1-stimulated capillary organization in vitro. Am J Pathol150:1835-1844, 1997

Parenti A, Cui X-L, Douglas JJ, Granger HJ, Ziche M. Nitric oxide is an upstream signal of VEGF-induced ERK 1/2 activation in postcapillary endothelium. J Biol Chem (in press).

Parenti A, Donnini S, Morbidelli L, Granger HJ, Ziche M: The effect of linomide on the migration and the proliferation of capillary endothelial cells elicited by vascular endothelial growth factor Brit J Pharmacol 1996;119:619-621.

Phillips GD, Whitehead RA, Knighton DR: Initiation and pattern of angiogenesis in wound healing in the rat. Am J Anat 1991;192:257-262.

Qu-Hong. Nagy JA. Senger DR. Dvorak HF, Dvorak AM: Ultrastructural localization of vascular permeability factor/vascular endothelial growth factor (VPF/VEGF) to the abluminal plasma membrane and vesiculovacuolar organelles of tumor microvascular endothelium. J Histochem Cytochem 1995;43:381-389.

Ratajska A, Torry RJ, Kitten GT, Kolker SJ, Tomanek RJ: Modulation of cell migration and vessel formation by vascular endothelial growth factor and basic fibroblast growth factor in cultured embryonic heart. Dev Dyn 1995;203:399.

Rathbone MP, Middlemiss PJ, Gysbers JW, DeForge S, Costello P, Del Maestro RF. Purine nucleosides and nucleotides stimulate proliferation of a wide range of cell types. In Vitro Cell Devel Biol 1992;28A:529-536.

Salzman AL, Menconi MJ, Unno N, Ezzell RM, Casey DM, Gonzalez PK, Fink MP. Nitric oxide dilates tight junctions and depletes ATP in cultured Caco-2BBe intestinal epithelial monolayers. Am J Physiol 1995;268:G361-G366.

Schiele JO, Schwabe U. Characterization of the adenosine receptor in microvascular coronary endothelial cells. Eur J of Pharmacol 1994;269:51-58.

Sexl V, Mancusi G, Baumgartner-Parzer S, Schutz W, Freissmuth M. Stimulation of human umbilical vein endothelial cell proliferation by A_2-adenosine and beta$_2$-adrenoceptors. Br J Pharmacol 1995;114:1577-1586.

Sexl V, Mancusi G, Holler C, Gloria-Maercker E, Schutz W, Freissmuth M. Stimulation of the mitogen-activated protein kinase via the A2$_A$-adenosine receptor in primary human endothelial cells. J Biol Chem. 1997;272:5792-5799.

Shweiki D, Itin A, Soffer D, Keshet E. Vascular endothelial growth factor induced by hypoxia may mediate hypoxia-initiated angiogenesis. Nature 1992;359:843-845.

Simionescu N, Simionescu M, Palade GE: Open junctions in the endothelium of the postcapillary venules of the diaphragm. J Cell Biol 1978;79:27-35.

Stanimirovic DB, Bertrand N, Merkel N, Bembry J, Spatz M. Interaction between histamine and adenosine in human cerebro-microvascular endothelial cells: modulation of second messengers. Metab Brain Dis 1994;9:275-289.

Takagi H, King GL, Ferrara N, Aiello LP: Hypoxia regulates vascular endothelial growth factor receptor KDR/Flk gene expression through adenosine A2 receptors in retinal capillary endothelial cells. Invest Ophthalmol Vis Sci 1996;37:1311-1321.

Takagi H, King GL, Robinson GS, Ferrara N, Aiello LP. Adenosine mediates hypoxic induction of vascular endothelial growth factor in retinal pericytes and endothelial cells. Invest Ophthalmol Vis Sci 1996;37:2165-2176.

Teuscher E, Weidlich V. Adenosine nucleotides, adenosine and adenine as angiogenesis factors. Biomed Biochim Acta. 1985;44:493-495.

Unemori EN, Ferrara N, Bauer EA, Amento EP. Vascular endothelial growth factor induces interstitial collagenase expression in human endothelial cells. J Cell Physiol 1992;153:557-562.

Vaisman N, Gospodarowicz D, Neufeld G. Characterization of the receptors for vascular endothelial growth factor. J Biol Chem 1990;265:19461-19470.

Wagner RC, Hossler FE: SEM of capillary pericytes prepared by ultrasonic microdissection: evidence for the existence of a pericapillary syncytium. Anat Rec 1992;234:249-255.

Ward KA, Jain RK, Zhou Q: Cancer cell interactions with the microvasculature. 1989; FASEB J 3:A543.

Warren BA, Shubik P: The growth of the blood supply to melonoma transplants in the hamster cheek pouch chamber. Lab Invest 1966; 15:464-476.

Wu G, Meininger CJ. Impaired arginine metabolism and NO synthesis in coronary endothelial cells of the spontaneously diabetic BB rat. Am J Physiol 1995;269:H1312-H1318.

Wu HM, Huang Q, Yuan Y, Granger HJ. VEGF induces NO-dependent hyperpermeability in coronary venules. Am J Physiol 1996;271:H2735-H2739.

Yuan Y, Granger H, Zawieja D, Chilian W. Flow modulates venular permeability by a nitric oxide-related mechanism. Am J Physiol 1992;263: H641-H646.

Yuan Y, Granger HJ, Zawieja DC, DeFily DV, Chilian WM. Histamine increases venular permeability via a phospholipase C-NO synthase-guanylate cyclase cascade. Am J Physiol 1993;264: H1734-H1739.

Ziche M, Parenti A, Ledda F, Dell'Era P, Granger HJ, Maggi CA, Presta M. Nitric oxide promotes proliferation and plasminogen activator production by coronary venular endothelium through endogenous bFGF. Circ Res 1997;80:845-852

Ziche M, Morbidelli L, Choudhuri R, Zhang H-T, Donnini S, Granger HJ, Bicknell R. Nitric oxide synthase lies downstream from vascular endothelial growth factor-induced but not basic fibroblast growth factor-induced angiogenesis. J Clin Invest 1997;99: 2625-2634.

Ziche M, Morbidelli L, Masini E, Amerini S, Granger HJ, Maggi CA, Geppetti P, Ledda P. Nitric oxide mediates angiogenesis in vivo and endothelial cell growth and migration in vitro promoted by substance P. J Clin Invest 1994;94:2036-2044.

Ziche M, Morbidelli L, Masini E, Granger H, Geppetti P, Ledda F. Nitric oxide promotes DNA synthesis and cyclic GMP formation in endothelial cells from postcapillary venules. Biochem Biophys Res Commun 1993;192: 1198-1203.

Ziche M, Morbidelli L, Parenti A, Amerini S, Granger HJ, Maggi CA. Substance P increases cyclic GMP levels on coronary postcapillary venular endothelial cells. Life Sci, Pharmacol Lett 1993;53: PL229-PL234.

ADENOSINE, K_{ATP} CHANNEL, AND CARDIOPROTECTION IN THE INTACT HEART

David W. Green and Gary J. Grover, Cardiovascular Drug Discovery, Bristol-Myers Squibb Pharmaceutical Research Institute, Rt. 206 and Provinceline Rd., Princeton, N.J. 08543-4000

I. SUMMARY

Recent breakthroughs in our understanding of the endogenous defense mechanisms of the heart to ischemia have made it possible to develop novel and efficacious therapeutics for treating myocardial ischemia. Of particular interest is the adenosine-K_{ATP} axis. Development of the interest in this area was based on the intersection of several independent lines of investigation. The interest in preconditioning brought the K_{ATP} and adenosine researchers together and it seems likely that there is a real interaction between the two systems. While we have learned much, we still have a long way to go. If one assumes that K_{ATP} activation is "downstream" of adenosine receptor activation, the mechanism by which K_{ATP} activation exerts cardioprotection is still unknown. This is probably the most important remaining question. Current progress is being made to identify the K_{ATP} subtype involved with cardioprotection and it may be mitochondrial. If this is the case, it will be imperative to determine the mechanism by which opening of these channels can exert cardioprotective and energy sparing effects. Determination of these mechanisms may potentially lead to the development of novel and efficacious therapeutics.

II. BACKGROUND

While acute myocardial ischemia has long been a major source of morbidity and mortality throughout the world, relatively little is known about the molecular mechanisms leading to myocyte injury and death. Until the advent of thrombolytics, little could be done except bed rest to protect severely ischemic myocardium. Unfortunately, there is still a large unmet medical need for agents which will protect severely ischemic myocardium. Pharmacologic therapy has been much more successful in treating chronic stable angina pectoris, although none of these antianginal agents (including nitrates, calcium antagonists, and β-adrenoceptor blockers) were designed with ischemia in mind and may not represent optimal cardioprotective (or even anti-anginal) therapeutics.

The idea of salvaging ischemic tissue (infarct size reduction) is certainly not new and much research has been done in this area, although little clinical success for these efforts can be claimed. Many classes of pharmacologic agents have been shown to salvage infarcting myocardium in experimental animal models, although clinical activity is still uncertain. These agents generally protect the ischemic/reperfused myocardium directly. Many of these agents protect ischemic myocardium by conserving energy, thereby enhancing the ability of the tissue to withstand ischemic stress and examples are calcium antagonists, K_{ATP} openers, adenosine receptor agonists, Na^+/H^+ exchange inhibitors and β-adrenoceptor blockers. Agents also can reduce ultimate infarction by directly inhibiting reperfusion injury and agents such as oxygen radical scavengers or anti-neutrophil treatment have shown promise in experimental models.

There has been a recent revival of interest in cardioprotection because of the discovery of preconditioning, which is the cardioprotective effect observed following a brief period of coronary occlusion (Murry *et al.*, 1990). Numerous mechanisms have been proposed, but two have received much attention. Adenosine receptors, most notably A_1, are thought to mediate preconditioning, but many investigators also find that K_{ATP} are involved (Schulz et al., 1994; Yao and Gross, 1994a). These studies were typically done using antagonists or agonists of the appropriate receptor or channel to either mimic or abolish preconditioning. Interestingly, selective blockers of either of these pathways alone completely abolished preconditioning. These results were reconciled (at least partially) by the findings that K_{ATP} and adenosine receptors might be functionally linked (Kirsch *et al.*, 1990; Grover *et al.*, 1992) . The remainder of this chapter will be devoted to the discussion of the interaction between these two systems in terms of cardioprotection. Before discussing the pharmacologic and molecular linkage between adenosine receptors and K_{ATP}, it would be useful to discuss the pharmacologic profile for these two systems individually. After this, we will link the activities of these two pathways and attempt to detail a general outline for how they fit into the mechanism of preconditioning and cardioprotection.

III. TEXT

A. Cardioprotective Profile of K_{ATP} Modulators

1. K_{ATP}: General Considerations

K_{ATP} are metabolically regulated ion channels found in brain, kidney, smooth muscle, heart, skeletal muscle, and insulin-secreting cells (Noma, 1983; Spruce *et al.*, 1985; de Weille *et al.*, 1988; Treherne and Ashford, 1991; Edwards and Weston, 1993). K_{ATP} are regulated by intracellular adenine nucleotides such that they are inhibited by ATP and this inhibition of K_{ATP} by ATP is reduced by ADP (although ADP alone will inhibit the channel) (Edwards and Weston, 1993). K_{ATP} are thought to serve as a mechanism for coupling metabolic activity to electrical activity in various cell types. Although ATP is an important modulator, K_{ATP} are regulated by other factors such as pH, arachidonic acid, acyl CoA, adenosine, NO, SH-redox state and other factors related to cellular metabolism (Kim and Clapham, 1989; Kirsch *et al.*, 1990; Edwards and Weston, 1993; Coetzee *et al.*, 1995; Ming *et al.*, 1997). K_{ATP} are the site of action of antidiabetic sulfonylureas which promote insulin release through blockade of this channel (glucose metabolism is linked to insulin secretion through K_{ATP}) (Edwards and Weston, 1993). K_{ATP} activation can relax vascular and nonvascular smooth muscle through hyperpolarization of the sarcolemmal membrane. The role of K_{ATP} in normal myocardium is not well understood, but agents which open cardiac sarcolemmal K_{ATP} will cause action potential duration (APD) shortening as they will enhance a repolarizing current. Since cardiac ventricular myocyte resting potential is close to K^+ equilibrium potential, little hyperpolarization would be expected for K_{ATP} opening.

K_{ATP} is a complex of two different proteins (Inagaki *et al.*, 1995; Ashcroft, 1996). One subunit is an inwardly-rectifying K^+ channel (Kir) subunit and it is thought that four of these form the channel pore. Two types of Kir (Kir6.1 and Kir6.2) have been identified at this time. The sulfonylurea receptor (SUR) is the protein which confers a regulatory role as well as sensitivity of the channel to pharmacologic agents and ATP (Ashcroft, 1996). SUR is a member of the ATP-binding cassette protein superfamily, also called ABC transporters and is related to CFTR channels (cystic fibrosis transmembrane receptor, also glyburide inhibitable) (Philipson and Steiner, 1995). This ATP transporter, CFTR, releases ATP, which interacts with a purinergic receptor subtype which then opens chloride channels. A similar ATP permeant protein has been termed the sulfonylurea receptor (SUR), which is thought to have a similar function to CFTR, but is linked with K_{ATP} (al-Awqati, 1995). It has been hypothesized that in smooth muscle cells and perhaps insulin secreting cells, K_{ATP} is tonically opened by ATP (or another factor) secreted by SUR through a purinergic receptor. SUR1 is highly expressed in β-cells while SUR2 is highly expressed in cardiac and skeletal muscle (Wellman and Quayle, 1997). It is unknown how many different ways these different Kirs and SURs can interact, but data suggest different combinations in different tissue types. Currently, it is thought that Kir6.2 and SUR2 form cardiac sarcolemmal K_{ATP} (Inagaki *et al.*, 1996). In the study by Inagaki et al. (1996), diazoxide did not activate the sarcolemmal cardiac channel (we will return to this concept later) while it did open pancreatic K_{ATP}, which is thought to be formed by Kir6.2 and SUR1 (Inagaki *et al.*, 1995). This suggests the exciting possibility of tissue selectivity. Pharmacologic evidence has suggested tissue selectivity for years and we will describe these data below.

Inhibition of K_{ATP} is produced not only by intracellular ATP, but its nonhydrolyzable analogs, suggesting phosphorylation is not critical (Edwards and Weston, 1993). ADP reduces the sensitivity of K_{ATP} for ATP and therefore the channel is modulated by the ratio of these nucleotides. It is thought that ADP binds to the ATP regulatory binding site and ADP alone inhibits channel activity, although another ADP site is thought to mediate weak agonist activity. This may be a nucleotide phosphate site which is stimulated by GDP.

In addition to the ATP regulatory site, a phosphorylation site has been hypothesized. K_{ATP} will "run down" in the absence of ATP and it is thought that ATP can prime K_{ATP} for channel opening as well as reactivate the channel after it has run down (Spruce *et al.*, 1987). This is thought to require phosphorylation and therefore, channel "run down" may be mediated by channel dephosphorylation. It has been proposed that a complex balance of phosphorylation and dephosphorylation of tyrosine and serine/threonine residues can modulate K_{ATP} activity (Kwak *et al.*, 1996). The phosphorylation site may be closely associated with SUR and the nucleotide diphosphate site. It has been proposed that K_{ATP} openers reduce the affinity of ATP at the regulatory (inhibitory) site. Occupation of the appropriate receptors associated with G_i protein releases GDP and enhances GTP binding. GTP bound a_i subunit of G_i has been shown to inhibit ATP binding affinity. Release of GDP will activate the nucleotide diphosphate site to increase K_{ATP} open probability. GTP is also known to increase K_{ATP} channel open probability. A number of ligands known to interact with G_i increase K_{ATP} open probability such as acetylcholine and adenosine. Ligands such as adenosine (through the A_1 receptor subtype) may also activate protein kinase C (PKC) and it is thought that PKC activates K_{ATP} in ventricular myocytes by reducing channel sensitivity to ATP (at the inhibitory site) (Hu *et al.*, 1996). Protein kinase A (PKA) is also thought to modulate K_{ATP} activity and data have suggested it is involved with K_{ATP} opening induced by calcitonin gene-related peptide, adenosine (adenosine A_2 receptor subtype), prostacyclin and b-adrenoceptor agonists (Jackson *et al.*, 1993; Kleppisch and Nelson, 1995; Ming *et al.*, 1997). Another regulatory role for SUR has been proposed. It has been hypothesized to be an ATP secreting protein and the secreted ATP may act extracellularly through a P_2 receptor to activate K_{ATP}. Recent data (Babenko and Vassort, 1997) show that extracellular ATP can enhance K_{ATP} current via a P_2 receptor, apparently via activation of adenylyl cyclase, although this was independent of PKA. Therefore, the signaling systems involved with K_{ATP} modulation appear complex and as yet, a clear picture has not been elucidated. We have included a rough diagram for proposed modulatory pathways for K_{ATP} (figure 3).

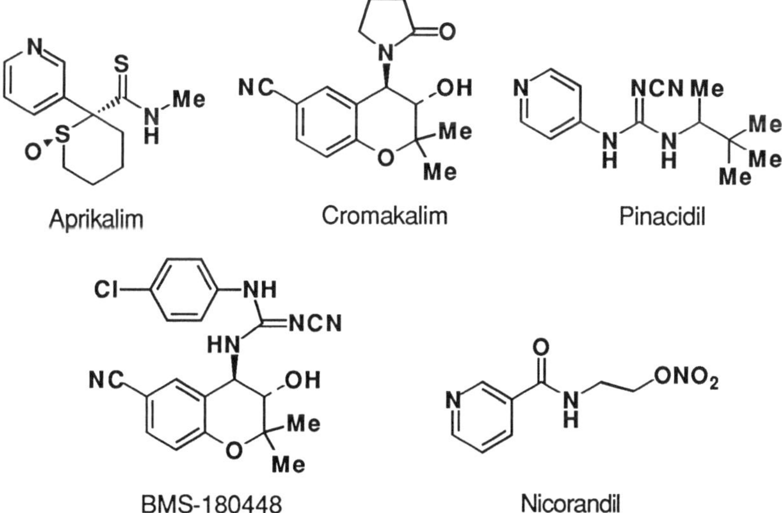

Figure 1. Chemical structure of K_{ATP} channel openers.

2. Cardioprotective Profile of K_{ATP} Openers

a. Cardioprotective Effects of K_{ATP} Openers *In Vitro*

We were fortunate to have selective inhibitors and activators of K_{ATP} to determine the role of this channel in myocardial ischemia (Atwal, 1992). The chemical structures of representative agents are shown in figure 1. Before testing in models of myocardial ischemia, it was unknown whether K_{ATP} openers would protect ischemic tissue or be pro-ischemic. Early studies were done in isolated heart or perfused heart strips in order to clearly observe any direct cardioprotective (or pro-ischemic) activity. At this time, all K_{ATP} openers were potent vasodilators which could cloud interpretation of data in whole animal models. In isolated heart models of ischemia, structurally diverse K_{ATP} openers exerted cardioprotective effects, usually within a 1-10 µM range (Cole *et al.*, 1991; Ohta *et al.*, 1991; Galinanes *et al.*, 1992; Grover, 1994; Grover *et al.*, 1995b; Grover and Atwal, 1995). The cardioprotective effects were expressed as increased post-ischemic functional recovery, reduced necrosis, reduced contracture formation, increased time to electrical uncoupling and improved energetic status (Cole *et al.*, 1991; Tan *et al.*, 1993; Grover, 1994; Monticello *et al.*, 1996). *In vitro*, addition of K_{ATP} openers only during reperfusion yields little cardioprotective effect, although the results of such studies must be interpreted with caution (Grover *et al.*, 1990). Most of the studies described above were done in isolated rat, guinea pig or rabbit hearts, although some of the studies were done in isolated muscle strips. In addition, cromakalim was shown to protect hypoxic/reoxygenated human atrial trabeculae (Speechly-Dick *et al.*, 1995), suggesting K_{ATP} to be relevant pharmacologic target.

The cardioprotective effects of K_{ATP} openers are universally abolished by K_{ATP} blockers such as glyburide and sodium 5-hydroxydecanoate (Grover, 1994). This strongly suggests K_{ATP} as the protective site of action. Both the vasodilator effects and cardioprotective effects of K_{ATP} openers are inhibited by glyburide (McCullough *et al.*, 1991). Interestingly, 5-HD does not block the vasodilator effect of K_{ATP} openers and appears to be relatively selective for inhibiting their cardioprotective activity (McCullough *et al.*, 1991). 5-HD, therefore, is a useful mechanistic tool and further discussion of this agent will be done later in this chapter.

b. Protective Effects of K_{ATP} Openers *in vivo*

Many investigators have shown K_{ATP} openers to protect ischemic/reperfused myocardium in whole animal models, although the results have been more variable compared to data *in vitro* . This may be related to the potent vasodilator effects of standard K_{ATP} openers which can induce hemodynamic changes and thereby cloud interpretation of data. We found that intracoronary infusion of K_{ATP} openers such as pinacidil and cromakalim significantly reduced infarct size and post-ischemic stunning in anesthetized dog models, although we could not achieve cardioprotection with these agents when they were administered systemically due to profound hypotension (Grover *et al.*,

1990; D'Alonzo *et al.*, 1992). Gross and colleagues showed that doses of bimakalim and aprikalim could be given i.v. or intracoronary in canine models of stunning and infarction and cardioprotection affected (Auchampach *et al.*, 1992a; Yao and Gross, 1994b). Studies by other investigators have shown similar results to those found by us or by Gross's group (Rohmann *et al.*, 1994). There have been several studies in which K_{ATP} openers were reported to lack protective effects *in vivo* (Imai *et al.*, 1988; Kitzen *et al.*, 1992). At the present time it is difficult to reconcile these studies with the studies showing beneficial activity.

c. Mechanism for the Cardioprotective effects of K_{ATP} Openers

In order to understand the possible role of K_{ATP} in preconditioning, it would be helpful to discuss current thoughts about the cardioprotective mechanism of action of K_{ATP} openers since they are said to mimic preconditioning. The original hypothesis was that K_{ATP} openers would shorten APD and protect hearts through a cardioplegic effect. Several laboratories showed K_{ATP} openers to conserve ATP, but interestingly they can do so at concentrations which do not depress cardiac function (or increase coronary flow for that matter) (Grover *et al.*, 1991; McPherson *et al.*, 1993). Several investigators showed that K_{ATP} openers exert additional protective effects over that conferred by depolarizing cardioplegia (completely arrested hearts), suggesting not only an effect independent of cardioplegia, but that changes in APD may also not be essential for K_{ATP} induced cardioprotection (Pignac *et al.*, 1994; Grover and Sleph, 1995).

We examined this in more detail in several studies. First, we determined the effect of doses of the delayed rectifier potassium channel blocker dofetilide on the cardioprotective activity (canine model of infarction) of cromakalim (Grover *et al.*, 1995a). The dose of dofetilide was titrated so that the APD shortening activity (as measured by epicardial monophasic action potential recordings) of cromakalim during ischemia was ablated. Despite a lack of APD shortening, cromakalim was equally cardioprotective in the presence of dofetilide compared to cromakalim alone. These results are in excellent agreement with previous work from Gross's group showing that doses of bimakalim could be given which reduced infarct size in dogs, but had no effect on monophasic APD (Yao and Gross, 1994b). These results were further confirmed by detailed structure-activity studies showing that cardioprotection and APD shortening were not correlated in compounds showing glyburide-reversible cardioprotective activity (Atwal *et al.*, 1993) Studies on one cromakalim analog, BMS-180448, showed that it was equi-cardioprotective compared to cromakalim, but unlike cromakalim was devoid of APD shortening activity in ischemic cardiac tissue (intracellular recordings) (Grover *et al.*, 1995b). Nevertheless, its cardioprotective effects were abolished by glyburide and 5-HD.

These data are difficult to reconcile with the idea that K_{ATP} openers are protecting hearts through enhancement of sarcolemmal potassium currents. While it is possible that structurally dissimilar compounds having the common activity of K_{ATP} activiation can work through a non-K_{ATP} related mechanism, this is doubtful. Recent studies have suggested that K_{ATP}s may be different in various tissues (see review by Ashcroft, 1996). On a molecular basis, cardiac sarcolemmal K_{ATP} is different from β-cell K_{ATP} and it is not surprising that the pharmacology of various compounds can also differ between tissues. A K_{ATP} has also been described in mitochondria (Inoue *et al.*, 1991) and the pharmacologic actions of various K_{ATP} openers on this channel seem different than sarcolemmal channels. Recent studies by Garlid's laboratory show that K_{ATP} openers such as cromakalim and bimakalim open mitochondrial K_{ATP} in concentrations which are similar to cardioprotective concentrations and their ability to open sarcolemmal cardiac K_{ATP} (Garlid *et al.*, 1996). These data are consistent with the *in vivo* pharmacology of these compounds showing equivalent cardioprotective and APD shortening potencies. Diazoxide, in contrast, only opens sarcolemmal K_{ATP} in the millimolar range while opening mitochondrial K_{ATP} in the low micromolar range. Diazoxide exerts cardioprotective effects within the same concentration range observed for mitochondrial K_{ATP} opening and both effects are inhibited by glyburide and 5-HD (Garlid et al., In Press). Further work will shed more light on the potential importance of mitochondrial K_{ATP} in cardioprotection.

3.   Role of K_{ATP} in Preconditioning

Brief periods of myocardial ischemia and reperfusion preceding a prolonged ischemic episode exert profound protective effects and this phenomenon is termed preconditioning. Much work has been devoted to determining the mechanism of this endogenous cardioprotective phenomenon. Suggested mechanisms include activation of adenosine receptors, K_{ATP}, PKC, α-adrenoceptors, and mitochondrial F_1F_0 ATPase inhibition. The multiple biochemical systems that seem to be involved with preconditioning makes it difficult to understand how these various pathways fit together. The most studied systems are adenosine and K_{ATP} and the remainder of this chapter will concentrate on describing how these two systems might fit together to mediate preconditioning.

Much of the work attempting to elucidate the mechanism of preconditioning has thus far been pharmacologic in nature. In the case of K_{ATP}, agonists of this channel "mimic" preconditioning because of their dramatic protective effects. While these data are consistent with K_{ATP} involvement with preconditioning, this is far from definitive proof. More definitive proof required the use of K_{ATP} blockers. Gross's laboratory was the first to show that glyburide abolished preconditioning in a canine model of preconditioning (Gross and Auchampach, 1992). The structurally distinct blocker 5-HD was also shown to abolish preconditioning in a similar model (Auchampach et al., 1992b). K_{ATP} blockers have been shown to abolish preconditioning in rabbits, rats, pigs, and man (see review by Grover, 1996). The ability of K_{ATP} blockers to abolish preconditioning in rats has been variable and may be model dependent, with preconditioning in isolated rat heart not abolished by 5-HD or glyburide (see review, Grover, 1996). Recent studies showed K_{ATP} blockers to abolish preconditioning in rat hearts *in vivo* (Schultz et al., 1997). An interesting proof of principle study on the role of K_{ATP} in meditating preconditioning in man was done by Tomai et al, 1994. In this study, human subjects were preconditioned using coronary angioplasty and the preconditioning effect was abolished by glyburide (Tomai *et al.*, 1994). Protection in this study was ascertained using ST-segment shifts and severity of cardiac pain. The ability of glyburide to abolish preconditioning in man was confirmed by Yellon's group using human atrial trabeculae (Speechly-Dick *et al.*, 1995). In this model, they also showed that K_{ATP} openers "simulated" preconditioning.

The studies outlined above strongly suggest the importance of K_{ATP} in mediating preconditioning, although further proof is necessary. It is possible that K_{ATP} blockers have nonspecific activities which could cloud data interpretation (probably a remote possibility). Also, merely showing that an agent protects ischemic myocardium does not mean that its pharmacologic mechanism is critical to preconditioning. Yao and Gross performed an interesting study suggesting that the cardioprotective effects of K_{ATP} openers actually do mimic preconditioning (Yao and Gross, 1994a). These authors used anesthetized dogs subjected to either 3 or 10 min preconditioning followed by 10 min of reperfusion. The prolonged ischemia was then insititued for 60 min followed by 4 hr reperfusion at which time infarct size was determined. The 3 min period of ischemia was not sufficient to precondition the hearts while 10 min was profoundly protective. A low, subthreshold, dose of bimakalim (0.3 μg/min, intracoronary) was found to be ineffective when given alone, but caused profound protection when combined with the subthreshold preconditioning protocol. This strongly suggests that the mechanism of cardioprotection for K_{ATP} openers and preconditioning are similar. Later studies by the same group (Mizumura *et al.*, 1995) showed K_{ATP} openers to simulate the effect of preconditioning on neutrophil function and adenosine release, although these data alone do not prove the involvement of K_{ATP} in preconditioning.

A study in whole-cell cardiac myocyte preparations showed that the potassium current elicited by pinacidil was potentiated by successive applications of this drug (Escande *et al.*, 1989). This would imply a "memory" phenomenon similar to that seen for preconditioning. This was also found for other structural classes of K_{ATP} openers. It is presently unknown if this effect of K_{ATP} openers on sarcolemmal K_{ATP} is relevant to preconditioning, although glyburide-reversible sarcolemmal potassium current can be enhanced by preconditioning (Schulz *et al.*, 1994). This study was done in a porcine model of 10 min preconditioning followed by 90 min of ischemia. The initial preconditioning period of 10 min of ischemia was associated with significant epicardial monophasic APD shortening, which was completely abolished by glyburide. Early into the prolonged ischemia, APD shortening was further enhanced in preconditioned hearts compared to sham treated hearts (glyburide-reversible), suggesting potentiation of a glyburide-inhibitable K^+ current. This is certainly consistent with the findings described above (Escande *et al.*, 1989), but this does not definitively prove that enhanced sarcolemmal K^+ currents mediate preconditioning. A study from our laboratories (Grover *et al.*, 1996) showed that a dose of dofetilide sufficient to abolish the APD shortening effect of the initial short period of ischemia in dogs, had no effect on cardioprotection. Additionally, preconditioning did not induce a more pronounced APD shortening during the prolonged ischemic episode, suggesting that enhanced sarcolemmal potassium efflux is not necessary for the cardioprotective effect of preconditioning to be observed. Nevertheless, glyburide completely abolished the cardioprotective effect of preconditioning in this model. These results are consistent with the pharmacology of K_{ATP} openers suggesting that an intracellular K_{ATP} may be mediating cardioprotection. These data also show that one must be cautious in interpreting data collected from sarcolemmal K_{ATP} and applying it to preconditioning.

B. Cardioprotective Profile of Adenosine Receptor Modulators
1. Adenosine Receptors: General Considerations

The cardioprotective effects of adenosine during myocardial ischemia have been recognized for several years (Berne, 1963; Lasley *et al.*, 1990). Although adenosine formation can occur via the metabolism of S-adenosylmethionine, during ischemia it is predominantly generated by the catabolism of ATP, specifically by the dephosphorylation of AMP by 5'nucleotidase (Mubagwa *et al.*, 1996). The conversion to nucleoside from nucleotide allows adenosine to freely diffuse to the intermembrane space of cardiomyocytes, where it is exported across the plasmalemmal membrane by a nucleoside transporter. Extracellular adenosine can also be generated by ectonucleotidase metabolism of plasma ATP that is released from vascular cells, thrombocytes and sympathetic nerves during ischemia.

The cardioprotective effects of adenosine are mediated through specific adenosine receptors (also known as purinergic P_1 receptors) also found in the respiratory, immune, and nervous systems. Adenosine receptors can be distinguished from ATP (P_2 purinergic) receptors by their affinity for adenosine over adenine nucleotides and their sensitivity to methyl xanthines (Rongen *et al.*, 1997). Binding studies with adenosine analogues have defined three distinct subtypes, A_1, A_2 and A_3 (Linden, 1994). Some of these analogs are illustrated in figure 2 and will be described in greater detail below. Subsequent cloning of the A_1, A_2 and A_3 receptors has confirmed the existence of subtypes. The A_2 receptors can be further subdivided into A_{2A} and A_{2B} isoforms that are genetically and pharmacologically distinct. Although isoforms of A_1 receptors have been proposed, there is no genetic evidence to support this. Amino acid sequence comparisons indicate adenosine receptors belong to the family of seven transmembrane domain receptors, and their interactions with guanine nucleotide-binding G proteins have been established by a number of studies (Mubagwa *et al.*, 1996). However, the subtypes do not all interact with the same type of G proteins, which will be discussed in greater detail.

Figure 2.

Adenosine Agonists:

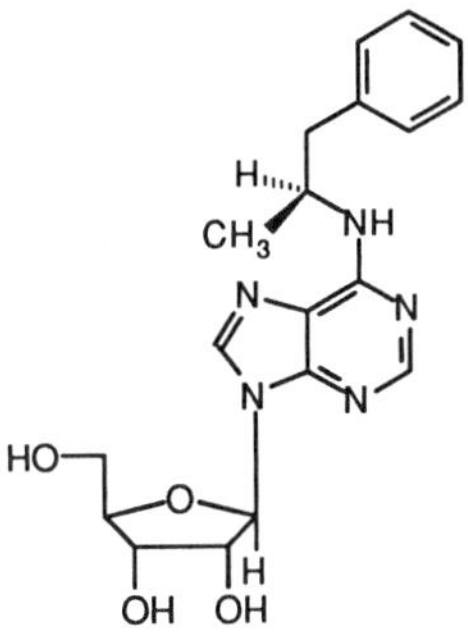

R-PIA (A_1)

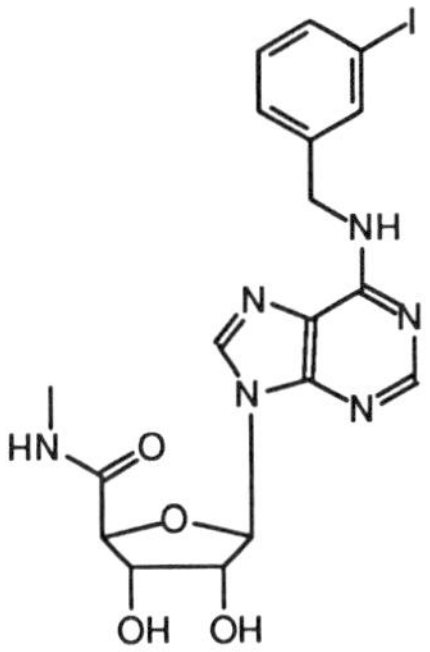

IB-MECA (A_3)

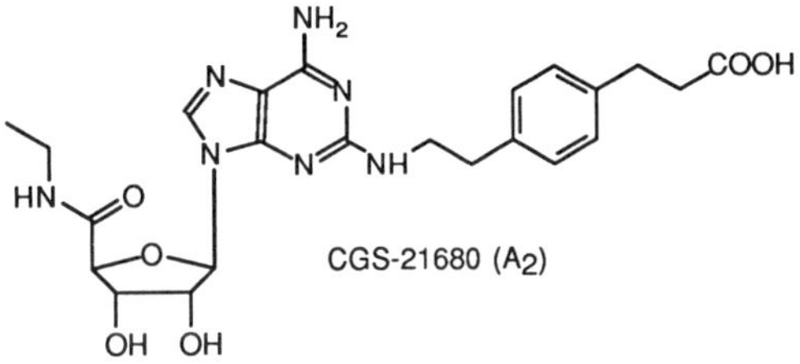

CGS-21680 (A_2)

Adenosine Antagonists:

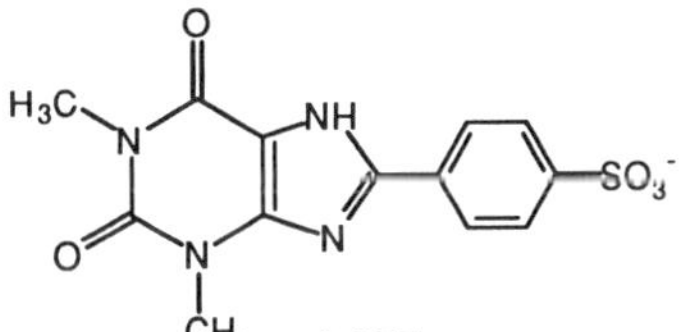

8-SPT

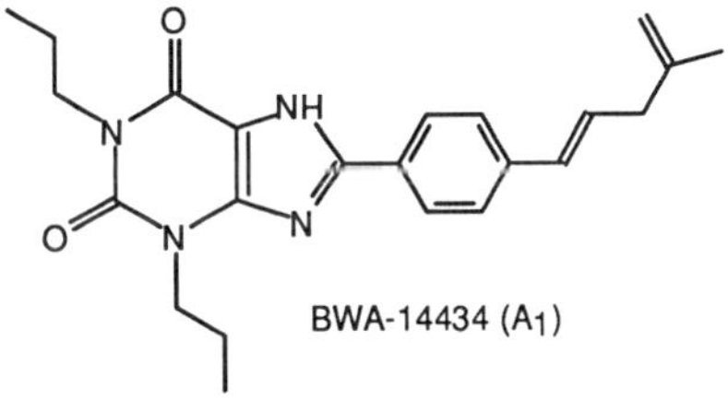

BWA-14434 (A_1)

Figure 2. Chemical structure of adenosine receptor agonists and antagonists.

C. Cardioprotective Profile of Adenosine Modulators

1. Cardioprotective Effects of Adenosine Modulators In Vitro

There is a wealth of information on adenosine receptor-mediated cardioprotection during ischemia and reperfusion in isolated hearts from various species using receptor subtype-selective agonists and antagonists. In the isolated rat heart model of global ischemia and reperfusion, the A_1 agonist (R)-N^6-phenylisopropyladenosine (R-PIA) was shown to increase ATP levels and the time to the onset of contracture during ischemia. These effects were blocked by the A_1-selective antagonist BW A1433U, which by itself caused a decrease in ATP levels and the time to contracture (Lasley *et al.*, 1990). In the same study, the A_2-selective agonist phenylaminoadenosine gave no cardioprotection. In a separate study using the same model, 2-chloro-N^6-cyclopentyladenosine, a highly selective A_1 adenosine receptor agonist, prevented the rise of diastolic pressure and coronary perfusion pressure during postischemic reperfusion (Monopoli *et al.*, 1994). Similar results have been found using A_1-selective agonists and antagonists in isolated perfused guinea pig and rabbit hearts (Lasley and Mentzer, 1995). A recent study where the A_1 receptor was overexpressed in mice resulted in increased time to the onset of contracture during global ischemia and improved functional recovery during reperfusion (Matherne *et al.*, 1997). Hearts from these transgenic mice also had a significantly lower heart rate. The adenosine A_3-selective agonist N^6-(3-iodobenzyl)-N-methyl-5'-carbamoyladenosine (IB-MECA) has also been found to be cardioprotective against ischemia-reperfusion injury in the isolated rabbit heart (Tracey *et al.*, 1997). Therefore, experiments using isolated heart models of ischemia and reperfusion demonstrate cardioprotection can be mediated through A_1 and A_3 adenosine receptors in a variety of species.

2. Protective Effects of Adenosine Modulators in vivo

As in isolated hearts, there is also ample evidence from several different animal models to support a cardioprotective role for adenosine that is mediated through adenosine receptors. In rat hearts *in situ*, the adenosine A_1 agonist BN-063 (1-cyclopropylisoguanosine) exerted antiarrhythmic and anti-infarct effects following left coronary artery occlusion and reperfusion (Lee *et al.*, 1995b). In dogs subjected to left coronary artery occlusion followed by reperfusion, exogenous intracoronary adenosine augmented postischemic recovery of function as assessed by a significant enhancement of systolic wall thickness (Randhawa *et al.*, 1993). In a similar model in pig, the A_1 agonist R-PIA was found to be antiarrhythmic.

A number of detailed studies have been performed in rabbits. In anesthetized rabbits, pretreatment with adenosine reduced infarct size during regional ischemia that was blocked with the nonselective antagonist 8-p-sulfophenyl theophylline (8-SPT) (Toombs *et al.*, 1992). In a similar open-chested rabbit model (Thornton *et al.*, 1992), adenosine A_1-selective agonists such as R-PIA and 2-chloro-N^6-cyclopentyladenosine (CCPA) were found to protect the heart from infarction during regional ischemia but the A_2-selective agonist CGS 21680 was not. R-PIA was also found to induce bradycardia, but still limited infarct size when the hearts were paced, indicating that protection was not the result of bradycardia. However, R-PIA given at reperfusion did not reduce infarct size, suggesting that receptors had to be occupied before ischemia in order to be cardioprotective.

In anesthetized dogs on total bypass, adenosine in cold blood cardioplegia improved postischemic left ventricular systolic function relative to cold blood alone or cold blood cardioplegia containing the nonselective adenosine antagonist 8-SPT (Hudspeth *et al.*, 1994). In a human study, adenosine infusion just prior to cardiopulmonary bypass resulted in an immediately improved post-bypass cardiac index. Forty hours postoperatively in the intensive care unit, patients treated with adenosine had improved cardiac index, maintained lowered resting heart rate and released significantly less creatine kinase during the first 24 hours of the postoperative period (Lee *et al.*, 1995a).

A recent study in conscious rabbits suggests adenosine-induced cardioprotection can also be mediated through A_3 receptors (Auchampach *et al.*, 1997). In a model of myocardial stunning and infarction, the A_3-selective agonist IB-MECA administered as an intravenous bolus before occlusion did not affect heart rate and reduced wall thickening and infarct size. The protective effect of the A_3 receptor agonist in this model was reversed by 8-SPT.

3. Mechanism for the Cardioprotective effects of Adenosine Modulators

The modulation of myocardial metabolism appears to be mediated through the stimulation of A_1 adenosine receptors in the heart. In contrast to A_2 receptors, adenosine binding to A_1 receptors activates the G_i or "inhibitable" G proteins (Lasley and Mentzer, 1993). The A_3 receptor is also linked to G_i (Zhou *et al.*, 1992), but its effects on heart rate and metabolism are not as well characterized as those of the A_1 receptor. G_i activation through A_1 receptors decreases cAMP levels through inhibition of adenylate cyclase following catecholamine stimulation. This decrease in cAMP attenuates the activity of L-type calcium channels, delayed rectifier potassium channels, chloride channels and pacemaker ion channels (Mubagwa *et al.*, 1996). G_i activation also affects heart rate through a cAMP-independent opening of muscarinic (K_{ACh}) channels, (Brodde *et al.*, 1992; Morton *et al.*, 1994). As has been mentioned in previous sections and will be discussed in greater detail, activation of G_i proteins is also thought to activate or open K_{ATP} channels, which can affect metabolic rates. Stimulation of A_1 receptors has been shown to inhibit glycolytic rates during reperfusion following ischemia (Finegan *et al.*, 1996), which may prevent acidosis caused by the production of lactate. Adenosine has also been shown to slow metabolism (rate of ATP depletion, glycogen utilization, and lactate accumulation) in isolated dog hearts during global ischemia (Vander Heide *et al.*, 1993). However, in a model of low-flow ischemia in isolated rabbit hearts, adenosine caused a stimulation of anaerobic glycolysis which could be blocked with the nonselective antagonist 8-phenyltheophylline (Janier *et al.*, 1993). This effect could be cardioprotective through increasing ATP, but would be accompanied by acidosis. The differences in these studies have not been reconciled, but may be model-dependent (low flow vs total flow cessation).

Adenosine-induced cardioprotection during ischemia may also be a result of decreasing cardiac work secondary to vasodilation, although this is doubtful. Vasodilation is associated with the A_2 receptors, which are found in vascular smooth muscle. Vasodilation is believed to result from increased cyclic AMP (cAMP) levels following A_2 receptor activation. Adenosine A_2 receptors interact with G_S or "stimulatory" G proteins, which increase adenylate cyclase activity and hence cAMP levels. While the vasodilating effects of adenosine could potentially offload ischemic myocardium, it is doubtful if this is an important contributory component of cardioprotection. Coronary vasodilation is also probably not important in mediating cardioprotection. The relative lack of importance of this mechanism is shown by the clear cardioprotective effects of A_1 selective agents.

4. Effect of Adenosine Modulators in Preconditioning

There has been much recent interest in the role of adenosine in mediating cardiac preconditioning. The majority of data suggests that adenosine does play an important role in preconditioning, although species differences seem to be important. Adenosine antagonists such as PD 115199 (Bunch *et al.*, 1992), 8-phenyltheophylline (nonselective) (Tsuchida *et al.*, 1992) and 8-SPT (Liu *et al.*, 1991) were found to block ischemic preconditioning in rabbits. The selective A_1 antagonist BW A1433 has also been shown to block preconditioning in isolated rabbit hearts (Liu *et al.*, 1994). PD 115199 and 8-cyclopentyl-1,3-dipropylxanthine (A_1 selective) blocked preconditioning in the hearts of anesthetized dogs (Auchampach and Gross, 1993) and a more recent study in dogs demonstrated PD 81723, an allosteric enhancer of A_1 receptors, was able to lower the threshold of ischemic preconditioning (Mizumura *et al.*, 1996). In humans the A_1 selective antagonist bamphylline was found to abolish preconditioning caused by single-vessel coronary angioplasty (Tomai *et al.*, 1996). A major limitation in intrepreting the role of adenosine A_1 and A_3 receptors in mediating the cardioprotective effect of preconditioning is that xanthines such as 8-cyclopentyl-1,3-dipropylxanthine and BWA1433 can also block the canine, human and rabbit A_3 receptor at sub-micromolar concentrations (Auchampach et al, 1997). Therefore, abolition of preconditioning by these xanthines may be due to blockade of not only the A_1 but also the A_3 receptors. Thus, preconditioning appears to involve adenosine receptors (A_1 and perhaps A_3) in a number of species with possible therapeutic utility in humans. A number of studies in rats have not been able to implicate adenosine as being involved in the cardioprotective effects of preconditioning. Although A_1 receptor agonists such as CCPA mimic preconditioning with respect to reducing infarct size (Liu and Downey, 1992) and anti-arrhythmic activity (Miyatake *et al.*, 1996) in the rat, this does not necessarily mean that adenosine is involved in preconditioning in this species. Adenosine receptor antagonists such as PD 115199 (nonselective A_1 and A_2) (Liu and Downey, 1992), BW A1433U (Asimakis *et al.*, 1993), and 8-SPT (nonselective) (Cave *et al.*, 1993) could not block the effects of preconditioning in isolated rat hearts.

D. Interaction Between K_{ATP} and Adenosine in Preconditioning

As reviewed above, a plethora of studies implicate both K_{ATP} and adenosine in mediating preconditioning, usually in the same species. It was surprising that antagonists of adenosine receptors and K_{ATP} completely abolished preconditioning in the same species when given separately. These results suggested the possibility of an interaction between them. The first study to show an interaction in which adenosine A_1 receptor activation caused K_{ATP} opening (sarcolemmal channels) was done in neonatal rat ventricular myocytes and demonstrated this link occurred via a G_i protein (Kirsch *et al.*, 1990). This has been confirmed in rabbit ventricular myocytes (Kim et al., 1997). The paper by Kirsch et al. (1990) prompted us to pharmacologically determine whether a link between K_{ATP} and adenosine A_1 receptors could be observed in models of ischemia. Since the role of sarcolemmal K_{ATP} in cardioprotection is unclear, it was important to determine a link in a model of ischemia. In an anesthetized canine model of ischemia and reperfusion, we showed that R-PIA significantly reduced infarct size and this effect was completely abolished by glyburide, suggesting that adenosine A_1 receptor activation can increase K_{ATP} activity (Grover *et al.*, 1992). Studies in a porcine model of infarction showed that the cardioprotective effects of R-PIA were abolished by 5-HD (Van Winkle *et al.*, 1994). The protective effects of adenosine in ischemic rabbit hearts have also been shown to be abolished by K_{ATP} blockers (Toombs et al., 1993). As usual, the isolated rat heart appears to be an outlier with no apparent pharmacologic linkage between adenosine A_1 receptors and K_{ATP} during ischemia (Fralix et al, 1993). Interestingly, rat cerebral preconditioning was abolished by glyburide (Heurteaux *et al.*, 1995). In this study, K_{ATP}-mediated protection appeared to be preceded and activated by adenosine A_1 receptor activation. Another study (Cleveland *et al.*, 1997) showed that adenosine preconditioning of human atrial trabeculae can be abolished by K_{ATP} blockers. These data are important in showing the relevance of the adenosine-K_{ATP} axis in man.

While there are numerous studies suggesting that adenosine A_1 receptor activation can stimulate K_{ATP}, there are several studies suggesting that the reverse pathway. It has been suggested that K_{ATP} openers are protecting ischemic myocardium by increasing adenosine production secondary to activation of ectosolic 5'-nucleotidase (Kitakaze *et al.*, 1994). A study in hypoxic rabbit ventricular myocytes showed that the K_{ATP} opener pinacidil was protective and this effect was abolished by the adenosine receptor antagonists SPT and DPCPX (Armstrong *et al.*, 1995). These data suggest that K_{ATP} activation in some way activates adenosine receptors which will then exert cardioprotection. It is presently difficult to reconcile these results with those showing adenosine receptor activation to activate K_{ATP}. A recent study showed that the protective effect of bimakalim was not abolished by DPCPX in dogs, suggesting that K_{ATP} openers do not protect through release of adenosine (Gross *et al.*, 1997). Hopefully, further research will help resolve this issue. Thus far, the weight of evidence seems to favor adenosine A_1 receptor-induced K_{ATP} activation.

Assuming that adenosine receptor activation is the stimulus for K_{ATP} activation, the signaling pathways linking the two need to be elucidated. Adenosine A_1 receptors are known to inhibit PKA and may also activate PKC. Both PKC and PKA are known to modulate K_{ATP} activity, although the nature of this interaction is still not clearly understood. The most likely cardioprotective pathway revolves around stimulation of PKC secondary to adenosine A_1 receptor activation. Several laboratories have shown inhibition of preconditioning by inhibitors of PKC and protective effects for PKC activation (see review, Gho et al., 1996). PKC has been shown to be involved with activation of sarcolemmal K_{ATP} in patch clamp studies and PKC activation has been shown to protect ischemic myocardium in a glyburide-reversible manner (Hu et al., 1996; Speechly-Dick et al., 1995). These studies suggest the schemes shown in figure 3 and figure 4 in which adenosine receptor activation causes PKC activation. PKC activation appears to phosphorylate a protein in or associated with K_{ATP}, resulting in activation. It should be remembered that much the work relating potassium currents to PKC activation were done using sarcolemmal currents and the relevance to the pertinent K_{ATP} (mitochondrial?) is not presently clear.

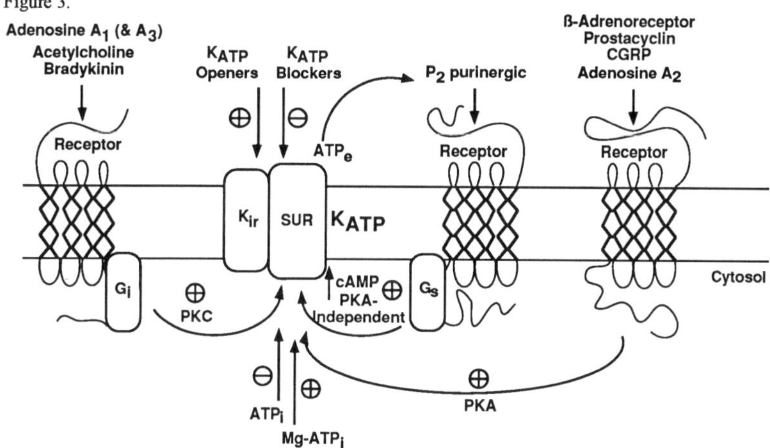

Figure 3. Mechanisms of sarcolemmal K_{ATP} channel modulation. The subunits of the K_{ATP} channel are indicated as "K_{ir}" for the inward rectifying potassium channel subunit and "SUR" for the sulfonylurea receptor. The seven-transmembrane receptor families are as indicated and their respective interactions with protein kinase A (PKA), the stimulatory (G_s) or inhibitory (G_i) G-coupled receptor proteins. Events or agents that open the K_{ATP} channel are labeled with an encircled "+"; those that close the channel are labeled with an encircled "-". The ATP subscripts "I" and "e" refer to intracellular and extracellular localization, respectively.

Figure 4. Schematic for possible mechanism of energy conservation during ischemia through adenosine receptor-mediated activation of the K_{ATP} channel via protein kinase C (PKC).

IV. REFERENCES

al-Aqati Q, Regulation of ion channels by ABC transporters that secrete ATP [comment]. Science 1995; 269: 805-806.

Armstrong S, Liu G, Downey J, Ganote C. Potassium channels and preconditioning of isolated rabbit cardiomyocytes: effects of glyburide and pinacidil. J Mol Cell Cardiol 1995; 27: 1765-1774.

Ashcroft F. Fresh insights into the interactions of drugs with ATP-sensitive K-channels. ID Research Alert 1996; 2: 43-37.

Asimakis G, Inners-McBride K, Conti V. Attenuation of postischaemic dysfunction by ischaemic preconditioning is not mediated by adenosine in the isolated rat heart. Cardiovasc Res 1993; 27: 1522-1530.

Atwal K. Modulation of potassium channels by organic molecules. Med Res Rev 1992; 12: 569-591.

Atwal K, Grover G, Ahmed S, Ferrara F, Harper T, Kim K, Sleph P, Dzwonczyk S, Russell A, Moreland S, and et, al. Cardioselective anti-ischemic ATP-sensitive potassium channel openers. J Med Chem 1993; 36: 3971-3974.

Auchampach JA, Grover GJ, Gross GJ. Blockade of ischemic preconditioning in dogs by the novel ATP dependent potassium channel antagonist sodium 5-hydroxydecanoate. Cardiovasc Res 1992b; 26:1054-1062.

Auchampach JA, Rizvi A, Qiu Y, Tang X.-L, Maldonado C, Teschner S, Bolli R. Selective activation of A_3 adenosine receptors with N6-(3-iodobenzyl)adenosine-5'-N-methyluronamide protects against myocardial stunning and infarction without hemodynamic changes in conscious rabbits Circulation Research 1997; 80: 800-809.

Auchampach JA, Jin X, Wan TC, Caughey GH, Linden J. Canine mast cell adenosine receptors: cloning and expression of the A_3 receptor and evidence that downregulation is mediated by the A_{2B} receptor. Mol Pharmacol 1997; 52:846-860.

Auchampach JA, Gross G. Adenosine A_1 receptors, K_{ATP} channels, and ischemic preconditioning in dogs. Am J Physiol 1993; 264: H1327-1336.

Auchampach JA, Maruyama M, Cavero I, Gross G. Pharmacological evidence for a role of ATP-dependent potassium channels in myocardial stunning. Circulation 1992a; 86: 311-319.

Babenko A, Vassort G. Enhancement of the ATP-sensitive K^+ current by extracellular ATP in rat ventricular myocytes. Involvement of adenylyl cyclase-induced subsarcolemmal ATP depletion. Circ Res 1997; 80: 589-600.

Berne RM. Cardiac nucleotides in hypoxia: possible role in regulation of coronary blood flow. Am J Physiol 1963; 204: 317-322.

Brodde, O. E., Broede, A., Daul, A., Kunde, K. and Michel, M. C. Receptor systems in the non-failing human heart. Basic Res Cardiol 1992; 87 Suppl 1: 1-14.

Bunch F, Thornton J, Cohen M, Downey J. Adenosine is an endogenous protectant against stunning during repetitive ischemic episodes in the heart. Am Heart J 1992; 124: 1440-1446.

Cave A, Collis C, Downey J, Hearse D. Improved functional recovery by ischaemic preconditioning is not mediated by adenosine in the globally ischaemic isolated rat heart. Cardiovasc Res 1993; 27: 663-668.

Cleveland JC, Meldrum D, Rowland R, Banerjee A, Harken A. Adenosine preconditioning of human myocardium is dependent upon the ATP-sensitive K^+ channel. J Mol Cell Cardiol 1997; 29: 175-182.

Coetzee W, Nakamura T, Faivre J. Effects of thiol-modifying agents on K_{ATP} channels in guinea pig ventricular cells. Am J Physiol 1995; 269: H1625-1633.

Cole W, McPherson C, Sontag D. ATP-regulated K+ channels protect the myocardium against ischemia/reperfusion damage. Circ Res 1991; 69: 571-581.

D'Alonzo A, Darbenzio R, Parham C, Grover G. Effects of intracoronary cromakalim on postischaemic contractile function and action potential duration. Cardiovasc Res 1992; 26: 1046-1053.

de Weille J, Schmid-Antomarchi H, Fosset M, Lazdunski M. ATP-sensitive K^+ channels that are blocked by hypoglycemia-inducing sulfonylureas in insulin-secreting cells are activated by galanin, a hyperglycemia-inducing hormone. Proc Natl Acad Sci U S A 1988; 85: 1312-1316.

Edwards G, Weston A. The pharmacology of ATP-sensitive potassium channels. Annu Rev Pharmacol Toxicol 1993; 33: 597-637.

Escande D, Thuringer D, Le Guern S, Courteix J, Laville M, Cavero I. Potassium channel openers act through an activation of ATP-sensitive K^+ channels in guinea-pig cardiac myocytes. Pflugers Arch 1989; 414: 669-675.

Finegan B, Lopaschuk G, Gandhi M, Clanachan A. Inhibition of glycolysis and enhanced mechanical function of working rat hearts as a result of adenosine A_1 receptor stimulation during reperfusion following ischaemia. Br J Pharmacol 1996; 118: 355-363.

Fralix TA, Steenbergen C, London RE, Murphy E. Glibenclamide does not abolish the protective effect of preconditioning on stunning in the isolated perfused rat heart. Cardiovasc Res 1993; 27:630-637.

Galinanes M, Shattock M, Hearse D. Effects of potassium channel modulation during global ischaemia in isolated rat heart with and without cardioplegia. Cardiovasc Res 1992; 26: 1063-1068.

Garlid K, Paucek P, Yarov-Yarovoy V, Sun X, Schindler P. The mitochondrial K_{ATP} channel as a receptor for potassium channel openers. J Biol Chem 1996; 271: 8796-8799.

Garlid KD, Paucek P, Yarov-Yarovoy B, Murray H N M, Darbenzio RB, D'Alonzo AJ, Lodge NJ, Smith MA, Grover GJ. Cardioprotective effect of diazoxide and its interaction with mitochondrial ATP-sensitive potassium channels: Possible mechanism of cardioprotection. in press, Circ Res.

Gho B C G, Eskildsen-Helmond Y E G, de Zeeuw S, Lamers J M J , Verdouw PD. Does protein kinase C play a pivotal role in the mechanisms of ischemic preconditioning. Cardiovasc. Drugs Ther. 1996; 10: 775-786.

Gross G, Auchampach JA. Blockade of ATP-sensitive potassium channels prevents myocardial preconditioning in dogs. Circ Res 1992; 70: 223 233.

Gross G, Mei D, Sleph P, Grover G. Adenosine A_1 receptor blockade does not abolish the cardioprotective effects of the adenosine triphosphate-sensitive potassium channel opener bimakalim. J Pharmacol Exp Ther 1997; 280: 533-540.

Grover G, Atwal KS. BMS-180448, a glyburide-reversible cardioprotective agent with minimal vasodilator activity. Cardiovasc Drug Rev 1995; 13: 123-136.

Grover G, Sleph P. Protective effect of K_{ATP} openers in ischemic rat hearts treated with a potassium cardioplegic solution. J Cardiovasc Pharmacol 1995; 26: 698-706.

Grover G. Protective effects of ATP-sensitive potassium-channel openers in experimental myocardial ischemia. J Cardiovasc Pharmacol 1994; 24 Suppl 4: S18-27.

Grover G, D'Alonzo A, Parham C, Darbenzio R. Cardioprotection with the K_{ATP} opener cromakalim is not correlated with ischemic myocardial action potential duration. J Cardiovasc Pharmacol 1995a; 26: 145-152.

Grover GJ, D'Alonzo AJ, Dzwonczyk S, Parham CS, Darbenzio RB. Preconditioning is not abolished by the delayed rectifier K^+ blocker dofetilide. Am J Physiol 1996; 271: H1207-H1214.

Grover G, D'Alonzo A, Hess T, Sleph P, Darbenzio R. Glyburide-reversible cardioprotective effect of BMS-180448 is independent of action potential shortening. Cardiovasc Res 1995b; 30: 731-738.

Grover G, Dzwonczyk S, Parham C, Sleph P. The protective effects of cromakalim and pinacidil on reperfusion function and infarct size in isolated perfused rat hearts and anesthetized dogs. Cardiovasc Drugs Ther 1990; 4: 465-474.

Grover G, Newburger J, Sleph P, Dzwonczyk S, Taylor S, Ahmed S, Atwal K. Cardioprotective effects of the potassium channel opener cromakalim: stereoselectivity and effects on myocardial adenine nucleotides. J Pharmacol Exp Ther 1991; 257: 156-162.

Grover G, Sleph P, Dzwonczyk S. Role of myocardial ATP-sensitive potassium channels in mediating preconditioning in the dog heart and their possible interaction with adenosine A_1-receptors. Circulation 1992; 86: 1310-1316.

Grover GJ "Role of the K_{ATP} channel in ischemic preconditioning." *In* Ischemia: Preconditioning and Adaptation, MS Marber and DM Yellon, ed, Oxford, UK : Bios Scientific Publishers, 1996.

Heurteaux C, Lauritzen I, Widmann C, Lazdunski M. Essential role of adenosine, adenosine A_1 receptors, and ATP-sensitive K^+ channels in cerebral ischemic preconditioning. Proc Natl Acad Sci U S A 1995; 92: 4666-4670.

Hu K, Duan D, Li G, Nattel, S. Protein kinase C activates ATP-sensitive K^+ current in human and rabbit ventricular myocytes. Circ Res 1996; 78: 492-498.

Hudspeth D, Nakanishi K, Vinten-Johansen J, Zhao Z, McGee D, Williams M, Hammon JW. Adenosine in blood cardioplegia prevents postischemic dysfunction in ischemically injured hearts. Ann Thorac Surg 1994; 58: 1637-1644.

Imai N, Liang C, Stone C, Sakamoto S, Hood WB J. Comparative effects of nitroprusside and pinacidil on myocardial blood flow and infarct size in awake dogs with acute myocardial infarction. Circulation 1988; 77: 705-711.

Inagaki N, Gonoi T, Clement JP, Namba N, Inazawa J, Gonzalez G, Aguilar-Bryan L, Seino S, Bryan J. Reconstitution of IK_{ATP}: an inward rectifier subunit plus the sulfonylurea receptor [see comments]. Science 1995; 270: 1166-1170.

Inagaki N, Gonoi T, Clement J, Wang C, Aguilar-Bryan L, Bryan J, Seino S. A family of sulfonylurea receptors determines the pharmacological properties of ATP-sensitive K+ channels. Neuron 1996; 16: 1011-1017.

Inoue I, Nagase H, Kishi K, Higuti T. ATP-sensitive K^+ channel in the mitochondrial inner membrane. Nature 1991; 352: 244-247.

Jackson W, Konig A, Dambacher T, Busse R. Prostacyclin-induced vasodilation in rabbit heart is mediated by ATP-sensitive potassium channels. Am J Physiol 1993; 264: H238-243.

Janier M, Vanoverschelde J, Bergmann S. Adenosine protects ischemic and reperfused myocardium by receptor-mediated mechanisms. Am J Physiol 1993; 264: H163-170.

Kim D, Clapham D. Potassium channels in cardiac cells activated by arachidonic acid and phospholipids. Science 1989; 244: 1174-1176.

Kim E, Han J, Ho W, Earm Y E. Modulation of ATP-sensitive K^+ channels in rabbit ventricular myocytes by adenosine A1 receptor activation. Am J Physiol 1997; 272: H325-H333.

Kirsch G, Codina J, Birnbaumer L, Brown A. Coupling of ATP-sensitive K^+ channels to A_1 receptors by G proteins in rat ventricular myocytes. Am J Physiol 1990; 259: H820-826.

Kitakaze M, Hori M, Morioka T, Minamino T, Takashima S, Sato H, Shinozaki Y, Chujo M, Mori H, Inoue M and et al. Infarct size-limiting effect of ischemic preconditioning is blunted by inhibition of 5'-nucleotidase activity and attenuation of adenosine release. Circulation 1994; 89: 1237-1246.

Kitzen J, McCallum J, Harvey C, Morin M, Oshiro G, Colatsky T. Potassium channel activators cromakalim and celikalim (WAY-120,491) fail to decrease myocardial infarct size in the anesthetized canine. Pharmacology 1992; 45: 71-82.

Kleppisch T, Nelson M. Adenosine activates ATP-sensitive potassium channels in arterial myocytes via A_2 receptors and cAMP-dependent protein kinase. Proc Natl Acad Sci U S A 1995; 92: 12441-12445.

Kwak Y, Park S, Cho K, Chae S. Reciprocal modulation of ATP-sensitive K^+ channel activity in rat ventricular myocytes by phosphorylation of tyrosine and serine/threonine residues. Life Sci 1996; 58: 897-904.

Lasley R D, Rhee J W, Van Wylen D G, Mentzer R M, Jr. Adenosine A_1 receptor mediated protection of the globally ischemic isolated rat heart. J Mol Cell Cardiol 1990; 22: 39-47.

Lasley RD, Mentzer R M Jr. Protective effects of adenosine in the reversibly injured heart. Ann Thorac Surg 1995; 60: 843-846.

Lasley RD, Mentzer R M Jr. Pertussis toxin blocks adenosine A_1 receptor mediated protection of the ischemic rat heart. J Mol Cell Cardiol 1993; 25: 815-821.

Lee H, LaFaro R, Reed G. Pretreatment of human myocardium with adenosine during open heart surgery. J Card Surg 1995a; 10: 665-676.

Lee Y, Sheu J, Yen M. BN-063, a newly synthesized adenosine A_1 receptor agonist, attenuates myocardial reperfusion injury in rats. Eur J Pharmacol 1995b; 279: 251-256.

Linden J. Cloned adenosine A_3 receptors: pharmacological properties, species differences and receptor functions. Trends Pharmacol Sci 1994; 15: 298-306.

Liu G, Richards S, Olsson R, Mullane K, Walsh R, Downey J. Evidence that the adenosine A_3 receptor may mediate the protection afforded by preconditioning in the isolated rabbit heart. Cardiovasc Res 1994; 28: 1057-1061.

Liu G, Thornton J, Van Winkle D, Stanley A, Olsson R, Downey J. Protection against infarction afforded by preconditioning is mediated by A_1 adenosine receptors in rabbit heart. Circulation 1991; 84: 350-356.

Liu G, Downey J. Ischemic preconditioning protects against infarction in rat heart. Am J Physiol 1992; 263: H1107-1112.

Matherne G P, Linden J, Byford A M, Gauthier N S, Headrick J P. Transgenic A_1 adenosine receptor overexpression increases myocardial resistance to ischemia. Proc. Natl. Acad. Sci. USA 1997; 94: 6541-6546.

McCullough J, Normandin D, Conder M, Sleph P, Dzwonczyk S, Grover G. Specific block of the anti-ischemic actions of cromakalim by sodium 5-hydroxydecanoate. Circ Res 1991; 69: 949-958.

McPherson C, Pierce G, Cole W. Ischemic cardioprotection by ATP-sensitive K^+ channels involves high-energy phosphate preservation. Am J Physiol 1993; 265: H1809-1818.

Ming Z, Parent R, Lavallee M. Beta 2-adrenergic dilation of resistance coronary vessels involves K_{ATP} channels and nitric oxide in conscious dogs. Circulation 1997; 95: 1568-1576.

Miyatake Y, Kusama Y, Kishida H, Hayakawa H. Adenosine mediates the antiarrhythmic effect of ischemic preconditioning in isolated rat hearts. Jpn Circ J 1996; 60: 341-348.

Mizumura T, Auchampach J, Linden J, Bruns R, Gross G. PD 81,723, an allosteric enhancer of the A_1 adenosine receptor, lowers the threshold for ischemic preconditioning in dogs. Circ Res 1996; 79: 415-423.

Mizumura T, Nithipatikom K, Gross G. Bimakalim, an ATP-sensitive potassium channel opener, mimics the effects of ischemic preconditioning to reduce infarct size, adenosine release, and neutrophil function in dogs. Circulation 1995; 92: 1236-1245.

Monopoli A, Conti A, Dionisotti S, Casati C, Camaioni E, Cristalli G, Ongini E. Pharmacology of the highly selective A_1 adenosine receptor agonist 2-chloro-N6-cyclopentyladenosine. Arzneimittelforschung 1994; 44: 1305-1312.

Monticello T, Sargent C, McGill J, Barton D, Grover G. Amelioration of ischemia/reperfusion injury in isolated rats hearts by the ATP-sensitive potassium channel opener BMS-180448. Cardiovasc Res 1996; 31: 93-101.

Morton M E, Brumwell C, Gartside C L, Hauschka S D, Nathanson N M, Characterization of muscarinic acetylcholine receptors expressed by an atrial cell line derived from a transgenic mouse tumor. Circ Res 1994; 74: 752-756.

Mubagwa K, Mullane K, Flameng W. Role of adenosine in the heart and circulation. Cardiovasc Res 1996; 32: 797-813.

Murry C, Richard V, Reimer K, Jennings R. Ischemic preconditioning slows energy metabolism and delays ultrastructural damage during a sustained ischemic episode. Circ Res 1990; 66: 913-931.

Noma A. ATP-regulated K^+ channels in cardiac muscle. Nature 1983; 305: 147-148.

Ohta H, Jinno Y, Harada K, Ogawa N, Fukushima H, Nishikori K. Cardioprotective effects of KRN2391 and nicorandil on ischemic dysfunction in perfused rat heart. Eur J Pharmacol 1991; 204: 171-177.

Philipson L, Steiner D. Pas de deux or more: the sulfonylurea receptor and K^+ channels [comment]. Science 1995; 268: 372-373.

Pignac J, Bourgouin J, Dumont L. Cold cardioplegia and the K^+ channel modulator aprikalim (RP 52891): improved cardioprotection in isolated ischemic rabbit hearts. Can J Physiol Pharmacol 1994; 72: 126-132.

Randhawa M P Jr., Lasley RD, Mentzer RM. Adenosine and the stunned heart. J Card Surg 1993; 8: 332-337.

Rohmann S, Weygandt H, Schelling P, Kie Soei L, Verdouw P, Lues I. Involvement of ATP-sensitive potassium channels in preconditioning protection. Basic Res Cardiol 1994; 89: 563-576.

Rongen GA, Floras JS, Lenders JW, Thien T, Smits P. Cardiovascular pharmacology of purines [editorial]. Clin Sci Colch 1997; 92: 13-24.

Schultz JEJ, Qian YZ, Gross GJ, Kukreja RC. The ischemia-selective K_{ATP} channel antagonist, 5-hydroxydecanoate, blocks ischemic preconditioning in the rat heart. J Mol Cell Cardiol 1997; 29: 1055-1060.

Schulz R, Rose J, Heusch G. Involvement of activation of ATP-dependent potassium channels in ischemic preconditioning in swine. Am J Physiol 1994; 267: H1341-1352.

Speechly-Dick M, Grover G, Yellon D. Does ischemic preconditioning in the human involve protein kinase C and the ATP-dependent K^+ channel? Studies of contractile function after simulated ischemia in an atrial in vitro model. Circ Res 1995; 77: 1030-1035.

Spruce A, Standen N, Stanfield P. Voltage-dependent ATP-sensitive potassium channels of skeletal muscle membrane. Nature 1985; 316: 736-738.

Spruce A, Standen N, Stanfield P. Studies of the unitary properties of adenosine-5'-triphosphate-regulated potassium channels of frog skeletal muscle. J Physiol (Lond) 1987; 382: 213-236.

Tan H, Mazon P, Verberne H, Sleeswijk M, Coronel R, Opthof T, Janse M. Ischaemic preconditioning delays ischaemia induced cellular electrical uncoupling in rabbit myocardium by activation of ATP sensitive potassium channels. Cardiovasc Res 1993; 27: 644-651.

Thornton J, Liu G, Olsson R, Downey J. Intravenous pretreatment with A_1-selective adenosine analogues protects the heart against infarction Circulation 1992; 85: 659-665.

Tomai F, Crea F, Gaspardone A, Versaci F, De Paulis R, Penta de Peppo, A, Chiariello L, Gioffre P. Ischemic preconditioning during coronary angioplasty is prevented by glibenclamide, a selective ATP-sensitive K^+ channel blocker. Circulation 1994; 90: 700-705.

Tomai F, Crea F, Gaspardone A, Versaci F, De Paulis R, Polisca P, Chiariello L, Gioffre P. Effects of A_1 adenosine receptor blockade by bamiphylline on ischaemic preconditioning during coronary angioplasty. Eur Heart J 1996; 17: 846-853.

Toombs CF, McGee DS, Johnston WE, Vinten-Johansen J. Protection from ischaemic-reperfusion injury with adenosine treatment is reversed by inhibition of ATP-sensitive potassium channels. Cardiovasc Res 1993; 27: 623-629.

Toombs C, McGee S, Johnston W, Vinten-Johansen J. Myocardial protective effects of adenosine. Infarct size reduction with pretreatment and continued receptor stimulation during ischemia. Circulation 1992; 86: 986-994.

Tracey W, Magee W, Masamune H, Kennedy S, Knight D, Buchholz R, Hill, R. Selective adenosine A_3 receptor stimulation reduces ischemic myocardial injury in the rabbit heart. Cardiovasc Res 1997; 33: 410-415.

Treherne J, Ashford M. The regional distribution of sulphonylurea binding sites in rat brain. Neuroscience 1991; 40: 523-531.

Tsuchida A, Miura T, Miki T, Shimamoto K, Iimura O. Role of adenosine receptor activation in myocardial infarct size limitation by ischaemic preconditioning. Cardiovasc Res 1992; 26: 456-461.

Van Winkle D, Chien G, Wolff R, Soifer B, Kuzume K, Davis R. Cardioprotection provided by adenosine receptor activation is abolished by blockade of the K_{ATP} channel. Am J Physiol 1994; 266: H829-839.

Vander Heide R, Reimer K, Jennings R. Adenosine slows ischaemic metabolism in canine myocardium in vitro: relationship to ischaemic preconditioning. Cardiovasc Res 1993, 27. 669-673.

Wellman G, Quayle J. ATP-sensitive potassium channels: Molecular structure and therapeutic potential in smooth muscle. ID Research Alert 1997; 2: 3-5.

Yao Z, Gross GJ. Activation of ATP-sensitive potassium channels lowers threshold for ischemic preconditioning in dogs. Am J Physiol 1994a; 267: H1888-H1894.

Yao Z, Gross G. Effects of the K_{ATP} channel opener bimakalim on coronary blood flow, monophasic action potential duration, and infarct size in dogs. Circulation 1994b; 89: 1769-1775.

Zhou QY, Li C, Olah ME, Johnson RA, Stiles GL, Civelli O. Molecular cloning and characterization of an adenosine receptor: the A_3 adenosine receptor. Proc Natl Acad Sci U S A 1992; 89: 7432-7436.

ADENOSINE RECEPTOR SUBTYPES AND CARDIOPROTECTION IN CARDIAC MYOCYTE AND TRANSGENIC MODELS

G. Paul Matherne, University of Virginia, USA; John P. Headrick, Griffith University, Australia ; Bruce T. Liang, University of Pennsylvania, USA

I. SUMMARY

Ischemic heart disease remains the leading cause of death in the U.S. and the Western World. Although much of the therapeutic efforts have largely focused on improving the blood supply to the ischemic myocardium, recent discoveries of myocardial adaptive protective mechanisms have re-focused therapeutic attention on the heart muscle. Adenosine is released in large amounts during ischemia and has been shown to mediate a number of important cardioprotective functions. Cardiac ventricular myocytes express multiple adenosine receptor subtypes including A_1, A_3 and A_{2A} receptors. Activation of the A_1 or the A_3 receptor can trigger as well as mediate the protective effect of ischemic preconditioning. Phospholipases, such as phospholipase C or D can be stimulated by A_1 or A_3 receptor agonists. The subsequent formation of diacylglyceride leads to protein kinase C activation which in turn activates or "primes" K_{ATP} channels. The primed channels can be more responsive to adenosine released during the ischemia and thus protect the myocytes. Although unproven, evidence is growing to support this sequence of signaling events in mediating the protective effect of ischemic preconditioning. Activation of the A_1 or A_3 receptor can also protect the heart against ischemia/reperfusion injury. Thus, acting at A_1 receptors adenosine has been shown to reduce ischemic injury, attenuate post-ischemic contractile dysfunction in reversibly injured myocardium, and reduce tissue necrosis in irreversibly injured hearts. To date most strategies designed to harness adenosinergic cardioprotection have involved classical pharmacological manipulations (eg. use of agonists, allosteric enhancers, uptake blockers). Unfortunately such studies yield mixed results, often due to undesirable systemic side effects. An alternate approach is to enhance receptor density. Transgenic enhancement of myocardial receptor expression may provide more efficacious and site-specific cardioprotection by the endogenous adenosine. Thus, a transgenic rather than traditional pharmacologic approach was used to enhance A_1 function in the murine myocardium. A_1 overexpression was shown to improve functional and metabolic tolerance to global ischemia reperfusion without compromising baseline contractile function or metabolism. Transgenic models such as this may prove valuable in unraveling the mechanisms of adenosine receptor mediated cardiac protection. The data provide support for the potential benefits of genetic therapy targeting the myocardial A_1 adenosine receptor. Targeting of the adenosine A_3 receptor is also potentially useful as both the A_3 receptor agonist and the receptor itself are possible novel anti-ischemic therapeutics.

II. BACKGROUND

<u>A. Adenosine and cardioprotection</u>

Adenosine can be formed intracellularly by the cytoplasmic 5'-nucleotidase that converts AMP to adenosine and by S-adenosylhomocysteine hydrolase (SAH) that converts SAH to adenosine. Adenosine can also be formed via a sequential dephosphorylation of ATP to ADP, AMP and then adenosine by ecto-5'-nucleotidases. Because adenosine is released in large amounts during ischemia, adenosine has been the focus of intense interest as a cardioprotective agent (Ely and Berne 1992). Adenosine has been shown to play an important role in triggering and mediating the cardioprotective effect of ischemic preconditioning (Murry et al., 1986; Li et al., 1990; Downey 1992; Liang 1996). Ischemic preconditioning or pre-ischemia conditioning, has been demonstrated in virtually every species studied including humans (Murry et al., 1986; Li et al., 1990; Downey 1992; Gross 1995; Leesar et al., 1997). In preconditioning, adenosine released during the first ischemic episode activates the adenosine receptor and triggers a protective effect that lasts well into the second period of ischemia. The basis for the memory of this cardioprotective effect is not known. Adenosine released during the second episode of ischemia is equally important in that activation of the adenosine receptor during that time is necessary to actually cause the cardioprotection. The cardioprotective effects of preconditioning include not only a reduction in the myocardial infarct size but also an attenuation of the ischemia-induced stunning, although these two effects may be related (Yao and Gross 1993; Sekili et al., 1995). Another observed beneficial effect of ischemic preconditioning is a decrease in the incidence of ventricular arrhythmias (Hagar et al., 1991). Taken together, these studies are extremely informative and important and clearly established the role of adenosine in this critical protective process. How adenosine acts to trigger and mediate the preconditioning effect remains poorly understood. The intact heart contains vascular cells, circulating blood cells, cardiac myocytes and neurons, all of which express one or another adenosine receptor subtype. Delineation of the role and signaling mechanisms of each adenosine receptor has been difficult due to the confounding effects arising from receptor activation in each cell type.

Another important cardioprotective function of adenosine is its role during the actual ischemia or during reperfusion following ischemia. For example, intracoronary administration of adenosine during reperfusion after prolonged no-flow ischemia can also limit infarct size in the intact heart (Olafsson et al., 1987, Babbitt et al., 1989). When adenosine is infused during low-flow ischemia and reperfusion, there is enhanced recovery of ventricular function (Reibel and Rovetto 1978; Ely and Berne., 1992). However, questions arise regarding whether the beneficial effect of adenosine is i) exerted during ischemia or during reperfusion, ii) due to activation of adenosine receptors because the adenosine concentrations used were high (50-100 mM), and iii) is mediated at the level of coronary vasculature, circulating neutrophils or cardiac myocytes.

<u>B. Cardioprotective roles of adenosine receptor subtypes</u>

1. Mediation of preconditioning.

The synthesis and availability of ligands selective at different adenosine receptor subtypes facilitated delineation of the cardioprotective function mediated by each receptor. Thus, activation of the adenosine A_1 or A_3 receptor can replace preconditioning ischemia and simulate the cardioprotective effect of preconditioning in the intact heart and in isolated cardiac myocytes (Headrick 1996; Strickler et al., 1996; Auchampach et al. 1997; Tracey et al., 1997; Wang et al., 1997; Carr et al., 1997). That is, adenosine A_1 or A_3 receptor agonists can induce preconditioning pharmacologically. On the other hand, activation of the adenosine A_{2A} receptor produces no preconditioning effect. These studies provided extremely valuable information on the role of different adenosine receptor subtypes in this important cardioprotective phenomenon.

2. Protection against ischemia/reperfusion injury.

Activation of A_1 receptors protects from injury during global ischemia (Lasley et al., 1990), attenuates post-ischemic contractile dysfunction (Lasley and Mentzer 1992; Randhawa et al., 1995; Sekili et al., 1995), and can reduce tissue necrosis in reperfused myocardium (Zhao et al., 1993; Lasley et al., 1995). Exogenous A_1 activation improves mechanical function and reduces injury in reperfused myocardium (Lasley et al., 1992, Zhao et al., 1993; Finegan et al., 1996) and has been used successfully as a myocardial protective agent in cardioplegia for heart surgery as well as in the catheterization lab during angioplasty (Leesar et al., 1997; Mentzer et al., 1997). Receptor activation by endogenous adenosine has also been shown to improve bioenergetic state during ischemia reperfusion (Angello et al., 1991; Headrick 1996). The role of the A_3 receptor in protecting against ischemia/reperfusion injury remains unclear, although preliminary data using a cardiac myocyte model suggest an important protective function of A_3 agonists even when ischemia has begun (Stambaugh et al, 1997).

Post-ischemic contractile dysfunction (stunning) is also reduced by endogenous and exogenous adenosine. Evidence indicates that the adenosine receptors involved in this effect must be activated prior to reperfusion in order to reduce stunning. When applied at the onset of reperfusion adenosinergic therapy is ineffective in attenuating contractile dysfunction (Ambrosio et al., 1989; Yao and Gross 1993; Randhawa et al., 1995) but is effective in reducing infarction size (Zhao et al., 1993). Once again mechanisms remained undetermined, although endogenous adenosine has been shown to improve metabolic and bioenergetic recovery from ischemia (Angello et al., 1991; Headrick 1996), and the functionally beneficial effects of exogenously applied adenosine have been associated with improved bioenergetic state *in vitro* and *in vivo* (Mentzer et al., 1993; Zhou et al., 1993). Improvement of ΔG_{ATP} may facilitate ionic recovery and improve sarcoplasmic reticular function, contributing to improved contractility. Reductions in ΔG_{ATP} depress contractile function in hypoxic myocardium (Kammermeier and Roeb 1982). Although most research has focused on the A_1 receptor, recent data demonstrate the importance of A_3 receptors in reduced stunning in conscious rabbit (Auchampach et al., 1997).

C. Importance of myocyte adenosine receptors

That activation of adenosine A_1 or A_3 receptors can simulate the protective effect of ischemic preconditioning pharmacologically raised the possibility that this cardioprotective effect is mediated at the level of cardiac myocyte in the intact heart. This notion is further supported by the observation that the A_{2A} receptor, the major subtype present in the coronary vasculature, is not able to mediate a preconditioning effect. Direct proof for this concept requires a number of experimental approaches; data already obtained thus far will be presented below.

<u>D. Mechanisms of cardioprotection</u>

The precise mechanism of adenosine mediated cardioprotection remains unclear, and there is some evidence that different protective effects of adenosine in ischemic and reperfused hearts may be mediated via differing mechanisms. For example, this is evidenced by the observation that adenosine-mediated reductions in post-ischemic contractile dysfunction require receptor activation during the ischemic insult itself whereas inhibition of infarct development is effective with stimulation of adenosine receptors during reperfusion alone (Zhao et al., 1993). In ischemic myocardium receptor activation by endogenous adenosine reduces injury (Lasley et al., 1990; Matherne et al., 1997), improves bioenergetic state (Angello et al., 1991; Headrick 1996), reduces acidosis (Headrick 1996), and inhibits myocardial Na^+ and Ca^{2+} accumulation (Fralix et al., 1993). These metabolic and ionic effects may be involved in enhanced functional tolerance. Beneficial effects are attenuated by A_1 antagonism and mimicked by A_1 receptor activation, implicating the A_1 receptor sub-type (Lasley et al., 1990; Lasley and Mentzer 1992; Yao and Gross 1993; Headrick 1996).

Recent evidence favors a role for inhibition of glycolysis in the A_1 mediated reduction in stunning (Finegan et al., 1996). This could improve recovery by reduction of H^+ generation. However, there is some controversy regarding the role of glycolysis in cardioprotection. Importantly, recent evidence shows that glycolytic metabolism is activated in post ischemic myocardium and plays a key role in restoring ion homeostasis and functional recovery during reperfusion (Jeremy et al., 1993; Tamm et al., 1994). Moreover, adenosine activates myocardial glycolysis in different models (Wyatt et al., 1989; Angello et al., 1993). Stimulation of glucose metabolism could enhance resistance to ischemic injury by providing crucial ATP under anaerobic conditions. Glycolytically derived ATP may preferentially support ion homeostasis and SR function. Myocardial injury and diastolic dysfunction during ischemic insult is reduced by glycolysis and exacerbated in its absence (Kingsley et al., 1991; Cross et al., 1996), and extent of ischemic acidosis has been shown to be unimportant in determining functional outcome (Cross et al., 1996; Cave et al., 1997). Thus, further research is needed to unmask the role of changes in glucose metabolism in effects of adenosine in ischemia reperfusion.

Overall, both A_1 and A_3 receptors have been implicated in protecting against ischemia/reperfusion, whether using the infarct size reduction or anti-stunning effect as endpoints. Although a number of possible mechanisms have been suggested, the precise cellular and molecular signaling events that underlie this receptor-mediated protective phenomenon are not known.

III. TEXT

<u>A. Identification of a novel adenosine A_3 receptor in the cardiac myocyte</u>

Studies carried out in human, rabbit, rat and chick ventricular myocytes provided important insight by indicating that the cardioprotective mechanism of preconditioning is exerted, at least in part, at the level of cardiac myocytes in the intact heart. Subsequent studies demonstrated that a non-A_1 receptor, possibly the A_3 subtype, may be involved in mediating preconditioning (Strickler et al., 1996). Direct proof that adenosine A_3 receptor is present on the cardiac myocyte was first obtained based on detailed pharmacological characterization studies on the chick ventricular myocyte (Strickler et al., 1996). It is well established that adenosine A_1

receptors are present and functional in the cardiac myocyte (Belardinelli et al., 1989; Olsson and Pearson 1990; Liang 1992). Thus, adenosine A_1 receptor agonist 2-chloro-N^6-cyclopentyladenosine (CCPA) can mediate a dose-dependent inhibition of the isoproterenol-stimulated cyclic AMP accumulation; such inhibition is completely abolished by the A_1 receptor-selective antagonist 8-cyclopentyl-1,3-dipropylxanthine (DPCPX) (Fig. 1A). The highly selective A_3 receptor agonist 2-chloro-N^6-(3-iodobenzyl)adenosine-5'-N-methyluronamide (Cl-IB-MECA) or N^6-(3-iodobenzyl)adenosine-5'-N-methyluronamide (IB-MECA) was also able to elicit a dose-dependent inhibition of isoproterenol-induced stimulation of cAMP accumulation, which, however, was not attenuated by DPCPX (Fig. 1B). These data provided the first demonstration that adenosine A_3 receptors are present and functional in the cardiac myocyte. Other data showed that Cl-IB-MECA or IB-MECA can also mediate a stimulation of the phospholipase D activity in the cardiac myocyte (Morley et al., 1996).

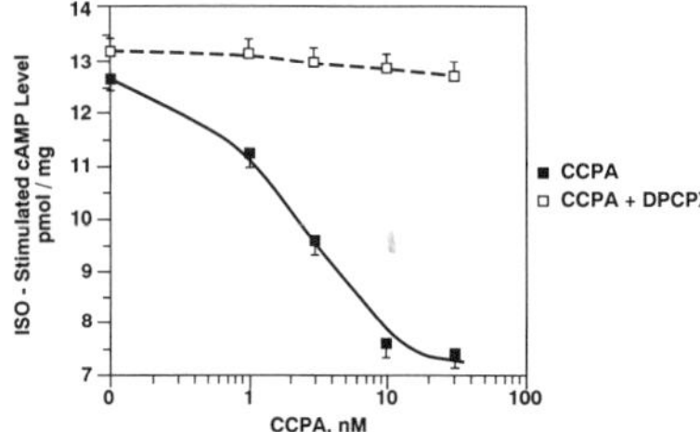
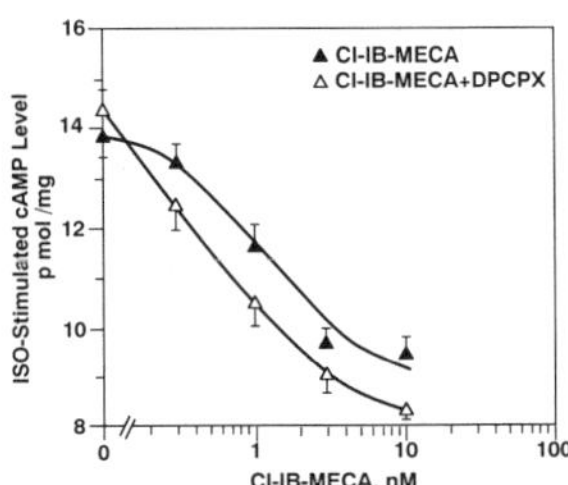

Figure1. <u>Effects of adenosine receptor agonists on isoproterenol-stimulated cAMP level</u>. The levels of cAMP (pmol/mg) were determined in the presence of isoproterenol and varying concentrations of (A) CCPA or (B) Cl-IB-MECA (closed symbols) or in the presence of isoproterenol, CCPA or Cl-IB-MECA plus DPCPX (open symbols). Reproduced from The Journal of Clinical Investigation, 1996, vol. 98, pp. 1777, by copy right permission of the American Society for Clinical Investigation.

<u>B. Cardioprotection functions mediated by the adenosine A_3 receptor</u>

1. Role in mediating preconditioning effect.

Having established the presence of a functional A_3 receptor on the cardiac myocyte, the next objective was to determine the potential cardioprotective function mediated by this receptor. To accomplish this objective, a cardiac myocyte model of ischemic preconditioning was established (Strickler et al., 1996). In this model, ischemia was simulated by exposure of the cardiac myocyte to hypoxia and glucose deprivation. A brief ischemic exposure was used to simulate preconditioning and a

second sustained period of ischemia was used to induce injury.   The protocol used
to precondition the myocyte was identical to that employed to precondition the
intact heart (Fig. 2A).   The basic characteristics of preconditioning of the myocyte
are similar to those of preconditioning of the intact heart (Murry et al., 1986; Li et
al., 1990; Downey 1992; Gross 1995; Headrick 1996; Leesar et al., 1997).
Adenosine triggers as well as mediates the preconditioning effect on the myocytes,
similar to its role in preconditioning of the intact heart.   These data suggest that
cultured myocytes are an excellent model to study the role of different adenosine
receptor subtypes in cardiac myocyte preconditioning.

<u>Preconditioning Protocol</u>

Fig. 2A

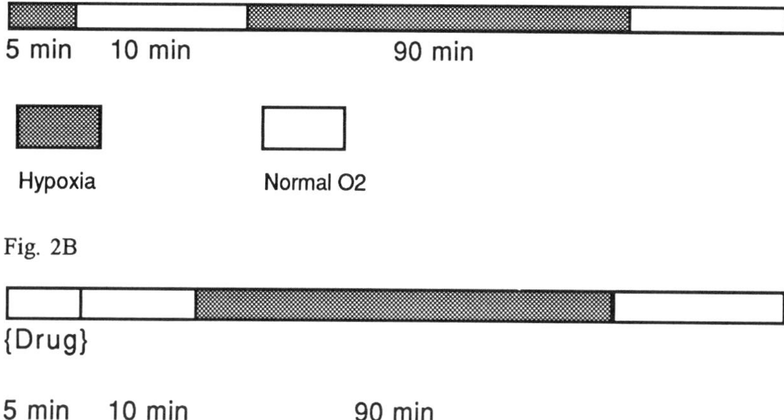

Two potential cardioprotective functions of the A_3 receptor were investigated.
The first cardioprotective function examined was its role in ischemic
preconditioning.   In determining this role of the adenosine A_3 receptor, myocytes
were exposed to A_3 agonist for five minutes prior to replacement with drug-free
media. Myocytes were incubated in this media for a further ten minutes before being
exposed to ninety minutes of simulated ischemia (Fig. 2D). Prior exposure to Cl-
IB-MECA (Fig. 3A) or IB-MECA (data not shown) was able to induce a DPCPX-
insensitive preconditioning response.   The data suggest that the response mediated
by the A_3 agonists is the result exclusively of a non-A_1 adenosine receptor, likely the
A_3 subtype.   The cAMP and preconditioning studies provide further evidence that a
functional adenosine A_3 receptor is present on the ventricular myocyte and that its
activation can also precondition the myocytes against ischemia-induced injury.   The
extent of cardioprotection afforded by A_3 receptor-mediated preconditioning is similar
to that of the protection in A_1 agonist-preconditioned myocytes (Fig. 3B).

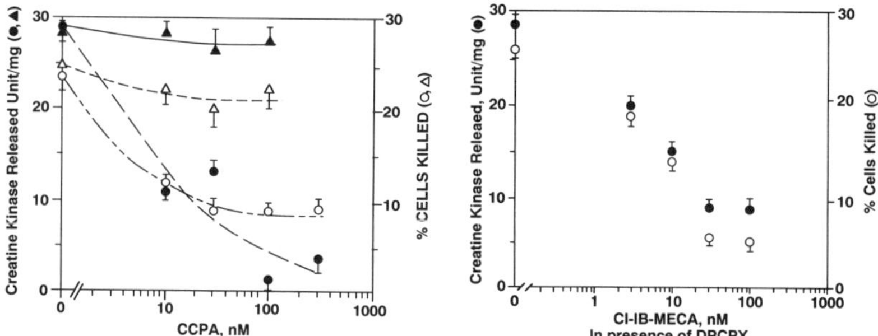

Figure 3. <u>Effects of prior exposure to A_1 or A_3 receptor agonists on the ischemia-induced cell injury</u>. Cells were pre-exposed to varying concentrations of (A) CCPA in the presence (open and closed triangles) or the absence of 0.1 µM DPCPX (open and closed circles) or (B) Cl-IB-MECA in the presence of 0.1 µM DPCPX (open and closed circles)., then incubated in the presence of fresh media lacking the adenosine analogs for 10 min, and finally exposed to 90 min of hypoxia. Data were plotted as percent cells killed or as amount of CK released vs. concentrations of the indicated agonists. Reproduced from The Journal of Clinical Investigation, 1996, vol. 98, pp. 1777, by copy right permission of the American Society for Clinical Investigation.

2. Cardioprotective role during prolonged ischemia.

The cardiac myocyte model of simulated ischemia provided an unique opportunity to investigate a potential protective function of the A_3 receptor during infarct-producing ischemia. In these studies, A_3 receptor-selective agonists or antagonists were included in media bathing the cells during the ninety-minute sustained ischemia. When present during the sustained simulated ischemia, IB-MECA or Cl-IB-MECA caused a dose-dependent reduction in the extent of ischemia-induced injury, as manifested by a decrease in the amount of creatine kinase released and the percentage of myocytes killed (Fig. 4A). The adenosine A_1 receptor-selective agonist CCPA, N^6-cyclohexyladenosine and adenosine amine congener were also able to attenuate myocyte injury when present during the ischemia (Fig. 4B). DPCPX completely abolished the A_1 agonist-mediated protection while it had no effect on the IB-MECA- or Cl-IB-MECA-mediated cardioprotection. Conversely, adenosine A_3 receptor-selective antagonist MRS1191 or 1097 completely blocked the IB-MECA- or Cl-IB-MECA-mediated protection. Neither MRS1191 nor MRS1097 had any significant effect on the protection exerted by the A_1 receptor agonists (Fig. 5). Therefore, the cardioprotective effects of adenosine A_1 and A_3 receptor agonists were mediated by their respective receptors.

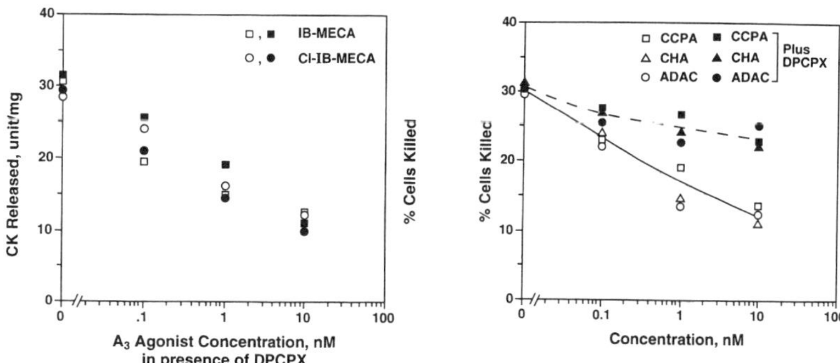

Figure 4. <u>Cardioprotective effects of adenosine A₁ and A₃ receptor agonists during sustained ischemia</u>. Cardiac ventricular myocytes were exposed to 90-min simulated ischemia. (A) The A₃ agonist, IB-MECA or Cl-IB-MECA, was present during the ischemia and the extent of myocyte injury determined. (B) The A₁ agonist, CCPA, CHA or ADAC, was present in the presence or the absence of the A₁ antagonist DPCPX during the ischemia. Reproduced from The American Journal of Physiology, 1996, vol. 273, pp. H502-H503, by copy right permission of The American Physiological Society.

Figure 5

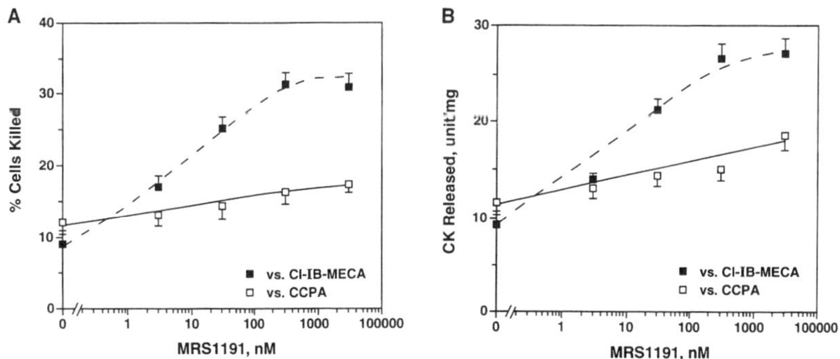

Figure 5. <u>Effect of adenosine A₃ receptor-selective antagonist MRS1191 on CCPA- and Cl-IB-MECA-induced cardioprotection</u>. 3-ethyl 5-benzyl-2-methyl-6-phenyl-4-phenylethynyl-1,4-(±)-dihydropyridine-3,5-dicarboxylate (MRS1191) was present at concentrations indicated with CCPA (10 nM) or Cl-IB-MECA (10 nM) during the

90-min ischemia. Reproduced from The American Journal of Physiology, 1996, vol. 273, pp. H503, by copy right permission of The American Physiological Society.

These studies define a novel cardioprotective function of the myocyte A_3 receptor and provide conclusive evidence that activation of both A_1 and A_3 receptors during ischemia can attenuate myocyte injury. Agonists selective at the A_1 or the A_3 receptors represent novel potent cardioprotective agents even when ischemia has begun. The data have important clinical implications in the treatment of ischemic heart disease and suggest that A_1 and A_3 receptor-selective agonists may reduce the size of myocardial infarction when given during infarct-producing ischemia.

C. Potential signaling mechanisms in the receptor-mediated cardioprotection

The preconditioning (PC) effect can be divided into two phases (Fig. 6). In the first phase, the preconditioning effect is initiated. Signaling mechanisms are activated and remain in an activated state during the sustained ischemia. During the sustained ischemia in which the second phase occurs, the actual cardioprotective effect of preconditioning is exerted. A series of signaling events take place during each phase as outlined in a hypothetical working model (Fig. 6). A growing body of evidence provides support for this model, which is summarized below.

Role of Ado receptor: a working model

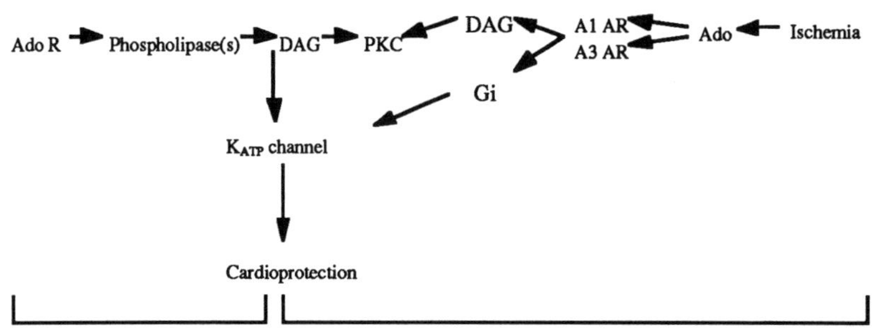

During initiation phase, adenosine is released from the ischemic myocardium which triggers the preconditioning by activating the adenosine receptor (Strickler et al., 1996). The receptor is coupled to stimulation of diacylglyceride (DAG) via either phospholipase C or D (Morley et al., 1996). A sustained increase in the level of DAG stimulates protein kinase C (PKC), which in turn, causes an activated K_{ATP} channel. Thus, inhibition of PKC blocks the preconditioning effect elicited by adenosine receptor activation (Liang 1997). K_{ATP} channel antagonist blocks the preconditioning effect elicited by adenosine agonist or phorbol ester, establishing the sequential activation of adenosine receptor, PKC and K_{ATP} channel as the sequence of signaling events during the initiation of preconditioning. During the second phase, activation of both adenosine receptor and K_{ATP} channel is required to exert the actual cardioprotective effect of preconditioning (Fig. 7). Central to this hypothesis is the concept that the K_{ATP} channel is first primed in an activated state during the initiation phase, and becomes more sensitive to the protective effect of adenosine during the second phase. Although definitive proof for this concept is yet to be provided, it is testable and deserves further study. Ultimately, activation of the channel is responsible for the protection against ischemia-induced injury. The precise mechanism by which K_{ATP} channel mediates the cardioprotective effect and the exact location of the channel in the cardiac myocyte remain to be determined.

Figure 7

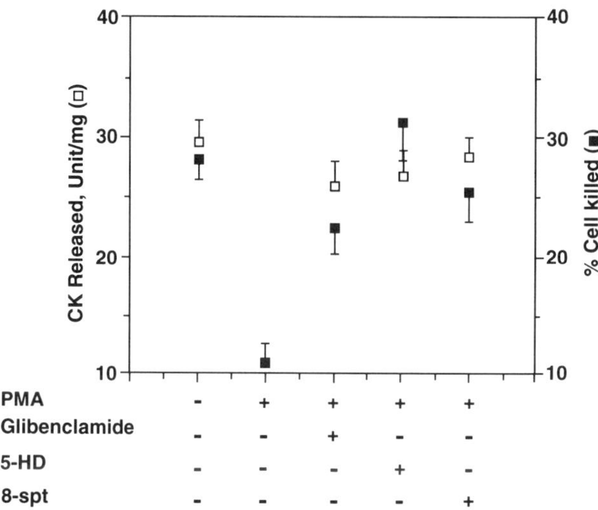

Figure 7. <u>Presence of glibenclamide, 5-HD or 8-SPT during the 90-min ischemia abolished PMA-induced preconditioning effect</u>. After prior exposure of myocytes to phorbol 12-myristate 13-acetate (PMA 100 nM) and PMA-free medium, 100 μM glibenclamide, 5-hydroxydecanoic acid (5-HD) or adenosine receptor antagonist 8-sulfophenyltheophylline (8-SPT) was added during the 90-min ischemic period. All three agents, when present individually during the ischemia, reversed the PMA-induced preconditioning effect. Reproduced from The American Journal of Physiology, 1997, vol. 273, pp. H851, by copy right permission of The American Physiological Society.

<u>D) Use of a Transgenic Approach to Determine the Function and Mechanism of A_1 Receptor-Mediated Cardioprotection</u>

1. A_1 receptor mediated protection: pharmacologic approach vs. altering receptor number.

Despite evidence in favor of cardioprotection by *endogenous* adenosine there remains controversy regarding the ability of *exogenous* adenosinergic therapy to reduce ischemic or post-ischemic injury (Vander Heide et al., 1996). One possibility which has received little attention is that it may be difficult to pharmacologically enhance an intrinsic response or mechanism that is normally maximal (or near maximal) during ischemia reperfusion. Extracellular adenosine levels in the heart exceed 1-10 μM during ischemia (Headrick et al., 1991; Van Wylen et al., 1992; Lasley et al., 1995), which should almost saturate myocardial A_1 receptors. An alternative to pharmacological therapy that could prove more effective is to increase the number of functional receptors present.

Because upregulation of A_1 receptors with antagonist is modest (10%-17%) (Rudolphi et al., 1989; Wu et al., 1989) and is not tissue or receptor subtype specific, we sought to increase A_1 adenosine receptors specifically in the heart using transgenic techniques. Transgenic models of G protein coupled receptors have been used to assess cardiac function and provide unique opportunities to study receptor signaling in heart (Koch et al., 1996). ß-Adrenoceptors have been overexpressed in the heart, enhancing myocardial function (Milano et al., 1994). This has been suggested as a possible approach for management of chronic heart failure when appropriate vectors for gene therapy become available. Transgenic techniques have also been used to improve the myocardial response to ischemia with overexpression of heat shock proteins (Plumier et al., 1995; Marber et al., 1995; Radford et al., 1996) and glutathione peroxidase (Yoshida et al., 1996) demonstrating that genetic manipulation can be used for a potentially therapeutic end result.

2. Development and characterization of the transgenic model of cardiac a_1 adenosine receptor overexpression.

Transgenic Construct: The full length rat A_1 cDNA (Reppert et al., 1991) was sub-cloned into a construct containing the α-MHC promoter (Adolph et al., 1993) for production of transgenic mice. This results in high level expression in atria and ventricles (Adolph et al., 1993). Founders were screened for the presence of the transgene by Southern analysis. Mouse genomic DNA was digested with EcoRI and probed with an α-MHC promoter fragment. EcoRI digestion of native α-MHC promoter results in a 2.6 kb fragment and digestion of the transgenic α-MHC A_1 construct results in a 1.6 kb fragment (Fig. 8a). Northern analysis was used to document increased message in transgene positive hearts using standard techniques (Matherne et al., 1996). Abundant message is present in transgenic hearts (Fig. 8b) while in the control animal, A_1 message was not detectable by standard northern analysis (Matherne et al., 1996).

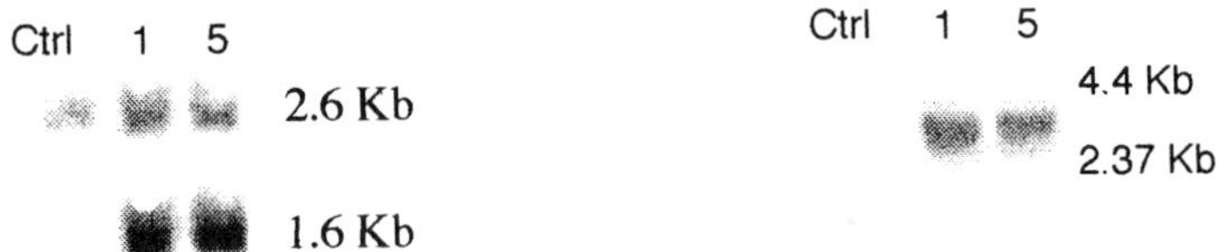

Figure 8 a) Southern   b) Northern

Figure 8. <u>Genotype and phenotype determination.</u> Demonstration of transgene detection by Southern analysis (a) and mRNA detection by northern analysis (b) in line 1 and line 5 mice. Permission to use Figures 8a was granted by National Academy of Sciences, U.S.A. Copyright 1987 (Matherne GP, Linden J, Byford AM, Gauthier NS, Headrick JP. Transgenic adenosine A1 receptor overexpression increases the resistance of the heart to ischemia. Proc Natl Acad Sci 94:6541-6546, 1997.).

Phenotype Determination Of Transgenic Lines: Founders were bred to C57Bl/6 mice to produce F1 pups. Tail biopsies were performed at 4 weeks and assessed for presence of transgene by Southern blotting. Transgenic lines were bred to either C57Bl/6 mice or transgene negative siblings to produce the F2 generation for phenotype determination. Phenotype was determined by: i) degree of receptor expression, ii) receptor coupling, and iii) receptor function.

Receptor Expression: A_1 adenosine receptor density was determined in whole heart membranes using standard radioligand binding techniques (Matherne et al., 1996) in 8 hearts from each of the transgenic lines. Although endogenous adenosine receptors can be more abundant in atria than ventricles (Musser et al., 1993), there is equal expression in atria and ventricles with the α-MHC promoter used (Adolph et al., 1993). Thus, whole heart membranes were used to quantitate receptor number. Antagonist (DPCPX) binding was used to measure the receptor number and receptor density (B_{max}) was reported as fmol receptor per mg protein. Transgene positive hearts expressed ~1000-fold greater levels of A_1 receptor than wild-type hearts (Fig. 9a). The control value of 8 fmol/mg protein is comparable to rat and rabbit (Musser et al., 1993; Matherne et al.. 1996). Binding was specific and saturable in transgenic tissue, and calculated dissociation constants for DPCPX were comparable in transgenic and control hearts (Line 1 - 0.92±0.09 nM, Line 5 - 1.12±0.09 nM, and Ctrl - 0.73±0.17 nM).

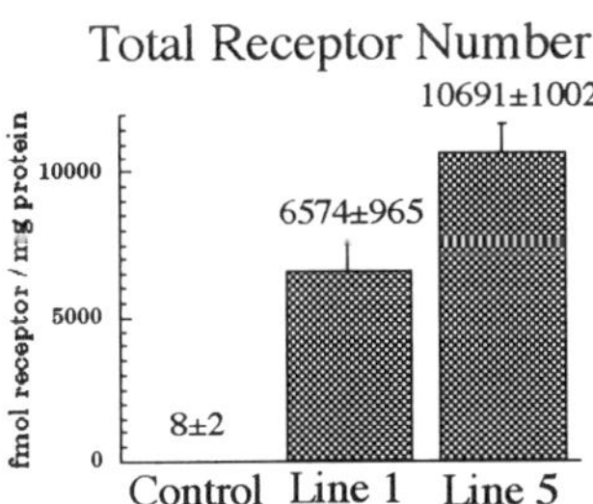

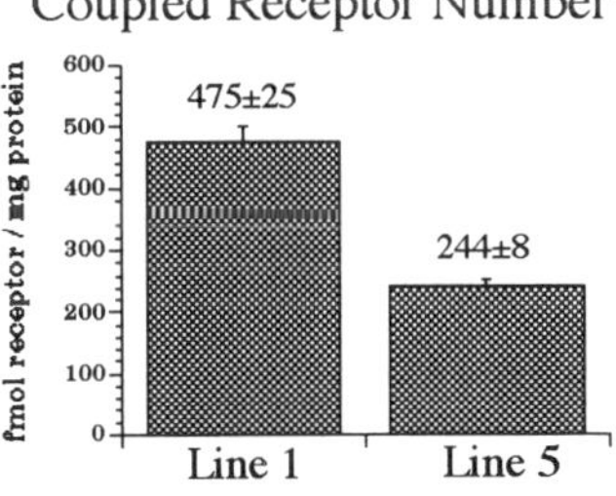

Figure 9: a) DPCPX binding   **b)** ABA binding

Figure 9. <u>Receptor number in transgenic hearts.</u> Determination of total receptor number by DPCPX binding (a) and coupled receptor number by ABA binding (b). Permission to use Figures 9a&b was granted by National Academy of Sciences, U.S.A. Copyright 1987 (Matherne GP, Linden J, Byford AM, Gauthier NS, Headrick JP. Transgenic adenosine A1 receptor overexpression increases the resistance of the heart to ischemia. Proc Natl Acad Sci 94:6541-6546, 1997.).

Receptor Coupling: Coupling of receptors to G-proteins was determined by high affinity agonist binding with ^{125}I ABA in 4 hearts from each transgenic line. GTP was added in some experiments to obliterate high affinity binding. % Receptor coupling was estimated by comparing coupled to total receptors ([^{3}H]DPCPX binding). Line 1 had more coupled receptors than line 5 (Fig. 9b) (line 1 - 7.7±1 %, line 5 - 2.4±3 %). Despite the low percentage of coupled receptors in transgenic hearts, the absolute number of coupled receptors (>240 fmol/mg) greatly exceeds the total receptor number in controls (8 fmol/mg).

Receptor Function: Receptor function was measured pharmacologically by determining the A_1 adenosine receptor profile with competition curves using 3 hearts from each transgenic line. The A_1 adenosine receptor profile was obtained with competition dissociation experiments and K_i's of competing ligands were calculated. Competition binding in all lines documented an A_1 receptor profile (Table 1) Thus the A_1 receptors appear pharmacologically and functionally normal.

TABLE 1: K_i data (nM)

	CPX	CPA	CGS	Theo
Line 1	0.4 ± 0.3	0.3 ± 0.2	400 ± 200	17000 ± 1800
Line 5	2 ± 0.6	0.3 ± 0.1	200 ± 100	10000 ± 2200

3. Cardiac function in transgenic mice.

Since A_1 receptor activation can inhibit myocardial inotropic state we determined the effects of A_1 overexpression on intrinsic contractility and the response to catecholamine stimulation (Gauthier et al., 1998). Basal heart rates were lower in transgenic (Trans) hearts than controls (~260 bpm for transgenic hearts n=3), but with pacing cardiac function and contractility were similar (Table 2).

TABLE 2. FUNCTIONAL PARAMETERS

	Output (ml/min/g)	LV pressure (mmHg)	+dP/dt (mmHg/s)	Heart Rate (bpm)
Control n=12	33±1	75±2	2440±154	330±10
Trans n=10	30±1	75±2	2628±81	324±2 (paced)

To assess intrinsic ventricular function, volume and pressure loading was performed (Starling curves shown in Fig. 10a and 10 b, n=12 per group). No differences were noted over the entire range of loads tested.

Figure 10

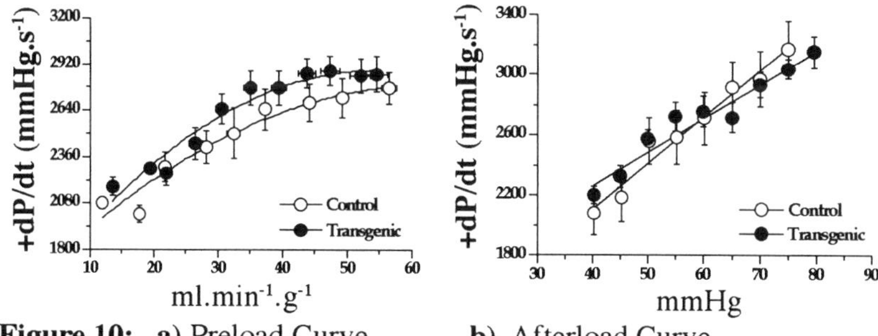

Figure 10: **a)** Preload Curve   **b)** Afterload Curve

Figure 10. <u>Ventricular function.</u> Starling type curves for volume loading, a)Preload Curve and pressure loading, b)Afterload Curve in control and transgenic mice.

No differences were observed in sensitivity to isoproterenol (EC_{50}= 1.6±0.7 x 10^{-8} M for controls, 1.2±0.5 x 10^{-8} M for transgenics), but at maximal doses there was a decrease in maximum +dP/dt in Trans compared to Ctrl hearts (maximum +dP/dt 152±6% baseline for Ctrl, 131±2% baseline for Trans, p<0.05). In summary, overexpression of A_1 receptors does not produce untoward effects on ventricular function or sensitivity to catecholamine stimulation, but dampens the contractile response at high doses of catecholamines (Gauthier et al., 1998). Permission was granted by Academic press Copyright 1998 (Gauthier NS, Headrick JP, Matherne GP. Myocardial function in the working mouse heart overexpressing cardiac A_1 adenosine receptors. J Mol Cell Cardiol 30:187-193(1998).

4. Cardioprotective effects of A_1 overexpression.

Severity of ischemic injury was assessed by time to onset of contracture (TIC) in control hearts, transgenic hearts and control hearts treated with either 50 µM 8-SPT (adenosine antagonist) or 20 nM CPA (A_1 adenosine agonist). As shown in Fig. 11a time to onset of contracture was ~10 min in control hearts. This was reduced to 7 min in adenosine blocked hearts (8-SPT) and was unaltered by pre-treatment with CPA presumably because adenosine released during ischemia saturated available A_1 receptors. In transgenic hearts, time to onset of contracture was significantly prolonged to 14 min.

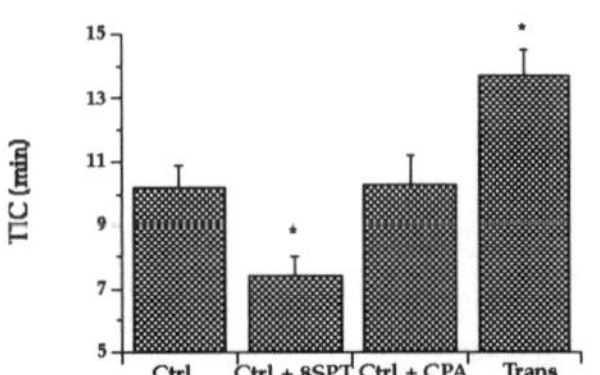 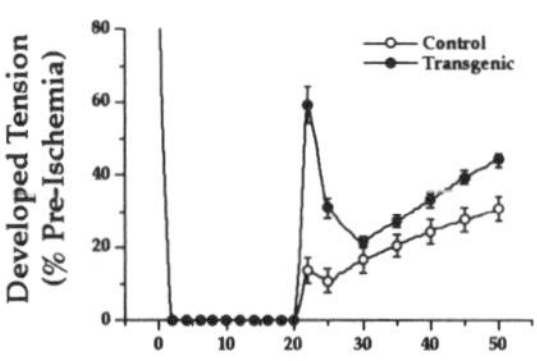

Figure 11: a) Time to Ischemic Contracture   **b)** Developed Tension during Ischemia

Figure 11. <u>Response to ischemia and reperfusion.</u> Time to ischemic contracture (TIC) (fig 11 a) in control (ctrl), adenosine receptor blocked (ctrl+8SPT), adenosine receptor stimulated (ctrl+CPA), and transgenic (trans) groups and developed tension (fig 11b b) in control and transgenic groups. *, different from control, p<0.05.

Functional parameters for control and transgenic hearts prior to, during and following 20 min of global ischemia are shown in Fig. 11b. Ischemia rapidly abolished contractile function. In the first 2 min of reperfusion there was an initial recovery of contractile function to 15% of pre-ischemia in control hearts but to 60% of pre-ischemia in transgenic hearts. Function then declined rapidly and stabilized. This profile indicates that A_1 overexpression reduces the ischemic injury together with subsequent reperfusion injury. At the end of reperfusion, diastolic tension remained elevated 25% above pre-ischemia in control hearts but recovered to baseline in transgenic hearts. Similarly, developed tension recovered to only 30% of pre-ischemia in control hearts and to 45% in transgenic hearts. Coronary flow did not differ between the two groups. Thus, functional recovery during reperfusion is improved significantly in transgenic hearts (Matherne et al., 1997). Adenosine receptor antagonism with 50 µM 8-SPT also reduced functional recovery on reperfusion in control and transgenic hearts (data not shown). Permission to use Figure 11a was granted by National Academy of Sciences, U.S.A. Copyright 1987 (Matherne GP, Linden J, Byford AM, Gauthier NS, Headrick JP. Transgenic adenosine A1 receptor overexpression increases the resistance of the heart to ischemia. Proc Natl Acad Sci 94:6541-6546, 1997.).

5. Effects of A_1 overexpression on bioenergetic and metabolic response to ischemia reperfusion.

Metabolic responses to 30 min global ischemia and 20 min reperfusion were compared in Langendorff perfused hearts from transgenic (n=7) and wild-type mice (n=8). Parallel bench studies were performed to study effects of ischemia reperfusion on contractile function (n=5 in transgenic, n=7 in wild-type). ^{31}P-NMR spectroscopy revealed that myocardial ATP was preserved to a much greater extent in transgenic hearts: 53±11% of pre-ischemic ATP was still present at the end of ischemia in transgenic hearts with only 4±4% present in wild-type hearts. Recovery

100

of ATP after 20 min reperfusion tended to be higher in transgenic (46±5%) versus wild-type hearts (37±12%) but did not reach statistical significance (Fig. 12a).

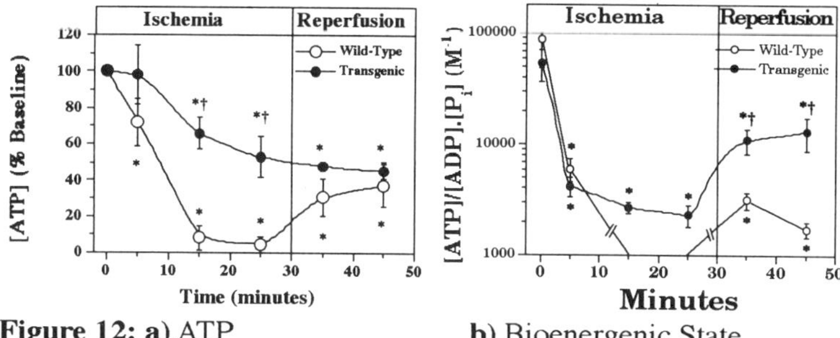

Figure 12: a) ATP   **b)** Bioenergenic State

Figure 12. <u>Metabolic response to ischemia.</u> Changes in ATP levels (a) and Bioenergetic state (b) during ischemia and recovery on control and transgenic hearts. *, different from control; + different from transgenic, both p<0.05. Permission was granted by by Academic press Copyright 1998 (Headrick JP, Gauthier NS, Berr SS, Matherne GP. Transgenic A1 adenosine receptor overexpression improves myocardial energy state during ischemia reperfusion. J Mol Cell Caradiol in press 1998).

The decline in phosphocreatine (PCr) during ischemia was similar in both groups but PCr recovery was higher in transgenic (67±8%) versus wild-type hearts (36±8%). Intracellular acidosis was similar in both groups during ischemia (minimum pH ~6.2), but recovery of pH during reperfusion was greater in transgenic hearts (pH=7.11±0.05) versus wild-type hearts (pH=6.90±0.02). Bioenergetic state, indexed by [ATP]/[ADP].[P$_i$], was higher in transgenic hearts during both ischemia and reperfusion (Fig. 5b). Functional studies verified that time to ischemic contracture was shorter in wild-type (10.4±0.3 min) versus transgenic hearts (13.6±0.8 min), degree of contracture was higher in wild-type versus transgenic hearts (270% vs. 215% of pre-ischemia), and recovery of diastolic tension was enhanced in transgenic versus wild type hearts (160% vs. 220% of pre-ischemia). These data indicate that A$_1$ overexpression reduces loss of ATP and improves bioenergetic state during ischemia, and improves metabolic and bioenergetic recovery during reperfusion. These pronounced changes may contribute to improved functional tolerance to ischemia reperfusion.

<u>E) Lessons Learned from the A$_1$ Receptor/Transgenic Model</u>

The primary goal of this work was to examine the impact of transgenic A$_1$ adenosine receptor overexpression on the myocardial response to ischemia and reperfusion. Transgenic manipulation permits substantial modification of protein levels in heart. Several studies demonstrate functionally beneficial effects of different transgenic manipulations in the setting of ischemia reperfusion, including overexpression of heat shock proteins, glutathione peroxidase, and superoxide

101

dismutase. The data verify that activation of the myocardial A_1 receptor by endogenous adenosine is indeed cardioprotective, reducing the degree of injury during ischemia and reperfusion. Although the mechanisms underlying the A_1 mediated cardioprotection remain unidentified, the data acquired to date provide some insight into potential mechanisms.

Interestingly, the primary effect of A_1 overexpression on contractile function appeared to be limited to diastolic function with lesser differences in recovery of systolic pressure. Factors impacting on diastolic tension include intracellular Ca^{2+} levels and sarcoplasmic reticular Ca^{2+} handling. These parameters may be modified by adenosine coupled transduction mecahnisms including activation of K_{ATP} channels and modification of glucose metabolism. Moreover, myocardial Ca^{2+} handling is sensitive to cellular energy state and the present data verify a profound improvement of cellular energy state by A_1 receptors.

The degree of preservation of ATP during ischemia in transgenic hearts is remarkable. A_1 receptors may activate anaerobic ATP production via glycolysis, and several studies have shown that receptor activation by exogenous and endogenous adenosine enhances myocardial glucose uptake (Angello et al., 1993; Murphy et al., 1993; Fang et al., 1997) and adenosine-mediated cardioprotection is glucose dependent in some models (Ganote et al., 1993). Glycolytically derived ATP may preferentially support metabolic and ionic homeostasis during ischemia due to co-localization of glycolytic enzymes and sarcolemmal and sarcoplasmic reticular ion channels. The data for ischemic ATP are not entirely consistent with inhibition of myocardial glycolysis and optimization of the coupling of myocardial glycolysis and glucose oxidation during ischemia, as proposed by Finegan et al. (1996), since there was no difference in cellular acidosis during the ischemic insult. We do note however, that transgenic hearts exhibited significantly lower $[H^+]$ during reperfusion whereas wild-type hearts exhibit sustained acidosis, consistent with the hypothesis of Finegan et al. (1996) in the reperfused (but not ischemic) myocardium. It may be that A_1 activation modifies different mechanisms at different stages of ischemia reperfusion.

IV. CONCLUSION AND FUTURE DIRECTIONS

Adenosine is released in large amounts during myocardial ischemia. Adenosine can serve two important cardioprotective functions. It can trigger as well as mediate the cardioprotective effect of ischemic preconditioning. Further, adenosine can also protect against ischemia-induced cell death and against reperfusion injury when it is present during ischemia and reperfusion, respectively. The preconditioning and the ischemia-resistant effects of adenosine are mediated via both the A_1 and A_3 receptors on the cardiac myocyte. Whether the myocyte A_1 or A_3 receptor can protect from reperfusion injury remains to be determined. The cardioprotective functions served by the A_1 and A_3 receptors are mediated by a sequential stimulation of diacylglyceride accumulation, of protein kinase C, and of K_{ATP} channel. In addition, direct activation of the channel by G_i subunit following receptor stimulation, perhaps in a membrane-delimited manner, may also play an important role in mediating these protective functions.

Future studies will need to address a number of important issues. First, remaining gaps of knowledge on the actions of adenosine receptor, phospholipases, protein kinase C and K_{ATP} channel during both the initiation and maintenance phases of ischemic preconditioning need to be closed. Second, the molecular basis of the interaction between the various receptors, enzymes and channels needs to be elucidated. For example, the G protein (both monomeric and heterotrimeric forms) that couples the adenosine receptor to phospholipases and the role of phosphorylation of the K_{ATP} channel by PKC need to be identified. Third, how activation of the K_{ATP} channel achieves the cardioprotective effect needs to be clarified. Whether it is the sarcolemmal or the mitochondrial channel that is important in mediating the cardioprotection needs to be determined. Finally, the question of whether apoptosis plays a quantitatively important role in the ischemia-induced cell death remains to be elucidated. If apoptosis contributes importantly to the cell death, whether adenosine receptor activation can protect against apoptosis and what the underlying mechanism(s) might be are the important questions.

The data from studies of transgenic hearts indicate that overexpression of myocardial A_1 adenosine receptors significantly increases functional and metabolic tolerance to ischemia reperfusion in murine myocardium. This provides additional evidence that A_1 receptor activation by endogenous adenosine improves myocardial functional and metabolic state during ischemia reperfusion. Improved tolerance to ischemia reperfusion with A_1 overexpression is not associated with untoward effects on baseline contractile function. These data collectively support the potential benefit of genetic therapy targeting the myocardial A_1 receptor. We are unaware of another stimulus producing such marked protection of myocardial ATP during severe ischemic insult. Transgenic overexpression of 70-kDa heat shock protein does not modify ischemic ATP depletion, and improves ATP repletion only following brief and not prolonged periods of ischemia (Radford et al., 1996). It has been postulated that A_1 receptor mediated cardioprotection is secondary to multiple mechanisms (Ely and Berne 1992; Sekili et al., 1995; Headrick 1996), and this is supported to some extent by data acquired here and elsewhere. Future studies should seek to identify the various signaling pathways involved in these effects, and the sequence of events involved in adenosine mediated cardioprotection.

REFERENCES.

Adolph EA, Subramaniam A, Cserjes P, Olson EN, Robbins J. Role of myocyte-specific enhancer-binding factor in transcriptional regulation of the alpha-cardiac myosin heavy chain gene. J Biol Chem 1993; 268:5349-5352.

Ambrosio G, Jacobus WE, Mitchell MC, Litt MR, Becker LC. Effects of ATP precursors on ATP and free ADP content and functional recovery of postischemic hearts. Am J Physiol 1989; 256: H560-H566.

Angello DA, Headrick JP, Coddington NM, Berne RM. Adenosine antagonism decreases metabolic but not functional recovery from ischemia. Am J Physiol 1991; 1260:H193-H200.

Angello DA, Berne RM, Coddington NM. Adenosine and insulin mediate glucose uptake in normoxic rat hearts by different mechanisms. Am J Physiol 1993; 265:H880-H885.

Auchampach JA, Rizvi A, Qiu Y, Tang X-L, Maldonado C, Teschner S, Bolli R. Selective activation of A3 adenosine receptors with N6-(3-iodobenzyl)adenosine-5'-N-methyluronamide protects against myocardial stunning and infarction without hemodynamic changes in conscious rabbits. Circ Res 1997; 80: 800-809.

Babbitt DG, Virmani R, Forman MB. Intracoronary adenosine administered after reperfusion limits vascular injury after prolonged ischemia in the canine model. Circ 1989; 80:1388-1399.

Belardinelli L, Linden J, Berne RM. The cardiac effects of adenosine. Prog Cardiovasc Dis 1989; 32:73-97.

Cave AC, Garlick PB. Ischemic preconditioning and intracellular pH: a ^{31}P-NMR study in the isolated rat heart. Am J Physiol 1997; 272: H544-H552.

Carr CS. Hill RJ, Masamune H, Kennedy SP, Knight DR, Tracey WR, Yellon DM. Evidence for a role for both the adenosine A_1 and A_3 receptors in protection of isolated human atrial muscle against simulated ischemia. Cardiovasc Res 1997; 36: 52-59.

Cross HR, Opie LH, Radda GK, Clarke K. Is a high glycogen content beneficial or detrimental to the ischemic rat heart? Circ Res 1996; 78:482-491.

Downey JM. Ischemic preconditioning. Nature's own cardio protective intervention. Trends Cardiovasc. Med. 1992; 2:170-176.

Ely SW, Berne RM. Protective effects of adenosine in myocardial ischemia. Circulation 1992; 85:893-904.

Fang HK, Sturgeon C, Segil LJ, Ripper RL, Law WR. Cardiac contractile function during coronary stenosis in dogs: association of adenosine in glycolytic dependence. Am J Physiol 1997; 272: H2195-H2203.

Finegan BA, Gandhi M, Lopaschuk GD, Clanachan AS. Antecedent ischemia reverses effects of adenosine on glycolysis and mechanical function of working hearts. Am J Physiol 1996; 271:H2116-H2125.

Fralix TA, Murphy E, London RE, Steenbergen C. Protective effects of adenosine in the perfused rat heart: changes in metabolism and intracellular ion homeostasis. Am J Physiol 1993; 264:C986-C994.

Ganote CE, Armstrong S, Downey JM. Adenosine and A_1 selective agonists offer minimal protection against ischaemic injury to isolated rat cardiomyocytes. Cardiovasc Res. 1993; 27:1670-1676.

Gauthier NS, Headrick JP, Matherne GP. Myocardial function in the working mouse heart overexpressing cardiac A_1 adenosine receptors. J Mol Cell Cardiol 1998; 30:187-193.

Gross GJ. ATP-sensitive potassium channels and myocardial preconditioning. Basic Res Cardiol. 1995; 90: 85-88.

Hagar JM, Hale SL, Kloner RA. Effect of preconditioning ischemia on reperfusion arrhythmias after coronary artery occlusion and reperfusion in the rat. Circ 1991; 68:61-68.

Headrick JP, Matherne GP, Berr SS, Berne RM. Effects of graded perfusion and isovolumic work on epicardial and venous adenosine and cytosolic metabolism. J Mol Cell Cardiol 1991; 23:309-324.

Headrick JP. Ischemic preconditioning: bioenergetic and metabolic changes and the role of endogenous adenosine. J Mol Cell Cardiol 1996; 28:1227-1240.

Headrick JP, Gauthier NS, Berr SS, Matherne GP. Transgenic A1 adenosine receptor overexpression improves myocardial energy state during ischemia reperfusion. J Mol Cell Caradiol 1998, in press.

Jeremy RW, Ambrosio G, Pike MM, Jacobus WE, Becker LC. The functional recovery of post-ischemic myocardium requires glycolysis during early reperfusion. J Mol Cell Cardiol 1993; 25:261-276.

Kammermeier H, Roeb E. Free energy change of ATP-hydrolysis: A causal factor of early hypoxic failure of myocardium? J Mol Cell Cardiol 1982; 14:267 277.

Kingsley PB, Sako EY, Yang MQ, Zimmer SD, Ugurbil K, Foker JE, From AHL. Ischemic contracture begins when anaerobic glycolysis stops: a 31P-NMR study of isolated rat hearts. Am J Physiol 1991; 261: H469-H478.

Koch WJ, Milano CA, Lefkowitz RJ. Transgenic manipulation of myocardial G protein coupled receptors and receptor kinases. Circ Res 1996; 78:511-516.

Lasley RD, Rhee JW, Van Wylen DGL, Mentzer RM. Adenosine A_1 receptor mediated protection of the globally ischemic isolated rat heart. J Mol Cell Cardiol 1990; 22:29-47.

Lasley RD, Mentzer RM. Adenosine improves recovery of postischemic myocardial function via an adenosine A_1 receptor mechanism. Am J Physiol 1992; 263:H1460-H1465.

Lasley RD, Konyn PJ, Hegge JO, Mentzer RM Jr. Effects of ischemic and adenosine preconditioning on interstitial fluid adenosine and myocardial infarct size. Am J Physiol 1995; 269(5): H1460-H1466.

Leesar MA, Stoddard M, Ahmed M, Broadbent J, Bolli R. Preconditioning of human myocardium with adenosine during coronary angioplasty. Circulation 1997; 95:2500-2507.

Li GC, Vasquez JA, Gallagher KP, Lucchesi BR. Myocardial protection with preconditioning. Circulation 1990; 82:609-619.

Liang BT. Adenosine receptors and cardiovascular function. Trends Cardiovasc Med 1992; 2:100-108.

Liang BT. Direct preconditioning of cardiac ventricular myocytes via adenosine A_1 receptor and KATP channel. Am J Physiol 1996; 271:H1769-H1777.

Liang BT. Protein kinase C-mediated preconditioning of cardiac myocytes: role of adenosine receptor and K_{ATP} channel. Am J.Physiol 1997; 273: H847-H853.

Marber MS, Mestril R, Chi S-H, Sayen MR, Yellon DM, Dillman WH. Overexpression of the rat inducible 70kD heat stress protein in a transgenic mouse increases the resistance of the heart to ischemic injury. J Clin Invest 1995; 95:1446-1456.

Matherne GP, Byford AM, Gilrain JT, Dalkin AC. Changes in myocardial A_1 adenosine receptor and message levels during fetal development and postnatal maturation. Biol Neonate 1996; 70:199-205.

Matherne GP, Linden J, Byford AM, Gauthier NS, Headrick JP. Transgenic adenosine A_1 receptor overexpression increases the resistance of the heart to ischemia. Proc Natl Acad Sci 1997, 94:6541-6546.

Mentzer RM, Bunger R, Lasley RD. Adenosine enhanced preservation of myocardial function and energetics. Possible involvement of the adenosine A_1 receptor system. Cardiovasc Res 1993; 27:28-35.

Mentzer RM, Rahko PS, Molina-Viamonte V, Canver CC, Chopra PS, Love RB, Cook TD, Hegge JO, Lasley RD. Safety, tolerance, and efficacy of adenosine as an additive to blood cardioplegia in humans during coronary artery bypass surgery. Am J Cardiol 1997; 79(12A):38-43.

Milano CA, Allen LF, Rockman HA, Dolber PC, McMinn TR, Chien KR, Johnson TD, Bond RA, Lefkowitz RJ. Enhanced myocardial function in transgenic mice overexpressing the ß2-adrenergic receptor. Science 1994; 264:582-586.

Morley JF, Jacobson KA, Liang BT. Differential coupling of adenosine A_1 and A_3 receptors to stimulation of diacylglycerol in chick ventricular myocytes. Circulation 1996; 94(8):268.

Murphy E, Fralix TA, London RE, Steenbergen C. Effects of adenosine antagonists on hexose uptake and preconditioning in perfused rat heart. Am J Physiol 1993; 265: C1146-C1155.

Murry CE, Jennings RB, Reimer KA. Preconditioning with ischemia: A delay of lethal cell injury in ischemic myocardium. Circulation 1986; 74:1124-1136.

Musser B, Morgan ME, Leid M, Murray TF, Linden J, Vestal RE. Species comparison of adenosine and β-adrenoceptors in mammalian atrial and ventricular. Eur J Pharmacol Mol Pharmacol 1993; 246:105-111.

Olafsson B, Forman MB, Puett DW, Pou A, Cates CU. Reduction of reperfusion injury in the canine preparation by intracoronary adenosine: Importance of the endothelium and the no-reflow phenomenon. Circulation 1987; 76:1135-1145.

Olsson RA, Pearson JD. Cardiovascular purinoceptors. Physiol Rev 1990; 70: 761-809.

Plumier J-CL, Ross BM, Currie RW, Angelidis CE, Kazlaris H, Kollias G, Pagoulatos GN. Transgenic mice expressing the human heat shock protein 70 have improved post-ischemic myocardial recovery. J Clin Invest 1995; 95:1854-1860.

Radford NB, Fina M, Benjamin IJ, Moreadith RW, Graves KH, Zhao P, Gavva S, Wiethoff A, Sherry AD, Malloy CR. Cardioprotective effects of 70-kDa heat shock protein in transgenic mice. Proc Natl Acad Sci USA 1996; 93: 2339-2342.

Randhawa MPS Jr, Lasley RD, Mentzer RM Jr. Salutary effects of exogenous adenosine on in vivo myocardial stunning. J Thorac Cardiovasc Surg 1995; 110:63-74.

Reibel DK, Rovetto MJ. Myocardial adenosine salvage rates and restoration of ATP content following ischemia. Am J Physiol 1979; 237: H247-H252.

Reppert SM, Weaver DR, Stehle JH, Rivkees SA. Molecular cloning and characterization of a rat A_1-adenosine receptor that is widely expressed in brain and spinal cord. Mol Endo 1991; 5:1037-1048.

Rudolphi KA, Keil M, Fastbom J, Fredholm BB. Ischaemic damage in gerbil hippocampus is reduced following upregulation of adenosine (A_1) receptors by caffeine treatment. Neurosci Lett 1989; 103:275-280.

Sekili S, Jeroudi MO, Tang X-L, Zughaib M, Sun J-Z, Bolli R. Effects of adenosine on myocardial stunning in the dog. Circ Res 1995; 76: 82-94.

Stambaugh K, Lacobson KA, Jiang J-L, Liang BT. A novel cardioprotective function of adenosine A_1 and A3 receptors during prolonged simulated ischemia. AM J Physiol 1997; 273: H501-H505.

Strickler J, Jacobson KA, Liang BT. Direct preconditioning of cultured chick ventricular myocytes: Novel functions of cardiac adenosine A_{2a} and A3 receptors. J Clin Invest 1996; 98:1773-1779.

Tamm C, Benzi R, Papageorgiou I, Tardy I, Lerch R. Substrate competition in postischemic myocardium. Circ Res 1994; 75:1103-1112.

Tracey WR, Magee W, Masamune H, Kennedy SP, Knight DR, Buchholz RA, Hill RJ. Selective adenosine A_3 receptor stimulation reduces ischemic myocardial injury in the rabbit heart. Cardiovascular Res 1997; 33: 410-415.

Vander Heide RS, Reimer KA. Effect of adenosine therapy at reperfusion on myocardial infarct size in dogs. Cardiovasc Res 1996; 31:711-718.

Van Wylen DGL, Schmit TJ, Lasley RD, Gingell RL, Mentzer RM Jr. Cardiac microdialysis in isolated rat hearts: interstitial purine metabolites during ischemia. Am J Physiol 1992; 262:H1934-H1938.

Wang J, Drake L, Sajjadi F, Firestein GS, Mullane KM, Bullough DA. Dual activation of adenosine A_1 and A3 receptors mediates preconditioning of isolated cardiac myocytes. Eur J Pharmacol 1997; 320:241-248.

Wu SN, Linden J, Visentin S, Boykin M, Belardinelli L. Enhanced sensitivity of heart cells to adenosine and up-regulation of receptor number after treatment with theophylline. Circ Res 1989; 65:1066-1077.

Wyatt DA, Edmunds MC, Rubio R, Berne RM, Lasley RD, Mentzer RM Jr. Adenosine stimulates glycolytic flux in isolated perfused rat hearts by A_1-adenosine receptors. Am J Physiol 1989; 257:H1952-H1957.

Yao Z, Gross GJ. Glibenclamide antagonizes adenosine A_1 receptor-mediated cardioprotection in the stunned myocardium. Circ 1993; 88:235-244.

Yoshida T, Watanabe M, Engelman DT, Engelman RM, Schley JA, Maulik N, Ho Y-S, Oberley TD, Das DK. Transgenic mice overexpressing glutathione peroxidase are resistant to myocardial ischemia reperfusion injury. J Mol Cell Cardiol 1996; 28:1759-1767.

Zhao ZQ, McGee DS, Nakanishi K, Toombs CF, Johnston WE, Ashar MS, Vinten-Johansen J. Receptor mediated cardioprotective effects of endogenous adenosine are exerted primarily during reperfusion after coronary occlusion in the rabbit. Circulation 1993; 88:709-719.

Zhou Z, Bunger M, Lasley RD, Hegge JO, Mentzer RM Jr. Adenosine pretreatment increases cytosolic phosphorylation potential and attenuates postischemic cardiac dysfunction in swine. Surg Forum 1993; 44: 249-252.

REGULATION OF ADENOSINE RECEPTOR SUBTYPES AND CARDIAC DYSFUNCTION IN HUMAN HEART FAILURE

Birgitt Stein PhD, Jasper Kiehn, and Joachim Neumann MD*

*Pharmakologisches Institut, Universitäts-Krankenhaus Eppendorf, Martinistraße 52, D-20246 Hamburg, *Institut für Pharmakologie und Toxikologie, Westfälische Wilhelms-Universität Münster, Domagkstraße 12, D-48129 Münster*

I. SUMMARY

A defective adenosine receptor-mediated autoregulation of the heart has been hypothesized to cause cardiomyopathy. Regulation of gene expression of the four adenosine receptor subtypes A_1, A_{2A}, A_{2B}, A_3 and the A_1-mediated antiadrenergic signal transduction pathway has been studied in nonfailing and failing human hearts. Gene expression of myocardial A_1-adenosine receptors and A_1-mediated inhibitory effects on cAMP levels or negative inotropic effects were not altered in failing compared to nonfailing human hearts, indicating that an impairment in the A_1-mediated pathway may not be involved in the pathogenesis of heart failure. Furthermore, A_3-adenosine receptor gene expression was unchanged in failing compared to nonfailing hearts, while A_{2B} mRNA was not detectable. However, A_{2A}-adenosine receptor gene expression was increased by about 60% in patients with dilated cardiomyopathy. Thus, the A_{2A}-upregulation may indicate a pathophysiological role for adenosine and adds therefore to the increasingly complex picture of molecular alterations in heart failure.

II. BACKGROUND

A. Effects of adenosine in the heart

Myocardial effects of adenosine were first described by Drury and Szént-György (1929), who observed a reduction of atrial and ventricular force of contraction as well as a decline in beating rate and an impairment of atrioventricular conduction in the guinea-pig heart in situ. Since that time the effects of adenosine on the heart have been intensively investigated in animals and, however to much lesser extent, in humans (reviewed in Belardinelli et al. 1989, 1995, Liang 1992, Tucker and Linden 1993, Olsson 1996, Müller and Stein 1996).

The main effects of adenosine in the heart are negative chronotropic, negative dromotropic, negative inotropic and vasodilatory (Drury and Szént-György 1929, Berne 1963, recently reviewed in Olsson 1996, Müller and Stein 1996). There is general agreement that adenosine can decrease ventricular automaticity (Pelleg and Kutalek 1997). However, in contrast there are recent data to suggest also an increase in ventricular automaticity (Hernández and Ribeiro 1996). It has been known for a long time that adenosine can decrease cardiac contractility. More recent data suggest, that at physiological relevant concentrations of adenosine a positive inotropic effect that at physiological relevant concentrations of adenosine a positive inotropic effect might be induced, if the antiadrenergic negative inotropic effect is inhibited (Liang 1992, Xu et al. 1992, 1996, Dobson and Fenton 1994, Liang and Haltiwanger 1995, Liang and Morely 1996). These effects are due to the action of adenosine on various cardiac cell types in different anatomic regions of the heart by activation of different adenosine receptors (Fig. 1). Four subtypes of adenosine receptors termed A_1, A_{2A}, A_{2B} and A_3 have been cloned and pharmacologically characterized, based on the rank order of potency of agonists and antagonists, and on second messenger coupling (Table 1). Whether all four adenosine receptor subtypes are expressed in the human heart, as well as their physiological and pathophysiological role remains to be elucidated.

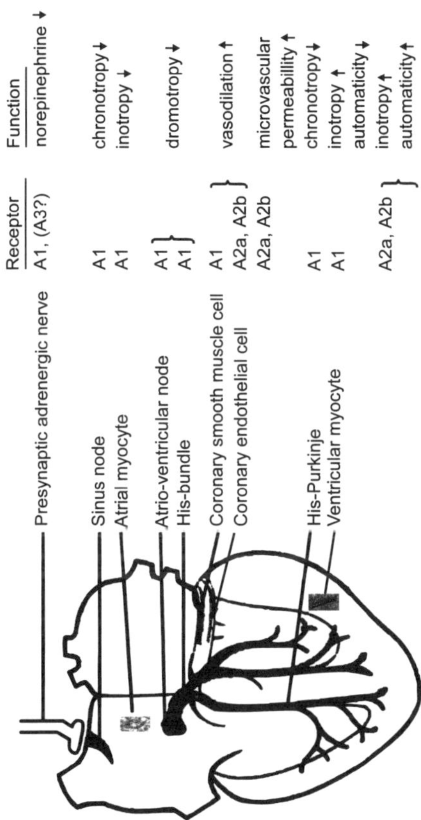

Figure 1. Cardiac action of adenosine receptor subtypes in the cardiovascular system.

<u>**Table 1.** Coupling of adenosine receptor subtypes to various effector systems</u>

	G-protein	**Effector system**
A_1	Gi	$AC\downarrow$, $K^+\uparrow$, $PLC\uparrow\downarrow$
A_{2A}	Gs	$AC\uparrow$
A_{2B}	Gs	$AC\uparrow$, $PLC\uparrow$
A_3	Gi	$AC\downarrow$, $PLC\uparrow$

Adenosine receptor subtypes couple to various effector systems via inhibitory (Gi) or stimulatory (Gs) GTP-binding proteins. AC = adenylyl cyclase, K^+ = potassium channel, PLC = phospholipase C, $\uparrow$ = activation, $\downarrow$ = inhibition.

The negative chronotropic and negative dromotropic effects of adenosine are mediated by A_1-adenosine receptors present on the sinus node and atrioventricular node, respectively. A_1-adenosine receptors present on atrial and ventricular cardiomyocytes mediate negative inotropic effects. The A_1-mediated negative inotropic effect in ventricular cardiomyocytes occurs only after previous stimulation with cAMP-elevating agents like ß-adrenoceptor agonists and has thus been termed as "indirect" or antiadrenergic effect. The "direct" negative inotropic effects of adenosine on the atria is due to an activation of the potassium outward current (Belardinelli and Isenberg 1983, Jochen and Nawrath 1983), whereas the "indirect" effect has been claimed to be mediated by an inhibition of adenylyl cyclase activity, hence decreasing cytosolic cAMP levels, which leads to a decrease in the L-type calcium current (Belardinelli et al. 1995). In contrast, others have presented evidence that at least in some species (guinea pigs), the ventricular effects of A_1-adenosine receptors occur independent of changes in cAMP content. That has been shown in isolated perfused ventricles (Neuman et al. 1995b), in papillary muscles (Böhm et al. 1988), and in isolated cardiomyocytes (Behnke et al. 1990, Gupta et al. 1993, Neumann et al. 1989, 1994, Stein et al. 1994). These A_1-adenosine receptor-mediated direct as well as indirect negative inotropic effects are mediated via inhibitory, pertussis toxin-sensitive, GTP-binding (Gi) proteins (Böhm et al. 1986, Brown et al. 1990, Stein et al. 1993, 1994, Neumann et al. 1994). We have demonstrated that A_1-adenosine receptor stimulation decreases phosphorylation of regulatory proteins (like phospholamban and the inhibitory subunit of troponin) in isolated perfused hearts (Neumann et al. 1995) and in isolated cardiomyocytes (Gupta et al. 1993; Neumann et al. 1994, Boknik et al. 1998). The effects of protein phosphorylation are also pertussis toxin sensitive (Neumann et al. 1994). Based on these findings we provided evidence that phosphatase activity is inhibited in the heart by cAMP-increasing agents (Neumann et al. 1991, Gupta et al. 1996). These inhibitory effects of cAMP-increasing agents on phosphatase activity could be reversed by A_1-adenosine receptor stimulation (Gupta et al. 1993). Fittingly, the effects of adenosine on contractility in the presence of cAMP-increasing agents could be blocked by fluoride, a phosphatase inhibitor (Neumann et al. 1995a). A2-adenosine receptors of the A_{2A}-subtype on ventricular cardiomyocytes may raise myocardial contractility, by an increase in cAMP content (rats, Xu et al. 1996) or via

110

controversy whether or not myocardial A_{2A}-adenosine receptors increase force of contraction (Behnke et al. 1990, Stein et al. 1993, 1994, Shryock et al. 1993, Boknik et al. 1998). It is clear that A2-adenosine receptor stimulation can increase cAMP content in cardiomyocytes (Behnke et al. 1990, Stein et al. 1993,1994, Xu et al. 1996, Boknik et al. 1998). However, unlike other cAMP-elevating agents like isoprenaline or isobutylmethylxanthine (IBMX) A2-adenosine receptor stimulation does not lead to phospholamban phosphorylation which could indicate that inotropically inactive compartments of cAMP do exist at least in guinea pigs (Boknik et al. 1998).

Furthermore, the inhibitory or excitatory effects of adenosine on cardiac ventricular automaticity have been suggested to be triggered by activation of the A_1- and A2- adenosine receptor subtype, respectively (reviewed in Hernández and Ribeiro 1996, Pelleg and Kubalak 1997). A2-adenosine receptors of the A_{2A} and, however, to a minor extent the A_{2B}-adenosine receptor subtypes on vascular endothelial and smooth muscle cells mediate coronary vasodilation (Keddi et al. 1996). Stimulation of A_3-adenosine receptors on mast cells, infiltrating myocardial tissue, caused a rapid degranulation and release of histamine and other mediators that induce vasodilation (Fozard et al. 1996).

B. Physiological role and pathophysiological adaptation

In the normal oxygenated ventricular (Berne et al. 1971, Rubio et al. 1973) and atrial (Thomas et al. 1975) myocardium, adenosine concentrations are in the range of 1-10 µM. At these low adenosine concentrations the antiadrenergic effects of adenosine are observed. The direct negative inotropic effects in the atria occur at somewhat higher concentrations (1-100 µM, de Gubareff et al. 1965, reviewed in Dobson and Fenton, 1993). Thus, based on the degree of negative inotropic effects produced by A_1-adenosine receptor activation and the amount of adenosine present in cardiac tissue it is conceivable, that the A_1-adenosine pathway may maintain some basal or tonic negative inotropic influence in working myocardium (Fredholm and Sollevi 1986).

Adenosine is released from the myocardium under certain physiological or pathophysiological conditions e.g. during ß-adrenergic catecholamine stimulation (Schrader et al. 1977, Bardenheuer et al. 1987, Headrick and Willis 1989, Fenton et al. 1990), cardiac ischemia (Fox et al. 1974), cardiac hypoxia (Berne 1963, Fenton and Dobson 1993), or increased workload (Mc Kenzie et al. 1974, recently reviewed in Schrader et al. 1998), conditions when the oxygen supply-demand ratio is unfavorable - situations present in failing human hearts from patients with dilated cardiomyopathy or ischemic cardiomyopathy. Indeed, myocardial adenosine levels have been found to be raised in a model of heart failure (Newman et al. 1984), indicating a regulatory role of adenosine in diseased hearts. Heart failure is functionally characterized by a diminished contractile response to ß-adrenergic stimulating catecholamines (Bristow et al. 1982, Böhm et al. 1988, Danielsen et al 1989, Bethke et al. 1991, Steinfath et al. 1992). Downregulation of myocardial ß1-adrenoceptors (Bristow et al. 1982, Steinfath et al. 1990) and the increase in Gi-proteins (Feldman et al. 1988, Neumann et al. 1989, Böhm et al. 1990) in myocardial tissue from failing compared to nonfailing human hearts may only be a part of the underlying mechanisms involved in the development of heart failure.

Adenosine antagonizes the stimulatory effects of ß-adrenergic agonists on contractile and metabolic function of the myocardium (Law et al. 1991, Romano et al. 1991). By reducing the adrenergic-elicited increase in contractility adenosine promotes a decrease in energy consumption at a time when energy production is limited because of reduced supply of oxygen. This antiadrenergic action, mediated by specific A_1-adenosine receptors, is therefore a negative feed back mechanism, that protects the heart against excessive stimulation by ß-adrenergic catecholamines. The pathophysiological role of these effects of adenosine in the human heart is not well established, but is of particular interest in the failing human heart where Gi-protein levels are elevated. It has been hypothesized that the increase in Gi- proteins in failing hearts could augment the antiadrenergic effect of adenosine, resulting in a more pronounced decrease in ß-adrenoceptor effects when adenosine levels are high. Thus, the increased effectiveness of adenosine could well contribute to the observed subsensitivity to ß-agonists in these patients. Furthermore, it has been hypothesized that a defective adenosine-mediated autoregulation of the heart might be an abnormality leading to cardiomyopathy (Watt et al. 1984). Therefore, the subtype of adenosine receptors, their density and coupling to effector systems have been studied in nonfailing and failing human hearts. The present overview focuses on the characterization and function of adenosine receptors in the human heart in vitro.

III. TEXT

A. A_1-adenosine receptors in human heart failure

1. Gene expression of A_1-adenosine receptors

In heart failure a number of proteins important for the regulation of myocardial contractility or development of hypertrophy have been described to be altered (reviewed in Mittmann et al. 1998). However, no data are available on the regulation of adenosine receptor gene expression in human heart failure. Therefore, we investigated gene expression of the four adenosine receptor subtypes in nonfailing and failing human hearts to answer the following questions: Which adenosine receptors are expressed in the human heart? Are myocardial adenosine receptors regulated on the mRNA level in failing human hearts?

In order to address these questions we performed ribonuclease (RNase) protection assays with total RNA isolated from ventricular myocardium from nonfailing and failing human hearts using specific adenosine receptor antisense probes. Figure 2 depicts a typical autoradiogram demonstrating the expression of the A_1-adenosine receptor mRNA in failing and nonfailing human hearts. The *in vitro* transcription and purification of ^{32}P-labeled RNA antisense probes by gel elution resulted in homogenous bands of A_1 (709 nt), and Gsα (150 nt) RNAs (Fig. 2, lane 2). Rat stimulatory Gsα, which is known to be unchanged in heart failure (Eschenhagen et al. 1992), was used as an internal standard. Hybridization with 20 µg total RNA extracted from left ventricular tissue of nonfailing or failing human hearts with A_1- and Gsα-antisense RNAs resulted in protection of a 680 nt A_1 fragment, and two Gsα fragments of 95 nt and 82 nt, respectively. The sizes of the probes and the protected fragments of sense-RNAs and total RNA, respectively, differed due to the multiple cloning sites of bacterial plasmid sequences in the RNAs and in the ^{32}P-labeled RNA probes.

RNase protection assay

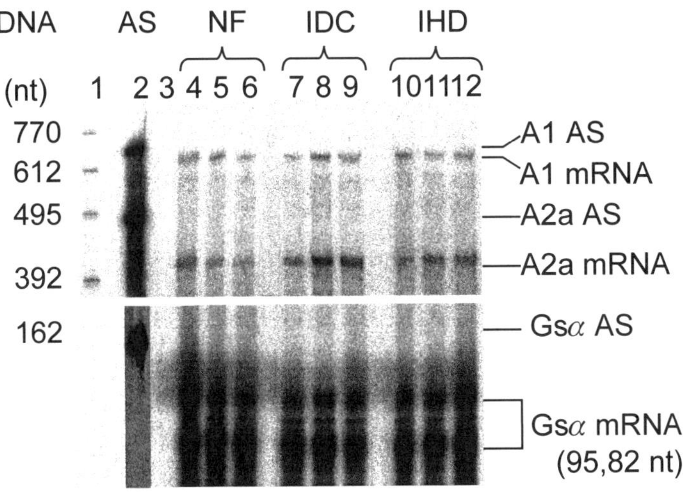

AS = antisense
NF = nonfailing
IDC = idiopathic dilated cardiomyopathy
IHD = ischemic cardiomyopathy

Figure 2. Determination of A_1- and A_{2A}-mRNA levels in relation to the mRNA-levels of Gsα using ribonuclease protection assays in ventricular tissue from nonfailing and failing human hearts from patients with dilated cardiomyopathy (IDC) or ischemic heart disease (IHD). Methods have been described in detail by Ihl-Vahl et al. 1996. Shown is a representative autoradiogram. Lanes indicate: (1) length marker, (2) A_1 (709 nt), A_{2A} (463 nt) or Gsα (105 nt) antisense RNA. A_1, A_{2A} and Gsα antisense RNA incubated with: (3) total RNA (10 _ g) from yeast (3; negative control), total RNA (20 µg) from ventricular myocardium from nonfailing (4-6; NF) and failing human hearts from patients with IDC (7-9) and IHD (10-12), respectively. Incubation of total RNA isolated from ventricular myocardial tissue with A_1- and A_{2A}-antisense RNA yielded in protected bands of the expected sizes (680 nt or 434 nt), indicating A_1- and A_{2A}-adenosine receptor gene expression in human ventricular myocardium.

Quantification of A_1-specific mRNA (pixel values) in nonfailing human hearts amounted to 230 ± 36 (n = 10, Fig. 3 A) and was unchanged in patients with idiopathic dilated cardiomyopathy (IDC, 232 ± 27; n = 13) or ischemic heart disease (IHD, 204 ± 19; n = 12). Thus, the present data demonstrate no alterations in the mRNA expression of the A_1-adenosine receptor in failing compared to nonfailing human hearts and show that no adaptation on mRNA level did occur.

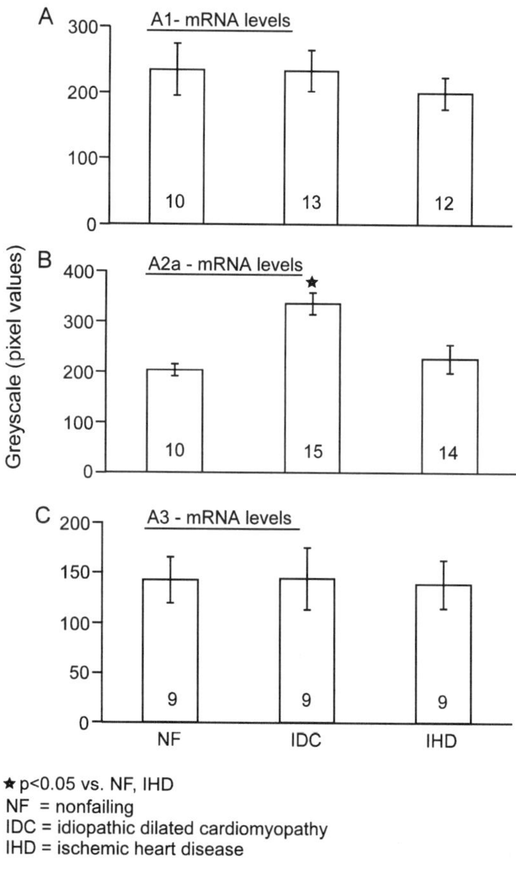

Figure 3. A_1-(A), A_{2A}-(B) or A_3-(C)mRNA levels (depicted in pixel values) in nonfailing (NF) and failing human hearts (F) from patients with dilated cardiomyopathy (IDC) and ischemic heart disease (IHD). Number in columns indicate number of hearts. A_1-mRNA levels were unchanged in failing compared to nonfailing human ventricular myocardium. However, A_{2A}-mRNA levels were increased in patients with IDC.

2. A_1-adenosine receptor density and affinity

Adenosine receptors on human myocardium were first identified by radioligand binding studies by Böhm et al. 1989a. Binding studies with the radiolabeled, selective A_1-adenosine receptor antagonist [3H] 8-cyclopentyl-1,3-dipropylxanthine (DPCPX) bound to human myocardial membranes with high affinity, indicating the presence of A_1-adenosine receptors in the human ventricle. Competition curves with the A_1-adenosine receptor agonist, R-(-)N6-phenylisopropyladenosine (R-PIA), revealed a high and a low-affinity state receptor.

Gpp(NH)p and pertussis toxin-pretreatment converted all high affinity receptors to a low affinity state (Böhm et al. 1989b). This indicated that the G-protein coupled to the human myocardial A_1-adenosine receptor was evidently a pertussis toxin substrate. Further investigations showed no alterations in either the density or affinity of myocardial A_1-adenosine receptors between failing and nonfailing human hearts. The maximal number (Bmax, fmol/mg protein) of A_1-adenosine receptors amounted to 21 ± 3 (n = 5) in nonfailing, or 17.5 ± 2 (n = 6) and 21 ± 2 (n = 6) in failing ventricular myocardium from patients with dilated cardiomyopathy, and ischemic heart disease, respectively (Böhm et al. 1989b, 1990, 1993). These data are in line with the unchanged A_1-adenosine receptor mRNA expression.

3. Effects of A_1-adenosine receptor agonists on cAMP levels

Additional studies focused on the A_1-mediated *signal transduction pathway* in heart failure. In ventricular myocardial membrane preparations from patients with dilated cardiomyopathy coupling of the A_1-adenosine receptor to a pertussis-toxin sensitive signal transducing Gi-protein has been demonstrated (Böhm et al. 1989a). Gi-proteins have been shown to be increased in ventricular myocardium from patients with heart failure (Feldman et al. 1988, Neumann et al. 1988, Böhm et al. 1990). Thus, it was conceivable that the increase in Gi protein in failing human hearts would lead to a facilitated coupling of A_1-adenosine receptors and hence to an augmentation in adenylyl cyclase coupling and a decrease in cAMP content.

Figure 4 depicts the effects of the A_1-adenosine receptor agonist R-PIA in the presence of forskolin on cAMP content and force of contraction in isolated human ventricular trabeculae from nonfailing and failing human hearts. Forskolin activates adenylyl cyclase directly, independent from receptor coupling. We used forskolin, instead of isoprenaline in these experiments to avoid alterations in cAMP levels, due to diminished ß-adrenoceptor density in ventricular myocardium from failing human hearts, which may interfere with the A_1-mediated decreases in cAMP content. R-PIA diminished forskolin-stimulated cAMP content significantly in ventricular trabeculae from nonfailing but not in failing human hearts. In ventricular myocardium from failing and nonfailing hearts the effects of R-PIA on adenylyl cyclase activity did not differ (Hershberger et al. 1991). These data might indicate that in the human heart like in the guinea-pig heart other pathways than cAMP decrease are important for the negative inotropic effect. For instance, it is obvious that the extent of the negative inotropic effect of A_1-adenosine receptors (60-70%) is larger than their effect to reduce the forskolin-stimulated cAMP content (25-40%), which is only significant in nonfailing hearts. Hence, one can speculate that the negative indirect inotropic effect of adenosine in the human heart is mediated at least in part by another pathway e.g. an activation of phosphatases. This pathway might be used more in failing than in nonfailing human myocardium. In agreement with this speculation we have noted that phosphatase expression and activity is elevated in the failing human heart (Neumann et al. 1997) and that the phospholamban phosphorylation in preparations from failing human hearts is lower than in preparations from nonfailing human hearts (Bartel et al. 1996).

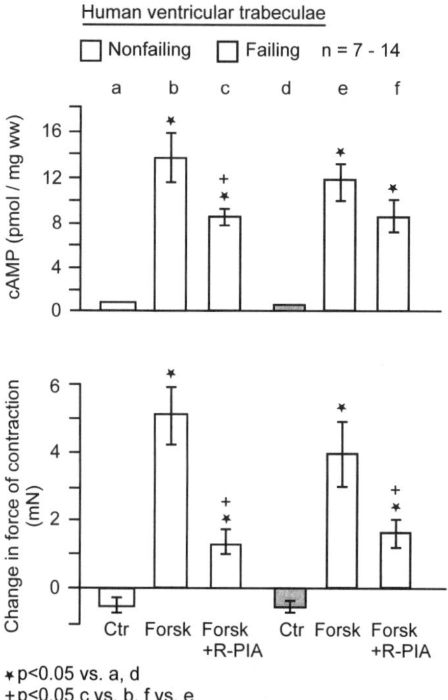

Figure 4. Effects of forskolin (Forsk, 1 µM, 30 min), R-(-)N6-phenyliso-propyl-adenosine (R-PIA, 1 µM, 5 min) in the presence of forskolin compared to control (Ctr, adenosine deaminase 1 µg/ml) on cAMP content (A, pmol/mg ww) and force of contraction (B, change in mN) in isolated electrically driven (0.5 Hz) ventricular trabeculae from nonfailing (open columns) and failing (hatched columns) human hearts. Methods are described in Danielsen et al. 1989. N denotes number of trabeculae from 4 nonfailing and 8 failing human hearts.

4. Contractile effects of adenosine in atrial and ventricular myocardium from humans

First evidence for the existence of myocardial adenosine receptors in human hearts was provided from functional studies (Böhm et al. 1985). Adenosine or the A_1-selective adenosine receptor agonist R-PIA added to isolated electrically driven preparations of ventricular heart muscles from failing human hearts antagonized the positive inotropic effect of isoprenaline. Thus, these data indicate an antiadrenergic action of adenosine in the human myocardium. Subsequent investigations on isolated human ventricular tissue (Böhm et al. 1989b, 1993, 1994) or isolated ventricular cardiomyocytes (Brown et al. 1990) as well as *atrial* preparations (Böhm et al. 1989b, 1993, Jakob et al. 1989) supported and extended these findings, demonstrating also "direct" negative inotropic effects of adenosine or adenosine analogues on human isolated *atrial* tissue, in addition to the "indirect" antiadrenergic effects of adenosine.

116

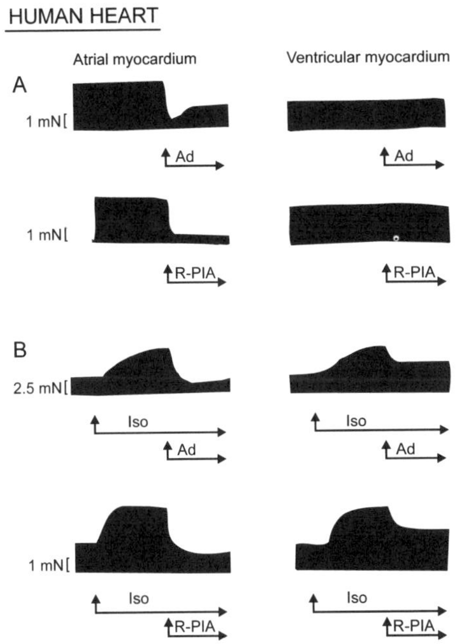

Figure 5. Original tracings illustrating the effects of adenosine (Ad) and the A_1-adenosine receptor agonist R-(-)N6-phenylisopropyladenosine (R-PIA) on force of contraction in isolated electrically driven right atrial trabeculae ("direct" negative inotropic effect; left panel) and papillary muscle strips ("indirect" negative inotropic or "antiadrenergic" effect; right panel) alone (A) and in the presence of isoprenaline (Iso, B). Note that adenosine and R-PIA reduced force of contraction in the atrial but not in the ventricular myocardium. In the presence of isoprenaline, adenosine and R-PIA exerted negative inotropic effects in both tissues. Reprinted by permission of Wiley-Liss, Inc., a subsidiary of John Wiley & Sons, Inc. (Adenosine Receptors in the Human Heart: Pharmacological Characterization in Nondiseased and Cardiomyopathic Tissue, by Michael Bohm, Martin Ungerer and Erland Erdmann, Drug Development Research, 1993, vol. 28. 268-276).

Figure 5 shows original traces illustrating the effect of the parent compound adenosine and R-PIA in human atrial and ventricular preparations (from Böhm et al. 1993). In atrial myocardium, adenosine and R-PIA exerted negative inotropic effects alone, while both agents failed to reduce force of contraction in ventricular tissue. However, when myocardial force of contraction was enhanced with the ß-adrenoceptor agonist isoprenaline, both adenosine and R-PIA exerted negative inotropic responses, indicating adenosine receptor-mediated antiadrenergic effects in human ventricular tissue. Moreover, the negative inotropic effects of adenosine and R-PIA could be antagonized by the A_1-adenosine receptor antagonist DPCPX, indicating that these effects are mediated by the A_1-adenosine receptor subtype.

Further studies investigated possible differences in the negative inotropic effects of adenosine in failing compared to nonfailing hearts, to evaluate whether alterations in the adenosine-mediated feedback control may contribute to or participate in the contractile dysfunctions observed in heart failure. However, in vitro studies on isolated human ventricular tissue failed to depict differences in the A_1-adenosine receptor-mediated antiadrenergic effects in failing compared to nonfailing hearts.

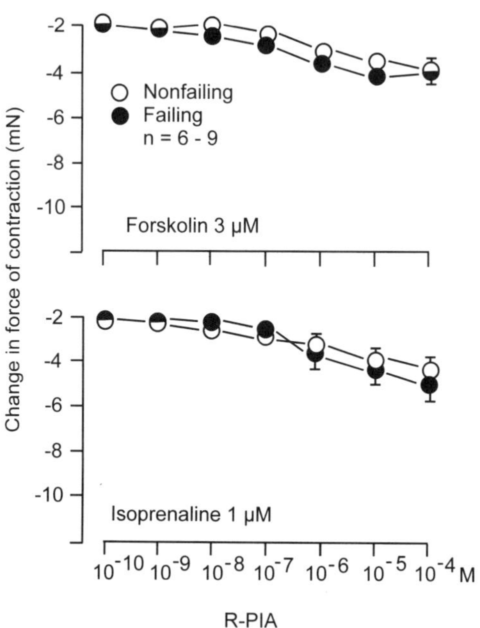

Figure 6. Concentration-dependent effects of R-(-)N6-phenylisopropyladenosine (R-PIA) in the presence of isoprenaline (0.1 µM) on force of contraction in isolated electrically driven (0.5 Hz) ventricular trabeculae from nonfailing (NF, open circles) and failing (F, closed) human hearts. N depicts number of trabeculae from 4 nonfailing and 8 failing human hearts. The negative inotropic effect of R-PIA did not differ in ventricular trabeculae from nonfailing and failing human hearts.

Figure 6 shows the concentration-dependent effects of R-PIA in the presence of forskolin (upper panel) or isoprenaline (lower panel) on force of contraction in isolated, electrically driven, human ventricular trabeculae from nonfailing and failing human hearts. It is clearly demonstrated that the A_1-adenosine receptor-mediated negative inotropic effect in failing heart is not different from that in nonfailing controls. This is in agreement with others, who investigated the A_1-mediated antiadrenergic effects in isolated papillary muscle strips from failing or nonfaling human hearts (Böhm et al. 1989b, 1993).

B. A_{2A}-adenosine receptor gene expression in human heart failure

Despite these investigations on the A_1-adenosine receptors and the A_1-adenosine receptor-mediated signal transduction pathway, no attention has been focused on the other adenosine receptor subtypes in failing human hearts. A_{2A}-adenosine receptors on cardiac vascular endothelial and smooth muscle cells mediate vasodilation. Furthermore, there is growing evidence for the expression and functionally coupling of A_{2A}-adenosine receptors in isolated ventricular cardiomyocytes from various species (rats, chicken, guinea pigs), indicating additional importance of this receptor subtype in the myocardium (recently reviewed in Olsson et al. 1996, Müller and Stein 1996). Therefore, we investigated whether A_{2A}-adenosine receptors are expressed in human myocardium and if so, whether the A_{2A} gene expression is altered in human heart failure.

Figure 2 depicts a representative autoradiogram demonstrating the expression of A_{2A}-adenosine receptor mRNA in failing and nonfailing human hearts. Hybridization of ^{32}P-labeled A_{2A}-RNA antisense probes with 20 μg total RNA extracted from left ventricular tissue of nonfailing or failing human hearts resulted in protection of a 434 nt A_{2A} fragment. Quantification of A_{2A}-specific mRNA (pixel values) in nonfailing human hearts amounted to 203 ± 11 (n = 10) and was not altered in hearts from patients with IHD (224.8 ± 23; n = 14, Fig. 3). However, A_{2A} mRNA was increased in failing human hearts from patients with IDC by 61% (328 ± 23, n = 15) compared to nonfailing hearts.

Thus, the present study shows for the first time gene expression of A_{2A}-adenosine receptors in human myocardium and interestingly, demonstrates an increase in A_{2A}-adenosine receptor mRNA in failing human hearts from patients with IDC compared to nonfailing human hearts. A_{2A}-adenosine receptors present on the vascular smooth muscle or endothelial cells could contribute to an improvement of the oxygen supply-demand ratio by increasing coronary vasodilation and/or angiogenesis. An adenosine-stimulated angiogenesis in vivo has been shown (Ziada et al. 1984). Furthermore, activation of A_{2A}-adenosine receptors on cardiomyocytes may lead to an increase in myocardial contractility, as has been shown before in cardiomyocytes from chicken and rats (Liang et al. 1995, Xu et al. 1996). Thus, it is conceivable that the upregulation of A_{2A}-adenosine receptors is a novel autoregulatory mechanism either to improve cardiac oxygen consumption or/and to overcome the diminished contractile response to ß-adrenergic agents observed in human heart failure. In addition, in other cell types a growth-promoting effect of the A_{2A}-adenosine receptor has been documented (Rozengurt 1986, Ledent et al. 1992, Sexel et al. 1997). In this respect one may speculate about a role of an increased A_{2A}-gene expression in morphological alterations such as hypertrophy as present in heart failure.

C. A_{2B}- and A_3-adenosine receptor gene expression

A_{2B}-adenosine receptor gene expression was not detectable in human myocardium as measured by RNase protections assay. However, quantification of A_3-specific mRNA (pixel values) in nonfailing human hearts amounted to 153 ± 28 (n = 9) and was not altered in hearts from patients with IDC (155.6 ± 35; n = 9) or IHD (152 ± 31; n = 9) as presented in figure 3c. Thus, the present study shows gene expression of A_3-adenosine receptors in human myocardium and demonstrates that A_3-adenosine receptor mRNA is not altered in failing compared to nonfailing human hearts.

IV. CONCLUSION

Impairment of the A_1-adenosine receptor-mediated feedback control has been hypothesized to cause cardiomyopathy (Watt et al. 1984). Results from the present study indicate that neither the A_1-adenosine receptor gene expression, nor the A_1-adenosine receptor-mediated antiadrenergic effects on force of contraction were altered in failing compared to nonfailing human hearts (Fig. 7). Thus, the data presented herein do not support the hypothesis, that a defective A_1-mediated autoregulation of myocardial contractile force could be the cause of cardiomyopathy. Furthermore, A_3-adenosine receptor gene expression did not differ in failing and nonfailing hearts.

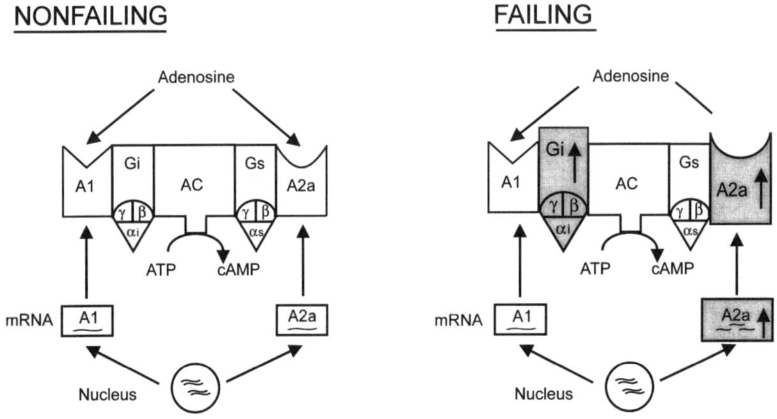

Figure 7. Scheme representing the regulation of A_1- and A_{2A}-adenosine receptor gene expression in failing hearts from patients with dilated cardiomyopathy (IDC) compared to nonfailing human hearts.

However, an increase in A_{2A}-adenosine receptor gene expression has been found in failing hearts from patients with IDC compared to nonfailing hearts (Fig. 7). This upregulation in myocardial A_{2A}-adenosine receptors may have significant general implications for regulation of basic cardiac function e.g. an increase in vasodilation mediated by A_{2A}-adenosine receptors on coronary endothelial and smooth muscle cells or an augmentation in contractile response mediated by A_{2A}-adenosine receptors on cardiomyocytes. Furthermore, A_{2A}-upregulation might be involved in the mechanisms underlying the morphological alterations such as hypertrophy as present in heart failure.

Stimulation of A_1-adenosine receptors led via an inhibitory GTP-binding protein (Gi protein) to an inhibition of adenylyl cyclase activity with a subsequent decrease in cAMP content. In contrast, stimulation of A_{2A}-adenosine receptors activates adenylyl cyclase activity and increases cAMP content. No alterations in the

A_1-adenosine receptor gene expression and the A_1-mediated inhibitory effects on cAMP content or contractile response were observed between failing and nonfailing human hearts. However, in failing hearts from patients with idiopathic dilated cardiomyopathy, besides an increase in Gi proteins, an upregulation of A_{2A}-adenosine receptor gene expression was found, which might be paralleled by an increase in A_{2A}-adenosine receptors and subsequent altered cellular responses e.g. augmentation in vasodilation and/or contractile response.

Even though the exact functional role and signal transduction mechanism in the myocardium remains largely unknown, the present finding of the existence of A_{2A}-adenosine receptors in human ventricular myocardium and its increase in failing human hearts, may indicate a pathophysiological role and adds therefore to the increasingly complex picture of molecular alterations in heart failure. The data also indicate that potential agonist effect on the A_{2A}-adenosine receptor should be considered when studying the cardiac action of adenosine in heart failure.

REFERENCES

Bardenheuer H, Whelton, B, Sparks HV Jr. Adenosine release by the isolated guinea pig heart in response to isoproterenol, acetylcholine, and acidosis: The minimal role of vascular endothelium. Circ Res 1987;61:594-600

Bartel S, Stein B, Eschenhagen T, Mende U, Neumann J, Schmitz W, Krause EG, Karczewski P, Scholz H. Impaired phosphorylation of phospholamban, troponin I and C-protein in the failing human heart. Mol Cell Biochem 1996;157:171-179

Behnke N, Müller W, Neumann J, Schmitz W, Scholz H, Stein B. Differential antagonsim by 1,3-dipropylxanthine-8-cyclopentylxanthine and 9-chloro-2-(2-furanyl)-5,6-dihydro-1,2,4-triazolo(1,5-c)quinazolin-5-imine of the effects of adenosine derivatives in the presence of isoprenaline on contractile response and cyclic AMP content in cardiomyocytes. Evidence for the coexistence of A_1- and A2-adenosine receptors on cardiomyocytes. J Pharmacol Exp Ther 1990;254:1017-1023

Belardinelli L, Isenberg G. Isolated atrial myocytes: Adenosine and acetylcholine increase in potassium conductance. Am J Physiol 1983;244:H291-H294

Belardinelli L, Linden J, Berne RM. The cardiac effects of adenosine. Prog Cardiovasc Res 1989;32:73-97

Belardinelli L, Shryock JC, Song Y, Wang D, Srinivas M. Ionic basis of the electrophysiological actions of adenosine on cardiomyocytes. FASEB J 1995;9:359-365

Berne RM. Cardiac nucleotides in hypoxia: Possible role in regulation of coronary blood flow. Am J Physiol 1983;204:317-322

Berne RM, Rubio R, Dobson JG Jr, Curnish RR. Adenosine and adenine nucleotides as possible mediators of cardiac and skeletal muscle blood flow regulation. Circ Res 1971;71:115-119

Bethke T, Klimkiewicz A, Kohl C, v d Leyen H, Mehl H, Mende U, Meyer W, Neumann J, Schmitz W, Scholz H, Starbatty J, Stein B, Wenzlaff H, Döring V, Kalmár P, Haverich A. Effects of isomazole on force of contraction and phosphodiesterase isoenzymes I-IV in nonfailing and failing human hearts. J Cardiovasc Pharmacol 1991;18:386-397

Böhm M, Brückner R, Neumann J, Schmitz W, Scholz H, Starbatty J. Role of guanine nucleotide binding protein in the regulation by adenosine of cardiac potassium conductance and force of contraction. Naunyn-Schmiedebergs Arch Pharmacol 1986;332:403-405

Böhm M, Beuckelmann D, Brown L, Feiler G, Lorenz B, Näbauer M, Kemkes B, Erdmann E. Reduction of ß-adrenoceptor density and evaluation of positive inotropic responses in isolated, diseased human myocardium. Eur Heart J 1988;9:844-852

Böhm M, Girschik P, Ungerer M, Erdmann E. Coupling of adenosine receptors to pertussis toxin-sensitive G protein in the human heart. Eur J Pharmacol 1989a;172:407-411

Böhm M, Meyer W, Mügge A, Schmitz W, Scholz H. Functional evidence for the existence of adenosine receptors in the human heart. Eur J Pharmacol 1985;116:323-326

Böhm M, Pieske M, Ungerer M, Erdmann E. Characterization of A_1-adenosine receptors in human atrial and ventricular myocardium from diseased human hearts. Circ Res 1989b;65:1201-1211

Böhm M, Gierschik P, Jakobs KH, Schnabel P, Ungerer M, Erdmann E. Increase in Gi-proteins in human hearts with dilated but not ischemic cardiomyopathy. Circulation 1990;82:1249-1265

Böhm M, Ungerer M, Erdmann E. Adenosine receptors in the human heart: Pharmacological characterization in nondiseased and cardiomyopathic tissue. Drug Dev Res 1993;28:268-276

Boknik P, Neumann J, Schmitz W, Scholz H, Wenzlaff H. Characterization of biochemical effects of CGS 21680C, an A2-adenosine receptor agonist, in the mammalian ventricle. J Cardiovasc Pharmacol 1998 (in press)

Bristow MR Gainsburg R, Minobe W, Cubicciotti RS, Sageman WS, Luri K, Billingham ME, Harrison DC Stinson EB. Decreased catecholamine activity and beta-adrenergic receptor density in failing human hearts. N Engl J Med 1982;307:205-211

Brown LA, Humphrey SM, Harding SE. The anti-adrenergic effect of adenosine and its blockade by pertussis toxin: a comparative study in myocytes isolated from guinea-pig, rat, and failing human hearts. Br J Pharmacol 1990;101:484-488

deGubareff T, Sleator WJ. Effects of caffeine on mammalian atrial muscle and its interaction with adenosine and calcium. J Pharmacol Exp Ther 1965;148:202-214

Dobson JG Jr. Reduction by adenosine of the isoproterenol-induced increase in cyclic adenosine 3′,5′-monophosphate formation and glycogen phosphorylase activity in rat heart muscle. Circ Res 1978;43:785-792

Dobson JG Jr. Adenosine reduces catecholamine contractile responses in oxygenated and hypoxic atria. Am J Physiol 1983;245:H468-H474

Dobson JG Jr, Fenton RA. "Antiadrenergic effects of adenosine in the heart". In: *Regulatory function of adenosine*, RM Berne, TW Rall, R Rubio, eds. Oak Park, L,1993, pp363-376

Dobson JG Jr, Fenton RA. Adenosine A2-receptor agonists elicit positive inotropic response and increase in adenylyl cyclase activity in rat ventricular myocytes. Drug Dev Res 1994;31:1232

Dobson JG Jr, Ordway RW, Fenton RA. Endogenous adenosine inhibits catecholmanine contractile responses in normoxic hearts. Am J Physiol 1986;251:H455-462

Drury AN, Szént-György. The physiological activity of adenine compounds with especial reference in their action upon the mammalian heart. J Physiol 1929;68:213-237

Feldman AM, Cates AE, Veazey WB, Hershberger RE, Bristow MR, Baughman KL, Baumgartner WA, Van Dop C. Increase of a 40,000 mol-wt pertussis toxin substrate (G protein) in the failing human heart. J Clin Invest 1988;82:189-197

Fenton RA, Dobson JG Jr. Measurement by fluorescence of intestinal adenosine levels in normoxic, hypoxic, and ischemic perfused rat hearts. Circ Res 1987;60:177-184

Fenton RA, Dobson JG Jr. Hypoxia enhances isopreterenol-induced increase in heart intestinal adenosine, depressing ß-adrenergic contractile responses. Circ Res 1993;72:571-578

Fenton RA, Tsimikas S, Dobson JG Jr. Influence of ß-adrenergic stimulation and contraction frequency on rat heart interstitial adenosine. Circ Res 1990;66:457-468

Fox AC, Reed GE, Glassma E, Kaltman AJ, Silk BB. Release of adenosine formation from human hearts during angina induced by rapid atrial pacing. J Clin Invest 1974;53:1447-1457.

Fozard JR, Pfannekuche H-J, Schurman H-J. Mast cell degranulation following adenosine A_3 receptor activation in rats. Eur J Pharmacol 1996;298:293-297

Fredholm BB, Abbracho MP, Burnstock G, Daly JW, Harden TK, Jacobson KA, Leff P, Williams M. Nomenclature and classification of purinoceptors. Pharmacol Rev 1994;46:143-156

Fredholm BB, Sollevi A. Cardiovascular effects of adenosine. Clin Physiol 1986;6:1-21

Gupta RC, Neumann J, Durant P, Watanabe AM. A_1-adenosine receptor-mediated inhibition of isoproterenol-stimulated protein phosphorylation in ventricular myocytes. Evidence against a cyclic AMP-dependent effect. Circ Res 1993;72:65-74

Gupta RC, Neumann J, Watanabe AM, Lesch M, Sabbah HN. Evidence for the existence and hormonal regulation of protein phosphatase inhibitor-1 in ventricular cardiomyocytes. Am J Physiol 1996;270:H1159-H1164

Headrick JP, Willis RJ. 5′Nucleotidase activity and adenosine formation in stimulated, hypoxic and underperfused rat heart. Biochem J 1989;261:541-550

Hernández J, Ribeiro JA. Excitatory actions of adenosine on ventricular automaticity. TIBS 1996;17:141-144

Hershberger RE, Feldman AM, Bristow M. A_1-adenosine receptor inhibition of adenylate cyclase in failing and nonfailing human ventricular myocardium. Circulation 1991;83:1343-1351

Ihl-Vahl R, Eschenhagen T, Kübler W, Marquetant R, Nose M, Schmitz W, Scholz H, Strasser RH. Differential regulation of mRNA specific for ß1- and ß2-adrenergic receptors in human failing hearts. Evaluation of the absolute cardiac mRNA levels by two independent methods. J Mol Cell Cardiol 1996;28:1-10

Jakob H, Oelert H, Rupp J, Nawrath H. Functional role of cholinoceptors and purinoceptors in human isolated atrial and ventricular heart muscle. Br J Pharmacol 1989;97:1199-1208

Jochen G, Nawrath H. Adenosine activates a potassium conductance in atrial heart muscle. Experienta Basel 1983;39:1347-1349

Keddi JR, Poucher SM, Shaw GR, Brooks R, Collis MG. In vivo characterization of ZM 241385, a selective adenosine A_{2A} receptor antagonist. Eur J Pharmacol 1996;301:107

Law WR, Carney PJ, McLane MP. Influence of adenosine on the stimulatory effects of isoprenaline and insulin on myocardial contractility in vivo. Cardiovasc Res 1991;25:151-157

Ledent C, Dumont JE, Vassart G, Parmentier M. Thyroid expression of an A2 adenosine receptor transgene induces thyroid hyperplasia and hyperthyroidism. EMBO J 1992;11:537-542

Liang BT. Adenosine receptors and cardiac function. Trends Cardiovasc Med 1992;2:100-108

Liang BT, Haltiwanger B. Adenosine A_{2A} and A_{2B} receptors in cultured fetal chick heart cells. High and low-affinity coupling to stimulation of myocyte contractility and cAMP accumulation. Circ Res 1995;76:242-251

Liang BT, Morley JF. A new cyclic AMP-independent, Gs-mediated stimulatory mechanism via the adenosine A_{2A} receptor in the intact cardiac cell. J Biol Chem 1996;2:18678-18658

Mittmann C, Eschenhagen T, Scholz H. Cellular and molecular aspects of contractile dysfunction in heart failure. Cardiovasc Res 1998 (submitted)

Müller CE, Stein B. Adenosine receptor antagonists: Structures and therapeutical applications. Curr Pharm Design 1996;2:501-530

Neumann J, Boknik P, Kaspareit G, Bartel S, Krause E-G, Pask HT, Schmitz W, Scholz H. Effects of the phosphatase inhibitor calyculin-A on the phosphorylation of C-protein in mammalian ventricular cardiomyocytes. Biochem Pharmacol 1995a;49:1583-1588

Neumann J, Boknik P, Bodor GS, Jones LR, Schmitz W, Scholz H. Effects of adenosine receptor and muscarinic cholinergic receptor on cardiac protein phosphorylation. Influence of pertussis toxin. J Pharmacol Exp Ther 1994;269:1310-1318

Neumann J, Eschenhagen T, Jones LR, Linck B, Schmitz W, Scholz H, Zimmermann N. Increased expression of cardiac phosphatases in patients with end-stage heart failure. J Mol Cell Cardiol 1997;29:265-272

Neumann J, Gupta RC, Jones LR, Bodor GS, Bartel S, Krause EG, Pask HT, Schmitz W, Scholz H, Watanabe AM. Interaction of ß-adrenoceptor and adenosine receptor agonists on phosphorylation. J Mol Cell Cardiol 1995b;27:1655-1667

Neumann J, Gupta RC, Schmitz W, Scholz H, Narin AC, Watanabe AM. Evidence for isoproterenol-induced phosphorylation of phosphatase inhibitor-1 in the intact heart. Circ Res 1991;61:1450-1457

Neumann J, Schmitz W, Scholz H, Stein B. Effects of adenosine analogues on contractile response and cAMP content in guinea-pig isolated ventricular myocytes. Naunyn-Schmiedebergs Arch Pharmacol 1989;340:689-695

Neumann J, Schmitz W, Scholz H, v Meyerinck L, Döring V, Kalmár P. Increase in myocardial Gi proteins in heart failure. Lancet 1988;19:936-937

Newman WH, Grossman SJ, Francis MB, Webb JG. Increased myocardial adenosine release in heart failure. J Mol Cell Cardiol 1984;16:577-580

Olsson RA.. Adenosine receptors in the cardiovascular system. Drug Dev Res 1996;3:133

Pelleg A, Kutalek SP. Adenosine in the mammalian heart: Nothing to get excited about. TIPS 1997;18:236-238

Romano FD, Naimi TS, Dobson JG Jr. Adenosine attenuation of catecholamine-enhanced contractility of rat heart in vivo. Am J Physiol 1991;260:H1635-H1639

Rozengurt E. Early signals in the mitogenic response. Science 1986;234:161-166

Rubio R, Berne RM, Dobson JG Jr. Sites of adenosine production in cardiac and skeletal muscle. Am J Physiol 1973;225:938-953

Schrader J, Baumann G, Gerlach E. Adenosine as inhibitor of myocardial effects of catecholamines. Pflügers Arch 1977;372:29-35

Schrader J, Deussen A, Decking UKM. "Adenosine metabolism and transport in the mammalian heart." In *Effects of extracellular adenosine and ATP on cardiac myocytes.* Amir Pelleg ed. Landes Bioscience Publisher, Austin, Texas, 1998 in press

Sexl V, Mancusi G, Höller C, Gloria-Maercker E, Schütz W, Freissmuth M. Stimulation of the mitogen-activated protein kinase via the A_{2A}-adenosine receptor in primary human endothelial cells. J Biol Chem 1997;272:5792-5799

Shryock J, Song Y, Wang D, Baker SP, Olsson R, Belardinelli L. Selective A2-adenosine receptor agonists do not alter action potential duration, twitch shortening, or cAMP accumulation in guinea pi, rat, or rabit isolated ventricular myocytes. Circ Res 1993;72:194-205

Stein B, Mende U, Neumann J, Schmitz W, Scholz H. Petussis toxin unmasks stimulatory myocardial A2-adenosine receptors on ventricular cardiomyocytes. J Mol Cell Cardiol 1993;25:655-659

Stein B, Schmitz W, Scholz H Seeland C. Pharmacological characterization of A2-adenosine receptors in guinea-pig ventricular cardiomyocytes. J Mol Cell Cardiol 1994;26:403-414

Steinfath M, Geertz B, Schmitz, W, Scholz H, Haverich A, Brel I, Hanrath P, Reupke C, Sigmund M, Loh B. Distinct downregulation of cardiac ß1 and ß2-adrenoceptors in different human heart diseases. Naunyn-Schmiedebergs Arch 1991;343:217-220

Steinfath M, Danielsen W, v d Leyen H, Mende U, Meyer W, Neumann J, Nose M, Reich T, Schmitz W, Scholz H, Starbatty J, Stein B, Döring V, Kalmár P, Haverich A. Reduced ß1- and ß2-adrenoceptor-mediated positive inotropic effects in human end-stage heart failure. Br J Pharmacol 1992;105,463-469

Thomas RA, Rubo R, Berne RM. Comparison of the adenine nucleotide metabolism of dog atrial and ventricular myocardium. J Mol Cell Cardiol 1975;7:115-123

Tucker A, Linden J. Cloned receptors and cardiovascular responses to adenosine. Cardiovasc Res 1993;27:62-67

Ungerer M, Böhm M, Schwinger RHG, Erdmann E. Antagonism of novel inotropic agents at A_1-adenosine receptors and cholinoceptors in human myocardium. Naunyn-Schmiedebergs Arch Pharmacol 1990;341:577-585

v der Leyen H, Mende U, Meyer W, Neumann J, Nose M, Schmitz W, Scholz H, Starbatty J, Stein J, Wenzlaff H, Döring V, Kalmár P, Haverich A. Mechanism underlying the reduced positive inotropic effects of the phosphodiesterase III inhibitors pimobendan, adibendan and saterinone in failing as compared to nonfailing human cardiac muscle preparations. Naunyn-Schmiedebergs Arch Pharmacol 1991;344:90-100

Watt AH. Hypertrophic cardiomyopathy: A disease of impaired adenosine-mediated autoregulation of the heart. Lancet 1984;9:1271-1273

Xu D, Kong H, Liang BT. Expression and pharmacological characterization of a stimulatory subtype of adenosine receptor in fetal chick ventricular myocytes. Circ Res 1992;70:56-65

Xu H, Stein B, Liang BT. Characterization of a stimulatory adenosine A_{2A} receptor in adult rat ventricular myocyte. Am J Physiol 1996;39:H1655-1661

ADENOSINE AND CARDIAC ARRHYTHMIAS

David J. Slotwiner, MD; Bruce B. Lerman, MD; Division of Cardiology, Department of Medicine, The New York Hospital - Cornell Medical Center, New York, USA.

I SUMMARY

The negative chronotropic and dromotropic properties of adenosine on the sinus and atrioventricular (AV) nodes have been know since 1929 [Drury et al, 1929]. These potent properties and its short half life in vivo have allowed adenosine to establish its place as a unique and invaluable tool for both elucidating the mechanism of cardiac arrhythmias and for treating supraventricular tachycardia.

The cardiac electrophysiologic effects of adenosine are mediated by two distinct cellular mechanisms. In supraventricular tissue the predominant action of adenosine results in hyperpolarization of the resting membrane voltage to the potassium equilibrium potential through stimulation of $I_{K,ACh,Ado}$ (an inward rectifying potassium current) and shortening of action potential duration. The second action of adenosine, ubiquitous throughout all regions of the heart, is mediated through adenosine's inhibition of adenylyl cyclase, thus antagonizing the stimulatory effects of catecholamine on I_{TI} (transient inward current), I_F (time- and voltage-dependent inward current activated by hyperpolarization), and I_{Ca} (inward calcium current).

This chapter will review the cellular and clinical electrophysiologic properties of adenosine that provide the basis for its emergence over the last 15 years as an antiarrhythmic agent.

II BACKGROUND

<u>A. Adenosine Sources:</u>

Endogenous adenosine is derived from two distinct pathways [Lerman et al, 1991]. The predominant pathway of adenosine synthesis during hypoxia (ATP pathway) is mediated by dephosphorylation of adenosine monophosphate (AMP), catalyzed by the cell surface enzyme 5'-nucleotidase. This enzyme is primarily bound to cell membranes in the heart but is also found in the cytosol in small quantities. Thus adenosine may be formed intracellularly or extracellularly. In addition to the ATP pathway, adenosine may be formed intracellularly by the degradation of *S*-adenosylhomocysteine hydrolase (SAH pathway). This pathway appears to predominate under normoxic conditions.

<u>B. Adenosine Receptor-Effector Complex:</u>

Adenosine receptors, present on most cells, have been divided into four subtypes based upon amino acid sequence. They are all glycoproteins with 7 transmembrane-spanning domains, and are all members of the G-protein-coupled receptor family. The cardiac electrophysiologic actions of adenosine are mediated exclusively by the cell surface adenosine A_1 receptor [Shryock et al, 1997]. These effects are either cAMP-independent and occur only at the level of nodal or atrial tissue, or are cAMP-dependent and inhibit the stimulatory effects of adenylyl cyclase in both atrial and ventricular myocytes (Fig. 1) [Lerman et al, 1991]. The adenosine A_1 receptor is coupled to its effector units via the inhibitory guanine nucleotide binding protein G_i [Belardinelli et al, 1989]. In nodal and atrial tissues binding of adenosine to the A_1 receptor results in activation of G_i which in turn activates the inward-rectifying potassium channel current $I_{K,ACh,Ado}$. In both ventricular and atrial myocytes binding of adenosine to the A_1 receptor in the presence of catecholamine stimulation (such as that achieved by the non-selective β-adrenergic receptor agonist isoproterenol) results in activation of G_i which inhibits adenylyl cyclase, thus decreasing the L-type calcium current ($I_{Ca,L}$), the transient inward current (I_{TI}) and the pacemaker current I_F.

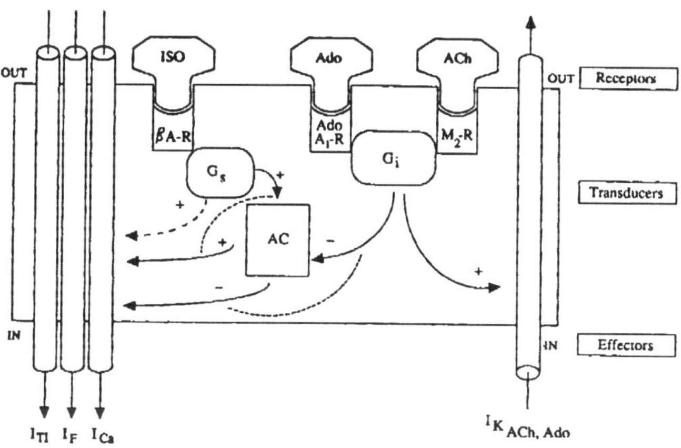

Figure 1: Adenosine receptor-effector coupling system. Details of schema are disclosed in text. ACh, acetylcholine; M_2-R, cardiac muscarinic receptor; Ado, adenosine; A1-R, cardiac adenosine receptor; ISO, isoproterenol; βA-R, β-adrenergic receptor; G_i and G_s, guanine nucleotide binding regulatory proteins; AC, adenylyl cyclase; +, activation or stimulation; -, attenuation or inhibition; $I_{K,ACh,Ado}$, ACh- and Ado-regulated inward rectifying potassium current; I_{Ca}, inward calcium current; I_F, time- and voltage-dependent inward current activated by hyperpolarization (pacemaker current); I_{Ti}, transient inward current; cAMP, cyclic adenosine monophosphate. (Reproduced with permission from Lerman BB, et al. Circulation 1991;83:1499-1509. Copyright 1991, The American Heart Association, Inc.)

<u>C. Removal of Adenosine:</u>

In contrast to adenosine synthesis, metabolism occurs exclusively within the cytosol by one of two enzymes. Deamination by adenosine deaminase produces the inactive metabolites inosine and hypoxanthine. Alternatively, phosphorylation of adenosine by adenosine kinase synthesizes AMP. Since both of these enzymes are cytosolic, adenosine must first be taken up by the cells. This occurs by either simple or facilitated diffusion via a nucleoside transport system. The transport system may be inhibited by nucleoside transport inhibitors such as dipyridamole.

Inhibition of this transport system leads to marked potentiation of the cardiac actions of adenosine. Alternatively, the cardiac actions of adenosine may be terminated by A_1 receptor antagonists, which include the alkylxanthines caffeine and theophylline.

III TEXT

A. Sinus Node

Bolus administration of adenosine has a biphasic effect on sinus node automaticity. The initial effects occur within 15 to 30 seconds following peripheral venous injection and are mediated by adenosine's stimulation of $I_{K,ACh,\ Ado}$, resulting in hyperpolarization of SA nodal cells from -60 mV to the potassium equilibrium potential ($\sim$ -90 mV). The extent of sinus node slowing correlates with the magnitude of the membrane hyperpolarization [West et al, 1985; West et al, 1985]. The negative chronotropic effect of adenosine usually lasts 5 to 10 seconds, and is then followed by reflex sinus tachycardia, which may be related to adenosine's stimulation of carotid body chemoreceptors, resulting in respiratory stimulation and secondary activation of pulmonary stretch receptors,[Biaggioni et al, 1987; Watt et al, 1986; Engelstein et al, 1994; Biaggioni et al, 1991] as well as peripheral vasodilation mediated by the adenosine A_2 receptor.

Adenosine also antagonizes cAMP-stimulated increases in $I_{Ca,L}$ and the pacemaker current I_F, a slow time and voltage-dependent inward current that is activated by hyperpolarization of sinus node cells [Belardinelli et al, 1988]. Adenosine markedly attenuates the increases in both currents observed in the presence of isoproterenol, but has no effect on $I_{Ca,L}$ in the absence cAMP stimulation. Preliminary evidence indicates, however, that adenosine may also inhibit unstimulated I_F in rabbit SA nodal and AV nodal cells, suggesting a second direct action of adenosine in slowing sinus node automaticity [Zaza et al, 1994; Wang et al, 1994].

Clinically, arrhythmias related to the sinus node include inappropriate sinus tachycardia (due to automaticity) and sinus node reentry. Adenosine, through its effects on $I_{K,ACh,Ado}$, terminates sinus node reentry (Fig. 2) [Griffith et al, 1989; Engelstein et al, 1994]. It may also have a causative role in mediating bradycardia observed in the sick sinus syndrome [Watt et al, 1985]. Preliminary data suggest that theophylline may be effective in treating bradycardia observed in these latter patients,[Alboni et al, 1991; Saito et al, 1993] and in increasing the ventricular response rate in patients with atrial fibrillation [Alboni et al, 1993].

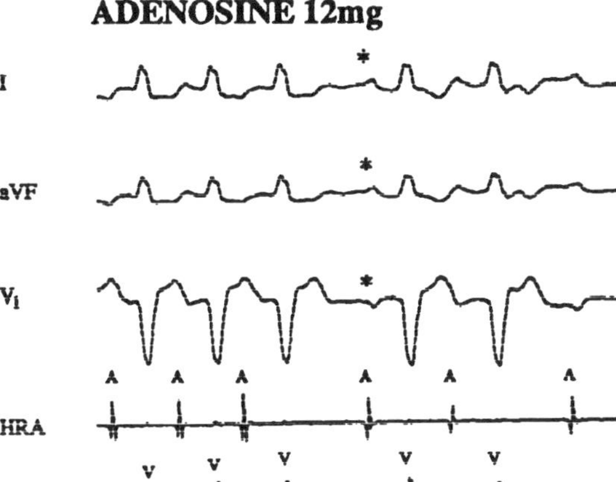

Figure 2: Termination of sinus node reentrant tachycardia with adenosine. Surface leads I, aVF, and V₁ and intracardiac recordings from the high right atrium (HRA) and His bundle (HBE) are shown. After a bolus injection of adenosine, sinus node reentrant tachycardia terminates and sinus rhythm is restored. The first sinus beat after termination of the tachycardia (*) has an atrial activation sequence identical to that recorded during sinus node reentry. Adenosine-induced transient atrioventricular block occurs after tachycardia termination. (Reproduced and modified with permission from Engelstein, et al. Circulation 1994;89:2645-2654. Copyright 1994, The American Heart Association, Inc.)

<u>B. Atrium</u>

The predominant effect of adenosine in the atria is mediated through activation of $I_{K,ACh,Ado}$. This results in shortening of the atrial action potential [Belardinelli et al, 1995; Haines et al, 1988] and also accounts for the observation that adenosine may precipitate atrial flutter and/or fibrillation, effects that may be blocked by nucleoside transport blockers such as aminophylline [Lerman et al, 1989]. Because the resting membrane potential in atrial myocytes approximates the E_K, adenosine has minimal effect on resting membrane potential. As in all myocardial tissue, adenosine has an antiadrenergic effect on atrial tissue.

Conventionally, adenosine was thought to have little effect in atrial tachycardia [Haines et al, 1990]. Although it is important to emphasize that most forms of atrial tachycardia do not respond to adenosine, including atrial flutter and fibrillation, [Engelstein et al, 1994] recently several studies have reported termination of specific forms of atrial tachycardias with adenosine [Shenasa et al, 1993; Chen et al, 1994; Kall et al, 1995]. Markowitz et. al. [Markowitz et al, 1997] proposed categorizing adenosine-sensitive atrial tachycardias into 3 groups. The first type consists of tachycardias originating from along the crista terminalis. These tachycardias are typically adenosine and verapamil sensitive and are thought to be due to either microreentry incorporating calcium-dependent sinus node like tissue, or cAMP-mediated triggered activity arising from atrial tissue. The second type typically presents with repetitive runs of monomorphic atrial tachycardia separated by one or more sinus beat. It arises from diverse focal sites, remote from the sinus node or crista terminalis, is initiated and terminated with programmed stimulation, and is sensitive to adenosine and verapamil. These tachycardias cannot be entrained and are thought to be due to triggered activity. The third type of adenosine-sensitive atrial tachycardia consists of a macroreentrant atrial tachycardia incorporating a zone of decremental atrial tissue. These findings are consistent with earlier studies showing that decremental accessory pathways (composed of working atrial myocardial tissue) are sensitive to adenosine (Fig. 3). Automatic atrial tachycardia, which cannot be initiated or terminated by programmed stimulation, and is facilitated by catecholamine stimulation and is sensitive to β-blockers, is transiently suppressed (less than 30 seconds) but not terminated by adenosine.

Figure 3: Termination of AV reciprocating tachycardia by adenosine in the retrograde accessory pathway after adenosine. Surface leads I, aVF, and V_1 and intracardiac recordings from the high right atrium (HRA), His bundle (HBE) and coronary sinus (CS) are shown. Adenosine caused a progressive increase in retrograde ventriculoatrial conduction time before termination. (Reproduced with permission from Lerman BB, et al. Circulation 1987;76:21-31 Copyright 1987, The American Heart Association, Inc.)

<u>C. AV Node</u>

The AV node is comprised of three regions of electrophysio-logically distinct cells. AN and N cells are typically located in the upper, and mid regions of the AV node respectively. Adenosine causes a concentration dependent decrease in duration and amplitude of action potentials in these cells, and in N cells it also causes a decrease in the upstroke velocity of the action potential [Belardinelli et al, 1995; Clemo et al, 1986]. At high concentrations adenosine completely abolishes the action potential of N cells [Clemo et al, 1986]. This effect stands in marked contrast to its effects in NH cells (located in the most distal region of the AV node), in which adenosine has no appreciable effect on the action potential (Fig. 4). et al Clemo, 1986]

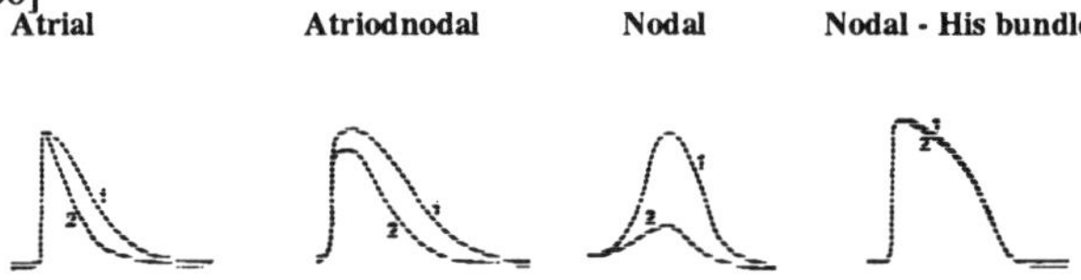

Figure 4: Effect of adenosine (15 μM) on the action potential of atrial, atrionodal, nodal and nodal-His bundle cells. Action potentials labeled (1) are controls and (2) are in the presence of adenosine. Action potential tracings have been retouched by the original authors. (Reproduced and modified with permission from Clemo HF, et al. Circ Res 1986;59:427-436. Copyright 1986, The American Heart Association, Inc.)

The mechanism by which adenosine affects the AV node remains to be elucidated. It is likely that $I_{K,ACh,Ado}$ plays a predominant role. Adenosine also has a small direct effect on $I_{Ca,L}$ and may attenuate I_F. [Clemo, et al 1986]   Therefore, adenosine induced hyperpolarization, shortening of the action potential duration and decreased automaticity contribute to the negative dromotropic effects observed in the AV node in response to adenosine.

Adenosine, in sufficient doses, will transiently block AV nodal conduction. As a result, it plays a important role in the diagnosis and treatment of reentrant arrhythmias in which the AV node is a critical component of the tachycardia circuit.

Typical AV nodal reentrant tachycardia, the single most common form of supraventricular tachycardia, consists of a circuit limited to the peri-AV nodal region, with anterograde conduction proceeding over a "slow" pathway and retrograde conduction traversing a "fast" pathway. While adenosine may terminate tachycardia in either limb, it terminates most often in the anterograde slow pathway [DiMarco et al, 1985; Belhassen et al, 1988].   In atypical AV nodal reentrant tachycardia, anterograde conduction precedes via a "fast" pathway and retrograde conduction via a "slow" pathway. The slow pathway typically demonstrates decremental conduction properties [Lerman et al, 1987].   Adenosine usually terminates tachycardia in the retrograde limb (Fig. 5) [Lerman et al, 1987].

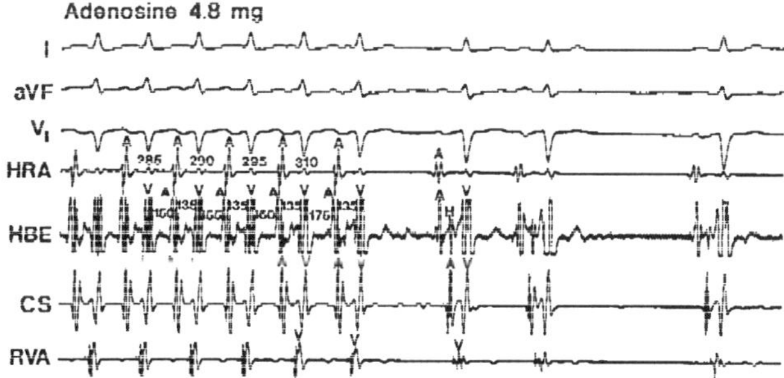

Figure 5:   Recording of atypical AV nodal reentrant tachycardia.   Following administration of adenosine, prior to termination of the tachycardia in the "slow" pathway, decremental conduction in the retrograde slow AV nodal pathway is induced by adenosine as indicated by the lengthening of the V-A interval.   A, atrium; H, His bundle; V, ventricle; HRA, high right atrium; HBE, His bundle electrogram; CS, distal coronary sinus; RVA, right ventricular apex. Surface leads I, aVF and V₁ are shown.

The effect of *endogenous* adenosine, augmented by dipyridamole, to slow or terminate AV nodal reentrant tachycardia and AV reciprocating tachycardia has also been demonstrated (Fig. 6) [Lerman et al, 1989]. As one might expect, aminophylline, a competitive antagonist of adenosine, completely reverses the negative dromotropic effects of dipyridamole on AV nodal conduction [Lerman et al, 1989].

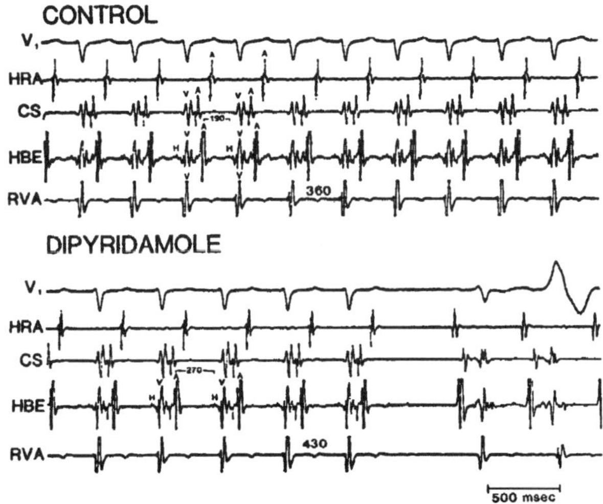

Figure 6:   Recording of termination of supraventricular tachycardia with dipyridamole. CONTROL Panel: AV reciprocating tachycardia in a patient with a left-sided accessory pathway. DIPYRIDAMOLE panel: In the same patient, ten minutes after bolus dose and continuous infusion of dipyridamole, cycle length of the tachycardia increased from 360 to 430 msec, an increment confided to the AV node. Tachycardia terminated in the AV node and could not be reinitiated. HRA, high right atrium; HBE, His bundle electrogram; CS, coronary sinus; RVA, right ventricular apex. (Reproduced with permission from Lerman BB, et al. Circulation 1989;80:1536-1541. Copyright 1989, The American Heart Association, Inc.)

Prior to the clinical availability of adenosine, intravenous verapamil was the drug of choice for diagnosing and treating AV nodal reentrant tachycardia. Adenosine has since become the preferred choice over verapamil due to its short half life, safety, reliability and because, unlike verapamil, it does not cause hypotension or negative inotropy.

D. His-Purkinje System and Ventricles
His-Purkinje System

Adenosine has little direct effect on the His-Purkinje system. Conduction from the His bundle to the ventricular myocardium is unaltered by adenosine [Favale et al, 1985]. Unlike observations in guinea pig [Wesley et al, 1985] and canine hearts, [Pelleg et al, 1986] adenosine does not diminish His-Purkinje automaticity in humans under basal conditions [Lerman et al, 1988]. However, adenosine does demonstrate potent antiadrenergic effects on the His-Purkinje system (Fig. 7) [Lerman et al, 1988]. Adenosine causes transient suppression of epinephrine-induced automatic rhythms that originate from Purkinje fibers at normal resting membrane potentials by decreasing the slope of phase 4 depolarization [Belardinelli et al, 1995]. However, in Purkinje fibers with depressed resting membrane potentials ($\leq$ -60 mV) adenosine has minimal effect on rate, indicating adenosine's effects on automatic rhythms are dependent on normal resting membrane potential [Rosen et al, 1983].

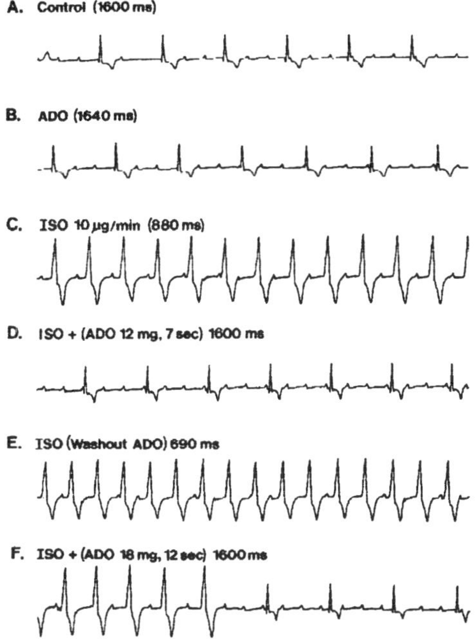

Figure 7: Antiadrenergic effect of adenosine on the His-Purkinje system in humans. (A) Complete heart block with an infra-Hisian escape rhythm cycle length of 1,600 ms. (B) Adenosine has minimal direct effect on the escape rhythm cycle length. (C) Isoproterenol increases the sensitivity of a more distal conduction site resulting in prolongation of the QRS complex and a marked decrease in the escape rhythm cycle length. (D) Adenosine completely antagonized the effects of isoproterenol and resulted in a proximal shift in pacemaker site identical to that observed during control (A). (E) After adenosine washout and during continued isoproterenol infusion (10 µg/min) the escape rhythm focus again shifted to a more distal site and the cycle length decreased to 690 ms. (F) The reproducible antiadrenergic effects of adenosine are shown. All recordings are from surface ECG lead 2. ADO adenosine, ISO isoproterenol. (Reproduced with permission from Lerman BB, et al. J Clin Invest 1988; 82:2127-2135. Copyright 1988, The American Society for Clinical Investigation, Inc.)

The electrophysiologic effects of adenosine on ventricular myocytes are species specific and dependent in large part to the presence or absence of $I_{K,ACh,Ado}$ [Belardinelli et al, 1995; Koumi et al, 1994]. Unlike ferret and rat ventricular myocytes, human, guinea pig and rabbit ventricular myocytes do not have $I_{K,ACh,Ado}$ channels [Koumi et al, 1994]. As a result, adenosine does not shorten the ventricular action potential in humans, nor does it alter ventricular inotropy [Belardinelli et al, 1995].

In the ventricle adenosine inhibits catecholamine induced stimulation of adenylyl cyclase and the subsequent increase in intracellular cAMP [Song et al, 1995]. Among the membrane currents stimulated by catecholamines are $I_{Ca,L}$, I_{Ti}, the delayed rectifier current I_K, and the chloride current I_{Cl}, all of which are attenuated by adenosine [Isenberg et al, 1984; Song et al, 1994; Harvey et al, 1990; Song et al, 1992; Belardinellli et al, 1995]. The effects of adenosine on ventricular action potential duration depend on the complex interplay of these various currents in response to catecholamines. For example, stimulation of $I_{Ca,L}$, increases action potential duration. However, catecholamine stimulation can also decrease action potential duration through its effects on I_K and I_{cl} [Song et al, 1995; Song et al, 1994; Harvey et al, 1990; Song et al, 1992; Belardinellli et al, 1995; Belardinelli et al, 1983]. Adenosine inhibition of catecholamine-stimulated $I_{Ca,L}$ therefore results in shortening of the action potential, whereas adenosine inhibition of catecholamine-stimulated I_K and I_{Cl} prolongs action potential duration. The overall effect of adenosine will reflect the algebraic summation of these effects.

Investigation of the role of adenosine in ischemic preconditioning suggest that adenosine, present in abundant quantities in ischemic myocardium, may facilitate the opening of $I_{K\text{-}ATP}$ channels [Kirsch et al, 1990; Ito et al, 1994]. These channels cause shortening of ventricular myocyte action potential. However, further work is required to substantiate this preliminary finding.

<u>E. Reentrant Ventricular Tachycardia</u>

In general, adenosine has no antiarrhythmic effect on reentrant ventricular tachycardia due to structural heart disease. It also has no effect on macroreentrant tachycardia due to bundle branch reentry and idiopathic verapamil-sensitive reentry originating from the region of the left posterior fascicle (intrafascicular ventricular tachycardia) [Lerman et al, 1996; Lerman et al, 1997; Lerman et al, 1993; Lerman et al, 1986].

<u>F. Triggered Ventricular Tachycardia</u>

Ventricular tachycardia due to delayed after depolarizations and triggered activity accounts for up to 80% of ventricular tachycardia occurring in individuals without structural heart disease. Clinically, these tachycardias may be classified into two entities, repetitive monomorphic ventricular tachycardia (RMVT) and paroxysmal stress-induced ventricular tachycardia [Lerman et al, 1995]. RMVT was initially described by Gallvardin and is characterized by frequent ventricular extrasystoles, ventricular couplets, and salvos of non sustained ventricular tachycardia with intervening sinus beats [Rahilly et al, 1982; Buxton et al, 1983; Gallavardin et al, 1922]. In contrast, paroxysmal stress-induced ventricular tachycardia occurs during exercise or emotional strain and is a sustained arrhythmia [Lerman et al, 1996]. Common to both is the usual absence of structural heart disease, similar tachycardia morphology (left bundle branch block, inferior axis), and similar site of origin, the right ventricular outflow tract. However, the tachycardia can also originate from deep within the interventricular septum and exit from the left side of the septum [Lerman et al, 1997]. It is important to recognize that there is considerable overlap between the clinical presentations of these two types of idiopathic ventricular tachycardia.

The resting ECG is usually normal, with the notable exception of T-wave abnormalities due to T-wave memory in subjects with frequent salvos of RMVT. Echocardiograms and ventriculo-graphy are normal, as are signal averaged ECG recordings and myocardial biopsy samples [Lerman et al, 1995].   Heart rate variability analysis in patients with RMVT shows evidence for transient increases in sympathetic tone (without parasympathetic withdrawal) that precede ventricular couplets and non sustained ventricular tachycardia [Lerman et al, 1995].

As discussed previously, adenosine is a potent inhibitor of delayed after depolarizations and triggered activity caused by cAMP mediated stimulation of I_{CaL} and I_{Ti}. Due to these effects and its lack of effect in other clinical forms of ventricular tachycardia, adenosine is thought to be a mechanism specific pharmacologic probe for identifying ventricular tachycardia due to triggered activity (Fig. 8). Other agents are less specific since they terminate VT due to more than one mechanism.   For example, verapamil, while frequently effective in terminating triggered arrhythmias, will also terminate intrafascicular reentrant ventricular tachycardia [Lerman et al, 1997].

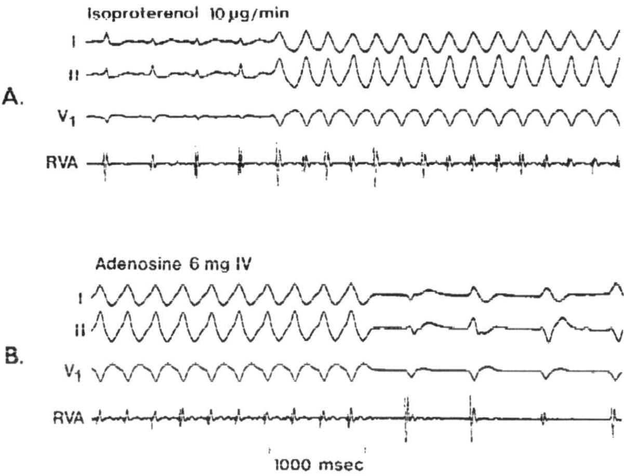

Figure 8:   (A) Initiation of ventricular tachycardia due to triggered activity during infusion of isoproterenol.   (B) Termination of the tachycardia initiated in (A) by adenosine in 7.5 seconds.   Note the slight slowing of tachycardia before termination. At termination of ventricular tachycardia there was temporary sinus slowing and AV nodal block caused by adenosine and an idioventricular escape rhythm that lasted 6 seconds.   RVA, right ventricular apex. (Reproduced with permission from Lerman BB, et al. Circulation 1986;74:270-280.   Copyright 1986, The American Heart Association, Inc.)

G. Automatic Ventricular Tachycardia

Adenosine's effects on automatic rhythms are dependent on resting membrane potential, with maximal effect observed at high potentials and minimal or no effect observed at low potentials [Rosen et al, 1983].   Adenosine causes transient suppression but does not terminate adrenergically-mediated automatic ventricular tachycardia (Fig. 9) [Lerman et al, 1993].

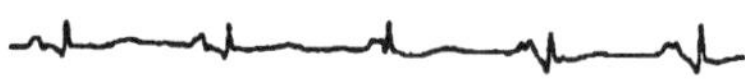

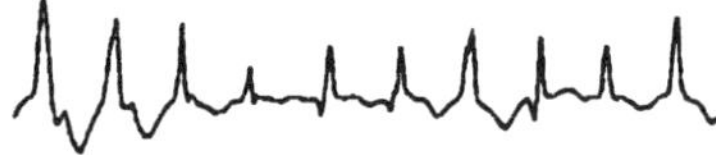

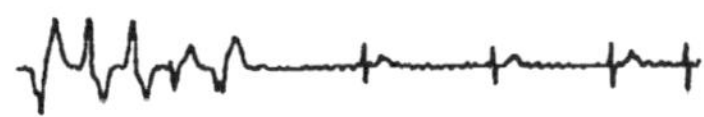

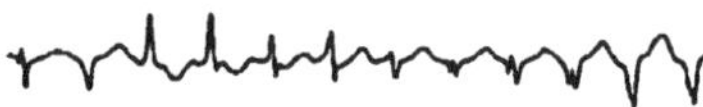

Figure 9: Initiation of idiopathic automatic ventricular tachycardia (VT) during isoproterenol infusion. (A) Sinus rhythm recording 60 seconds after beginning isoproterenol infusion (µg/min). (B) Sustained polymorphic VT (identical to the patient's clinical arrhythmia) developed 95 seconds after initiation of isoproterenol infusion (1 µg/min). (C) Adenosine transiently suppressed automatic VT (for approximately 5 seconds). This was followed by the emergence of multiform ventricular extrasystoles until sustained VT resumed. (D) Resumption of automatic VT 20 seconds after initial effects of adenosine were demonstrated. (Reproduced and modified with permission from Lerman BB, Circulation 1993; 87:382-90. Copyright 1993, The American Heart Association, Inc.)

<u>H. Atrioventricular Accessory Pathways</u>

Reentrant tachycardia involving an AV accessory pathway consists of anterograde conduction over the AV node and His-Purkinje system with retrograde conduction to the atrium proceeding via an accessory pathway. Occasionally, the direction of the circuit may be reversed with anterograde conduction across the accessor pathway and retrograde conduction over the AV node. Typically, AV accessory pathways are not sensitive to adenosine, likely due to the fact that they consist primarily of normal atrial tissue with resting membrane potentials near E_k (~ -90 mV). Therefore, adenosine has little direct effect in these tissues [Lerman et al, 1991] and thus terminates AV reciprocating tachycardia by transiently blocking conduction in the AV node.

Rarely, some accessory pathways may demonstrate adenosine sensitivity. In general, these pathways demonstrate decremental conduction properties (Fig. 3) [Lerman et al, 1987]. Decremental accessory pathways may be divided into two groups based upon their electrophysiologic characteristics. Pathways that respond only to adenosine and not verapamil are likely comprised of partially depolarized atrial tissue (> -60 mV to -70 mV) [Lerman et al, 1987]. In contrast, decremental accessory pathways that respond to both verapamil (modulation of the slow-inward calcium current) and to adenosine, have pharmacologic properties similar to that of

the AV node, and may be comprised of more completely depolarized atrial fibers
with resting membrane potentials of -60 mV or less.

Because the majority of accessory pathways are not sensitive to adenosine,
inducing transient AV block with bolus administration of adenosine has become a
useful tool for diagnosing latent preexcitation, localizing the region of the accessory
pathway, and in assessing the immediate efficacy of accessory pathway ablation
[Wilbur et al, 1997; Walker et al, 1995; Engelstein et al, 1994].

I. Adenosine and Heart Block

Endogenous adenosine may contribute to AV block observed during inferior
wall myocardial infarction. High degree AV block, observed in up to one third of
patients experiencing acute inferior wall myocardial infarction, has been attributed to
increased parasympathetic tone [Scheinman et al, 1975; Adgey et al, 1968].
However, atropine is frequently ineffective in abolishing heart block complicating an
inferior wall infarction [Adgey et al, 1968; Feigl et al, 1984]. An alternative
explanation for infarct related AV block is that related to the increased release of
adenosine during ischemia, resulting in a negative dromotropic effect on AV nodal
conduction. Treatment with the adenosine antagonist aminophylline can reverse
complete heart block in this setting [Sutton et al, 1968; Wesley et al, 1986].

J. Pharmacology
Dose

Effective dosing varies widely between individuals and is dependent on
baseline autonomic tone. To observe the cardiac electrophysiologic effects of
adenosine it must be administered intravenously as a bolus. The recommended
initial dose for adults is 6 mg, followed by a flush of normal saline. If the desired
effect is not obtained, doses of 12 mg or mor may be required. If administered via a
central vein, lower starting doses such as 3 mg are appropriate. The onset of action
is usually within 20 seconds of administration into a peripheral vein . Because the
half-life of adenosine ranges from 0.6 to 10 seconds [Klabunde et al, 1983; Ontyd et
al, 1984; Möser et al, 1989] its electrophysiologic effects in the myocardium
typically resolve within 20 seconds.

Adenosine's use during pregnancy [Afridi et al, 1992; Mason et al, 1992]
and in children [Till et al, 1989] appears to be safe and effective in terminating
supraventricular tachycardia. There is one report in which adenosine successfully
terminated a supraventricular tachycardia in a pregnant woman without affecting the
fetal heart rate [Harrison et al, 1992].

Recipients of orthotopic heart transplants appear to be extremely sensitive to
the negative chronotropic and dromotropic effects of adenosine [Ellenbogen et al,
1990]. Sinus bradycardia occurs frequently in this population, perhaps as a result of
adenosine release along with other inflammatory mediators [Madara et al, 1993].
Theophylline has been demonstrated to be effective in reversing bradycardia observed
in these patients [Redmond et al, 1993].

K. Adverse Reactions

Adenosine produces uncomfortable side effects in approximately 30% of
patients [DiMarco et al, 1990]. These include flushing (due to vasodilation),
dyspnea (due to stimulation of carotid body chemoreceptors [Biaggioni et al, 1987],

transient bronchospasm [Mann et al, 1985; Cushley et al, 1985]), and chest pressure (which may be due to direct sympathetic nerve afferent stimulation or to the presence of adenosine pain receptors located on the heart) [Crea et al, 1990; Sylvén et al, 1987; Montano et al, 1995]. Adenosine may precipitate severe bronchospasm in patients with reactive airway disease [Mann et al, 1985; Cushley et al, 1985].

<u>L. Proarrhythmic Effects</u>

Atrial fibrillation is the most common arrhythmia induced following administration of adenosine, and is usually associated with a "long-short" atrial sequence [Strickberger et al, 1997]. As noted previously, atrial fibrillation is potentiated as a result of shortening of the atrial action potential. In most cases atrial fibrillation is brief in duration and resolves spontaneously. AV block, sinus bradycardia and sinus pauses are expected with bolus administration of adenosine. However, these effects may be prolonged if the patient is taking dipyridamole [Watt et al, 1986]. Caution should be exercised when administering adenosine to patients with an anterogradely conducting accessory pathways since it may precipitate atrial fibrillation with a rapid ventricular response over the pathway. Rapid conduction across the AV node may develop in patients with atrial flutter (280/min) after the negative dromotropic effects of adenosine subside [Brodsky et al, 1995; Rankin et al, 1993; Slade et al, 1993]. This is related to adenosine's sympathoexcitation.

Sporadic cases of adenosine induced ventricular arrhythmias have been reported. This usually occurs in the setting of a prolonged pause following adenosine termination of supraventricular tachycardia and results in pause-dependent polymorphic ventricular tachycardia or ventricular fibrillation [Ben-Sorek et al, 1993]. There have also been cases of adenosine causing torsade de pointes in subjects with normal QT intervals [Harrington et al, 1993] and in those with the congenital long QT syndrome [Wesley et al, 1992].

IV CONCLUSIONS AND FUTURE DIRECTIONS

Adenosine has developed an important role for use in the diagnosis and treatment of both supraventricular and ventricular tachyarrhythmias. Adenosine's mechanism and specific effects have greatly facilitated the acute treatment of supraventricular tachycardias involving the AV node. In addition, adenosine has played a crucial role in identifying triggered activity as the mechanism responsible for most forms of idiopathic ventricular tachycardia. If one appreciates the relatively few contraindications for its use, then adenosine offers considerable utility as a safe and effective antiarrhythmic agent as well as a unique mechanistic probe into the pathogenesis of clinical arrhythmias.

Acknowledgments

This work was supported in part by a grant from the National Institutes of Health (RO1-HL56139)

REFERENCES

Adgey AAJ, Geddes JS, Mulholland HC, Keegan DAJ, Pantridge JF. Incidence, significance, and management of early bradyarrhythmia complicating acute myocardial infarction. Lancet 1968; 2:1097-1101.

Afridi I, Moise KJJ, Rokey R. Termination of supraventricular tachycardia with intravenous adenosine in a pregnant woman with Wolff-Parkinson-White syndrome. Obstet Gynecol 1992; 80:481-3.

Alboni P, Paparella N, Cappato R, Pirani R, Yiannacopulu P, Antonioli GE. Long-term effects of theophylline in atrial fibrillation with a slow ventricular response. Am J Cardiol 1993; 72:1142-5.

Alboni P, Ratto B, Cappato R, Rossi P, Gatto E, Antonioli GE. Clinical effects of oral theophylline in sick sinus syndrome. Am Heart J 1991; 122:1361-7.

Belardinelli L, Giles W, West A. Ionic mechanisms of adenosine actions in pacemaker cells from rabbit heart. J Physiol (Lond) 1988; 405:615-633.

Belardinelli L, Isenberg G. Actions of adenosine and isoproterenol on isolated mammalian ventricular myocytes. Circ Res 1983; 53:287-297.

Belardinelli L, Linden J, Berne RM. The cardiac effects of adenosine. Prog Cardiovasc Dis 1989; 32:73-97.

Belardinelli L, Shryock JC, Song Y, Wang D, Srinivas M. Ionic basis of the electrophysiologic actions of adenosine on cardiomyocytes. FASEB J 1995; 9:359-365.

Belardinellli L, Song Y. Adenosine and ATP regulated ion currents in cardiomyocytes. In: Dodfraind T, Mancia O, Abbracchio MP, eds. Pharmacological control of calcium and potassium homeostasis: Biological, therapeutical and clinical aspects. Dordrecth: Kluwer Academic Publishers, 1995:65-72.

Belhassen B, Glick A, Laniado S. Comparative clinical and electrophysiologic effects of adenosine triphosphate and verapamil on paroxysmal reciprocating junctional tachycardia. Circulation 1988; 77:795-805.

Ben-Sorek ES, Wiesel J. Ventricular fibrillation following adenosine administration. A case report. Arch Intern Med 1993; 153:2701-2.

Biaggioni I, Killian TJ, Mosqueda-Garcia R, Robertson RM, Robertson D. Adenosine increases sympathetic nerve traffic in humans. Circulation 1991; 83:1668-1675.

Biaggioni I, Olafsson B, Robertson RM, Hollister AS, Robertson D. Cardiovascular and respiratory effects of adenosine in conscious man: Evidence for chemoreceptor activation. Circ Res 1987; 61:779-786.

Brodsky MA, Hwang C, Hunter D, Chen PS, Smith D, Ariani M, Johnston WD, Allen BJ, Chun JG, Gold CR. Life-threatening alterations in heart rate after the use of adenosine in atrial flutter. Am Heart J 1995; 130:564-71.

Buxton AE, Waxman HL, Marchlinski FE, Simson MB, Cassidy D, Josephson ME. Right ventricular tachycardia: clinical and electrophysiologic characteristics. Circulation 1983; 68:917-27.

Chen S-A, Chiang C-E, Yang C-J, Cheng C-C, Wu T-J, Wang S-P, Chiang BN, Chang M-S. Sustained atrial tachycardia in adult patients: electrophysiological characteristics, pharmacologic response, possible mechanism, and effects of radiofrequency ablation. Circulation 1994; 90:1262-2178.

Clemo HF, Belardinelli L. Effect of adenosine on atrioventricular conduction: I. Site and characterization of adenosine action in the guinea pig atrioventricular node. Circ Res 1986; 59:427-436.

Crea F, Pupita G, Galassi AR, el-Tamimi H, Kaski JC, Davies G, Maseri A. Role of adenosine in pathogenesis of anginal pain. Circulation 1990, 81:164-72.

Cushley MJ, Holgate ST. Adenosine-induced bronchoconstriction in asthma: role of mast cell-mediator release. J Allergy Clin Immunol 1985; 75:272-8.

DiMarco JP, Miles W, Akhtar M, Milstein S, Sharma AD, Platia E, McGovern B, Scheinman MM, Govier WC. Adenosine for paroxysmal supraventricular tachycardia: dose ranging and comparison with verapamil. Assessment in placebo-controlled, multicenter trials. The Adenosine for PSVT Study Group. Ann Intern Med 1990; 113:104-10.

DiMarco JP, Sellers TD, Lerman BB, Greenberg ML, Berne RM. Diagnostic and therapeutic use of adenosine in patients with supraventricular tachyarrhythmias. J Am Coll Cardiol 1985; 6:417-425.

Drury AN, Szent-Gyorgyi A. The physiologic activity of adenine compounds with special reference to their action upon the mammalian heart. J Physiol 1929; 68:213-237.

Ellenbogen KA, Thames MD, DiMarco JP, Sheehan H, Lerman BB. Electrophysiological effects of adenosine in the transplanted human heart. Evidence of supersensitivity. Circulation 1990; 81:821-8.

Engelstein E, Lippman N, Stein KM, Lerman BB. Mechanism-specific effects of adenosine on atrial tachycardia. Circulation 1994; 89:2645-2654.

Engelstein ED, Lerman BB, Somers VK, Rea RF. Role of arterial chemoreceptors in mediating the effects of endogenous adenosine on sympathetic nerve activity. Circulation 1994; 90:2919-26.

Engelstein ED, Wilber D, Wadas M, Stein KM, Lippman N, Lerman BB. Limitations of adenosine in assessing the efficacy of radiofrequency catheter ablation of accessory pathways. Am J Cardiol 1994; 73:774-9.

Favale S, DiBiase M, Rizzo V, Belardinelli L, Rizzon P. Effect of adenosine and adenosine-5'-triphosphate on atrioventricular conduction in patients. J Am Coll Cardiol 1985; 5:1212-1219.

Feigl D, Ashkenazy J, Kishon Y. Early and late atrioventricular block in acute inferior myocardial infarction. J Am Coll Cardiol 1984; 4:35-38.

Gallavardin L. Extrasystolic ventriculaire a paroysmes tachycardiques prolonges. Arch Mal Coeur Vaiss 1922; 15:298-306.

Griffith MJ, Garrett CJ, Ward DE, Camm AJ. The effects of adenosine on sinus node reentrant tachycardia. Clin Cardiol 1989; 12:409-411.

Haines DE, DiMarco JP. Sustained intraatrial reentrant tachycardia: clinical, electrocardiographic and electrophysiologic characteristics and long-term follow-up. J Am Coll Cardiol 1990; 15:1345-1354.

Haines DE, Lerman BB, DiMarco JP. Intravenous adenosine shortens atrial refractoriness in man (abstract). Circulation 1988:II-153.

Harrington GR, Froelich EG. Adenosine-induced torsades de pointes. Chest 1993; 103:1299-301.

Harrison JK, Greenfield RA, Wharton JM. Acute termination of supraventricular tachycardia by adenosine during pregnancy. Am Heart J 1992; 123:1386-8.

Harvey RD, Clark CD, Hume JR. Chloride current in mammalian cardiac myocytes. Novel mechanism for autonomic regulation of action potential duration and resting membrane potential. J Gen Physiol 1990; 95:1077-1102.

Isenberg G, Belardinelli L. Ionic basis for the antagonism between adenosine and isoproterenol on isolated mammalian ventricular myocytes. Circ Res 1984; 55:309-325.

Ito H, Vereecke J, Carmeliet E. Mode of regulation by G protein of the ATP-sensitive K+ channel in guinea-pig ventricular cell membrane. J Physiol (London) 1994; 478:101-108.

Kall JG, Kopp D, Olshansky B, Kinder C, O'Connor M, Cadman CS, Wilber D. Adenosine-sensitive atrial tachycardia. PACE 1995; 18:300-306.

Kirsch GE, Codina J, Birnbaumer L, Brown AM. Coupling of ATP-sensitive K+ channels to A1 receptors by G proteins in rat ventricular myocytes. Am J Physiol 1990; 259:H820-H826.

Klabunde R. Dipyridamole inhibition of adenosine metabolism in human blood. Eur J Pharmacol 1983; 93:21-26.

Koumi SI, Wasserstrom JA. Acetylcholine-sensitive muscarinic K+ channels in mammalian ventricular myocytes. Am J Physiol 1994; 266:H1812-H1821.

Lerman BB, Belardinelli L, West GA, Berne RM, DiMarco JP. Adenosine-sensitive ventricular tachycardia: evidence suggesting cyclic AMP-mediated triggered activity. Circulation 1986; 74:270-80.

Lerman BB, Greenberg M, Overholt ED, Swerdlow CD, Smith RT, Sellers TD, DiMarco JP. Differential electrophysiologic properties of decremental retrograde pathways in long RP' tachycardia. Circulation 1987; 76:21-31.

Lerman BB, Stein K, Engelstein ED, Battleman DS, Lippman N, Bei D, Catanzaro D. Mechanism of repetitive monomorphic ventricular tachycardia. Circulation 1995; 92:421-9.

Lerman BB, Stein KM, Markowitz SM. Adenosine-sensitive ventricular tachycardia: a conceptual approach. J Cardiovasc Electrophysiol 1996; 7:559-69.

Lerman BB, Stein KM, Markowitz SM. Mechanisms of idiopathic left ventricular tachycardia. J Cardiovasc Electrophysiol 1997; 8:571-83.

Lerman BB, Wesley RC, Belardinelli L. Electrophysiologic effects of dipyridamole on atriventricular nodal conduction and supraventricular tachycardia: Role of endogenous adenosine. Circulation 1989; 80:1536-1543.

Lerman BB, Wesley RC, DiMarco JP, Haines DE, Belardinelli L. Antiadrenergic effects of adenosine on His-Purkinje automaticity: Evidence for accentuated antagonism. J Clin Invest 1988; 82:2127-2135.

Lerman BB. Response of nonreentrant catecholamine-mediated ventricular tachycardia to endogenous adenosine and acetylcholine. Evidence for myocardial receptor-mediated effects. Circulation 1993; 87:382-90.

Lerman BL, Belardinelli L. Cardiac electrophysiology of adenosine. Circulation 1991; 83:1499-1509.

Madara JL, Patapoff TW, Gillece-Castro B, Colgan SP, Parkos CA, Delp C, Mrsny RJ. 5'-adenosine monophosphate is the neutrophil-derived paracrine factor that elicits chloride secretion from T84 intestinal epithelial cell monolayers. J Clin Invest 1993; 91:2320-2325.

Mann JS, Cushley MJ, Holgate ST. Adenosine-induced bronchoconstriction in asthma. Role of parasympathetic stimulation and adrenergic inhibition. Am Rev Respir Dis 1985; 132:1-6.

Markowitz SM, Stein KM, Mittal S, Lerman BB. Mechanistic spectrum of adenosine-sensitive atrial tachycardia. J Am Coll Cardiol 1997; 25:345A.

Mason BA, Ricci-Goodman J, Koos BJ. Adenosine in the treatment of maternal paroxysmal supraventricular tachycardia. Obstet Gynecol 1992; 80:478-80.

Montano N, Gnecchi-Ruscone T, Lombardi F, Malliani A. Excitatory effects of adenosine on cardiac sympathetic afferent fibers. In: Belardinelli L, Pelleg A, eds. Adenosine and adenine nucleotides: From molecular biology to integrative physiology. Vol. I. Boston Dordrecht London: Kluwer Academic Publishers, 1995:307-314.

Möser G, Schrader J, Deussen A. Turnover of adenosine in plasma of human and dog blood. Am J Physiol 1989; 256:C799-806.

Ontyd J, Schrader J. Measurement of adenosine, inosine, and hypoxanthine in human plasma. J Chromatogr 1984; 307:404-9.

Pelleg A, Mitamura H, Misuoka T, Michelson EL, Dreifus LS. Effects of adenosine and adenosine 5'-triphosphate on ventricular escape rhythm in the canine heart. J Am Coll Cardiol 1986; 8:1145-1151.

Rahilly GT, Prystowsky EN, Zipes DP, Naccarelli GV, Jackman WM, Heger JJ. Clinical and electrophysiologic findings in patients with repetitive monomorphic ventricular tachycardia and otherwise normal electrocardiogram. Am J Cardiol 1982; 50:459-68.

Rankin AC, Rae AP, Houston A. Acceleration of ventricular response to atrial flutter after intravenous adenosine. Br Heart J 1993; 69:263-5.

Redmond JM, Zehr KJ, Gillinov MA, Baughman KL, Augustine SM, Cameron DE, Stuart RS, Acker MA, Gardner TJ, Reitz BA, et al. Use of theophylline for treatment of prolonged sinus node dysfunction in human orthotopic heart transplantation. J Heart Lung Transplant 1993; 12:133-8; discussion 138-9.

Rosen MR, Danilo P, Weiss RM. Actions of adenosine on normal and abnormal impulse initiation in canine ventricile. Am J Physiol 1983; 83:H715-H721.

Saito D, Matsubara K, Yamanari H, Obayashi N, Uchida S, Maekawa K, Sato T, Mizuo K, Kobayashi H, Haraoka S. Effects of oral theophylline on sick sinus syndrome. J Am Coll Cardiol 1993; 21:1199-204.

Scheinman MM, Thorburn D, Abbott JA. Use of atropine in patients with acute myocardial infarction and sinus bradycardia. Circulation 1975; 52:627-633.

Shenasa H, Kanter RJ, Hamer ME, Sorrentino RA, Sanders WE, Greenfield RA, Page RL, Wharton JM. Reappraisal of the efficacy of adenosine for termination of extopic atrial tachycardia. J Am Coll Cardiol 1993; 88:I-396.

Shryock JC, Belardinelli L. Adenosine and adenosine receptors in the cardiovascular system: Biochemistry, physiology, and pharmacology. Am J Cardiol 1997; 79:2-10.

Slade AK, Garratt CJ. Proarrhythmic effect of adenosine in a patient with atrial flutter. Br Heart J 1993; 70:91-2.

Song Y, Shryock J, Belardinelli L. Modulation of cardiomyocyte membrane currents by A1 adenosine receptors. In: Belardinelli L, Pelleg A, eds. Molecular Biology to Integrative Physiology. Norwell, Massachusetts: Kluwer Academic Publishers, 1995:On Call From Library.

Song Y, Srinivas M, Belardinelli L. The effects of adenosine on isoproterenol-induced outward currents in cardiac ventricular myocytes. Drug Dev Res 1994; 31:324.

Song Y, Thedford S, Lerman BB, Belardinelli L. Adenosine-sensitive after depolarizations and triggered activity in guinea pig ventricular myocytes. Circ Res 1992; 70:743-753.

Strickberger SA, Man KC, Daoud EG, Goyal R, Brinkman K, Knight BP, Weiss R, Bahu M, Morady F. Adenosine-induced atrial arrhythmia: a prospective analysis. Ann Intern Med 1997; 127:417-22.

Sutton R, Davies M. The conduction system in acute myocardial infarction complicated by heart block. Circulation 1968; 38:987-992.

Sylvén C, Jonzon B, Brandt R, Beermann B. Adenosine-provoked angina pectoris-like pain--time characteristics, influence of autonomic blockade and naloxone. Eur Heart J 1987; 8:738-43.

Till J, Shinebourne EA, Rigby ML, Clarke B, Ward DE, Rowland E. Efficacy and safety of adenosine in the treatment of supraventricular tachycardia in infants and children. Br Heart J 1989; 62:204-11.

Walker KW, Silka MJ, Haupt D, Kron J, McAnulty JH, Halperin BD. Use of adenosine to identify patients at risk for recurrence of accessory pathway conduction after initially successful radiofrequency catheter ablation. PACE 1995; 18:441-446.

Wang D, Belardinelli L. Effects of adenosine on phase 4 depolarization and pacemaker current (IF) in single rabbit atrioventricular nodal myocytes. FASEB J 1994; 8:A611.

Watt AH, Bernard MS, Webster J, Passani SL, Stephens MR, Routledge PA. Intravenous adenosine in the treatment of supraventricular tachycardia: a dose-ranging study and interaction with dipyridamole. Br J Clin Pharmacol 1986; 21:227-30.

Watt AH, Routledge PA. Transient bradycardia and subsequent sinus tachycardia produced by intravenous adenosine in healthy adult subjects. Br J Clin Pharmacol 1986; 21:533-536.

Watt AH. Sick sinus syndrome: An adenosine-mediated disease. Lancet 1985; 1:786-788.

Wesley RC, Belardinelli L. Role of adenosine on ventricular overdrive suppression in isolated guinea pig hearts and Purkinje fibers. Circ Res 1985; 57:517-531.

Wesley RCJ, Lerman BB, DiMarco JP, Berne RM, Belardinelli L. Mechanism of atropine-resistant atrioventricular block during inferior myocardial infarction: possible role of adenosine. J Am Coll Cardiol 1986; 8:1232-4.

Wesley RCJ, Turnquest P. Torsades de pointe after intravenous adenosine in the presence of prolonged QT syndrome. Am Heart J 1992; 123:794-6.

West GA, Belardinelli L. Correlation of sinus slowing and hyperpolarization caused by adenosine in sinus node. Pflugers Arch 1985; 403:75-81.

West GA, Belardinelli L. Sinus slowing and pacemaker shift caused by adenosine in rabbit SA node. Pfleugers Arch 1985; 403:66-74.

Wilbur SL, Marchlinski FE. Adenosine as an antiarrhythmic agent. Am J Cardiol 1997; 79:30-37.

Zaza A, Rocchetti M, DiFrancesco D. Modulation of the hyperpolarization-activated current (IF) by adenosine in rabbit sino-atrial myocytes. J Physiol 1994; 475P:84P.

ADENOSINE AND CARDIAC AGING

Richard A.Fenton, Mojca Lorbar and James G. Dobson, Jr., Department of Physiology, University of Massachusetts Medical School and Graduate School of Biomedical Sciences, Worcester, Massachusetts U.S.A.

I. SUMMARY

Basal and ß-adrenergic-induced interstitial adenosine levels are significantly higher in the aged compared to the young adult myocardium. This results in a greater manifestation of the antiadrenergic action of adenosine reducing ß-adrenergic responsiveness. Results suggest that despite similar adenosine production by young adult and aged myocardium, elevated basal interstitial adenosine levels in aged myocardium result from depressed *in vivo* activity of adenosine kinase coupled to a reduced inward adenosine transport. With ß-adrenergic stimulation, adenosine production is greater in the young compared to aged heart. However, efficient recycling of adenosine in the young heart minimizes the enhanced adenosine release. This results in a net adenosine release which is still greater in the aged heart.

II. BACKGROUND

The ß-adrenergic responsiveness of the myocardium is known to become depressed with advancing age (Weisfeldt, 1980; Lakatta and Yin, 1982; Guarnieri, et al., 1980; Lakatta, et al., 1975; Kronenberg and Drage, 1973; Petrofsky and Lind, 1975; Conway, et al., 1971; Vestal, et al., 1979). To investigate this phenomenon, considerable attention has been placed directly on the ß-adrenergic signaling pathway. Receptor binding assays in myocardial membrane preparations suggest either an age-related decrease (Ferrara, et al., 1997; Scarpace and Abrass, 1986) or no difference (Guarnieri, et al., 1980) in ß-receptor affinity or number of receptors. Other studies have been conducted investigating the influence of aging on ß-adrenergic-elicited changes in myocardial cyclic AMP (cAMP) levels and protein kinase A activities. It was reported that when senescent and adult rat heart septa were stimulated with a level of isoproterenol eliciting a maximal contractile increase, there was no age-dependent difference in the level of cAMP or activation of cAMP-dependent protein kinase attained (Guarnieri, et al., 1980). However, other studies indicate that the isoproterenol-induced formation of myocardial cAMP is reduced in isolated perfused hearts obtained from aged rats (Dobson and Fenton, 1993; Robberecht, et al., 1986). This reduced responsiveness to adrenergic stimulation may reflect an age-dependent decrease in hormone-sensitive adenylyl cyclase activity (Gao, et al., 1997; Cai, et al., 1997; Narayanan and Tucker, 1986; Scarpace, 1986; Scarpace, 1990) resulting from an alteration in the catalytic subunit of the enzyme (O'Connor, et al., 1983; Narayanan and Tucker, 1986; Tobise, et al., 1994), the coupling of receptor and the adenylyl cyclase complex, or both (Scarpace, 1990; O'Connor, et al., 1983; Shu and Scarpace, 1994). Further down the signalling cascade, aging also appears to reduce norepinephrine-induced activation of protein kinase A and enhanced incorporation of phosphate into proteins involved in contractile responses to adrenergic stimulation (Robberecht, et al., 1986; Jiang, et al., 1993; Sakai, et al., 1989). Aged hearts retain responsiveness to elevations in perfusate Ca^{2+} equal to that of young adult counterparts (Guarnieri, et al., 1980; Sakai, et al., 1989; Lakatta, et al., 1975; Ferrara, et al., 1997; Dobson, et al., 1990).

Adenosine manifests an antiadrenergic action in the mammalian heart (Dobson, et al., 1987) which is expressed as an attenuation of contractile and metabolic responses to ß-adrenergic stimulation (Dobson, 1983; Dobson and Fenton, 1983; Fenton, et al., 1991; Fenton and Dobson, 1984; LaMonica, et al., 1985). This topic is presented in greater detail in an earlier chapter of this volume. Briefly, in the catecholamine-stimulated heart, adenosine attenuates the increased formation of cAMP (Dobson, 1978), activation of protein kinase A (Dobson, 1983), phosphorylation of cellular proteins (Fenton and Dobson, 1984; George, et al., 1991), activation of glycogen phosphorylase activity (Dobson, 1983) and ventricular contractility (Dobson, 1978; Dobson, 1983; Dobson and Fenton, 1983; Romano, et al., 1991). These inhibitory effects are observed when adenosine is at a concentration normally found in the interstitial compartment of the heart (0.1-10 µM; Fenton, et al., 1990; Fenton and Dobson 1987). The antiadrenergic property of adenosine allows it to be antiarrhythmogenic as well (Fenton, 1996; Fenton, et al., 1991), preventing development of adrenergic-induced arrhythmic activity. The antiadrenergic action of adenosine is mediated by sarcolemmal adenosine A_1-receptors (Gilman, 1987; Romano, et al., 1989) with signal transduction accomplished by an inhibitory guanine nucleotide binding protein, G_i (Kubalak, et al., 1991; Gilman, 1987).

Although adenosine is normally present in the interstitial fluid of the well-oxygenated heart (Fenton and Dobson, 1987; Fenton, et al., 1990), the production of adenosine by the myocardium increases with isoproterenol stimulation, ischemia and hypoxia (Fenton, et al., 1990; Fenton and Dobson, 1993; Bardenheuer, et al., 1987). Increased energy demand in response to adrenergic stimulation (Achterberg, et al., 1986) and reduced oxidative phosphorylation occurring with hypoxia or ischemia (Mentzer, et al., 1988) increases cytosolic AMP (Achterberg, et al., 1986). Adenosine derives from the cardiomyocyte (Bardenheuer, et al., 1987) upon dephosphorylation of AMP (Kroll, et al., 1993; Achterberg, et al., 1986; Bunger and Sobolle, 1986) by cytosolic (Smolenski, et al., 1991) or sarcolemmal-bound (Imai, et al., 1989; Headrick and Willis, 1989) 5'-nucleotidase activity, and from the hydrolysis of S-adenosylhomocysteine, a metabolite of the transmethylation reaction utilizing S-adenosylmethionine (Lloyd and Schrader, 1993). Subsequent to production, adenosine is deaminated to inosine by adenosine deaminase, is recycled into the nucleoside pool by the action of adenosine kinase, or released into the interstitial space where it may interact with cell surface adenosine specific receptors.

Adenosine-induced modulation of myocardial contractile function has been studied at different stages of development from juvenile to aged adult. The results have been provocative, though not always consistent. In most studies, aging appears to modify adenosine signal transduction. Maturation of the rat heart from an immature (23 days) to mature (80 days) stage has been associated with a decrease in adenosine A_1-receptor sensitivity as determined by attenuation of ß-adrenergic-induced ventricular and atrial contractile responses, and adenylyl cyclase activities (Sawmiller, et al., 1998). However, others have reported that the antiadrenergic properties of adenosine do not change with maturation (Gao, et al., 1997). Progression from young adult to aged adult has been reported to increase Fischer 344 rat heart adenosine A_1-receptor density (Janczewski and Lakatta, 1993). Juvenile-like sensitivity in adenosine A_1-receptor activity has been reported to return with aging as indicated by greater inhibition of ß-adrenergic stimulated adenylyl cyclase (Romano and Dobson 1996) and stronger negative dromotropic actions (Headrick, 1996), possibly as a result of increased adenosine A_1-receptor sensitivity (Janczewski and Lakatta, 1993) and greater $G_{i\alpha}$ activity in the aged myocardium (Ferrara, et al., 1997). A reduced chronotropic response to ß-adrenergic stimulation in aging humans was attributed to higher levels of adenosine production (Suteparuk, et al., 1995). An aging-elicited decrease in adenosine A_2-receptor activity was apparent as a reduction in vasodilation and smooth muscle relaxation with isolated perfused hearts and aortic rings, respectively, obtained from aged adult rats (Headrick, 1996). However, others have reported a marked attenuation of antiadrenergic action (Gao, et al., 1997) with aging as a result of a decrease in adenosine A_1-receptor/G_i coupling (Cai, et al., 1997) and a significantly reduced negative chronotropic effect of adenosine with aging (DiGennaro, et al., 1987). These latter results may have been influenced by uncontrolled high levels of endogenous adenosine reported to occur with aging (see below). In experiments to be described, evidence suggests that the antiadrenergic action of adenosine plays an important role in the diminished responsiveness of the aged adult heart to ß-adrenergic stimulation (Dobson, et al., 1990; Dobson and Fenton, 1993; Dobson, et al., 1991; Headrick, 1996).

III. TEXT

Removing the Influence of Endogenous Adenosine Restores ß-Adrenergic Contractile Responsiveness of Aged Hearts.

Hearts obtained from young adult (3-4 mo) and aged (17-20 mo) adult rats of Sprague-Dawley strain were perfused at a constant pressure of 70 cm H_2O and paced at 270/min. The maximum rates of left ventricular pressure development (+dP/dt $_{max}$) and relaxation (-dP/dt $_{max}$), indices of left ventricular function, were determined using a latex balloon placed in the left ventricular lumen. Basal values for +dP/dt$_{max}$ and -dP/dt$_{max}$ were similar between the young adult and aged adult groups, ranging from 2682 to 2912 mm Hg/sec and 1425 to 1609 mm Hg/sec for + and -dP/dt$_{max}$, respectively. Isoproterenol, a ß$_1$-adrenoceptor agonist, administered at 10^{-8} M for one min elicited significant elevations in both contractile function variables (Fig. 1). The values for +dP/dt$_{max}$ and -dP/dt$_{max}$ were 27 to 30% lower in the aged adult hearts, confirming findings reported in the literature (Lakatta, et al., 1975; Abrass, et al., 1982; Lakatta and Yin, 1982). Administration of the adenosine receptor antagonist, 8-sulphophenyl-theophylline (8-SPTheo) at 5×10^{-5} M to block the effects of endogenous adenosine resulted in an increase in adrenergic contractile responsiveness of aged hearts to a level similar to that of young adult hearts. The antagonist did not influence the $+dP/dt_{max}$ and $-dP/dt_{max}$ in the absence of isoproterenol. In addition the basal levels of $+dP/dt_{max}$ and $-dP/dt_{max}$ were not different between young and aged adult hearts. A similar observation was made in hearts treated with adenosine deaminase to metabolize endogenous adenosine to the biologically inactive inosine. In these experiments the ß-adrenergic contractile response was lower by 16-17% in the aged when compared to young adult hearts. Adenosine deaminase restored contractile responsiveness of aged hearts to a level not different from young hearts. The results suggest that the antiadrenergic action of adenosine plays a role in the aging-dependent reduction in myocardial ß-adrenergic sensitivity (Dobson and Fenton, 1993).

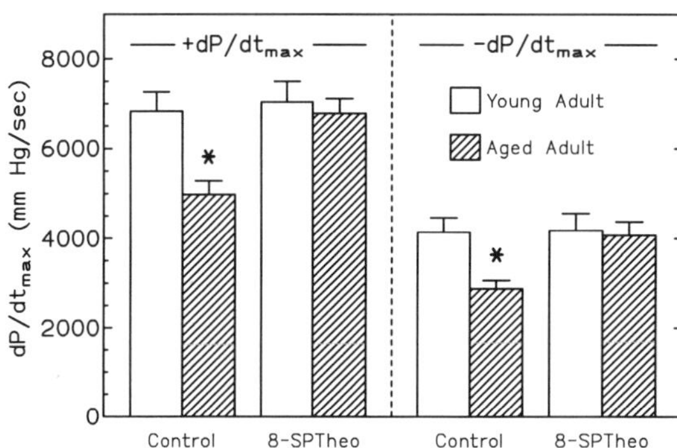

Figure 1: Effect of ISO in the absence or presence of 8-sulpho-phenyltheophylline (8-SPTheo) on the maximal rates of left ventricular pressure development ($+dP/dt_{max}$) and relaxation ($-dP/dt_{max}$) of isolated young adult and aged rat hearts. Values are mean ± SE for 12 hearts. *denotes signi-ficance from corresponding young value.

Blockade of Adenosine Receptor Activity Restores ß-Adrenergic Metabolic Responsiveness in the Aged Heart.

ß-Adrenergic stimulation is manifest as an elevation of cardiomyocyte levels of cAMP which subsequently elicit changes in a multitude of cellular functions. The effect of aging on changes in cAMP content and glycogen phosphorylase activity elicited by ISO stimulation was investigated by freeze-clamping hearts after stimulation with 10^{-8} M ISO for 1 min. Glycogen phosphorylase activities were determined in the absence or presence of AMP. An increase in the ratio of these activities indicates activation by an increase in conversion of phosphorylase *b* to *a*. Basal levels of cAMP and enzyme activity ratios for young and aged adult hearts were similar, ranging from 6.16 to 6.47 pmol cAMP/mg protein, and were not affected by the presence of adenosine blockade with 50 μM 8-SPTheo. ISO stimulation increased cAMP levels in young hearts 105% above preISO levels (Fig. 2). However, the ISO response with aged hearts was 33% less than that with young myocardium. Treatment with 8-SPTheo eliminated this difference. ISO stimulation of young adult hearts increased the glycogen phosphorylase activity 144% from a preISO ratio of 0.180 ± 0.032. However, aged hearts were less responsive in that ISO-stimulated enzyme ratios were 30% less than ratios of young hearts despite similar preISO values

(0.171 ± 0.019). Administration of 8-SPTheo eliminated the difference between young and aged hearts, for ratios with ISO stimulation were 0.464 ± 0.032 and 0.443 ± 0.029 for young and aged hearts, respectively (Dobson and Fenton, 1993).

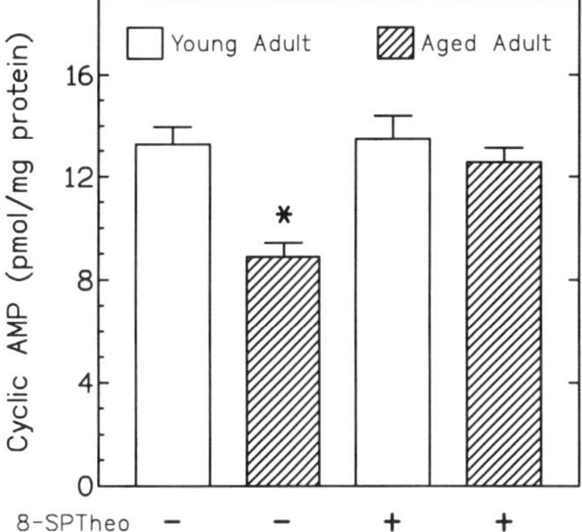

Figure 2: Effect of ISO on cAMP content of isolated perfused Sprague-Dawley rat hearts in the absence or presence of 8-SPTheo. Values are mean ± SE for 6-12 hearts. * denotes difference from corresponding young heart value.

Adenosine Levels in Coronary Effluent and Interstitial Fluid are Greater in Aged than Young Adult Hearts.

Coronary effluent was obtained from the pulmonary arteries of isolated perfused young adult and aged hearts and analyzed for adenosine content (Fenton and Dobson, 1985). Adenosine levels were 57% to 66% higher in effluent issuing from aged than from young adult hearts, whether of Sprague-Dawley or Fischer 344 strain (Fig. 3). For both strains this difference was maintained during ISO stimulation when coronary effluent adenosine concentrations were found with aged hearts to be 41% to 49% higher than values for young adult hearts as compared to adenosine values prior to isoproterenol administration. It is interesting that the increase in adenosine release with aging is a continuation of a progressive change in adenosine metabolism apparently beginning at the juvenile stage. Sawmiller, et al., (1996) have found that as the heart progresses from an immature (25 days) to mature (79 days) state, adenosine levels as determined in the coronary effluent increase.

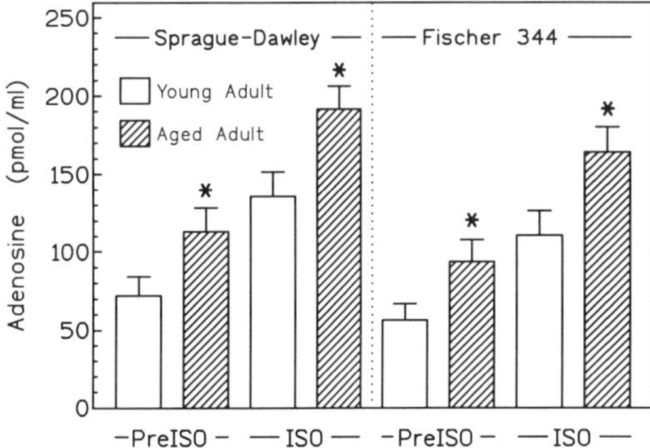

Figure 3: Effect of isoproterenol (ISO) on coronary effluent adenosine concentration of isolated young and aged adult hearts from Sprague-Dawley and Fischer 344 rats. Values are means ± SE of 10 to 11 hearts before (PreISO) and during the 1 min administration of 10^{-8} M ISO. * denotes significance from corresponding young values.

A similar pattern of concentration differences was observed in epicardial surface transudates obtained from Sprague-Dawley rat hearts (Fig. 4). When the perfused heart is inverted and the pulmonary effluent is directed away from the heart, droplets of fluid accumulate on the myocardial surface. These droplets, or transudates, are thought to be representative of the interstitial fluid bathing the cardiomyocytes of the heart (Fenton and Dobson 1987; Fenton, et al., 1990). Transudate samples of approximately 5 microliters volume were removed from ventricular surfaces of inverted heart preparations and the

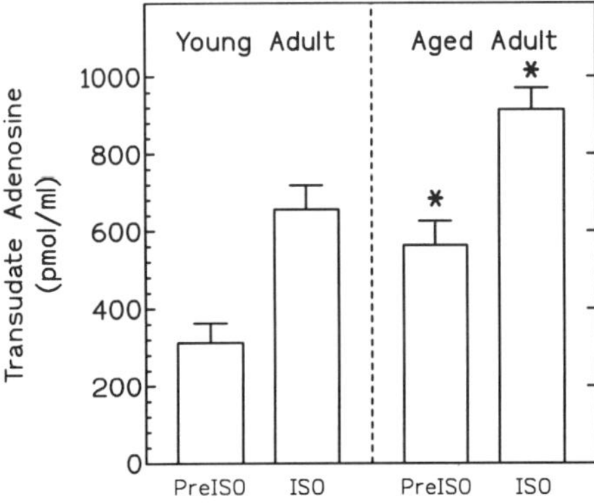

Figure 4: Effect of isoproterenol (ISO) on the concentration of adenosine in epicardial surface transudates obtained from isolated young and aged adult rat hearts. Transudates were collected immediately prior to (PreISO) or for 15 sec toward the end of a 1 min administration of 10^{-8} M ISO. Values represent mean ± SE from 12 Sprague-Dawley rat hearts. * denotes significance from the corresponding young adult value.

adenosine contained was derivatized with chloroacetaldehyde (Fenton and Dobson, 1993; Fenton and Dobson 1992; Fenton, et al., 1990; Fenton and Dobson 1987). Fluorometric analysis revealed that interstitial levels of adenosine in the aged heart, as estimated from the transudate values, were 80% higher than those found in the young adult heart. While stimulation with 10^{-8} M ISO enhanced transudate adenosine levels, these levels remained a significant 37% greater in aged compared to young adult hearts. Thus, significantly higher coronary effluent levels of adenosine in aged versus young adult hearts reflected a similar situation existing in the interstitial fluid (Dobson and Fenton, 1993). These findings have recently been confirmed by another laboratory (Headrick, 1996).

Activities of Adenosine Metabolism Enzymes are Similar Between Homogenates of Aged and Young Rat Hearts.

Studies were initiated to ascertain mechanism(s) which would manifest a level of adenosine greater in the aged than young adult heart. In that differences in myocardial oxygen consumption do not appear to be involved in the enhanced adenosine production in the aged heart (Dobson, et al., 1990), activities of enzymes participating in adenosine metabolism were determined. Young adult and aged hearts were frozen with liquid N_2-cooled aluminum clamps. To determine the activities of 5'-nucleotidase, the enzyme catalyzing the dephosphorylation of AMP to adenosine, frozen ventricles were homogenized in Tris buffer (pH 7.5). Enzyme activity was determined by measuring [^{14}C] adenosine formed from [U-^{14}C]AMP. AOPCP (α,β-methyleneadenosine diphosphate), a potent inhibitor of 5'-nucleotidase, was used to determine nonspecific phosphatase activities, which amounted to 1-6% of total activity under the conditions of the assay (Dobson and Fenton, 1993). The 5'-nucleotidase activities in homogenates obtained from young or aged hearts were not found to be significantly different. Enzyme activity with 400 µM AMP as substrate in the absence of AOPCP was 30.10 ± 1.41 nmol/min/mg protein for young hearts and 30.42 ± 0.95 nmol/min/mg protein for aged hearts. Centrifugation of the homogenates at 100,000 g yielded supernatants which contained, at most, 9% of the homogenate enzyme activity. Once again, 5'-nucleotidase activities in young and aged myocardial supernatants were not different at 2.51 ± 0.53 and 1.7 ± 0.50 nmol/min/mg homogenate protein, respectively. In addition V_{max} values of 37.6 ± 0.8 and 39.4 ± 0.9 nmol/min/mg protein for young and aged heart homogenates, respectively, were also not different. The K_m value for young hearts (53.7 ± 3.9 µM) was less than that for aged hearts (71.9 ± 5.1 µM). These data suggest that, as determined from *in vitro* analysis of heart homogenates, high levels of adenosine in aged hearts do not derive from 5'-nucleotidase activity enhanced with aging (Dobson and Fenton, 1993).

To determine the activities of adenosine deaminase metabolizing adenosine to inosine, frozen samples of young and aged myocardium were homogenized in phosphate buffer and centrifuged at 14,400 g. Adenosine deaminase activities were assessed in the whole homogenate and supernatant samples by measuring the rate of conversion of adenosine to inosine using a spectrophotometer set to an absorbance wavelength of 265 nm. Adenosine deaminase activities in whole heart homogenates were 7.93 ± 1.19 and 6.81 ± 1.02 nmol/min/mg protein for young and aged rat hearts, respectively. These values were not significantly different. The respective supernatant values were 18.62 ± 1.63 and 20.14 ± 1.76 nmol/min/mg protein. Supernatant fractions were used to determine the K_m and V_{max} values for the enzyme. Values for K_m and V_{max} were 33.4 ± 3.3 and 34.7 ± 5.8 µM, and 20.8 ± 0.6 and 20.4 ± 1.0 nmol/min/mg protein for young and aged hearts, respectively. It is apparent that, based on these data obtained from myocardial homogenates, elevated adenosine levels in aged hearts do not derive from reduced degradation of adenosine to inosine (Dobson and Fenton, 1993).

Adenosine Uptake is Reduced in the Aged Heart.

Extracellular adenosine is transported into cardiomyocytes or endothelial cells by a nucleoside-specific facilitated diffusion carrier (Plagemann, et al., 1988). Inhibition of this carrier results in an increase in extracellular levels of adenosine (Nagy, 1992). Experiments were conducted to determine if adenosine uptake transport was limited in the aged heart, resulting in increased interstitial and coronary adenosine levels. Adenosine uptake was assessed in young (3 mo) and aged (20 mo) Fischer 344 rat hearts constant-flow perfused for 20 min with [U-^{14}C]adenosine and nonradioactive adenosine in a concentration of either 100 nM or 1.08 nM. The ventricular myocardium was subsequently solubilized with NaOH. Aliquots of heart solution and coronary perfusate were counted to determine contained radioactivity. Intracellular ^{14}C-adenosine was determined after correction for extracellular ^{14}C contamination. Whether determined with low or high specific activity, uptake of adenosine by the ventricular myocardium was found to be reduced by 16% to 24% in response to aging (Fig. 5). These experiments confirm those of others utilizing a different methodology, wherein it was reported that the apparent V_{max} for the combined transport and catabolism of adenosine was significantly reduced in aged hearts (Headrick, 1996).

Carrier-mediated transport of adenosine across the sarcolemma into the cardiomyocyte is coupled to intracellular phosphorylation of adenosine by adenosine kinase (Plagemann, et al., 1988). The activity of this enzyme is the rate determining factor for transport because of the high affinity of adenosine kinase for the substrate (Plagemann and Wohlhueter, 1983). Thus, while it is possible that transporter efficiency may differ between young and aged cardiomyocytes, it is equally possible that *in vivo* adenosine kinase activity may differ between the two groups. Reduced adenosine recycling in the aged heart would result in more adenosine released provided adenosine production in the two age groups was similar.

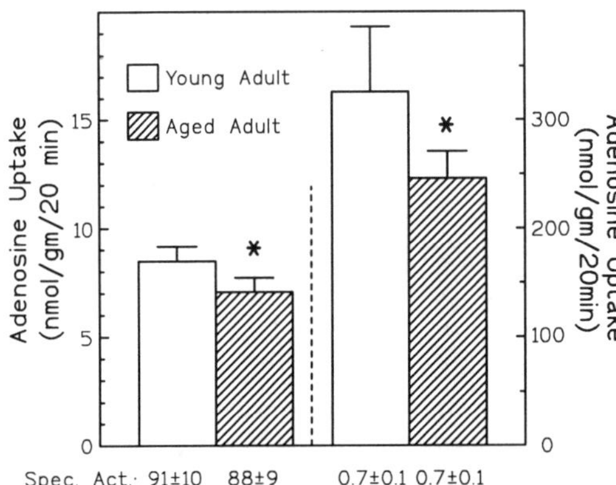

Figure 5: Effect of aging on the uptake of adenosine by the ventricular myocardium. Specific activities (Spec.Act.), given as 10^{12} dpm/mol adenosine, were varied by changing the perfusion fluid content of nonradioactive adenosine. Values are means ± SE for 5-7 hearts. * denotes significance from the appropriate young adult values. Note differences in scaling of left and right vertical axes.

Basal Adenosine Production by Young Adult and Aged Hearts are Similar, and ß-Adrenergic-stimulated Adenosine Production by Young Adult Exceeds That of Aged Hearts.

Levels of adenosine present in the interstitial fluid and coronary effluent are the net result of adenosine produced by dephosphorylation of AMP by 5'-nucleotidase and adenosine deaminated to inosine or rephosphorylated to AMP by adenosine deaminase and adenosine kinase, respectively. Inhibition of the latter two enzymatic activities should allow determination of total adenosine production by the myocardium (Kroll, et al., 1992). Hearts from young adult (3 months) and aged (20 months) Sprague-Dawley rats were perfused at a constant pressure of 65 mm Hg and paced at 360/min. After sampling of the coronary effluent, infusion of iodotubercidin (ITC; adenosine kinase inhibitor; 2 μM) and erythro-9-(2-hydroxy-3-nonyl)adenosine (EHNA; adenosine deaminase inhibitor; 50 μM) was initiated for 10 min, whereupon the coronary effluent again was sampled. Hearts were then stimulated with a 2 min administration of isoproterenol at 10^{-8} M. Coronary effluent was sampled continuously every minute for 7 minutes beginning with the appearance of ß-adrenergic-induced enhancement of contractile function. As observed previously, adenosine release prior to enzyme inhibition was found to be greater by 136% in aged compared to young adult hearts (Fig. 6). With the inhibition of adenosine deaminase and adenosine kinase, adenosine release in the young adult heart increased more than 16-fold from the pre-inhibition value of 2.24 ± 0.51 nmol/min/gdw. Because aged heart values increased only 7-fold after similar treatment with inhibitors, adenosine release by young hearts was not significantly different from that of aged hearts. This suggests that total

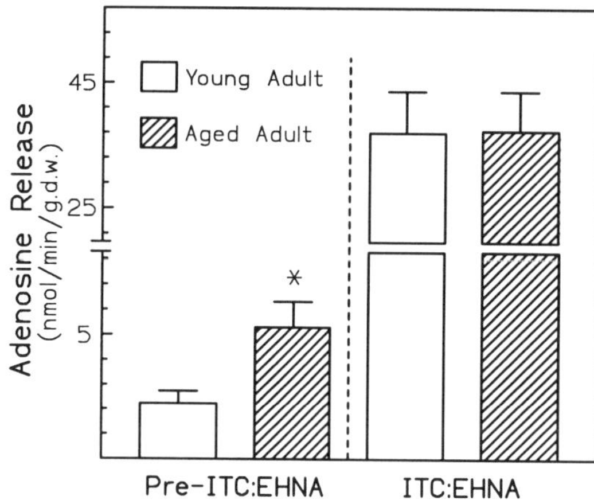

Figure 6: Effect of iodo-tubercidin (ITC) and erythro-9-(2-hydroxy-3-nonyl)adenosine (EHNA) on basal adenosine release from isolated perfused hearts of young or aged Sprague-Dawley rats. Values are means ± SE for 4 and 5 hearts, young and aged, respectively. * denotes significance from the young adult heart values.

productions of adenosine by young adult and aged hearts are not different. This conclusion is different from that of Headrick (1996) who has proposed a greater adenosine production in the aged heart as a result of enhanced cytosolic levels of adenosine 5'-monophosphate as calculated using ^{31}P-NMR.

More pronounced effects of adenosine kinase and adenosine deaminase inhibition are observed upon ß-adrenergic stimulation. With ITC and EHNA, adenosine production determined during ß-adrenergic stimulation was significantly greater in young adult versus aged hearts, increasing 21- and 8-fold over basal values of 5.7 and 7.6 nmol/min/gdw for young and aged hearts, respectively (Fig. 7). Coronary effluent inosine increased only 97% and 34% from basal levels of 17.2 and 14.5 nmol/min/gdw in young and aged hearts, respectively. This indicates that the difference of adenosine production observed with aging is not a manifestation of enhanced deamination of adenosine in the aged heart. A lower adenosine release by aged hearts does not derive from a lower contractile response to adrenergic stimulation. Although basal levels of contractile function declined 36 to 38% after treatement with ITC and EHNA (Fig. 8, left), hearts of both age groups responded similarly to the 2 min administration of ISO (Fig. 8, right).

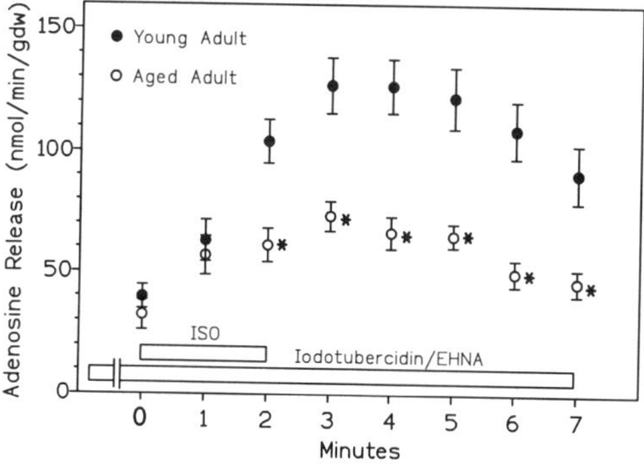

Figure 7: Total adenosine released during, and for 5 min after, a 2 min administration of isoproterenol (ISO; 10^{-8} M). Hearts were treated with iodotubercidin (2 µM) and EHNA (50 µM for 10 min prior to and through-out the coronary effluent collection period. ISO infusion was initiated immediately after the first effluent sample was collected at zero time. Values represent mean ± SE for 4 young and 5 aged hearts. * denotes significance from young values.

151

Thus, when adenosine metabolism is intact, adenosine release from aged hearts, whether basal or ß-adrenergic-induced, exceeds that of young hearts. However, with inhibition of enzymes catalyzing the removal of adenosine, basal adenosine release by the two age groups are similar, and adenosine release in response to ß-adrenergic stimulation by the young heart exceeds that of the aged heart. It is proposed that while the inherent basal production of adenosine by young and aged myocardium is not different, the recycling of adenosine by adenosine kinase is reduced in the aged heart thereby allowing basal release from aged hearts to exceed that of the young. With ß-adrenergic stimulation, even though adenosine production by the young heart exceeds that of the aged, efficient adenosine recycling in the young heart prevents the greater production to be manifest as a greater release. The net result is again a greater adenosine release from the ß-adrenergic stimulated aged heart.

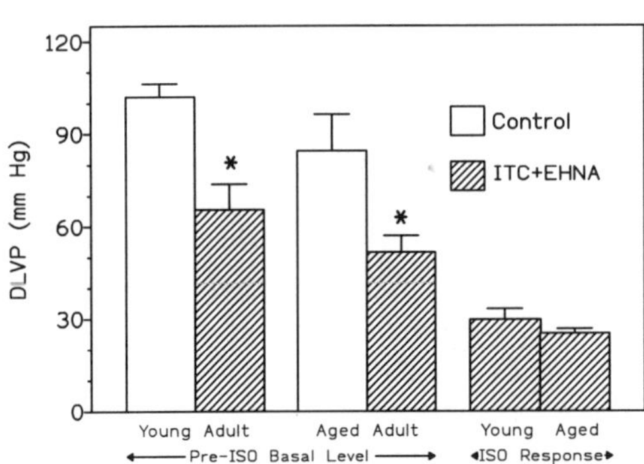

Figure 8: Effect of iodotubercidin (ITC) and EHNA on developed left ventricular pressure (DLVP) of young and aged hearts before and during isoproterenol (ISO) stimulation. ITC (2 µM) and EHNA (50 µM) were administered beginning 10 min before and during ISO administration (10^{-8} M). The ISO Response, depicted with the right columns, is the peak response to ISO occurring above the basal DLVP values, shown with the left columns. Values are mean ± SE for 4 young and 5 aged hearts. * denotes significance from the corresponding control value.

Adenosine Release in Response to Reduced Perfusion is Less in Aged than Young Adult Hearts.

The supply-to-demand ratio for oxygen is one determinant of adenosine release from the heart (Bardenheuer and Schrader, 1986). Thus, a reduction in coronary perfusion would reduce the delivery of oxygen to the myocardium, thereby lowering the ratio. An enhanced release of adenosine would result. It is interesting to inspect the relationship between the perfusion rate existing in young and aged hearts at a constant perfusion pressure of 65 mm Hg and the coronary effluent adenosine levels attained (Fig. 9). In well oxygenated hearts, adenosine levels are higher in the aged than young adult heart. However, assuming that differences in the perfusion rate at constant pressure resulted only from differences in vascular reactivity, it is suggested by these experiments that the young adult heart has a much greater ability than the aged heart to increase adenosine release when challenged with a reduction in perfusion. The ramifications of this observation are important with respect to hypoperfusion and ischemia. Attention has recently been focused on adenosine-induced cardioprotection offered the heart (Ely and Berne, 1992; Mullane and Bullough, 1995) by mechanisms activated during ischemia (Lasley and Mentzer, 1992; Ely, et al., 1985) and preconditioning (Wang, et al., 1997). Preconditioning is a phenomenon whereby brief ischemic episodes provide protection against functional and metabolic injury elicited by a subsequent ischemic episode of greater duration. Preconditioning appears to involve adenosine A_1-receptor activation (Liang, 1996; Liu, et al., 1991; Rice, et al., 1996) with signal transduction mediated, at least in part, by G_i protein activity (Lasley and Mentzer, 1993) and activation of protein kinase C (Perlini, et al., 1997). Final target mechanisms may include the short-term attenuation of the cardiotoxic actions of endogenously released ß-adrenergic neurotransmitters (Fenton, et al., 1995), persistent antiadrenergic "memory" initiated by short-term exposure to adenosine (Perlini, et al., 1997), and the activation of ATP-dependent potassium channels (Liang, 1996; Kirsch, et al., 1990). Although an in-depth discussion of mechanisms by which adenosine affords cardioprotection is not within the scope of this chapter, it is clear that the beneficial attributes of adenosine to the aged ischemic myocardium would be sacrificed should adenosine release in response to attenuated coronary flow become reduced. It is suggested that aged ischemic heart would be at greater risk than the younger heart experiencing a similar flow deprivation were it not for an enhanced responsiveness of the aged heart adenosine A_1-receptors to adenosine stimulation (Romano and Dobson 1996).

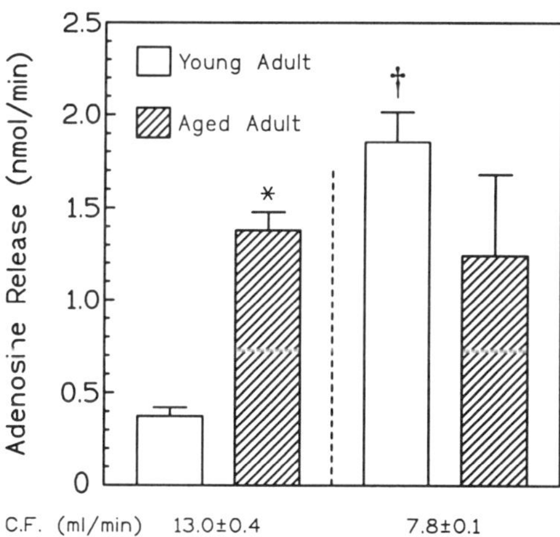

Figure 9: Relationship between coronary flow (C.F.) of isolated young and aged Sprague-Dawley rat hearts and coronary effluent adenosine release. Perfusion was conducted at a constant pressure of 65 mm Hg and hearts were paced at 360/min. Values are mean ± SE for 8 young and 6 aged hearts. *denotes significance from the corresponding young value. † denotes significance from the corresponding high C.F. value.

IV. CONCLUSIONS AND FUTURE DIRECTIONS

In the adult myocardium, contractile responsiveness to ß-adrenergic stimulation diminishes with age. Significant reduction results from the manifestation of an antiadrenergic action by greater levels of adenosine present in the interstitial fluid of aged as compared to the young myocardium (Fig. 10). The data suggest that basal levels of adenosine production (a) and adenosine deamination (c) are equivalent in young and aged hearts. These data also suggest a restricted uptake which may enhance the accumulation of extracellular adenosine (d). It could be hypothesized that rephosphorylation of adenosine in the aged heart (b) is also reduced, allowing more adenosine to leave the cardiomyocyte. The net result would be a greater adenosine release from the aged heart. With ß-adrenergic stimulation, adenosine production by the young heart exceeds that of the aged heart. However, a more efficient recycling of adenosine in the young myocardium prevents an enhanced adenosine release. Thus, in the aged heart, net adenosine release is still greater than that of the young adult heart.

The studies presented herein describe the beginning steps toward understanding the impact aging imposes on adenosine production and release, and adenosine-induced modulation of heart function. Conclusions regarding the effect of aging on adenosine kinase activity are based on indirect evidence obtained from investigations of adenosine transport. Direct determination of adenosine kinase activity is required to strengthen the hypotheses. Enzymatic activities must be determined not only with homogenates, but *in vivo* as well to insure the inclusion of all cellular mechanisms which may influence the final expression of adenosine kinase action. Cellular mechanism(s) which negatively impact on the activity of this kinase in the aged heart must be identified. The consequences of aging on adenosine transporter characteristics require studies of less than 5 sec duration to eliminate the confounding influence of adenosine kinase kinetics on interpretation of results. Transporter proteins in young and aged hearts must be isolated and compared to understand how aging-induced changes in protein expression modifies functioning of the adenosine transporter. Finally, more needs to be learned regarding the reduced ability of the aged heart to enhance adenosine release during a reduction in perfusion rate, especially in the context of coronary artery disease and ischemic-induced myocardial injury. A better understanding of the effects of aging on adenosine metabolism and adenosine-induced modulation of contractile and metabolic function will aid in developing methodologies to preserve cardiac function with advancing age.

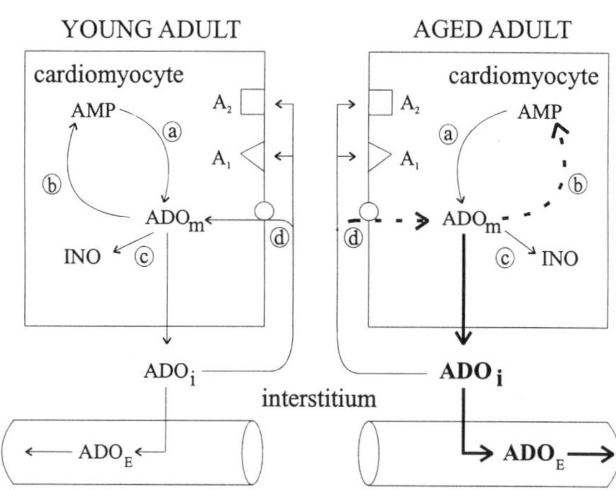

Figure 10: Schematic of adenosine metabolism in the young adult and aged adult heart. ADO_m, ADO_i, ADO_E: adenosine levels in the cell, interstitial fluid, and coronary effluent, respectively; a: production of ADO via 5'-nucleotidase; b: rephosphorylation of ADO via adenosine kinase; c: deamination of ADO via adenosine deaminase; d: ADO transport; INO: inosine; A_1, A_2: ADO receptor subtypes.

V. REFERENCES

Abrass IB, Davis JL, Scarpace PJ. Isoproterenol responsiveness and myocardial β-adrenergic receptors in young and old rats. *J Gerontol, 37:*156-160, 1982.

Achterberg W, Stroeve J, DeJong JW. Myocardial adenosine cycling rates during normoxia and under conditions of stimulated purine release. *Biochem J, 235:*13-17, 1986.

Bardenheuer H, Schrader J. Supply-to-demand ratio for oxygen determines formation of adenosine by the heart. *Am J Physiol, 250:*H173-H180, 1986.

Bardenheuer H, Whelton B, Sparks, Jr. Adenosine release by the isolated guinea pig heart in response to isoproterenol, acetylcholine, and acidosis: The minimal role of vascular endothelium. *Circ Res, 614:*594-600, 1987.

Bunger R, Sobolle S. Cytosolic adenylates and adenosine release in perfused working heart: Comparison of whole tissue with cytosolic nonaqueous fractionation analyses. *Eur J Biochem, 159:*203-213, 1986.

Cai G, Wang H-Y, Gao E, Horwitz J, Snyder DL, Pelleg A, Roberts J, Friedman E. Reduced adenosine A_1 receptor and G_α protein coupling in rat ventricular myocardium during aging. *Circ Res, 81:*1065-1071, 1997.

Conway J, Wheeler R, Sannerstedt R. Sympathetic nervous activity during exercise in relation to age. *Cardiovasc Res, 5:*577-581, 1971.

DiGennaro M, Bernabei R, Sgadari A, Carosella L, Carbonin PU. Age-related differences in isolated rat sinus node function. *Basic Res Cardiol, 82:* 530-536, 1987.

Dobson JG, Jr. Reduction by adenosine of the isoproterenol-induced increase in cyclic adenosine 3',5'-monophosphate formation and glycogen phosphorylase activity in rat heart muscle. *Circ Res, 43:* 785-792, 1978.

Dobson JG, Jr. Mechanism of adenosine inhibition of catecholamine-induced elicited responses in heart. *Circ Res, 52:* 151-160, 1983.

Dobson JG, Jr. Adenosine reduces catecholamine contractile responses in oxygenated and hypoxic atria. *Am J Physiol, 245:* H468-H474, 1983.

Dobson JG, Jr., Fenton RA. Antiadrenergic effects of adenosine in the heart. In: Berne RM, Rall TW, and Rubio R (eds.), *Regulatory Function of Adenosine.* Boston: Nijhoff, 1983, pp. 363-376.

Dobson JG, Jr., Fenton RA. Adenosine inhibition of β-adrenergic induced responses in aged hearts. *Am J Physiol, 265:* H494-H503, 1993.

Dobson JG, Jr., Fenton RA, Romano FD. The antiadrenergic actions of adenosine in the heart. In: Gerlach E and Becker BF (eds.), *Topics and Perspectives in Adenosine Research.* Berlin: Springer-Verlag, 1987, pp. 356-368.

Dobson JG, Jr., Fenton RA, Romano FD. Increased myocardial adenosine production and reduction of β-adrenergic contractile response in aged hearts. *Circ Res, 66:* 1381-1390, 1990.

Dobson JG, Jr., Fenton RA, Romano FD. Adenosine and the reduced responsiveness of the aged heart to adrenergic stimulation. In: Imai S and Nakazawa M (eds.), *Role of Adenosine and Adenine Nucleotides in Biological Systems.* Amsterdam: Elsevier, 1991, pp. 377-386.

Ely SW, Berne RM. Protective effects of adenosine in myocardial ischemia. *Circulation, 85:* 893-904, 1992.

Ely SW, Mentzer RM, Lasley RD, Lee BK, Berne RM. Functional and metabolic evidence of enhanced myocardial tolerance to ischemia and reperfusion with adenosine. *J Thorac Cardiovasc Surg, 90:* 549-556, 1985.

Fenton RA. Purines and ventricular arrhythmias. In: Abd-Elfattah AS and Wechsler AS (eds.), *Purines and Myocardial Protection*. Boston: Kluwer, 1996, pp. 383-394.

Fenton RA, Dobson JG, Jr. Adenosine and calcium alter adrenergic-induced intact heart protein phosphorylation. *Am J Physiol, 246:* H559-H565, 1984.

Fenton RA, Dobson JG, Jr. Nicotine increases heart adenosine release, oxygen consumption, and contractility. *Am J Physiol, 249:* H463-H469, 1985.

Fenton RA, Dobson JG, Jr. Measurement by fluorescence of interstitial adenosine levels in normoxic, hypoxic and ischemic perfused rat hearts. *Circ Res, 60:* 177-184, 1987.

Fenton RA, Dobson JG, Jr. Fluorometric quantitation of adenosine concentration in small samples of extracellular fluid. *Anal Biochem, 207:* 134-141, 1992.

Fenton RA, Dobson JG, Jr. Hypoxia enhances isoproterenol-induced increase in heart interstitial adenosine depressing β-adrenergic contractile responses. *Circ Res, 72:* 571-578, 1993.

Fenton RA, Galeckas KJ, Dobson JG, Jr. Endogenous adenosine reduces depression of cardiac function induced by β-adrenergic stimulation during low flow perfusion. *J Mol Cell Cardiol, 27:* 2373-2383, 1995.

Fenton RA, Moore EDW, Fay FS, Dobson JG, Jr. Adenosine reduces the Ca^{2+} transients of isoproterenol-stimulated rat ventricular myocytes. *Am J Physiol, 261:* C1107-C1114, 1991.

Fenton RA, Tsimikas S, Dobson JG, Jr . Influence of β-adrenergic stimulation and contraction frequency on heart interstitial adenosine. *Circ Res, 66:* 457-468, 1990.

Ferrara N, Bohm M, Zolk O, O'Gara P, Harding SE. The role of G_i-proteins and β-adrenoceptors in the age-related decline of contraction in guinea-pig ventricle myocytes. *J Mol Cell Cardiol, 29:* 439-448, 1997.

Gao E, Snyder DL, Johnson MD, Friedman E, Roberts J, Horwitz J. The effect of age on adenosine A_1 receptor function in the rat heart. *J Mol Cell Cardiol, 29:* 593-602, 1997.

George EE, Romano FD, Dobson JG, Jr. Adenosine and acetylcholine reduce isoproterenol-induced protein phosphorylation of rat myocytes. *J Mol Cell Cardiol, 23:* 749-764, 1991.

Gilman AG. G proteins: transducers of receptor generated signals. *Ann Rev Biochem, 56:* 615-649, 1987.

Guarnieri T, Filburn CR, Zitnik G, Roth GS, Lakatta EG. Contractile and biochemical correlates of β-adrenergic stimulation of the aged heart. *Am J Physiol, 239:* H501-H508, 1980.

Headrick JP. Impact of aging on adenosine levels, A_1/A_2 responses, arrhythmogenesis, and energy metabolism in rat heart. *Am J Physiol, 270:* H897-H906, 1996.

Headrick JP, Willis RJ. 5'-nucleotidase activity and adenosine formation in stimulated, hypoxic and underperfused rat heart. *Biochem J, 261:* 541-550, 1989.

Imai S, Chin W-P, Jin H, Nakazawa M. Production of AMP and adenosine in the interstitial fluid compartment of the isolated perfused normoxic guinea pig heart. *Pflugers Arch, 414:* 443-449, 1989.

Janczewski AM, Lakatta EG. Thapsigargin inhibits Ca^{2+} uptake, and Ca^{2+} depletes sarcoplasmic reticulum in intact cardiac myocytes. *Am J Physiol, 265:* H517-H522, 1993.

Jiang MT, Moffat MP, Narayanan N. Age-related alterations in the phosphorylation of sarcoplasmic reticulum and myofibrillar proteins and diminished contractile response to isoproterenol in intact rat ventricle. *Circ Res, 72:* 102-111, 1993.

Kirsch GE., Codma J, Birnbaumer L, Brown AM. Coupling of ATP-sensitive K^+ channels to A_1 receptors by G proteins in rat ventricular myocytes. *Am J Physiol, 259:* H820-H826, 1990.

Kroll K, Decking UKM, Dreikorn K, Schrader J. Rapid turnover of the AMP-adenosine metabolic cycle in the guinea pig heart. *Circ Res, 73:* 846-856, 1993.

Kroll K, Deussen A, Sweet IR. Comprehensive model of transport and metabolism of adenosine and S-adenosylhomocysteine in the guinea pig heart. *Circ Res, 71:* 590-604, 1992.

Kronenberg RS, Drage CW. Attenuation of the ventilatory and heart rate responses to hypoxia and hypercapnia with aging in normal men. *J Clin Invest, 52:* 1812-1819, 1973.

Kubalak SW, Newman WH, Webb JG. Differential effect of pertussis toxin on adenosine and muscarinic inhibition of cyclic AMP accumulation in canine ventricular myocytes. *J Mol Cell Cardiol, 23:* 199-205, 1991.

Lakatta EG, Gerstenblith G, Angell CS, Shock NW, Weisfeldt ML. Diminished inotropic response of aged myocardium to catecholamines. *Circ Res, 36:* 262-269, 1975.

Lakatta EG, Yin FCP. Myocardial aging: Functional alterations and related cellular mechanisms. *Am J Physiol, 242:* H927-H941, 1982.

LaMonica DA, Frohloff N, Dobson JG, Jr. Adenosine inhibiton of catecholamine-stimulated cardiac membrane adenylate cyclase. *Am J Physiol, 248:* H737-H744, 1985.

Lasley RD, Mentzer RM. Adenosine improves recovery of postischemic myocardial function via an adenosine A_1 receptor mechanism. *Am J Physiol, 32:* H1460-H1465, 1992.

Lasley RD, Mentzer RMJ. Pertussis toxin blocks adenosine A_1 receptor mediated protection of the ischemic rat heart. *J Mol Cell Cardiol, 25:* 815-821, 1993.

Liang BT. Direct preconditioning of cardiac ventricular myocytes via adenosine A_1 receptor and K_{ATP} channel. *Am J Physiol, 271:* H1769-H1777, 1996.

Liu GS, Thornton J, Van Winkle DM, Stanley AWH, Olsson RA, Downey JM. Protection against infarction afforded by preconditioning is mediated by A_1 adenosine receptors in rabbit heart. *Circulation, 84:* 350-356, 1991.

Lloyd HGE, Schrader J. Adenosine metabolism in the guinea pig heart: the role of cytosolic S-adenosyl-L-homocysteine hydrolase, 5'-nucleotidase and adenosine kinase. *Eur Heart J, 14 (Suppl.I):* 27-33, 1993.

Mentzer RMJ, Ely SW, Lasley RD, Berne RM. The acute effects of AICAR on purine nucleotide metabolism and postischemic cardiac function. *J Thorac Cardiovasc Surg, 95:* 286-293, 1988.

Mullane K, Bullough D. Harnessing an endogenous cardioprotective mechanism: Cellular sources and sites of action of adenosine. *J Mol Cell Cardiol, 27:* 1041-1054, 1995.

Nagy LE. Ethanol metabolism and inhibition of nucleoside uptake lead to increased extracellular adenosine in hepatocytes. *Am J Physiol, 262:* C1175-C1180, 1992.

Narayanan N, Tucker L. Autonomic interactions in the raging heart: Age-associated decrease in muscarinic cholinergic receptor mediated inhibition of β-adrenergic activation of adenylate cyclase. *Mech Age Dev, 34:* 249-259, 1986.

O'Connor SW, Scarpace PJ, Abrass IB. Age-associated decrease in the catalytic unit activity of rat myocardial adenylate cyclase. *Mech Age Dev, 21:* 357-363, 1983.

Perlini S, Khoury E, Chung ES, Fenton RA, Dobson JG, Jr., Meyer TE. Adenosine mediates sustained antiadrenergic depression via activation of protein kinase C in the rat heart. *Circulation, 96:* I-449, 1997 (Abstract).

Petrofsky JS, Lind AR. Isometric strength, endurance, and the blood pressure and heart rate responses during isometric exercise in healthy men and women, with special reference to age and body fat content. *Pflugers Arch, 360:* 49-61, 1975.

Plagemann PGW, Wohlhueter RM. Nucleoside transport in mammalian cells and interaction with intracellular metabolism. In: Berne RM, Rall TW, and Rubio R (eds.), *Regulatory Function of Adenosine*. Boston: Martinus Nijhoff, 1983, pp. 179-201.

Plagemann PGW, Wohlhueter RM, Woffendin C. Nucleoside and nucleobase transport in animal cells. *Biochim Biophys Acta, 947:* 405-443, 1988.

Rice PJ, Armstrong SC, Ganote CE. Concentration-response relationships for adenosine agonists during preconditioning of rabbit cardiomyocytes. *J Mol Cell Cardiol, 28:* 1355-1365, 1996.

Robberecht P, Gillard M, Waelbroek M, Camus JC, Neef P. Alterations of rat cardiac adenylate cyclase activity with age. *Eur J Pharmacol, 126:* 91-95, 1986.

Romano FD, Dobson JG, Jr. Adenosine attenuation of isoproterenol-stimulated adenylyl cyclase activity is enhanced with aging in the adult heart. *Life Sci, 58:* 493-502, 1996.

Romano FD, Macdonald SG, Dobson JG, Jr. Adenosine receptor coupling to adenylate cyclase of rat ventricular myocyte membranes. *Am J Physiol, 257:* H1088-H1095, 1989.

Romano FD, Naimi TS, Dobson JG, Jr. Adenosine attenuation of catecholamine-enhanced contractility of rat heart in vivo. *Am J Physiol, 260:* H1635-H1639, 1991.

Sakai M, Danziger RS, Staddon JM, Lakatta EG, Hansford RG. Decrease with senescence in the norepinephrine-induced phosphorylation of myofilament proteins in isolated rat cardiac myocytes. *J Mol Cell Cardiol, 21:* 1327-1336, 1989.

Sawmiller DR, Fenton RA, Dobson JG, Jr. Myocardial adenosine A_1 and A_2 receptor activities during juvenile and adult stages of development. *Am J Physiol, 271:* H235-H243, 1996.

Sawmiller DR, Fenton RA, Dobson JG, Jr. Myocardial adenosine A_1-receptor sensitivity during juvenile and adult stages of maturation. *Am J Physiol, 274:* H627-H635, 1998.

Scarpace PJ. Decreased β-adrenergic responsiveness during senescence. *Fed Proc, 45:* 51-54, 1986.

Scarpace PJ. Forskolin activation of adenylate cyclase in rat myocardium with age: Effects of guanine nucleotide analogs. *Mech Age Dev, 52:* 169-178, 1990.

Scarpace PJ, Abrass IB. Beta-adrenergic agonist-mediated densensitization in senescent rats. *Mech Age Dev, 35:* 255-264, 1986.

Shu Y, Scarpace PJ. Forskolin binding sites and G-protein immunoreactivity in rat hearts during aging. *J Cardiovasc Pharmacol, 23:* 188-193, 1994.

Smolenski RT, Schrader J, deGroot H, Deussen A. Oxygen partial pressure and free intracellular adenosine of isolated cardiomyocytes. *Am J Physiol, 260:* C708-C714, 1991.

Suteparuk S, Nies AS, Andros E, Gerber JG. The role of adenosine in promoting cardiac β-adrenergic subsensitivity in aging humans. *J Gerontol: Series A, 5017:* B128-B134, 1995.

Tobise K, Ishikawa Y, Holmer SR, Im M-J, Newell JB, Yoshie H, Fujita M, Susannie EE, Homcy CJ. Changes in type VI adenylyl cyclase isoform expression carrelate with a decreased capacity for cAMP generation in the aging ventricle. *Circ Res, 74:* 596-603, 1994.

Vestal RE, Wood AJJ, Shand DG. Reduced β-adrenoceptor sensitivity in the elderly. *Clin Pharmacol Ther, 26:* 181-186, 1979.

Wang J, Drake L, Sajjadi F, Firestein GS, Mullane KM, Bullough DA. Dual activation of adenosine A_1 and A_2 receptors mediates preconditioning of isolated cardiac myocytes. *Eur J Pharmacol, 320:* 241-248, 1997.

Weisfeldt ML. Left Ventricular Function. In: Weisfeldt ML (ed.), *Aging*. New York: Raven Press, 1980, pp. 297-316.

MOLECULAR BIOLOGY OF P2X PURINOCEPTORS.

Brian F. King, Autonomic Neuroscience Institute, Royal Free Hospital School of Medicine, Rowland Hill Street, Hampstead, London NW3 2PF, United Kingdom.

I. SUMMARY

A major class of ATP receptor, the P2X purinoceptor, operates an ion-channel intrinsic to the quaternary structure of the receptor protein. ATP-gated ion-channels are permeable to monovalent (Na^+ and K^+) and divalent cations (Ca^{2+}), but charge selection precludes the passage of anions. The channel pore is approximately 9Å in diameter. Estimates of P_{Ca}/P_{Na} ratios vary widely (0.3 to 10) amongst known examples of recombinant and native P2X receptors, whereas the pore shows no such selectivity for monovalent cations (P_K/P_{Na} ~1). Thus, P2X ion-channels permit physiologically-significant levels of Ca^{2+}-influx into cells at voltages near the resting membrane potential. Since 1994, seven types of P2X receptor subunits ($P2X_{1-7}$) have been cloned and the pharmacology of homomeric $P2X_{1-7}$ receptors studied. However, little is known with certainty about the ways P2X subunits combine to form an ion-channel - in terms of receptor stoichiometry, the lining of the ion pore, means by which the ion-channel opens and closes, allosteric modulatory sites, the ligand docking site and reasons for selectivity of nucleotidic analogues. The situation has been compounded by the discovery of splice variants of P2X subunits, which alter the operational profile of recombinant P2X receptors. However, advances have been made in understanding these basic properties by carefully studying the operational features of cloned P2X receptors expressed as either homomeric or heteromeric assemblies of P2X subunits in mammalian and non-mammalian host cells. The predicted amino acid sequences of the seven known P2X subunits varies significantly, showing between 32% and 52% sequence identity between subunit proteins. However, the topology of $P2X_{1-7}$ subunits is unerringly alike and key domains have been identified as having a bearing on operational features of the P2X purinoceptor. The possibility exists that $P2X_{1-7}$ subunits and their splice variants can produce a large number of permutations of heteromeric P2X purinoceptors. Nonetheless, emerging evidence indicates that structural diversity between P2X subunits actually limits their potential for union and, if polymerisation does occur, a single P2X subunit tends to dominate by conferring key phenotypic properties to native P2X purinoceptors.

II. BACKGROUND

A. Heterogeneity of P2 purinoceptors

Twenty years ago, pharmacological and biochemical criteria were established to distinguish between extracellular receptors for adenosine (P1 purinoceptors) and ATP (P2 purinoceptors) in cardiovascular tissues (Burnstock, 1978). Chemical alterations to the P2 purinoceptor agonist ATP (at the adenine and ribose moieties or at the triphosphate chain itself) yielded analogues which showed preferential activity at some but not all P2 purinoceptors in whole tissues (Burnstock & Kennedy, 1985; Gordon, 1986). Thus, structure-activity relationships for ATP-related agonists revealed that members of the P2 purinoceptor class were heterogeneous. This realisation led initially to a proposal for two pharmacologically-distinct classes of ATP receptor, the P2X and P2Y subtypes, in vascular and visceral smooth muscles (Burnstock & Kennedy, 1985). This seminal paper was closely followed by a proposal for another two subtypes, the P2T subtype of blood platelets and the P2Z subtype of mast cells and lymphocytes (Gordon, 1986). Subsequently, two classes of P2 purinoceptors activated by pyrimidine nucleotides were included - a UTP-selective pyrimidinoceptor in phagocytic blood cells and walls of blood vessels (in both endothelium and smooth muscle) (Siefert & Schultz, 1989) and the UTP/ATP-activated P2U subtype in the endothelial lining of blood vessels (O'Connor *et al.*, 1991). Another major P2 purinoceptor subtype was proposed, the P2D purinoceptor in neuronal and cardiac tissues, to take into account those nucleotide receptors with a strong preference for diadenosine polyphosphates, particularly Ap_4A (Hilderman *et al.*, 1991). More thorough reviews on the heterogeneity of P2 purinoceptors are available elsewhere (Burnstock, 1991; Dubyak &El-Moatassim, 1993; Dalziel & Westfall, 1994; Harden *et al.*, 1995).

<u>B. The P2X purinoceptor subtype</u>

1. *Pharmacological properties of native P2X purinoceptors*
On pharmacological grounds, the P2X purinoceptor was originally distinguished by the agonist potency order: α,β-meATP $>$ β,γ-meATP$>>$ATP$\gg$2-MeSATP. α,β-meATP was approximately three orders of magnitude more potent than ATP, but all agonists were excitatory at visceral and vascular smooth muscles and cardiac muscle (Burnstock & Kennedy, 1985). The enantiomer, L-β,γ-meATP, was considered to be a potent and selective agonist of the P2X subtype (Hourani *et al.*, 1985) and, like all methylene phosphonate ATP derivatives, was resistant to break-down by ectonucleotidases (Welford *et al.*, 1986, 1987). A fuller survey of agonist activities at the P2X subtype (and other P2 subtypes) can be found elsewhere (Cusack, 1993). The ATP analogue ANAPP$_3$ was claimed to be a selective yet irreversible antagonist of the P2X subtype which also shows marked desensitisation following prolonged exposure to either ATP or α,β-meATP (Burnstock & Kennedy, 1985). The trypanocide, suramin, was found to be a reversible antagonist of P2X purinoceptors (Dunn & Blakeley, 1988), although it is not selective for the P2X subtype (Den Hertog *et al.*, 1989; Hoyle *et al.*, 1990; Kennedy, 1990).

2. *Pharmacological differences between native P2X and P2Y purinoceptors*
In addition to the above pharmacological features, native ATP receptors have been tentatively identified as belonging to the P2X subtype through a process of exclusion - *i.e.*, when they did not show the pharmacological phenotype of the P2Y subtype. Thus, the P2Y subtype was distinguished from the P2X subtype by an identifying agonist potency order: 2-MeSATP$>$ATP$>>\alpha,\beta$-meATP $=\beta,\gamma$-meATP. 2-MeSATP was 100-1000 fold more potent than ATP and 3-5 orders of magnitude more potent than methylene phosphonate ATP derivatives. Nonetheless, all agonists caused relaxation of visceral and vascular smooth muscle (Burnstock & Kennedy, 1985). ADP-β-F was shown to be a potent and selective agonist of the P2Y subtype, but is only moderately resistant to breakdown by ectonucleotidases (Hourani *et al.*, 1988). The P2Y subtype was weakly antagonised by ANAPP$_3$ and, depending on the tissue studied, P2Y purinoceptors show a variable degree of desensitisation to either ATP or α,β-meATP (Burnstock & Kennedy, 1985). The anthroquinone-sulphonate, Reactive blue-2 (RB2), was claimed to be a selective nonsurmountable antagonist at the P2Y subtype over a limited concentration range, but high concentrations (~100 μM) also blocked P2X purinoceptors (Burnstock & Warland, 1987; Houston *et al.*, 1987; Reilly *et al.*, 1987). Suramin was a reversible but weak antagonist at the P2Y subtype (Den Hertog *et al.*, 1989; Hoyle *et al.*, 1990; Kennedy, 1990). Reactive red 2 was

reported to antagonise P2 purinoceptors, with a 15-fold selectivity for the P2Y subtype over the P2X subtype (Bültmann & Starke, 1995).

3. *Problems in characterising native P2X purinoceptors by agonist activity*

Although distinct agonist profiles for P2X and P2Y subtypes have helped identify P2 purinoceptors present in cardiovascular tissues (and elsewhere), there have been occasions when these defining phenotypic features have been misleading. Thus, the view that the P2X subtype is selectively and potently activated by α,β-meATP , while the P2Y is selectively and potently activated by 2-MeSATP, is true only under limited circumstances. Evans and Kennedy (1994) showed that the agonist potency order differed significantly for P2X-mediated responses in dissociated smooth muscle cells of rat tail artery (ATP=2-MeSATP_α,β-meATP) and in smooth muscle rings from the same tissue (α,β-meATP >>2-MeSATP>ATP), although all three agonists showed cross-desensitisation in both preparations. These differences in agonist potency were explained as the influence of ectonucleotidases that rapidly degrade ATP and 2-MeSATP but cause little breakdown of methylene phosphonate compounds. This point was demonstrated effectively where ectonucleotidase activity in vascular tissue was blocked by FPL67156 and FPL66301 to enhance the potency of ATP and 2-MeSATP but not α,β-meATP , so changing the agonist potency order relative to α,β-meATP (Crack *et al.*, 1995; Kennedy & Leff, 1995). Of the seven P2X subunits cloned to date, only 2 subunits (P2X$_1$ and P2X$_3$) are activated by α,β-meATP and, even then, 2-MeSATP is still more potent than α,β-meATP at these homomeric P2X receptors (Humphrey *et al.*, 1998). In fact, all known P2X$_{1-7}$ subunits are activated by 2-MeSATP which is often as potent as ATP. Confusion over agonist properties and relative potencies of 2-MeSATP, ATP and α,β-meATP at P2X and P2Y subtypes led to a proposal for ionotropic P2Yα purinoceptors in central neurons (Illes & Nörenberg, 1993). In hindsight however, these central P2 purinoceptors are now known to belong to the P2X class and not the P2Y class (Illes *et al.*, 1996).

4. *Further problems in characterising P2X purinoceptors*

In addition to problems with agonist potency (see review by Humphrey *et al.*, 1995), other defining characteristics of P2X and P2Y subtypes have been called into question. PPADS was reported to be a selective antagonist of peripheral P2X purinoceptors in vascular smooth muscle (Ziganshin *et al.*, 1994). PPADS is now known to inhibit ATP-responses at the P2Y$_1$ receptor (pA$_2$=6) and is a weak antagonist at P2Y$_4$ and P2Y$_6$ receptors, while it is a slowly-reversible antagonist at P2X$_{1,2,3,5}$ subunits (IC$_{50}$,1-2μM) and a weak antagonist at P2X$_{4,6,7}$ subunits (IC$_{50}$, 50-500μM) (Boarder & Hourani, 1998; also see Table 2). RB2 was once considered to be a relatively-selective antagonist for P2Y purinoceptors (Burnstock & Warland, 1987). RB2 is now known to block native P2X purinoceptors (Humphrey *et al.*, 1995; Evans & Surprenant, 1996) and recombinant P2X$_2$ receptors (Brake *et al.*, 1994) at concentrations similar to or lower than those known to block the P2Y subtype (King *et al.*, 1997b). Suramin was thought to be a reversible antagonist at P2X and P2Y purinoceptors (for references, see above). Suramin is now known to be relatively inactive at a number of native P2X and P2Y purinoceptors (Humphrey *et al.*, 1995), recombinant P2X$_{4,6}$ subunits (Humphrey *et al.*, 1998) and P2Y$_4$ and P2Y$_6$ receptors (Boarder & Hourani, 1998). Additionally, suramin has blocking actions at sites other than ATP receptors (Voogd *et al.*, 1993).

Initially, all P2X purinoceptors were believed to desensitise rapidly in the presence of an agonist (Burnstock and Kennedy, 1985). However, it gradually became clear that operational properties (both desensitisation and rundown of ATP- and α,β-meATP -responses) of native P2X purinoceptors show considerable variability which depends as much on their site of origin as the stage of development of tissues. For example, P2X purinoceptors in DRG cells show fast desensitisation in neonatal tissues but slower rates of desensitisation in older animals (Evans & Surprenant, 1996). Recombinant $P2X_{1-7}$ subunits also show different rates of desensitisation and rundown, but this is unrelated to age or tissue of origin for any particular P2X subunit (Humphrey *et al.*, 1998).

<u>C. P2X purinoceptors as calcium channels</u>

Electrophysiological evidence for ATP-activated ion-channels was first presented 15 years ago for chick embryonic skeletal muscle and rat sensory neurons and, subsequently, were reported in cardiac muscle, visceral and vascular smooth muscles, central and peripheral neurons, sensory neurons and secretory cells (see review by Dubyak & El-Moatassim, 1993). The discovery of ATP-gated channels at other sites has grown with the general application of whole-cell patch-clamp recording techniques to ATP signalling. In many cases, ATP-activated ion-channels do not show selectivity for the supposedly P2X-selective agonist α,β-meATP . However, two pharmacological phenotypes of native P2X purinoceptor can be distinguished (Surprenant *et al.*, 1995). 2-MeSATP, ATP and α,β-meATP were equi-effective at some P2X ion-channels, particularly in visceral and vascular smooth muscles, sensory neurons and cortical neurons. A second type of P2X ion-channel was selectively activated by ATP and 2-MeSATP, but not by α,β-meATP , and found on secretory cells, cochlear hair cells, brainstem neurons, autonomic neurons and embryonic skeletal muscle. This latter type does not desensitise rapidly, if at all, to prolonged exposure to nucleotide agonists.

Regardless of pharmacological profile, most ATP-activated ion-channels are permeable to calcium ions. Estimates of P_{Ca}/P_{Na} ratios vary considerably: 10 in rat medial habenula neurons (Edwards *et al.*, 1997); 4 in PC12 cells (Nakazawa *et al.*, 1990); 3 in rabbit ear artery smooth muscle (Benham & Tsien, 1987); 0.3 in rat sensory neurons (Bean *et al.*, 1990). However, some P2X channels show negligible Ca^{2+} permeability (Edwards *et al.*, 1997). Subunit composition of native P2X purinoceptors may have considerable bearing on Ca^{2+} permeability, because homomeric assemblies of $P2X_{1-7}$ subunits do not show the same P_{Ca}/P_{Na} values (Humphrey *et al.*, 1998). About 6-15% of the total ionic current is carried by calcium at native P2X-gated ion-channels (Benham, 1989; Rogers & Dani, 1995; Edwards *et al.*, 1997). Such levels of ATP-gated Ca^{2+}-influx can achieve peak intracellular Ca^{2+} concentrations of ~500nM which is sufficient to activate contractile proteins in vascular smooth muscle (Benham, 1989). When Ca^{2+}-influx is abolished, P2X purinoceptor-mediated release of catecholamines from PC12 cells is reduced by 80% (Rhoads *et al.*, 1993).

In addition to carrying calcium inside cells, the P2X purinoceptor is involved in fast postjunctional excitation of vascular and visceral smooth muscles (see review by Burnstock, 1997). P2X purinoceptors are involved in fast synaptic transmission in central (Edwards *et al.*, 1992), brainstem (Nieber *et al.*, 1997), dorsal horn (Bardoni *et al.*, 1997) sympathetic (Evans *et al.*, 1992) and enteric neurons (Galligan & Bertrand, 1994). In all excitable tissues showing fast purinergic signalling, macroscopic currents carried by the P2X purinoceptor are reduced by extracellular calcium ions (Nakazawa and Hess, 1993; and references therein). Thus, Na^+ permeability is significantly reduced as extracellular Ca^{2+} levels are raised, and similarly reduced by other divalent cations (where $Cd^{2+}>Mn^{2+}>Mg^{2+}\gg Ca^{2+}>Ba^{2+}$). The K_d value for Ca^{2+} block of P2X purinoceptors in PC12 cells is 6 μM but is much higher (100μM) for P2X purinoceptors in sensory neurons (Nakazawa & Hess, 1993). Intracellular Ca^{2+}

and Ba^{2+} also reduced single channel currents at concentrations lower than that causing blockade by external Ca^{2+} and Ba^{2+} (Nakazawa & Hess, 1993). The external and internal blocking sites for Ca^{2+} (and Ba^{2+}) may not be in the same location. Thus, permeant Ca^{2+} may have a secondary role in limiting the depolarising action of ATP-activated P2X purinoceptors, in addition to providing trigger calcium for contraction, secretion *etc.*.

D. The P2Z purinoceptor subtype

Extracellular ATP, at near millimolar levels, can permeabilize a number of cell types, including: Ehrlich ascites tumour cells; canine erythrocytes; renal epithelia; macrophages; mast cells; fibroblasts; lymphocytes; erythroleukaemic cells; salivary acinar cells (see review by Wiley *et al.*, 1996). Cockcroft and Gomperts (1980) named the activated extracellular receptor an ATP^{4-} receptor, since solutions lacking Mg^{2+} and Ca^{2+} gave greater agonist activity with free ATP (ATP^{4-}) and lower activity when divalent cations were present to form $Mg(H_X)ATP$ and $Ca(H_X)ATP$ (where X=0,1,2). The ATP^{4-} receptor was later renamed the P2Z purinoceptor (Gordon, 1986), the Z subtype being considered a logical extension to the X and Y subtypes of P2 purinoceptors already described by Burnstock & Kennedy (1985). ATP^{4-} activation of P2Z purinoceptors, for example in human lymphocytes, opens an intrinsic ion-channel that is highly selective for Ca^{2+}, with a $P_{Ca}:P_{Na}$ ratio of the order of 30-35:1 (Bretschneider *et al.*, 1995; Wiley *et al.*, 1996). However, extracellular Na^+ appears to limit the permeability of the P2Z purinoceptors which, therefore, may be even more permeable to Ca^{2+} than is now assumed (Wiley *et al.*, 1996). The activation of P2Z purinoceptors leads to the formation of large pores in the membrane, although the means by which pores form is not yet understood. The size-pass of permeants varies with cell type; pores in macrophages are permeable to molecules £900Da, whereas pores in lymphocytes will allow the influx of molecules £400Da (Wiley *et al.*, 1996). Benzoylbenzoyl ATP (BzATP) is approximately 10-fold more potent than ATP at native P2Z purinoceptors, while antagonists include: oxidised ATP; suramin and DIDS; amiloride and its hexamethylene derivatives; KN62 and KN04 (Wiley *et al.*, 1996; Gargett & Wiley, 1997).

The P2Z subtype was considered from the outset to be distinct from P2X purinoceptors in terms of operational and pharmacological profiles (Gordon, 1986). However, the beginning of the 1990s saw a revision of P2 purinoceptor subtypes into two general classes: ligand-gated ion-channels (ionotropic ATP receptors or the P2X family, including P2X and P2Z subtypes) and G protein coupled receptors (metabotropic ATP receptors or the P2Y family, including P2Y, P2U, P2T and P2D subtypes) (Kennedy, 1990; Dubyak, 1991; Dubyak & El-Moatassim, 1993; Abbracchio & Burnstock, 1994; Fredholm *et al.*, 1994). The predictive powers of this revision were fully vindicated after the outcome of cloning experiments. Thus, the native P2Z purinoceptor is now thought to be a homomeric assembly of the cloned $P2X_7$ subunit (Surprenant *et al.*, 1996; Rassendren *et al.*, 1997b).

III. TEXT

A. Structural features of P2X subunits

1. *Subunit topology*

P2X subunits were first cloned in 1994, two encoding cDNAs being derived from rat vas deferens for $P2X_1$ (Valera *et al.*, 1994) and rat PC12 cells for $P2X_2$ (Brake *et al.*, 1994). The open reading frame for the $P2X_1$ subunit yields a protein containing 399aa (M_r 45kDa), while the P2X$_2$ subunit has 472aa (M_r 52.5kDa). The deduced amino acid sequences of these two P2X receptor proteins differ considerably (38% sequence homology). Both P2X$_1$ and P2X$_2$ receptors form functional homomeric ATP-gated ion-channels, although their respective operational profiles are markedly different (see below). Coexpression of P2X$_1$ and P2X$_2$ gives a P2X receptor population with an operational profile similar to the homomeric P2X$_1$ receptors (Lewis *et al.*, 1995). The hydropathy profile (using either Kyte-Doolittle or GES scales) for P2X$_1$ and P2X$_2$ proteins shows two regions (M1 and M2) sufficiently lipophilic to insert into and cross the cell membrane, plus another short hydrophobic region (H5 or P segment) which loops in and out of the membrane (Brake *et al.*, 1994; Valera *et al.*, 1994). All of the known P2X$_{1-7}$ proteins share this hydropathy profile, although the amino acid sequences for these proteins are only 32-52% alike (Humphrey *et al.*, 1998). Thus, the topology of P2X subunits reveals six key segments: *i*) intracellular N-terminus; *ii*) first transmembrane spanning region (M1); *iii*) large extracellular loop; *iv*) H5 or P segment, adjacent to M2; *v*) second transmembrane spanning region (M2); *vi*) intracellular C-terminus (see Table 1). This overall transmembrane topology is shared by other types of non-purinergic ion-channels, including FMRFamide-gated ion-channel subunits, inward rectifying and pH-sensitive K$^+$-channel subunits, amiloride-sensitive Na$^+$-channel subunits, and mechanosensory ion-channel subunits of the degenerin family (see North, 1996a). The recently-cloned acid-sensing ion-channel subunit (ASICs) (Waldmann *et al.*, 1997) also shows a similar topology to that of the P2X subunits. The topology of P2X receptor proteins is quite different from that of the nicotinic superfamily of intrinsic ion-channels gated by acetylcholine, 5-HT$_3$, GABA$_A$ and glycine or that of the excitatory amino acid superfamily of intrinsic ion-channels gated by AMPA, kainate and NMDA (Evans *et al.*, 1998).

2. *Extracellular loop*

Comparison of the amino acid sequences of P2X$_{1-7}$ subunits reveals 76 conserved positions - including 14 glycine residues, 10 cysteine residues and 6 lysine residues (Buell *et al.*, 1996a; North, 1996a). The cysteine residues found in the central section of the P2X subunits are thought to stabilise the extracellular loop by forming a ladder of Cys-Cys disulphide bridges (Brake *et al.*, 1994). Do cysteine residues link and hold together adjacent extracellular loops? It has been shown that solubilization of rat vas deferens membranes yields a protein containing a binding site for [^{3}H]α,β-meATP with a similar K_d value to that of the native P2X purinoceptor of this tissue (Bo *et al.*, 1992). The solubilized protein (62kDa) has a molecular weight similar to glycoslyated P2X$_1$ subunits (60kDa) (Valera *et al.*, 1995) and, so, appears to be a single P2X$_1$ subunit containing a high-affinity binding site for α,β-meATP . The formation of intermolecular Cys-Cys disulphide bridges seems unlikely. It is presumed that this would also be true for P2X$_{2-7}$ subunits.

Short segments of the extracellular loop in P2X$_1$ and P2X$_2$ subunits were initially thought to bear some similarity to the Walker type-A ATP-binding motif (Brake *et al.*, 1994; Valera *et al.*, 1994). However, doubt has been cast over the presence of such binding motifs in P2X$_{1-7}$ subunits (Buell *et al.*, 1996a). Localisation of the ATP-binding site within the extracellular loop is based on data from single-point mutations of P2X$_2$, P2X$_4$ and P2X$_6$ that change the inhibitory activity of the P2 purinoceptor antagonist PPADS on ATP-responses (Buell *et al.*, 1996a). Exchanges of large segments of the extracellular loop between human and rat P2X$_4$ receptors alter the potency of the P2 purinoceptor antagonists suramin and PPADS at chimeric (h/r)P2X$_4$ receptors (Garcia-Guzman *et al.*, 1997a). Chimeric P2X$_{1/2}$ receptors change their responsiveness to α,β-meATP in a predictable manner when extracellular loops

Table 1. Physical properties of rP2X$_{1-7}$ subunits and recombinant P2X$_{1-7}$ receptors.

	[a]P2X$_1$	[b]P2X$_2$	[c]P2X$_3$	[d]P2X$_4$	[e]P2X$_5$	[e]P2X$_6$	[f]P2X$_7$
Genbank N$^\circ$	X80477	U14414	X90651	X91200	X92069	X92070	X95882
Amino acids	399	472	397	388	455	379	595
M_r	45,000	52,557	45,000	43,500	51,000	42,500	66,500
N-terminus	Met^1-Lys^{28}	Met^1-Met^{35}	Met^1-Arg^{28}	Met^1-Arg^{33}	Met^1-Lys^{28}	Met^1-Trp^{29}	Met^1-Gly^{27}
M1 segment	Val^{29}-Val^{50}	Val^{36}-Gln^{52}	Arg^{29}-Glu^{46}	Ala^{34}-Glu^{52}	Val^{29}-Ile^{51}	Val^{30}-Ala^{52}	Thr^{28}-Leu^{51}
E/c loop	Tyr^{51}-Pro^{330}	Lys^{53}-Lys^{308}	Lys^{47}-Lys^{300}	Lys^{53}-Pro^{334}	Lys^{52}-Thr^{335}	Lys^{53}-Thr^{329}	Tyr^{52}-Gln^{332}
H5 segment	Thr^{331}-Gly^{338}	Ala^{309}-Ala^{322}	Ala^{301}-Ala^{314}	nd	nd	nd	nd
M2 segment	Ile^{339}-Pro^{358}	Phe^{324}-Leu^{352}	Phe^{317}-Leu^{345}	Thr^{335}-Met^{360}	Val^{336}-Ile^{358}	Ala^{330}-Leu^{352}	Leu^{333}-Asn^{356}
C-terminus	Pro^{359}-Ser^{399}	Thr^{353}-Leu^{472}	Asn^{346}-His^{397}	Lys^{361}-Gln^{388}	Tyr^{359}-Thr^{455}	Tyr^{353}-Ala^{379}	Thr^{357}-Tyr^{595}
[g]P_{Ca}/P_{Na}	4.8	2.2	4	4.2	nd	nd	nd
[h]**Unitary Cond.**	19pS	21pS	"flickery"	9pS	nd	nd	nd
E_{rev}	0mV	-5mV	0mV	-10mV	-1mV	0mV	-2mV
Desensitisation	very fast	very slow	very fast	slow	very slow	slow	very slow / none

References: [a]Valera *et al.*, 1994; [b]Brake *et al.*, 1994; [c]Chen *et al.*, 1995; [d]Bo *et al.*, 1995; [e]Collo *et al.*, 1996; [f]Surprenant *et al.*, 1996; [g]Evans *et al.*, 1996; [h]Evans, 1996.

are exchanged between $P2X_1$ and $P2X_2$ receptors (North, 1996b). The Hill coefficient (n_H) is 2 for agonist dose-response curves for homomeric $P2X_2$ receptors, indicating cooperativity between at least two ATP-binding sites to open the ion-channel (King *et al.*, 1997b). Values for n_H are generally >1 for other recombinant $P2X_{1,3-7}$ receptors, again suggesting the rate limiting step for channel activation is the binding of more than one agonist molecule. Nakazawa and Hess (1993) reported sub-conductance states for native P2X purinoceptors in PC12 cells, but there is no evidence as yet to link this phenomenon with occupancy of a certain number of ATP-binding sites per P2X receptor. Subconductance states have not yet been reported for homomeric P2X receptors (see Evans, 1996).

3. *Pore-lining region*

A number of the amino acids in the M2 segment of $P2X_{1-7}$ subunits are polar and it has been suggested that M2 could form an amphiphatic α-helix to line the wall of an ion-channel pore (Brake *et al.*, 1994, Buell *et al.*, 1996a; North, 1996b). Van Rhee and colleagues (1998) have put forward an alternative proposal, still involving the M2 segment, where the pore lining region has been modelled on the same lines as voltage-dependent K^+-channels. This proposal involves a β-strand and extended chain embedded in the membrane to from "pore loops" thought to be common to a number of ion-channels (MacKinnon, 1995). Voltage-dependent K^+-channels, such as the inward rectifier, have a predicted stoichiometry of 4 subunits (Kubo *et al.*, 1993). A tetrameric assembly has been tentatively proposed for $P2X_3$ receptors (Burnstock & Wood, 1996), with M2 segments lining the pore and M1 segments lying behind the M2 ring to stiffen the structure. Earlier suggestions that P2X receptors may be pentameric assemblies were based on comparisons with intrinsic ion-channels of nicotinic and $GABA_A$ receptors (which are pentamers). However, the transmembrane topology of nicotinic and $GABA_A$ subunits are quite different from that of P2X subunits (see Evans *et al.*, 1998) and any inferences on like stoichiometry are questionable.

The clearest indication that the M2 segment forms the pore lining region comes from single point mutations of residues 316-354 on the $P2X_2$ subunit (Rassendren *et al.*, 1997a). Each residue was individually substituted with cysteine which, in these constructs, had no bearing on ATP potency and efficacy. Aqueous sulphydryl reagents of various sizes (4.8-5.8Å diameter) and ionic charge (+/-) were used to crosslink with cysteine residues and physically block the P2X ion-channel (9Å diameter). This technique is referred to as SCAM (substituted cysteine accessibility method). Whole-cell currents were inhibited by positively-charged MTSEA and MTSET as well as the negatively-charged MTSES for residues Ile[328], Asn[333] and Thr[336]. Thus, these residues may be situated in a pore vestibule where the ionic charge of sulphydryl reagents is unimportant. Residues between Thr[336] and Asp[349] were only affected by MTSEA, the smallest (4.8Å) of the positively-charged sulphydryl reagents. Thus, residues 336-349 may line the pore which, for native and recombinant P2X receptors, is both size- and charge-selective. Asp[349] was only accessible to MTSEA when the P2X ion-channel was opened by ATP. This aspartic acid residue may demarcate the inner limit of an intracellular (cytoplasmic) vestibule or channel gate. A theoretical model of the P2X pore lined by four M2 segments appears elsewhere (North & Barnard, 1997). This model has been validated in part by Egan and colleagues (1998), who confirmed by SCAM the involvement of the M2 segment in the pore but have suggested that Gly[342] may be the site of the ion-channel gate. Egan *et al.* also raised doubts over the M2 segment forming either an a-helix or b-sheet.

The M1 segment of P2X subunits has also been proposed to form a β-strand that lies adjacent to the M2 β-hairpin structure (Van Rhee *et al.*, 1998). Recombinant P2X$_3$ receptors function poorly where the M1 segment has been deleted by truncating the P2X$_3$ protein at Met[75] (King *et al.*, 1997a). A splice variant of P2X$_4$ (devoid of the M1 segment) fails to respond to ATP, as do P2X$_4$ constructs missing either residues 1-94 or 1-50 (Dhulipala *et al.*, 1998). Splice variants of P2X$_2$ subunits (P2X$_{2C}$ and P2X$_{2D}$), where short sections of the M1 segment are deleted, similarly fail to respond to agonists (Simon *et al.*, 1997). Thus, the presence of the M1 segment and its position relative to the M2 segment seems critical for full agonist activity, perhaps by anchoring the extracellular loop and influencing the shape of the ligand docking pocket. Another splice variant, P2X$_{2-1}$, possessing the M1 segment and lacking most of the M2 segment occurs in cochlear and pituitary tissues but this P2X subunit has not been tested for functionality (Housley *et al.*, 1995).

4. *Intracellular termini*

The N-terminus is the shorter of the two intracellular segments (see Table 1), comprising some 27-35 amino acid residues of which only are four conserved among the P2X$_{1-7}$ subunits (see sequence alignments by Buell *et al.*, 1996a). The C-terminus is much longer and shows greater variability in its primary structure, ranging from 28-248 amino acid residues of which only two are conserved amongst the P2X$_{1-7}$ subunits. There is no obvious correlation between the relative lengths of the C-terminus and rates of desensitisation of recombinant P2X$_{1-7}$ receptors. However, it is now clear that intracellular termini do play a role in the process of ion-channel inactivation.

The C-terminus of the P2X$_{2-2}$ (or P2X$_{2B}$) splice variant lacks 69 of 118aa (Brändle *et al.*, 1997; Simon *et al.*, 1997). This splice variant shows an accelerated rate of desensitisation to ATP (t = 11s, compared to 57s in wild-type P2X$_2$ (or P2X$_{2a}$)), yet there is no in agonist potency. Another P2X$_2$ splice variant, P2X$_{2-1}$, shows a 10aa deletion in the M2 region and further 101aa deletion in the C-terminus (Housley *et al.*, 1995). Although P2X$_{2-1}$ has not been tested functionally, it has been implicated with a persistent non-desensitising current to ATP in the sensory hair cells of the cochlea from which P2X$_{2-1}$ was cloned. We have generated a chimeric P2X$_{2/3}$ subunit where the M2 segment and C-terminus of the P2X$_2$ subunit is replaced by that of the P2X$_3$ subunit. Under these conditions, the chimeric P2X$_{2/3}$ receptor is non-desensitising and remains fully open for periods of 20-30 minutes in the presence of ATP (King & Townsend-Nicholson, unpublished data).

Deletion of the residues 419-595 from the C-terminus of P2X$_7$ subunits (generating the P2X$_7$DC subunit) prevents the development of sustained inward currents seen with repetitive challenges to BzATP (Surprenant *et al.*, 1996). Also, the P2X$_7$DC receptor no longer shows the cytolytic response nor significant uptake of YO-PRO-1 that is normally associated with agonist activation of a P2Z purinoceptor. Amino acid sequences of the C-termini of rat and human P2X$_7$ subunits differ at 52 of 248 residues and, accordingly, the pattern of desensitisation and degree of YO-PRO-1 uptake are different for these two P2X$_7$ orthologues (Rassendren *et al.*, 1996b). An exchange of C-termini (cut at position 347) between human and rat orthologues confers the appropriate rate of desensitisation and uptake of YO-PRO-1 to the respective chimeric P2X$_7$ receptor (Rassendren *et al.*, 1996b).

The N-terminus also appears have an impact on desensitisation of P2X$_3$ receptors. Removal of the N-terminus and M1 segment converts the truncated P2X$_3$ receptor from a fast desensitising ion-channel (t = 1s) to a slowly desensitising receptor (t > 300s) (King *et al.*, 1997a). On a similar vein, two substitutions to the slowly-desensitising P2X$_2$ subunit - by inserting the M1 segment and 15 residues of the N-terminus plus M2 segment and 11 residues of the C-terminus from the P2X$_3$ subunit - generate a fast-desensitising chimeric P2X$_2$ receptor (Werner *et al.*, 1996). The same two substitutions, but taking the equivalent P2X$_1$ segments instead, have the same outcome on P2X$_2$ receptor desensitisation (Werner *et al.*, 1996). Conversely, substitutions made to the P2X$_1$ subunit - using either one or both of the above segments from P2X$_2$ subunits -generates a non-desensitising receptor (Werner *et al.*, 1996). Further substitutions revealed that the cytoplasmic portions (pre-M1 and post-M2) of the replaced segments were critical for regulating desensitisation, while substituted M1 and M2 portions (particularly when taken from P2X$_3$) appeared to have an impact of the rate of recovery from desensitisation.

5. *Calcium influx*

The ion-channels of homomeric P2X$_{1-7}$ receptors are permeable to calcium. This is particularly evident where P2X receptors are expressed in the *Xenopus* oocyte which possesses an endogenous calcium-activated chloride channel (Barish, 1983). Accordingly, Ca^{2+}-entry into oocytes via the P2X ion-channel causes the secondary activation of oscillatory inward currents (I$_{Cl,Ca}$). Therefore, P2X receptors expressed in oocytes are normally studied under calcium-free conditions - substituting Ca^{2+} with isomolar Ba^{2+} which fails to activate chloride channels. The P$_{Ca}$/P$_{Na}$ ratio is between 3.9-4.8 for P2X$_1$ receptors (Valera *et al.*, 1994; Evans *et al.*, 1996), 2.2 for P2X$_2$ (Evans *et al.*, 1996), approximately 4 for P2X$_3$ (Lewis *et al.*, 1995) and 4.2 for P2X$_4$ (Buell *et al.*, 1996b). It is notable that P2X$_2$ receptors show the lowest P$_{Ca}$/P$_{Na}$ ratio, least evidence for Ca^{2+}-activation of endogenous I$_{Cl,Ca}$ when expressed in *Xenopus* oocytes (and extracellular Ca^{2+} is present) and, subsequently, the slowest rate of desensitisation. King and colleagues (1997a) showed that complete withdrawal of extracellular Ca^{2+} and buffering of residual/bound Ca^{2+} with EGTA converts the fast-desensitising P2X$_3$ receptor into a non-desensitising ion-channel. Extracellular cyclosporin (an inhibitor of calcineurin) and intracellular injection of calcineurin auto-inhibitory peptide (CaN A457-481) also slowed down the rate of desensitisation of P2X$_3$ receptor (King *et al.*, 1997a). These results suggest that Ca^{2+}-entry into the cell during P2X$_3$ activation can trigger a calcineurin-mediated dephosphorylation of the P2X$_3$ receptor to initiate channel inactivation. Truncating P2X$_3$ at Met[75] yielded a poorly functioning P2X receptor where rapid desensitisation was abolished (King *et al.*, 1997a); the serine-rich N-terminus of the P2X$_3$ subunit may be the locus for the calcineurin-mediated dephosphorylation event. The direct involvement of intracellular Ca^{2+} in the desensitisation of other P2X subtypes has not been addressed. Cook & McCleskey (1997) have shown that transiently-elevated extracellular Ca^{2+} accelerates the rate of recovery from desensitisation for P2X$_3$-like receptors on sensory neurons, without affecting the desensitisation process during subsequent agonist activation. This phenomenon, together with the above data, indicates that the Ca^{2+} level on either side of the channel pore has a role to play in operational features of recombinant P2X receptors. About 8% of the total current at P2X$_4$ receptors is carried by Ca^{2+} ions (Garcia-Guzman *et al.*, 1997a). This value is in agreement with estimations of Ca^{2+} influx at native P2X purinoceptors in SCG neurons (Rogers & Dani, 1995) from which P2X$_4$ was cloned (Buell *et al.*, 1996b).

1. *ATP agonism*

A summary of the pharmacological activity of key agonists at homomeric P2X$_{1-7}$ receptors is given in Table 2. P2X$_{1-7}$ receptors are foremost ATP receptors and, by far, ATP is the most potent of the naturally-occurring nucleotides. The fast desensitising P2X$_1$ and P2X$_3$ receptors are the most sensitive to extracellular ATP, which is active over a working range of 0.01-30µM (threshold-to-maximum) with EC$_{50}$ values in the region of 0.5-1.5µM. This sensitivity to ATP can be matched at the P2X$_2$ receptor if extracellular pH (pH$_e$) is lowered to pH5.5. At normal pH$_e$ levels (7.4±0.1) however, P2X$_{2,4,5,6}$ receptors are approximately 10-fold less sensitive to extracellular ATP which is active over a working range of 0.3-300µM with EC$_{50}$ values of the region of 5-15µM. The P2X$_7$ receptor stands out as being the least sensitive to ATP, which is active over a range of 3-1000µM with an EC$_{50}$ value in the region of 100µM. ATP sensitivity of the P2X$_7$ receptor is influenced by experimental design. Petrou and colleagues (1997) showed under certain conditions (Ca^{2+}$_o$, 0.1mM; Mg^{2+}$_o$, 0mM; BAPTA$_i$, ~1mM) that the apparent EC$_{50}$ for ATP is 10µM - equivalent to 3.3µM for free ATP (ATP^{4-}). It is difficult, however, to envisage such Ca^{2+}- and Mg^{2+}-buffering conditions in the vicinity of the P2Z purinoceptor under normal physiological circumstances.

2. *Other nucleoside triphosphates*

Whereas P2X$_{1-7}$ receptors are activated by ATP, other nucleoside triphosphates appear to be relatively inert. UTP has weak agonist activity at the α,β-meATP -activated P2X$_1$ and P2X$_3$ receptors, but requires concentrations greater than 100µM for significant agonist activity. [^{3}H]α,β-meATP binding at smooth muscle cells is displaced by nucleoside triphosphates with the following potency order (IC$_{50}$ values, µM): ATP(8)>UTP(20)>CTP(54)>TTP(63)>ITP(68)>GTP(101) (Bo & Burnstock, 1993). However, binding displacement data bear little relationship to the activity of these compounds at either native P2X purinoceptors in smooth muscle or homomeric P2X$_1$ receptors. UTP activates P2X$_3$ receptors with an EC$_{50}$ value of ~245µM (Chen *et al.*, 1995) and shows a similar potency at the native P2X purinoceptor in neonatal rat DRG (Rae *et al.*, 1998), from which the P2X$_3$ subunit was cloned (Chen *et al.*, 1995; Lewis *et al.*, 1995). Another pyrimidine nucleotide, CTP, is more potent at human P2X$_3$ receptors (EC$_{50}$, 18µM) (Garcia-Guzman *et al.*, 1997b) than rat P2X$_3$ receptors (EC$_{50}$, 63µM) (Chen *et al.*, 1995). Rat and human P2X$_3$ receptors show 93% sequence homology, differing by 14 of 217 residues in the extracellular loop. CTP activates native P2X purinoceptors in sensory neurons of rat nodose ganglia (Krishtal *et al.*, 1988).

3. *ADP activation*

While ADP is found to be weakly active at recombinant P2X$_{1-7}$ receptors, such data should be treated with caution. ADP is approximately 20-fold less potent than ATP at P2X$_1$ receptors and shows a lower efficacy (lower maximal activity) than ATP. Similar results are seen at P2X$_{2-7}$ receptors where ADP is 30-100 fold less potent than ATP. Commercial ADP stocks are heavily contaminated with ATP (4-10%) and, accordingly, the observed ADP activity may be due to the presence of contaminating ATP. ADP solutions can be purified by using hexokinase which, in the presence of glucose, converts ATP into ADP. This purification process is now used to study recombinant P2Y

Table2. Pharmacological properties of recombinant $P2X_{1-7}$ receptors.

	[a]$P2X_1$	[c]$P2X_2$	[d]$P2X_3$	[g]$P2X_4$	[k]$P2X_5$	[k]$P2X_6$	[n]$P2X_7$
Agonists (EC_{50})							
2-MeSATP	0.4µM	7.1µM	0.2µM	74µM	20µM	9µM	300µM
ATPγS	3.1µM	7.4µM	9µM	23µM	9.3µM	16µM	ia
ATP	0.5µM	4.6µM	1.5µM	10µM	15.4µM	12µM	100µM
ADP	10.5µM	ia	ia	ia	ia	ia	ia
α,β-meATP	1.5µM	ia	3µM	ia	[m]ia	ia	ia
L-β,γ-meATP	[b]2µM	ia	ia	nd	[m]ia	nd	ia
BzATP	[b]0.8µM	[b]25µM	[e]10µM	[e]25µM	[e]40µM	[e]25µM	3.7µM
CTP	nd	ia	[f]18µM	[h]250µM	[m]ia	nd	nd
UTP	>100µM	ia	245µM	ia	[m]ia	nd	ia
Antagonists (IC_{50})							
PPADS	[b]1µM	1.6µM	[f]1.7µM	>500µM	1.3µM	>500	[p]51µM
Suramin	[b]1µM	10.4µM	[f]15µM	>500µM	4µM	>500	[p]78µM
Reactive blue-2	nd	0.36µM	nd	[j]128µM	nd	nd	nd

(ia = inactive; nd = not determined)

References: [a]Valera *et al.*, 1994; [b]Evans *et al.*, 1995; [c]King *et al.*, 1997b; [d]Chen *et al.*, 1995; [e]Evans *et al.*, 1998; [f]Garcia-Guzman *et al.*, 1997b; [g]Bo *et al.*, 1995; [h]Soto *et al.*, 1996; [j]Garcia-Guzman *et al.*, 1997a; [k]Collo *et al.*, 1996; [m]Garcia-Guzman *et al.*, 1996; [n]Surprenant *et al.*, 1996; [p]Rassendren *et al.*, 1997b.

receptors that are activated selectively by nucleoside diphosphates and not by the corresponding triphosphate (Lazarowski *et al.*, 1997). Purified ADP has not been tested on any of the P2X$_{1-7}$ receptors. However, inorganic diphosphates and triphosphates displaced [^{3}H]-α,β-meATP binding from smooth muscle cells with same avidity as ATP, whereas inorganic monophosphates, adenosine, adenine and xanthine moieties did not (Bo & Burnstock, 1993). Potentially, purified nucleoside diphosphates could be weak agonists of P2X$_{1-7}$ receptors. AMP and adenosine do not activate P2X$_{1-7}$ receptors. Thus, P2X$_{1-7}$ subunits are all relatively selective for the purine triphosphate, ATP.

4. α,β-meATP *agonism*

The prototypic P2X agonist, α,β-meATP , is active at the fast-desensitising P2X$_1$ and P2X$_3$ receptors - over the same concentration range as ATP and with comparable potency and efficacy (see Table 2). The reputedly P2X-selective L-β,γ-meATP is active only at P2X$_1$ receptor with the same potency and efficacy as α,β-meATP and ATP. L-β,γ-meATP is virtually inactive at P2X$_3$ receptors (Chen *et al.*, 1995) and at P2X$_3$-like receptors in rat DRG cells (Rae *et al.*, 1998). Thus, L-β,γ-meATP could be used to distinguish between P2X$_1$ and P2X$_3$ subunits which otherwise possess identical operational profiles. Trezise and colleagues (1995) have already reported that L-β,γ-meATP can discriminate between smooth muscle P2X purinoceptors (which are P2X$_1$-like) and vagal nerve P2X purinoceptors (which may be P2X$_{2/3}$ heteromers). At high concentrations (>100μM), α,β-meATP and L-β,γ-meATP are weak partial agonists at P2X$_2$ receptors (Evans *et al.*, 1996). Similarly, α,β-meATP (100μM) is a weak partial agonist at P2X$_4$ receptors (Bo *et al.*, 1995; Buell *et al.*, 1996b; Soto *et al.*, 1996a) and can cause a partial suppression (<50%) of submaximal ATP-responses (Bo *et al.*, 1995). [^{35}S]-ATPgS binding is displaced by α,β-meATP at P2X$_1$ (pIC$_{50}$, 7.2) and P2X$_2$ receptors (pIC$_{50}$, 7.1) (Michel *et al.*, 1996) as well as P2X$_4$ receptors (pIC$_{50}$, 7.8) (Michel *et al.*, 1997); ATPγS is a full and potent agonist at all three recombinant P2X receptors (see Table 2). [^{35}S]-ATPγS binding is also displaced by L-β,γ-meATP at: P2X$_1$, 5.4; P2X$_2$, 5.2; P2X$_4$, 5.7 (subtype, pIC$_{50}$) (Michel *et al.*, 1996, 1997). Thus, P2X$_{1,2,4}$ subunits recognise and bind α,β-meATP and L-β,γ-meATP but, paradoxically, these two compounds only are agonists at P2X$_1$ subunits.

5. *2-MeSATP agonism*

The supposedly P2Y-selective agonist, 2-MeSATP, is a full agonist at P2X$_{1,2,3,5,6}$ receptors and partial agonist at P2X$_{4,7}$ receptors. 2-MeSATP is as potent as ATP where it is a full agonist (particularly at P2X$_1$ and P2X$_3$), and less potent than ATP (by 3-7 fold) where it is a partial agonist. 2-MeSATP is either 10-fold more potent than ATP (Chen *et al.*, 1995) or seemingly equipotent (Lewis *et al.*, 1995) at rP2X$_3$ receptors, and 2.5-fold more potent than ATP at hP2X$_3$ receptors (Garcia-Guzman *et al.*, 1997b). Similar variations in relative potency of these two agonists have been reported for native P2X purinoceptors in sensory neurons. 2-MeSATP is 10-fold more potent than ATP at rat nodose neurons (Khakh *et al.*, 1995a) or seemingly equipotent at rat vagal nerve (Humphrey *et al.*, 1995) and rat DRG cells (Robertson *et al.*, 1996). 2-MeSATP is more potent than α,β-meATP at P2X$_1$ and P2X$_3$ receptors, thus matching the potency order of these agonists at native P2X purinoceptors in dissociated smooth muscle cells (P2X$_1$-like) and sensory neurons (P2X$_3$-like) (Evans & Kennedy, 1994; Khakh *et al.*, 1995a,b; Robertson *et al.*, 1996).

6. *BzATP agonism*

In the past, benzoylbenzoyl ATP (BzATP) has been regarded as a selective agonist of the native P2Z purinoceptor. This premise has been laid bare by the identification of the homomeric $P2X_7$ receptor as the P2Z purinoceptor (Surprenant *et al.*, 1996) and subsequent realisation that all $P2X_{1-7}$ subunits are activated to some extent by BzATP. From available evidence, BzATP appears to be at least as potent at $P2X_1$ as $P2X_7$ receptors (see Table 2). EC_{50} values for BzATP of either $0.8\mu M$ or $3\mu M$ can be determined for $hP2X_1$ receptors expressed in either HEK293 cells or *Xenopus* oocytes, respectively (Evans *et al.*, 1995). The EC_{50} value for BzATP is $3.7\mu M$ at $rP2X_7$ (Surprenant *et al.*, 1996) and $45\mu M$ for $hP2X_7$ (Rassendren *et al.*, 1997b) when expressed in HEK293 cells. However, BzATP is a full agonist at $P2X_7$ receptors and only a partial agonist (~50% maximal activity of ATP) at $P2X_1$ receptors. $P2X_1$ has been associated with apoptosis in murine thymocytes (Chvatchko *et al.*, 1996) as has $P2X_7$ receptors elsewhere (Surprenant *et al.*, 1996), although it has been suggested that these P2X receptors might cause necrosis rather than apoptosis (Chow *et al.*, 1997). The intimate relationship between BzATP-stimulated cell death and BzATP activation of native P2Z purinoceptors as well as $P2X_1$ and $P2X_7$ subtypes suggests that all reported examples of native P2Z purinoceptors may not necessarily correspond to the $P2X_7$ receptor subtype alone. BzATP also activates $P2X_{2-6}$ receptors (EC_{50}, $10-40\mu M$).

7. *Extracellular pH and Zn^{2+} on agonist activity*

Extracellular pH has a strong bearing on ATP-sensitivity at recombinant and native P2X purinoceptors. This phenomenon was reported first for homomeric $P2X_2$ receptors, where lower extracellular pH (7.4-6.5) displaced the ATP dose/ response curve leftwards without altering maximal agonist activity (King *et al.*, 1996). Raising extracellular pH (7.4-8.0) had the opposite effect and reduced ATP-sensitivity. ATP-sensitivity is increased 10-fold at pH5.5, and reduced 5-fold at pH8.0 (King *et al.*, 1997b). This phenomenon was also shown to occur at native P2X purinoceptors in rat nodose ganglia (Li *et al.*, 1996a,b). P2X purinoceptors in nodose ganglia are reported to be heteromeric $P2X_{2/3}$ receptors (Lewis *et al.*, 1995; but see Thomas *et al.*, 1998) and subsequently, acid sensitisation has been observed where $P2X_{2/3}$ receptors were coexpressed in HEK293 (Stoop *et al.*, 1997). ATP-sensitivity is reduced under acidic conditions at $P2X_3$ subunits, as well as at $P2X_{1,4,7}$ subunits (Stoop *et al.*, 1997; Virginio *et al.*, 1997). Acidic conditions reduced neuronal apoptosis (Xu *et al.*, 1998), although the involvement of either $P2X_1$ or $P2X_7$ receptors was not established. Thus, proton sensitisation of native P2X receptors implicates the presence of $P2X_2$ subunits while proton desensitisation points to their absence. Small changes in pH (±0.03 pH-units) were shown to have an effect on ATP-sensitivity at $P2X_2$ receptors (Wildman *et al.*, 1997). Mild acidification (pH6.95) significantly enhances the bone-resorption capacity of ATP-activated osteoclasts (Morrison *et al.*, 1998).

Extracellular Zn^{2+} potentiates ATP activity at native P2X receptors in sympathetic and sensory ganglia (Cloues *et al.*, 1993; Li *et al.*, 1993). Zn^{2+} increases the opening frequency of single ATP-gated ion-channels and increases the burst duration of the open conductance state at neuronal P2X purinoceptors (Cloues 1995; Wright & Li, 1995). Extracellular Zn^{2+} also has marked effects at homomeric P2X receptors. Zn^{2+}-potentiation of ATP-responses occurs at $P2X_4$ receptors (Garcia-Guzman *et al.*, 1997a) and is even more pronounced at $P2X_2$ receptors (Wildman *et al.*, 1998a). This potentiating effect is time-dependent at $P2X_2$ receptors and time-independent at $P2X_4$ receptors (Wildman *et al.*, 1998b). Zn^{2+} has a mild excitatory action on ATP-responses at $P2X_3$ receptors but an inhibitory action on ATP-responses at $P2X_1$ receptors (Wildman *et al.*, 1998b). The potentiating actions of Zn^{2+} and H^+ were shown to be independent of each other at P2X$_2$ receptors (Wildman *et al.*, 1998a) and P2X$_4$ receptors (Wildman & King; unpublished observations). Extracellular Cu^{2+} appears to mimic Zn^{2+} by potentiating ATP-responses at neuronal P2X purinoceptors (Li *et al.*, 1996c). Little is known about the modulatory role of Zn^{2+}, Cu^{2+} and extracellular pH at non-neuronal P2X purinoceptors in cardiovascular tissues.

C. Antagonist activity at P2X$_{1-7}$ receptors

1. *PPADS blockade*

Pyridoxal-a^5-phosphate -6-azophenyl-2′,4′-disulphonic acid, PPADS, inhibits ATP-responses at recombinant rP2X$_{1,2,3,5,7}$ receptors but is relatively inactive at rP2X$_{4,6}$ receptors (Table 2). PPADS blockade is generally nonsurmountable and, with the exception of P2X$_3$ receptors, difficult to reverse on washout. It has been proposed that the aldehyde group on PPADS reacts irreversibly with lysine residues in some P2X subunits to form a Schiff's base (Buell *et al.*, 1996a,b). Scrutiny of the amino acid sequences of P2X$_{1-6}$ subunits drew attention to a lysine residue on P2X$_{1,2,5}$ subunits, at a site equivalent to Glu^{249} on P2X$_4$ and Leu^{251} on P2X$_6$ subunits. A threonine residue occurs at the equivalent position in P2X$_3$ receptors. Substitution of glutamic acid with lysine (E249K) on the P2X$_4$ subunit enhanced PPADS potency by 30-fold (Buell *et al.*, 1996b) and a similar substitution of leucine with lysine (L251K) at P2X$_6$ receptors also enhanced PPADS potency (Collo *et al.*, 1996). A reciprocal mutation of the P2X$_2$ subunit (lysine with glutamic acid, K246E) did not lower potency although PPADS blockade became rapidly reversible (Buell *et al.*, 1996b). These results support, but do not confirm, the notion of Schiff base formation.

Although missing Lys^{249}, the human orthologue of the rP2X$_4$ receptor still shows a greater sensitivity to PPADS (IC$_{50}$, 28μM) and other P2 purinoceptor antagonists (Garcia-Guzman *et al.*, 1997a). Antagonist activity at a series of chimeric human/rat P2X$_4$ receptors indicated that residues 81-183 on the extracellular loop were crucial for the enhanced PPADS-sensitivity at the human P2X$_4$ receptor. The human and rat orthologues differ by 22 residues over this PPADS-sensitive domain but only one lysine residue (Lys^{127}) is present and is unique to the human sequence. Substitution of the residue (N127K) in the rat sequence failed to enhance PPADS sensitivity. This finding, coupled with the absence of Lys^{249} in the human orthologue, indicates the complex nature of the PPADS binding pocket.

PPADS blockade of P2X subunits appears to be use-dependent, where long application times significantly increase the potency of this P2 receptor antagonist. For example, PPADS (10μM, 30-60min) inhibited ATP-responses by 60% at the supposedly PPADS-insensitive P2X$_6$ receptor (Collo *et al.*, 1996). Similarly, the potency of PPADS and isoPPADS are increased by greater than 10-fold at P2X$_1$ and P2X$_3$ receptors when exposed to these antagonists for long periods (3h) (Jacobson *et al.*, 1998). PPADS and P5P derivatives, where the aldehyde is replaced by a cyclic phosphate moiety, show a reduction in potency but gain by showing selectivity for P2X over P2Y subtypes and becoming rapidly reversible antagonists (Jacobson *et al.*, 1998).

2. *Suramin blockade*

Like PPADS, suramin also inhibits ATP-responses at homomeric rP2X$_{1,2,3,5,7}$ receptors and is relatively inactive at rP2X$_{4,6}$ receptors (Table 2). Unlike PPADS however, suramin blockade is reversible with washout and is not affected by inserting a strategic lysine residue into rP2X$_4$ and rP2X$_6$ subunits

(Buell *et al.*, 1996b; Collo *et al.*, 1996). In general, suramin is less potent than PPADS at antagonising $P2X_{1-7}$ receptors (Table 2). The exception is the $P2X_2$ receptor, where suramin-blockade is dependent on extracellular pH while PPADS is not. Under acidic conditions, suramin potently inhibits ATP-responses at $P2X_2$ receptors (IC_{50}, 78nM at pH5.5 *vs* 10μM at pH7.4) (King *et al.*, 1997b). Extracellular Zn^{2+} mimics H^+ to enhance the blocking activity of suramin at $P2X_2$ receptors (Wildman *et al.*, 1998). Suramin blocks $P2X_1$ at low concentrations (IC_{50}, 1μM), but is not particularly selective for $P2X_1$ over $P2X_{2,3,5,7}$ subunits. Suramin (1μM) also blocks some recombinant P2Y receptors (Boarder & Hourani, 1998).

3. *Other antagonists*

Preliminary experiments have shown that di-inosine pentaphosphate (Ip_5I) is a potent antagonist of $P2X_1$ subtypes (pA_2, 8.1) (King & Liu; unpublished data). The anthroquinone-sulphonate, Reactive blue-2 (RB2), is potent inhibitor of $P2X_2$ receptors (IC_{50}, 0.36μM) (King *et al.*, 1997b). Trinitrophenyl-ATP (TNP-ATP) also antagonises $P2X_2$ receptors (IC_{50}, 1μM) (King *et al.*, 1997b), but is 330-fold more potent at the $P2X_3$ subtype (IC_{50}, 3nM) (Thomas *et al.*, 1998). There are no potent antagonists of $P2X_4$ and $P2X_6$ receptors. Technical problems with expressing recombinant $P2X_5$ receptors have hindered antagonist studies at this subtype. Calmidazolium blocks $P2X_7$ receptors (IC_{50}, ~10nM) but only antagonises ATP-responses and not BzATP-responses (Virginio *et al.*, 1997).

D. P2X subunit distribution

The distribution of $P2X_{1-7}$ receptor proteins has been assessed primarily by Northern analysis to detect tissue contents of P2X transcripts, and by hybridisation histochemistry to reveal the cellular localisation of these transcripts. Two commonly-used radioligands for P2X receptors, $[^3H]\alpha,\beta$-meATP and $[^{35}S]$-ATPγS, are not selective for particular P2X subunits and, at best, can only suggest the presence of P2X purinoceptors. $[^3H]\alpha,\beta$-meATP also appears to bind to 5-nucleotidase on endothelial cells and, accordingly, may give false-positive results for P2X purinoceptors in blood vessels (Michel *et al.*, 1995). There is a small number of papers describing the localisation of P2X subunit-specific immuno-histochemistry, although selective antibodies for P2X subunits are not yet freely available to the community at large.

1. *P2X₁ subunit*

$P2X_1$ transcripts were first located in vascular smooth muscle of several small blood vessels (lingual, trapezius muscle, carotid and umbilical arteries) (Valera *et al.*, 1994; Collo *et al.*, 1996). Subsequently, mRNA transcripts for $P2X_1$ subunits have been found in aortic, pulmonary, internal and external iliac, renal, femoral and coronary arteries but are absent in superior mesenteric artery (Nori *et al.*, 1998). $P2X_1$-specific signals were detected in vas deferens, urinary bladder, small intestine and colon (Valera *et al.*, 1994; Longhurst *et al.*, 1996). $P2X_1$ transcripts found in kidney, liver, lung, placenta and spleen may be due to their vascularisation (Valera *et al.*, 1995). $P2X_1$ transcripts occur in adrenal glands, gonads, heart, pancreas and various immune cells (Valera *et al.*, 1994; Buell *et al.*, 1996a; Longhurst *et al.*, 1996, Bogdanov *et al.*, 1998). Recently, mRNA for $P2X_1$ was found in human blood platelets and the cloned cDNA showed 100% sequence identity to the encoding cDNA from human bladder $P2X_1$ (Scase *et al.*, 1998).

$P2X_1$-like immunoreactivity ($P2X_1$-ir) was found in the smooth muscle of vas deferens and urinary bladder, blood vessel walls of submucosal arteries and in intercalated disks of cardiac muscle (Vulchanova *et al.*, 1996). $P2X_1$-ir is found in some blood vessels in kidney (intrarenal, arcuate and interlobular arteries, and afferent arterioles) but is not found in efferent arterioles, glomeruli, or renal tubules (Chan *et al.*, 1998). $P2X_1$-ir is found in some cerebral blood vessels (Bo *et al.*, 1998).

$P2X_1$ and $P2X_3$ receptors are activated by α,β-meATP . The prevailing evidence suggests $P2X_1$ subunits are concentrated in smooth muscle (see above) whereas $P2X_3$ subunits are restricted mainly to sensory neurons (see below). However, $[^3H]\alpha,\beta$-meATP binding to smooth muscle is not necessarily an indicator of the presence of $P2X_1$ subunits. Transcripts for both $P2X_2$ and $P2X_4$ subunits colocalise with $P2X_1$ transcripts in blood vessels (Nori *et al.*, 1998). Both $P2X_2$ and $P2X_4$ receptors bind $[^3H]\alpha,\beta$-meATP with the same avidity as $P2X_1$ receptors (Michel *et al.*, 1996, 1997), although only $P2X_1$ receptors are activated by the bound methylene phosphonate compound.

2. *P2X₂ subunit*

$P2X_2$ transcripts are found mainly in the neuraxis, but in a more restricted distribution than $P2X_4$- and $P2X_6$-specific signals (Collo *et al.*, 1996). mRNA for splice variants of $P2X_{2a,b,c}$ colocalise in the same regions of the CNS (Simon *et al.*, 1997). mRNA for the $P2X_2$ subunit and splice variants and $P2X_2$-ir are found in peripheral ganglia (DRG, nodose ganglion, SCG, myenteric plexus) (Brake *et al.*, 1994; Kidd *et al.*, 1995; Lewis *et al.*, 1995; Collo *et al.*, 1996; Vulchanova *et al.*, 1996; Simon *et al.*, 1997). $P2X_2$ transcripts are found in heart and colocalise with $P2X_1$ and $P2X_4$ transcripts in aortic, pulmonary, internal and external iliac, renal, femoral and coronary arteries (Nori *et al.*, 1998). $P2X_2$ transcripts are found in vas deferens (Brake *et al.*, 1994), but $P2X_2$-ir appears to be restricted to sympathetic nerves within this tissue (Vulchanova, *et al.*, 1996). $P2X_2$ transcripts and $P2X_2$-ir are present in PC12 cells, anterior pituitary and adrenal medulla (Brake *et al.*, 1994; Collo *et al.*, 1996; Vulchanova *et al.*, 1996). $P2X_2$ mRNA and $P2X_2$-ir are seen in the retina, in the outer layer near photoreceptors and inner layer in the plexus of dendrites formed by retinal ganglion cells (Greenwood *et al.*, 1997). Thus, $P2X_2$ subunits are located primarily in neural tissues in the CNS and PNS (and neurosecretory tissues therein), but are also present in cardiovascular tissues.

3. *P2X₃ subunit*

$P2X_3$ transcripts are concentrated in sensory neurons (dorsal root, nodose and trigeminal ganglia) particularly in capsaicin-sensitive nociceptors (Chen *et al.*, 1995; Lewis *et al.*, 1995; Collo *et al.*, 1996; Cook *et al.*, 1997). Transcripts for $P2X_{1,2,4,5,6}$ are also found in DRG (Collo *et al.*, 1996), while $P2X_{2,4}$ occur in nodose ganglia (Lewis *et al.*, 1995), and $P2X_5$ is present in trigeminal ganglia (Collo *et al.*, 1996). $P2X_1$-ir is not found in nodose and trigeminal ganglia (Vuchanova *et al.*, 1996; Cook *et al.*, 1997). mRNA for $P2X_3$ is 40-fold more abundant than $P2X_2$ transcripts in DRG (Chen *et al.*, 1995). Both mRNA for $P2X_3$ and $P2X_3$-ir have been located in the dorsal spinal horn, nucleus tractus solitarius (NTS) and spinal trigeminal nucleus - *i.e.*, at the central termini of dorsal root, nodose and trigeminal sensory neurons (Collo *et al.*, 1996; Garcia-Guzman *et al.*, 1997b; Vulchanova *et al.*, 1997). $P2X_3$ transcripts have not been found in vascular tissue but do occur in heart tissue in unspecified locations (Garcia-Guzman *et al.*, 1997b; Bogdanov *et al.*, 1998).

4. *P2X₄ subunit*

$P2X_4$ transcripts and $P2X_4$-ir occur extensively throughout the neuraxis (Collo *et al.*, 1996; Lê *et al.*, 1998) and are detected in peripheral neurons (dorsal root, coeliac, nodose, superior cervical and trigeminal ganglia) as well

(Lewis *et al.*, 1995; Buell *et al.*, 1996b). P2X$_4$ transcripts are found in the adrenal gland, gastrointestinal tract, gonads, kidney, liver, lung, pancreas and islets of Langerhans, salivary gland, skeletal muscle, thymus, trachea, vas deferens and urinary bladder (Bo *et al.*, 1995; Buell *et al.*, 1996b; Soto *et al.*, 1996a; Wang *et al.*, 1996; Dhulipala *et al.*, 1998). mRNA for P2X$_4$ is present in heart tissue (Soto *et al.*, 1996a; Wang *et al.*, 1996; Bogdanov *et al.*, 1998; Dhulipala *et al.*, 1998), being confined to coronary arteries and intramural ganglia but absent in myocytes (Nori *et al.*, 1998). P2X$_4$ transcripts (colocalised with P2X$_1$ and P2X$_2$ transcripts) are present in major blood vessels including aortic, pulmonary, internal and external iliac, renal, femoral arteries and vena cava (Soto *et al.*, 1996a; Nori *et al.*, 1998). P2X$_4$ transcripts are expressed heavily in smooth muscle-containing tissues, including cultured (human airway) smooth muscle cells (Dhulipala *et al.*, 1998). Splice variants of P2X$_4$ subunits also occur in human smooth muscle, but do form functional homomeric P2X receptors (Dhulipala *et al.*, 1998).

5. *P2X$_5$ subunit*

Of the known P2X subunits, P2X$_5$ transcripts has the most restricted distribution (at least in rat). mRNA for P2X$_5$ occurs mainly in rat mesencephalic nucleus of the trigeminal nerve (MeN5), and is found sparingly in sensory neurons in rat dorsal root and trigeminal ganglia (Collo *et al.*, 1996). P2X$_5$-ir has been associated with proprioceptive sensory neurons in MeN5 (Cook *et al.*, 1997). P2X$_5$ transcripts are found in ventral horn motoneurons of rat cervical spinal cord (Collo *et al.*, 1996). A P2X$_5$-like peptide (62% sequence homology with rP2X$_5$) was cloned from human tissue (Lê *et al.*, 1997). Most of the M2 segment and a short section of the C-terminus is missing from the hP2X$_5$ protein and the homomeric receptor fails to respond to ATP. A chimeric human/rat P2X$_5$ receptor spliced at Ile^{318} does respond to ATP but strongly inactivates and fails to recover. A non-functional splice variant of the hP2X$_5$ subunit occurs where the domain Gly^{97}-Glu^{120} is deleted. Transcripts for the shorter and longer forms of hP2X$_5$ are widely spread throughout the human body and occur mainly in brain and immune system (lymphocytes, leukocytes, thymus and bone marrow). There is no evidence for functional P2X$_5$ receptors in human and rat cardiovascular tissues although P2X$_5$ transcripts do occur in rat heart (Garcia-Guzman *et al.*, 1996).

6. *P2X$_6$ subunit*

P2X$_6$ transcripts are distributed heavily throughout the neuraxis in both neurons and astrocytes, and are also detected in peripheral neurons (dorsal root, trigeminal and coeliac ganglia) (Collo *et al.*, 1996; Soto *et al.*, 1996b). P2X$_6$-specific signal was found in bronchial and uterine epithelia but not in salivary epithelia. mRNA for P2X$_6$ was found in adrenal gland, gonads, heart, pituitary, lung and trachea but was undetected in aorta, vena cava, urinary bladder and vas deferens. In general, the distribution of P2X$_6$ transcripts match that of P2X$_4$-specific signal, except P2X$_6$ is absent in smooth muscle.

7. *P2X$_7$ subunit*

P2X$_7$ transcripts were detected first in macrophages, microglia, brain, lung and spleen in rat (Surprenant *et al.*, 1996). Subsequently, P2X$_7$ transcripts were detected in human heart, liver, pancreas, thymus with lower levels found in brain, leukocytes, lung, placenta, prostate, skeletal muscle and testis (Rassendren *et al.*, 1997). The presence of P2X$_7$ in some tissues should be treated with caution and may reflect infiltration of immune cells. Thus, P2X$_7$ mRNA and P2X$_7$-ir in brain is located in microglia and ependymal cells but not neurons (Collo *et al.*, 1997). The isolation of the P2X$_7$ cDNA from SCG and medial habenula libraries may be due to contaminating RNA from microglia

(Collo *et al.*, 1997). P2X$_7$ transcripts are prominent in haemopoietic tissues, and occur in myeloid stem cells (monocytes, granulocytes) and lymphoid stem cells (B cells and T cells). There is no evidence as yet for P2X$_7$ transcripts in cardiovascular tissues.

<u>E. P2X subunit coexpression</u>

The colocalisation of P2X transcripts in many tissues raises two questions. Are native P2X purinoceptors composed of heteromeric assemblies of P2X subunits? And, do populations of homomeric assemblies of P2X subunits occur in the same tissue? The answers to both questions seems a guarded "yes".

The strongest evidence for heteromeric P2X receptors stems from the coexpression of P2X$_2$ and P2X$_3$ subunits to yield a recombinant receptor with operational features similar to the native P2X purinoceptor in nodose and dorsal root ganglia (Lewis *et al.*, 1995; Radford *et al.*, 1997). In these circumstances, the P2X$_{2/3}$ receptor shows slowly-desensitising α,β-meATP -responses which are enhanced when extracellular pH is lowered (Lewis *et al.*, 1995; Stoop *et al.*, 1997). Thus, α,β-meATP agonism is thought to be conferred by the P2X$_3$ subunit whereas the reduced desensitisation rate and H$^+$-potentiation are properties of the P2X$_2$ subunit. However, recent results with the antagonist TNP-ATP (which prefers P2X$_3$ over P2X$_2$ receptors) indicate that both homomeric P2X$_2$ receptors and heteromeric P2X$_{2/3}$ receptors occur in individual nodose ganglion cells (Thomas *et al.*, 1998). ATP-responses at some nodose ganglion cells (42 of 188 cells, 22%) were not potentiated by low pH levels (Li *et al.*, 1996a,b), indicating a subset of nodose neurons possess neither homomeric P2X$_2$ receptors nor heteromeric P2X$_{2/3}$ receptors (see Section III, B6 for rationale).

Vulchanova and colleagues (1997) showed that P2X$_2$-ir and P2X$_3$-ir are colocalised in many (but not all) sensory neurons in nodose and dorsal root ganglia. Additionally, some sensory nerves contained either P2X$_2$-ir or P2X$_3$- ir or none at all. The central terminals of vagal sensory neurons showed colocalisation of P2X$_2$-ir and P2X$_3$-ir within rat NTS (but not monkey NTS), whereas central terminals of dorsal roots showed P2X$_2$-ir and P2X$_3$-ir in different dorsal horn laminae. Thus, vagal sensory neurons and their axons in the rat possess mainly heteromeric P2X$_{2/3}$ receptors, but spinal sensory neurons possess homomeric P2X$_2$ and P2X$_3$ receptors and their fibres seek different targets in the spinal cord. The operational features of P2X purinoceptors in DRG more closely match the P2X$_3$ receptor than the P2X$_2$ receptor (Rae *et al.*, 1998), perhaps for the reason that the levels of transcripts in DRG are 40-fold higher for P2X$_3$ than P2X$_2$ subunits (Chen *et al.*, 1995). It has been suggested that neonatal DRG (1-4 days old) express homomeric P2X$_3$ receptors while adult DRG express heteromeric P2X$_{2/3}$ receptors (Evans & Surprenant, 1996). However, desensitisation of ATP-responses is still fast (t = 2.5-3s) in DRG of cat (5-10 days old) and rat (4-60 days old) (Krishtal *et al.*, 1983, 1988). The rate of desensitisation is much slower for homomeric P2X$_2$ receptors (t = >60s) and native P2X purinoceptors of PC12 cells (t = >180s) from which P2X$_2$ was cloned (see Surprenant *et al.*, 1995).

Apart from the example P2X$_{2/3}$ receptors (and controversy over its presence in DRG cells and all nodose neurons), the evidence is weak for other heteromeric P2X receptors elsewhere. Although mRNA for P2X$_{1,2,4}$ colocalise in smooth muscle of blood vessels (Nori *et al.*, 1998), the operational profile of native P2X purinoceptors in vascular tissues is predominantly that of the P2X$_1$

177

phenotype (*i.e.*, α,β-meATP -sensitive, PPADS- and suramin-sensitive, fast desensitising responses). What happens when $P2X_{1,2,4}$ subunits are coexpressed? $P2X_1$ and $P2X_2$ subunits were coexpressed in HEK293 cells, resulting in an α,β-meATP sensitive P2X receptor with the same kinetics of activation and inactivation as homomeric $P2X_1$ receptors (Lewis *et al.*, 1995). Coexpression of $P2X_1$ and $P2X_4$ subunits also resulted in a P2X receptor with the kinetics of a $P2X_1$ receptor (Lewis *et al.*, 1995). Where rundown of ATP-responses was studied, coexpression of $P2X_1$ with either $P2X_2$ and $P2X_4$ resulted in independent populations of ion-channels (Lewis *et al.*, 1995). Thus, $P2X_{1,2,4}$ transcripts may occur in vascular tissues but the available evidence points to mixed populations of homomeric P2X receptors rather than heteromeric $P2X_{1,2,4}$ receptors.

$P2X_1$ and $P2X_2$ transcripts also occur in PC12 cells (Valera *et al.*, 1994, Brake *et al.*, 1994) as does $P2X_1$-ir and $P2X_2$-ir (Vulchanova *et al.*, 1996). However, the native P2X purinoceptor in PC12 cells has the same pharmacological profile as homomeric $P2X_2$ receptors (see Rhoads *et al.*, 1993; King *et al.*, 1996, 1997b). $P2X_2$-ir occurs in the myenteric plexus (Vulchanova *et al.*, 1996) as does $P2X_4$-ir (Burnstock, unpublished data). The ATP receptors in the myenteric plexus show *either* a $P2X_2$-like phenotype (Zhou & Galligan, 1996) *or* a $P2X_4$-like phenotype (Barajas-Lopez *et al.*, 1996) but not a mixed phenotype. $P2X_4$ and $P2X_6$ transcripts colocalise throughout the neuraxis. Coexpression of $P2X_4$ and $P2X_6$ subunits resulted in a P2X receptor with an operational profile identical to homomeric $P2X_4$ receptors (Soto *et al.*, 1996). Thus, the prevailing evidence suggests that homo-meric P2X receptors occur more often than heteromeric P2X receptors. Possible reasons for such circumstances may rest with differences in the structure and surface charge of the M2 segment of P2X subunits. Little is known with certainty about the "molecular glue" that binds subunits together, in terms of weak and strong electrical forces or formation of stable covalent bonds.

IV. CONCLUSIONS AND FUTURE DIRECTIONS

We are currently aware of seven structurally-distinct P2X subunits which preferentially select ATP amongst all the naturally-occurring nucleotides and nucleosides. An eighth P2X subunit (P2XM) is similar in structure to the $P2X_6$ subunit (80% identity), but this particular subtype has not been functionally characterised (Urano *et al.*, 1997). There is no certainty that every P2X subunit has been discovered. Splice variants of some P2X subunits have been identified and, where shown to be functional, possess a different operational profile from the corresponding wild-type P2X receptor. Little is known about the roles for splice variants and the consequences of coexpression with wild-type P2X subunits. Homomeric and, possibly, heteromeric P2X receptors exist in the same tissue if not the same cell. The ways in which heterogeneous P2X purinoceptors can interact is not understood. It is likely that a heterogeneous population of homomeric P2X receptors would broaden cell responsiveness to a wide range of concentrations of extracellular ATP (from nanomolar to micromolar). Additionally, heterogeneous P2X receptors represent a heterogeneous population of Ca^{2+}-channels that can open and close at different rates.

We are beginning to understand the structure of the P2X subunit and can identify six key regions that influence the operational profile of homomeric P2X receptors. The M2 segment is involved with lining the pore of the ion-channel, but the means by which a tetramer of M2 segments are assembled remains illusive. The extracellular loop forms a binding pocket for ATP but the key amino acid residues anchoring ATP have yet to be defined. Allosteric modulatory sites for inorganic ions (H^+, Zn^{2+}, Cu^{2+}) occur near the ATP-binding site and also remain to be identified. The intracellular N- and C-termini are involved in the process of receptor inactivation and channel closure. Phosphorylation sites occur on both termini but their role in intracellular signalling beyond closing the ion-channel has not been addressed. Similarly, the consequences of activating P2Y receptors as well as non-purinoceptors in the vicinity of P2X receptors has been ignored so far.

A key area for investigation is the cellular localisation of P2X subunits and levels of expression of P2X subunits following gene up/down-regulation. A role for heteromeric P2X receptors can only be determined if the building blocks of these P2X purinoceptors are known to us. Additionally, the appearance of particular P2X subunits during tissue development, growth, ageing and disease will advance our understanding of the physiological purposes of P2 receptor signalling. Possible cross-talk between heterogeneous P2X purinoceptors remains a fertile area for investigation. Such studies would benefit from selective agonists and antagonists for homomeric P2X receptors. Drug discovery represents an enormous challenge but there are encouraging signs that novel selective chemical entities are within reach. The application of such drugs in the clinical field represents a logical extension to the process of drug discovery, to complete the journey from gene to disease.

V. REFERENCES

Abbracchio MP, Burnstock G. Purinoceptors: are there families of P2X and P2Y purinoceptors? Pharmacol Ther 1994; 64:445-475.

Barajas-López C, Huizinga JD, Collins SM, Gerzanich V, Espinoza-Luna R, Peres LA. P2X-purinoceptors of myenteric neurones from the guinea-pig ileum and their unusual pharmacological properties. Brit J Pharmacol 1996; 119:1541-1548.

Bardoni R, Goldstein PA, Lee CJ, Gu JG, MacDermott AB. ATP P2X receptors mediate fast synaptic transmission in the dorsal horn of the rat spinal cord. J Neurosci 1997; 17:5297-5304.

Barish M. A transient calcium-dependent chloride current in the immature *Xenopus* oocyte. J Physiol 1983; 342:309-325.

Bean BP, Williams CA, Ceelen PW. ATP-activated channels in rat and bullfrog sensory neurons: current-voltage relation and single-channel behaviour. J Neurosci 1990; 10:11-19.

Benham, CD. ATP-activated channels gate calcium entry into smooth muscle cells dissociated from rabbit ear artery. J Physiol 1989; 419:689-701.

Benham CD, Tsien RW. A novel receptor-operated Ca^{2+}-permeable channel activated by ATP in smooth muscle. Nature 1987; 328:275-278.

Bo X, Burnstock G. Triphosphate, the key structure of the ATP molecule responsible for interaction with P2X purinoceptors. Gen Pharmac 1993; 24:637-640.

Bo X, Karoon P, Nori SL, Bardini M, Burnstock G. P2X purinoceptors in postmortem human cerebral arteries. J Cardiovasc Pharmacol 1998; in press.

Bo X, Simon J, Burnstock G, Barnard EA. Solubilization and molecular size of the P2X purinoceptor from rat vas deferens. J Biol Chem 1992; 267:15581-17587.

Bo X, Zhang Y, Nassar M, Burnstock G, Schoepfer R. A P2X purinoceptor cDNA conferring a novel pharmacological profile. FEBS Lett 1995; 375:129-133.

Boarder MR, Hourani SMO. The regulation of vascular function by P2 receptors: multiple sites and multiple receptors. Trends Pharmacol Sci 1998; 19:99-107.

Bogdanov Y, Rubino A, Burnstock G. Characterisation of subtypes of the P2X and P2Y families of ATP receptors in the foetal human heart. Life Sci 1998; 62:697-703.

Brake AJ, Wagenbach MJ, Julius D. New structural motif for ligand-gated ion channels defined by an ionotropic ATP receptor. Nature 1994; 371:519-523.

Brändle U, Spielmanns P, Osteroth R, Sim J, Surprenant A, Buell G, Ruppersberg JP, Plinkert PK, Zenner HP, Glowatzki E. Desensitization of the $P2X_2$ receptor controlled by alternative splicing. FEBS Lett 1997; 404:294-298.

Bretschneider F, Klapperstück M, Löhn M, Markwardt F. Nonselective cation currents elicited by extracellular ATP in human B-lymphocytes. Pflügers Arch 1995; 429:691-698.

Buell G, Collo G, Rassendren F. P2X receptors: an emerging channel family. Eur J Neurosci 1996a; 8:2221-2228.

Buell G, Lewis C, Collo G, North RA, Surprenant A. An antagonist-insensitive P2X receptor expressed in epithelia and brain. EMBO J 1996b; 15:55-62.

Bültmann R, Starke K. Reactive red 2: a P2Y selective purinoceptor antagonist and an inhibitor of ectonucleotidase. Naunyn-Schmiedeberg's Arch Pharmacol 1995; 352:477-482.

Burnstock G. "A basis for distinguishing two types of purinergic receptor." In *Cell Membrane Receptors for Drugs and Hormones: A Multidisciplinary Approach*, RW Straub & L Bolis, eds. New York: Raven Press, pp.107-118, 1978.

Burnstock G. Distribution and roles of purinoceptor subtypes. Nucleos Nucleot 1991; 10:917-930.

Burnstock G. The past, present and future of purine nucleotides as signalling molecules. Neuropharmacol 1997; 36:1127-1139.

Burnstock G, Kennedy C. Is there a basis for distinguishing two types of P2 purinoceptor? Gen Pharmac 1985; 16:433-440.

Burnstock G, Warland JJI. P2 purinoceptors of two subtypes in the rabbit mesenteric artery; reactive blue 2 selectively inhibits responses mediated via the P2Y but not the P2X purinoceptor. Brit J Pharmacol 1987; 90:383-390.

Burnstock G, Wood JN. Purinergic receptors: their role in nociception and primary afferent neurotransmission. Curr Opin Neurobiol 1996; 6:526-532.

Chen CC, Akopian AN, Sivilotti L, Colquhoun D, Burnstock G, Wood JN. A P2X purinoceptor expressed by a subset of sensory neurons. Nature 1995; 377:428-431.

Chow SC, Kass GEN, Orrenius S. Purines and their roles in apoptosis. Neuropharm 1997; 36:1149-1156.

Chvatchko Y, Valera S, Aubry JP, Renno T, Buell G, Bonnefoy JY. The involvement of an ATP-gated ion channel, P(2X1), in thymocyte apoptosis. Immunity 1996; 5:275-283.

Cloues R. Properties of ATP-gated channels recorded from rat sympathetic neurons; voltage dependence and regulation by Zn^{2+}. J Neurophysiol 1995; 73:312-319.

Cloues R, Jones S, Brown DA. Zn^{2+} potentiates ATP-activated currents in rat sympathetic neurons. Pflügers Arch 1993; 424:152-158.

Cockcroft S, Gomperts BD. The ATP^{4-} receptor of rat mast cells. Biochem J 1980; 188:789-798.

Collo G, North RA, Kawashima E, Merlo-Pich E, Neidhart S, Surprenant A, Buell G. Cloning of $P2X_5$ and $P2X_6$ receptors and the distribution and properties of an extended family of ATP-gated ion channels. J Neurosci 1996; 16:2495-2507.

Collo G, Neidhart S, Kawashima E, Kosco-Vilbois M, North RA, Buell G. Tissue distribution of the $P2X_7$ receptor. Neuropharmacol 1997; 36:1277-1285.

Cook SP, Vulchanova L, Hargreaves KM, Elde R, McCleskey EW. Distinct ATP receptors on pain-sensing and stretch-sensing neurons. Nature 1997; 387:505-508.

Cook SP, McCleskey EW. Inactivation, recovery and Ca^{2+}-dependent modulation of ATP-gated P2X receptors in nociceptors. Soc Neurosci Abst 1997; 23:151.12.

Crack BE, Beukers MW, McKechnie KCW, IJzerman AP, Leff P. Pharmacological analysis of ecto-ATPase inhibition: evidence for combined enzyme inhibition and receptor antagonism in P2X purinoceptor ligands. Brit J Pharmacol 1995; 113:1432-1438.

Cusack, NJ. P2 receptor: subclassification and structure-activity relationship. Drug Devel Res 1993; 28:244-252.

Dalziel HH, Westfall DP. Receptors for adenine nucleotides and nucleosides: subclassification, distribution, and molecular characterization. Pharmacol Rev 1994; 46:449-466.

Den Hertog A, Nelemans A, Van Den Akker J. The inhibitory action of suramin on the P2 purinoceptor response in smooth muscle cells of guinea-pig taenia caeci. Eur J Pharmacol 1989; 166:531-534.

Dhulipala PDK, Wang YX, Kotlikoff MI. The human $P2X_4$ receptor gene is alternatively spliced. Gene 1998; 207:259-266.

Dubyak GR. Signal transduction by P2 purinergic receptors for extracellular ATP. Am J Respir Cell Mol Biol 1991; 4:295-300

Dubyak GR, El-Moatassim C. Signal transduction via P2 purinergic receptors for extracellular ATP and other nucleotides. Am J Physiol 1993; 265:C577-C606.

Dunn PM, Blakeley AGH. Suramin: a reversible P2 purinoceptor antagonist in the mouse vas deferens. Brit J Pharmacol 1988; 93:243-245.

Edwards FA, Gibb AJ, Colquhoun D. ATP receptor-mediated synaptic currents in the central nervous system. Nature 1992; 359:144-147.

Edwards FA, Robertson SJ, Gibb AJ. Properties of ATP receptor-mediated synaptic transmission in the rat medial habenula. Neuropharmacol 1997; 36:1253-1268.

Egan TM, Haines WR, Voigt MM. A domain contributing to the ion channel of ATP-gated $P2X_2$ receptors identified by the substituted cysteine accessibility method. J Neurosci 1998; 18:2350-2359.

Evans RJ. Single channel properties of ATP-gated cation channels (P2X receptors) heterologously expressed in Chinese hamster ovary cells. Neurosci Lett 1996; 212;212-214.

Evans RJ, Derkach V, Surprenant A. ATP mediates fast synaptic transmission in mammalian neurons. Nature 1992; 357:503-505.

Evans RJ, Kennedy C. Characterization of P2 purinoceptors in the smooth muscle of rat tail artery: a comparison between contractile and electrophysiological responses. Brit J Pharmacol 1994; 113:853-860.

Evans RJ, Lewis C, Buell G, Valera S, North RA, Surprenant A. Pharmacological characterization of heterologously expressed ATP-gated cation channels (P2X purinoceptors). Mol Pharmacol 1995; 48:178-183.

Evans RJ, Lewis C, Virginio C, Lundstrom K, Buell G, Surprenant A, North RA. Ionic permeability of, and divalent effects on, two ATP-gated cation channels (P2X receptors) expressed in mammalian cells. J Physiol 1996; 497:413-422.

Evans RJ, Surprenant A. P2X receptors in autonomic and sensory neurons. Seminars Neurosci 1996; 8:217-223.

Evans RJ, Surprenant A, North RA. "P2X receptors: Cloned and Expressed". In *The P2 Nucleotide Receptors*, JT Turner, GA Weisman & JS Fedan eds, Human Press, New Jersey, Ch.2, pp43-62, 1998.

Fredholm BB, Abbracchio MP, Burnstock G, Daly JW, Harden TK, Jacobson KA, Leff P, Williams M. Nomenclature and classification of purinoceptors. Pharmacol Rev 1994; 46:143-156.

Galligan JJ, Bertrand PP. ATP mediates fast synaptic potentials in enteric neurons. J. Neurosci 1994; 14:7563-7571.

Garcia-Guzman M, Soto F, Gomez-Hermandez JM, Lund PE, Stühmer W. Characterization of recombinant human P2X$_4$ receptor reveals pharmacological differences to the rat homologue. Mol Pharmacol 1997a; 51:109-118.
Garcia-Guzman M, Soto F, Laube B, Stühmer W. Molecular cloning and functional expression of a novel rat heart P2X purinoceptor. FEBS Lett 1996; 388:123-127.

Garcia-Guzman M, Stühmer W, Soto F. Molecular characterization and pharmacological properties of the human P2X$_3$ purinoceptor. Mol Brain Res 1997b; 47:59-66.

Gargett CE, Wiley JS. The isoquinoline derivative KN-62 a potent antagonist of the P2Z-receptor of human lymphocytes. Brit J Pharmacol 1997; 120:1483-1490.

Gordon JL. Extracellular ATP: effects, sources and fate. Biochem J 1986; 233:309-319.

Greenwood D, Yao WP, Housley GD. Expression of the P2X2 receptor subunit of the ATP-gated ion channel in the retina. NeuroReport 1997; 8:1083-1088.

Harden TK, Boyer JL, Nicholas RA. P2 purinergic receptors: subtype-associated signalling responses and structure. Ann Rev Pharmacol Toxicol 1995; 35:541-579.

Hilderman RH, Martin M, Zimmermn JK, Pivorun EB. Identification of a unique membrane receptor for adenosine 5′,5′′′-P1,P4-tetraphosphate. J Biol Chem 1991; 266:6915-6918.

Hourani SMO, Cusack NJ, Welford LA. L-AMP-PCP, an ATP receptor agonist in guinea pig bladder, is inactive on taenia coli. Eur J Pharmacol 1985; 108:197-200.

Hourani SMO, Welford LA, Loizou GD, Cusack NJ. Adenosine-5′-(2-flourodiphosphate) is a selective agonist at P2 purinoceptors mediating relaxation of smooth muscle. Eur J Pharmacol 1988; 147:131-136.

Housley GD, Greenwood D, Bennett T, Ryan AF. Identification of a short form of the P2XR-1 purinoceptor subunit produced by alternative splicing in the pituitary and cochlea. Biochem Biophys Res Comm 1995; 212:501-508.

Houston DA, Burnstock G, VanHoutte PM. Different P2 purinergic receptor subtypes of endothelium and smooth muscle cells in canine blood vessels. J Pharmacol Exp Ther 1987; 241:501-506.

Hoyle CHV, Knight GE, Burnstock B. Suramin antagonizes responses to P2 purinoceptor agonists and purinergic nerve stimulation in the guinea-pig urinary bladder and taenia coli. Brit J Pharmacol 1990; 99:617-621.

Humphrey PPA, Buell G, Kennedy I, Khahk BS, Michel AD, Surprenant A, Tresize DJ. New insights on P2X purinoceptors. Naunyn-Schmiedeberg's Arch Pharmacol 1995; 352:585-596.

Humphrey PPA, Khakh BS, Kennedy C, King BF, Burnstock G. "Classification of ionotropic (P2X) receptors for adenosine triphosphate (ATP)." In *IUPHAR Receptor Compendium*, IUPHAR Media, 1998.

Illes P, Nieber K, Nörenberg W. Electrophysiological effects of ATP on brain neurones. J Auton Pharmacol 1996; 16:407-411.

Illes P, Nörenberg W. Neuronal ATP receptors and their mechanism of action. Trends Pharmacol Sci 1993; 14:50-54.

Jacobson KA, Kim YC, Wildman SS, Mohanram A, Harden K, Boyer JL, King BF, Burnstock G. A pyridoxine cyclic phosphate and its 6-azoaryl-derivative selectively potentiate and antagonize activation of P2X$_1$ receptors. J Med Chem 1998; in press.

Kennedy C. P1 and P2 purinoceptor subtypes - an update. Arch Int Pharmacodyn 1990; 303:30-50.

Kennedy C, Leff P. How should P2X purinoceptors be classified pharmacologically? Trends Pharmacol Sci 1995; 16:168-174.

Khakh BS, Humphrey PPA, Surprenant A. Electrophysiological properties of P2X purinoceptors in rat superior cervical, nodose and guinea-pig coeliac neurones. J Physiol 1995a; 484:385-395.

Khakh BS, Surprenant A, Humphrey PPA. A study of P2X purinoceptors mediating the electrophysiological and contractile effects of purine nucleotides in rat vas deferens. Brit J Pharmacol 1995b; 115:177-185.

Kidd EJ, Grahames CBA, Simon J, Michel AD, Barnard EA, Humphrey PPA. Localization of P2X purinoceptor transcripts in the rat nervous system. Mol Pharmacol 1995; 48:569-573.

King B, Chen CC, Akopian AN, Burnstock G, Wood JN. A role for calcineurin in the desensitization of the P2X$_3$ receptor. NeuroReport 1997a; 8:1099-1102.

King BF, Wildman SS, Ziganshina LE, Pintor J, Burnstock G. Effects of extracellular pH on agonism and antagonism at a recombinant P2X$_2$ receptor. Brit J Pharmacol 1997b; 121:1445-1453.

King BF, Ziganshina LE, Pintor J, Burnstock G. Full sensitivity of P2X$_2$ purinoceptor to ATP revealed by changing extracellular ATP. Brit J Pharmacol 1996; 117:1371-1373.

Krishtal OA, Marchenko SM, Pidoplichko VI. Receptor for ATP in the membrane of mammalian sensory neurones. Neurosci Lett. 1983; 35:41-45.

Krishtal OA, Marchenko SM, Obukhov AG, Volkova TM. Receptors for ATP in rat sensory neurones: the structure-function relationship for ligands. Brit J. Pharmacol 1988; 95:1057-1062.

Kubo Y, Bladwin TJ, Jan YN, Jan LY. Primary structure and functional expression of a mouse inward rectifier potassium channel. Nature 1993; 362:127-133.

Lazarowski ER, Paradiso AM, Watt WC, Harden TK, Boucher RC. UDP activates a mucosal-restricted receptor on human nasal epithelial cells that is distinct from the P2Y$_2$ receptor. Proc Nat Acad Sci USA 1997; 94:2599-2603.

Lê KT, Paquet M, Nouel D, Babinski K, Séguéla P. Primary structure and expression of a naturally truncated human P2X ATP receptor subunit from brain and immune system. FEBS Lett 1997; 418:195-199.

Lê KT, Villeneuve P, Ramjaun AR, McPherson PS, Beadet A, Séguéla P. Sensory presynaptic and widespread somatodendritic immunolocalization of central ionotropic P2X ATP receptors. Neurosci 1998; 83:177-190.

Lewis C. Neidhart S, Holy C, North RA, Buell G, Surprenant A, Coexpression of P2X$_2$ and P2X$_3$ receptor subunits can account for ATP-gated currents in sensory neurones. Nature 1995; 377:432-435.

Li C, Peoples RW, Li Z, Weight FF. Zn^{2+} potentiates excitatory action of ATP on mammalian neurons. Proc Nat Acad Sci USA 1993; 90:8264-8267.

Li C, Peoples RW, Weight FF. Proton potentiation of ATP-gated ion channel responses to ATP and Zn^{2+} in rat nodose ganglion neurons. J. Neurophysiol 1996a; 76:3048-3058.

Li C, Peoples RW, Weight FF. Acid pH augments excitatory action of ATP on a dissociated mammalian sensory neuron. NeuroReport 1996b; 7:2151-2154.

Li C, Peoples RW, Weight FF. Cu^{2+} potently enhances ATP-activated current in rat nodose ganglion neurons. Neurosci Lett 1996c; 219:45-48.

Longhurst PA, Schwegel T, Folander K, Swanson R. The human P2X$_1$ receptor; molecular cloning, tissue distribution, and localization to chromosome 17. Biochim Biophys Acta 1996; 1308:185-188.

MacKinnon R. Pore loops - an emerging theme in ion channel structure. Neuron 1995; 14:889-892.

Michel AD, Chau NM, Fan TPD, Frost EE, Humphrey PPA. Evidence that [^{3}H]-_,_-meATP may label an endothelial-derived cell line 5´-nucleotidase with high affinity. Brit J Pharmacol 1995; 115:767-774.

Michel AD, Lundström K, Buell GN, Surprenant A, Valera S, Humphrey PPA. A comparison of the binding characteristics of recombinant P2X$_1$ and P2X$_2$ purinoceptors. Brit J Pharmacol 1996; 118:1806-1812.

Michel AD, Miller KJ, Lundstrom K, Buell GN, Humphrey PPA. Radiolabeling of the rat $P2X_4$ purinoceptor: evidence for allosteric interaction of purinoceptor antagonists and monovalent cations with P2X purinoceptors. Mol Pharmacol 1997; 51:524-532.

Morrison MS, Turin L, King BF, Burnstock G, Arnett TR. ATP is a potent stimulator of the activation and formation of rodent osteoclasts. J Physiol 1998; in press.

Nakazawa K, Fujimori K, Takanaka A, Inoue K. An ATP-activated conductance in phaeochromocytoma cells and its suppression in extracellular calcium. J Physiol 1990; 428:257-272.

Nakazawa K, Hess P. Block by calcium of ATP-activated channels in pheochromocytoma cells. J Gen Physiol 1993; 101:377-392.

Nieber K, Poelchen W, Illes P. Role of ATP in fast excitatory synaptic potentials in locus coeruleus neurones of the rat. Brit J Pharmacol 1997; 122:423-430.

Nori S, Fumagalli L, Bo X, Bogdanov Y, Burnstock G. Coexpression of mRNAs for $P2X_1$, $P2X_2$ and $P2X_4$ receptors in rat vascular smooth muscle: an *in situ* hybridization and RT-PCR study. J Vasc Res 1998; 35:179-185.

North RA. Families of ion channels with two hydrophobic segments. Curr Opin Cell Biol 1996a; 8:474-483.

North RA. P2X purinoceptor plethora. Seminars Neurosci 1996b; 8:187-194.

North RA, Barnard EA. Nucleotide receptors. Curr Opin Neurobiol 1997; 7:346-357.

O'Connor SE, Dainty IA, Leff P. Further subclassification of ATP receptors based on agonist studies. Trends Pharmacol Sci 1991; 12:137-141.

Petrou S, Ugur M, Drummond RM, Singer JJ, Walsh Jr. JV. $P2X_7$ purinoceptor expression in *Xenopus* oocytes is not sufficient to produce a pore-forming P2Z-like phenotype. FEBS Lett 1997; 411:339-345.

Radford KM Virginio C, Surprenant A, North RA, Kawashima E. Baculovirus expression provides direct evidence for heteromeric assembly of $P2X_2$ and $P2X_3$ receptors. J Neurosci 1997; 17:6529-6533.

Rae MG, Rowan EG, Kennedy C. Pharmacological properties of $P2X_3$-receptors present in neurones of the rat dorsal root ganglia. Brit J Pharmacol 1998; 124:176-180.

Rassendren F, Buell G, Newbolt A, North RA, Surprenant A. Identification of amino acid residues contributing to the pore of a P2X receptor. EMBO J 1997a; 16:3446-3454.

Rassendren F, Buell GN, Virginio C, Collo G, North RA. The permeabilizing ATP receptor, $P2X_7$. Cloning and expression of a human cDNA. J Biol Chem 1997b; 272:5482-5486.

Reilly WM, Saville VL, Burnstock G. An assessment of the antagonistic activity of reactive blue 2 at P1 and P2 purinoceptors: supporting evidence for purinergic innervation of the rabbit portal vein. Eur J Pharmacol 1987: 140;47-53.

Rhoads AR, Parui R, Vu ND, Cadogan R, Wagner PD. ATP-induced secretion in PC12 cells and photoaffinity labeling of receptors. J Neurochem 1993; 61:1657-1666.

Robertson SJ, Rae M, Rowan EG, Kennedy C. Characterization of a P2X purinoceptor in cultured neurones of the rat dorsal root ganglia. Brit J. Pharmacol 1996; 118:951-956.
Rogers M, Dani JA. Comparison of quantitative calcium flux through NMDA, ATP, and ACh receptor channels. Biophys J 1995; 68:501-506.

Scase TJ, Heath MF, Allen JM, Sage SO, Evans RJ. Identification of a $P2X_1$ purinoceptor expressed in human platelets. Biochim Biophys Res Comm 1998; 242:525-528.

Seifert R, Schultz G. Involvement of pyrimidinoceptors in the regulation of cell functions by uridine and uracil nucleotides. Trends Pharmacol Sci 1989; 10:365-369.

Simon J, Kidd EJ, Smith FM, Chessell IP, Murrell-Lagnado R, Humphrey PPA, Barnard EA. Localization and functional expression of splice variants of the $P2X_2$ receptor. Mol Pharmacol 1997; 52:237-248.

Soto F, Garcia-Guzman M, Gomez-Hernandez JM, Hollmann M, Karschin C, Stühmer W. P2X$_4$: an ATP-activated ionotropic receptor cloned from rat brain. Proc Nat Acad Sci USA 1996a; 93:3684-3688.

Soto F, Garcia-Guzman M, Karschin C, Stühmer W. Cloning and tissue distribution of a novel P2X receptor from rat brain. Biochem Biophys Res Comm 1996; 223:456-460.

Stoop R, Surprenant A, North RA. Different sensitivities to pH of ATP-induced currents at four cloned P2X receptors. J Neurophysiol 1997; 78:1837-1840.

Surprenant A, Buell G, North RA. P2X receptors bring new structure to ligand-gated ion channels. Trends Neurosci 1995; 18:224-229.

Surprenant A, Rassendren F, Kawashima E, North RA, Buell G. The cytolytic P2Z receptor for extracellular ATP identified as a P2X receptor (P2X$_7$). Science 1996; 272:735-738.

Thomas S, Virginio C, North RA, Surprenant A. The antagonist trinitrophenyl-ATP reveals co-existence of distinct P2X receptor channels in rat nodose ganglion. J Physiol 1998; 509:411-417.

Trezise DJ, Michel AD, Grahames CB, Khakh BS, Surprenant A, Humphrey PP. The selective P2X purinoceptor agonist, b,g-methylene-L-adenosine 5'-triphosphate, discriminates between smooth muscle and neuronal P2X purinoceptors. Naunyn-Schmiedeberg's Arch Pharmacol 1995; 351:603-609.

Urano T, Nishimori H, Han HJ, Furuhata T, Kimura Y, Nakamura Y, Tokino T. Cloning of P2XM, a novel P2X receptor gene regulated by p53. Cancer Res 1997; 57:3281-3287.

Valera S, Hussy N, Evans RJ, Adami N, North RA, Surprenant A, Buell G. A new class of ligand-gated ion channel defined by P2X receptor for extracellular ATP. Nature 1994; 371:516-519.

Valera S, Talabot F, Evans RJ, Gos A, Antonarakis SE, Morris MA, Buell GN. Characterization and chromosomal localization of a human P2X receptor from the urinary bladder. Recept Chann 1995; 3:283-289.

Van Rhee AM, Jacobson KA, Garrad R, Weisman GA, Erb L. "P2 receptor modeling and identification of ligand binding sites. In *The P2 Nucleotide Receptors*, JT Turner, GA Weisman & JS Fedan eds, Human Press, New Jersey, Ch.6, pp135-166, 1998.

Virginio C, Church D, North RA, Surprenant A. Effects of divalent cations, protons and calmidazolium at the rat P2X$_7$ receptor. Neuropharmacol 1997; 36:1285-1294.

Voogd TE, Vansterkenburg EL, Wilting J, Janssen LH. Recent research on the biological activity of suramin. Pharmacol Rev 1993; 45:177-203.

Vulchanova L, Arvidsson U, Riedl M, Wang J, Buell G, Surprenant A, North RA, Elde R. Differential distribution of two ATP gated ion channels (P2X receptors) determined by immunohistochemistry. Proc Nat Acad Sci USA 1996; 93:8063-8067.

Vulchanova L, Riedl MS, Shuster SJ, Buell G, Surprenant A, North RA, Elde R. Immunohistochemical study of the P2X$_2$ and P2X$_3$ receptor subunits in rat and monkey sensory neurons and their central terminals. Neuropharmacol 1997; 36:1229-1242.

Waldmann R, Champigny G, Bassilana F, Heurteux C, Lazdunski M. A proton-gated cation channel involved in acid-sensing. Nature 1997; 386:173-177.

Wang CZ, Namba N, Gonoi T, Inagaki N, Seino S. Cloning and pharmacological characterization of a fourth P2X receptor subtype widely expressed in brain and peripheral tissues including various endocrine tissues. Biochem Biophys Res Comm 1996; 196-202.

Welford LA, Cusack NJ, Hourani SMO. ATP analogues and the guinea-pig taenia coli: a comparison of the structure-activity relationships of ectonucleotidases with those of the P2 purinoceptor. Eur J Pharmacol 1986; 129:217-224.

Welford LA, Cusack NJ, Hourani SMO. The structure-activity relationships of ectonucleotidases and of excitatory P2 purinoceptors: evidence that dephosphorylation of ATP analogues reduces pharmacological potency. Eur J Pharmacol 1987; 141:123-130.

Werner P, Seward EP, Buell GN, North RA.   Domains of P2X receptors involved in desensitization. Proc Nat Acad Sci 1996; 93:15485-15490.

Wildman SS, King BF, Burnstock G. Potentiation of ATP-responses at recombinant P2X$_2$ receptor by neuro-transmitters and related substances. Brit J Pharmacol 1997; 120:221-224.

Wildman SS, King BF, Burnstock G. Zn^{2+} modulation of ATP-responses at recombinant P2X$_2$ receptors and its dependence on extracellular pH. Brit J Pharmacol 1998a; 123:1214-1220.

Wildman SS, King BF, Burnstock G. Zn^{2+} modulation of ATP-responses at recombinant P2X receptors: dependence on extracellular pH. Drug Devel Res 1998b; in press.

Wiley JS, Snook MS, Gargett CE, Jamieson GP. "Transduction mechanisms of P2Z purinoceptors". In *P2 purinoceptors: localization, function and transduction mechanisms* (Ciba Foundation Symposium 198), Wiley & Sons, Chichester, pp149-165, 1996.

Wright JM, Li C. Zn^{2+} potentiates steady-state ATP activated currents in rat nodose ganglion neurons by increasing the burst duration of a 35pS channel. Neurosci Lett 1995; 193:177-180.

Xu L, Glassford AJM, Giaccia AJ, Giffard RG.   Acidosis reduces neuronal apoptosis. NeuroReport 1998; 9:875-879.

Zhou X, Galligan JJ. (1996). P2X purinoceptor in cultured myenteric neurons of guinea-pig small intestine. J Physiol 1996; 496:719-729.

Ziganshin AU, Hoyle CHV, Lambrecht G, Mutschler E, Bäumert HG, Burnstock G.   Selective antagonism by PPADS at P2X purinoceptor in rabbit isolated blood vessels. Brit J Pharmacol 1994; 111:923-929.

THE G PROTEIN-COUPLED P2Y RECEPTORS

T. Kendall Harden, Department of Pharmacology, University of North Carolina
School of Medicine, Chapel Hill, NC 27599

SUMMARY

A family of G protein-coupled receptors has been identified that respond to
extracellular adenine and uridine nucleotides. These receptors are designated the
P2Y receptors and are distinguished individually by a numerical subscript. This
family includes five phospholipase C-activating mammalian receptors, the $P2Y_1$,
$P2Y_2$, $P2Y_4$, $P2Y_6$, and $P2Y_{11}$ receptors, that have been cloned and unambiguously
shown to be activated by nucleotides after heterologous expression. Several
mammalian P2Y receptors remain to be cloned, including an ADP-activated
receptor(s) that couples to G proteins of the Gi family and inhibits adenylyl cyclase.
Three non-mammalian P2Y receptors also have been cloned. The P2Y receptors
exhibit a broad tissue distribution. However, lack of availability of P2Y receptor
subtype selective agonists and antagonists has made difficult the alignment of
nucleotide-promoted physiological/ pharmacological responses of various tissues
with a specific receptor. The most encouraging progress in drug development has
been in the synthesis of selective agonists and antagonists for the $P2Y_1$ receptor.
Mutational and modeling analyses have provided some insight into identification of
the binding pocket for $P2Y_1$ and $P2Y_2$ receptors.

TEXT

Extracellular nucleotides interact with a group of G protein-coupled
metabotropic receptors collectively termed the P2Y receptors. The "P2Y receptor"
terminology follows from and also supplants that coined in 1985 by Burnstock and
Kennedy in their initial conception of a subdivision of P2 receptors into P2X- and
P2Y-purinergic receptor subtypes (Burnstock and Kennedy, 1985). P2Y purinergic
receptors were identified originally in studies with intact tissues as receptor(s) that
were potently activated by the ATP analogue 2MeSATP and not by α,βMeATP and
β,γMeATP. As was introduced in the previous chapter by Brian King, more recent
terminology places nucleotide-activated receptors that are ligand-gated ion channels
into the P2X receptor class of signaling proteins; G protein-coupled receptors that
are activated by nucleotides comprise the P2Y receptor class of signaling proteins
(Fredholm et al., 1994; Fredholm et al., 1997a). An important distinction in the new
classification scheme is that "purinergic" has been deleted from the P2X and P2Y
receptor designations. Many of these receptors, in particular several in the P2Y class
of signaling proteins, are activated by pyrimidines rather than, or in addition to,
purines. The original pharmacological distinction of P2X and P2Y purinergic
receptors also no longer holds since in many cases 2MeSATP is more potent than

α,βMeATP or β,γMeATP for activation of P2X receptors (Brake et al., 1994; Valera et al., 1994). Moreover, 2MeSATP is not an agonist at the majority of the P2Y receptors that have been cloned to date. The resolution of P2X and P2Y receptors ultimately depends on molecular cloning of individual members of the two classes, and the discussion in this chapter will focus almost entirely on proteins of defined sequence rather than on the extensive pharmacological studies that partially resolved P2Y receptors prior to their molecular cloning. The distinguishing properties of P2X receptors versus P2Y receptors are summarized in Table 1.

	P2X receptor	P2Y receptor
General mechanism	ionotropic receptor: ligand-gated cation channel	metabotropic: G protein-coupled receptor
Signaling response	depolarization and Ca^{2+} influx	activation of phospholipase C release of internal Ca^{2+} stores activation of protein kinase C
Timeframe of activation	msec	100s of msec to sec
Natural agonists	ATP	ATP, UTP, ADP, UDP
Cloned mammalian receptors	seven ($P2X_1$-$P2X_7$)	five ($P2Y_1$, $P2Y_2$, $P2Y_4$, $P2Y_6$, $P2Y_{11}$)

Table 1. Properties of P2X and P2Y receptors.

A. G protein-mediated signaling

The G protein-coupled receptors by definition couple to heterotrimeric G proteins, which in turn couple to effector enzymes, ion channels, or other proteins. In most cases signal transduction is effected by receptor-promoted exchange of GTP for GDP on the G protein α-subunit followed by dissociation of the heterotrimer into GTP-bound α-subunit plus free βγ-subunit (Gilman, 1995). The activated α-subunit then interacts with and activates its cognate effector protein. Activation is reversed by hydrolysis of GTP in the active GTP·α-subunit in a catalytic step that is promoted by some effector proteins, e.g. phospholipase C-β (Biddlecome et al., 1996), and by a group of GTPase-stimulating proteins named the RGS proteins (Dohlman and Thorner, 1997). The identity of the G protein heterotrimer is established by its α-subunit. Thus, GTP·$α_s$ released from heterotrimeric Gs, GTP·$α_q$ released from heterotrimeric G_q, and GTP·$α_t$ released from G_t (transducin) activate adenylyl cyclase, phospholipase C-β, and cyclic GMP phosphodiesterase, respectively. G protein β- and γ-subunits remain tightly associated under non-denaturing conditions. Although G protein βγ-subunits were originally thought primarily to subserve a chaperone role in maintaining α-subunits in a heterotrimeric inactive state, the βγ-subunit component of the G protein heterotrimer now is known to subserve more complex functions (Clapham and Neer, 1993). For example, G protein βγ-subunits help define receptor-G protein coupling specificity, and release of G protein βγ-subunits accounts entirely for activation of some effector proteins, e.g. K^+ channels (Reuveny et al., 1994), and accounts partially for activation of some effector isoenzymes, e.g. phospholipase C-β2 (Camps et al., 1992; Boyer et al., 1992). At least 5 β-subunit genes and eleven γ-subunit genes exist in mammals and most combinations of β- and γ-subunits form a functional dimer (Clapham and Neer,

188

1997). Thus, it is likely that even more complex functional roles for G protein $\beta\gamma$-subunits will be elucidated.

The five mammalian P2Y receptors that have been cloned to date all activate the inositol lipid signaling pathway as their primary signaling response. Activation involves Gq-mediated stimulation of one of the four mammalian phospholipase C-β isoenzymes (PLC-β1, PLC-β2, PLC-β3, and PLC-β4), which in turn hydrolyze phosphatidylinositol $(4,5)P_2$ to inositol $(1,4,5)P_3$ and diacylglycerol (Harden et al., 1996). Inositol $(1,4,5)P_3$ interacts with a specific receptor on intracellular storage sites for Ca^{2+}, which results in a marked increase in release of Ca^{2+} and in intracellular Ca^{2+} concentration (Berridge and Irvine, 1987). Diacylglycerol in concert with Ca^{2+} and/or phosphatidylserine activates one of the approximately dozen isoenzymes of protein kinase C (Newton, 1995). Although all five of the cloned mammalian P2Y receptors couple to Gq to activate phospholipase C, pertussis toxin partially blocks the inositol lipid responses to $P2Y_2$ and $P2Y_4$ receptor activation. Therefore, these two receptors may also couple to members of the Gi class of heterotrimeric G proteins since these are the only α-subunits that contain the NH_2-terminal cysteine that is the site of ADP-ribosylation catalyzed by pertussis toxin (Gilman, 1995). A Gi-linked P2Y receptor that inhibits adenylyl cyclase in response to adenine nucleotides has been studied for years in platelet membranes (Cooper et al., 1979; Hourani et al., 1994) and more recently in several cultured cell lines (Boyer et al., 1993; Feolde et al., 1995). This receptor(s) has not yet been cloned.

B. Cloned Mammalian P2Y Receptors

The P2Y receptors are members of the superfamily of G protein-coupled receptors of which hundreds have been cloned. These receptors have a predicted seven transmembrane spanning topology as originally described for rhodopsin (Nathans et al., 1983) and the β-adrenergic receptor (Dixon et al., 1986; Yarden et al., 1986). The highest extent of homology among these signaling relatives exists in the α-helical transmembrane spanning domains.

receptor	$P2Y_1$	$P2Y_2$	$P2Y_4$	$P2Y_6$	$P2Y_{11}$
ligand	ADP/ATP	ATP/UTP	UTP	UDP	ATP
signaling response	Gq/PLC-β	Gq/Gi/PLC-β	Gq/PLC-β	Gq/PLC-β	Gq/PLC-β Gs/AC

Table 2. Cloned Mammalian G Protein-Coupled P2Y Receptors

All G protein-coupled receptors that are activated by nucleotides are designated P2Y receptors. Mammalian P2Y receptors that have been identified unambiguously by molecular cloning and conferrence of nucleotide-promoted signaling responses after heterologous expression also are designated by a subscript number (Fredholm et al., 1997a; Fredholm et al., 1997b). Five mammalian G protein-coupled receptors ($P2Y_1$, $P2Y_2$, $P2Y_4$, $P2Y_6$, and $P2Y_{11}$) have been cloned that share up to 51% identity (Table 2) and when expressed in a null cell line confer unambiguous second messenger signaling responses to extracellular nucleotides. These receptors are discussed in detail below. An alignment of their amino acid sequences is presented in Figure 1, and the percent identities of these receptors is compared in Table 3. The properties of three non-mammalian (two avian and a Xenopus receptor) P2Y receptors also are discussed. It is unclear whether these non-mammalian receptors are species homologues of an already identified mammalian receptor, are species homologues of a yet-to-be-identified mammalian receptor, or are non-mammalian P2Y receptors for which a corresponding homologue does not exist in the human (or other mammalian) genome. The number designation of the five mammalian receptors is not consecutive because number designations have been given to several G protein-coupled receptors that may or may not prove to be P2Y receptors. The receptor referred to as a p2y3 receptor has been cloned from chicken and turkey but not from mammalian species. This receptor is discussed in detail below. A G protein-coupled receptor originally cloned from activated chick T-lymphocytes and referred to as the 6H1 gene product (Kaplan et al., 1993; Owman et al., 1996), was later proposed to be the $P2Y_5$ receptor based on radioligand binding assays (Webb et al., 1996b). Others have stably expressed this receptor (Li et al., 1997) and another G protein-coupled receptor (Janssens et al., 1997) with high homology to this receptor and have failed to observe nucleotide-promoted second messenger signaling responses. Thus, the 6H1 G protein-coupled receptor is unlikely to be a P2Y receptor. A G protein-coupled receptor cloned from a human erythroleukemia cell library was initially proposed (Akbar et al., 1996) to be a P2Y receptor (the $P2Y_7$ receptor), but was later illustrated to be a receptor for leukotriene B4 (Yokomizo et al., 1997). A *Xenopus* P2Y receptor (see below) has been referred to as the p2y8 receptor. The sequence of two additional G protein-coupled receptors have been entered into the GenBank data base and given p2y9 and p2y10 monikers based on relatively low homology to the defined P2Y receptors. It is unlikely that these receptors will prove to be bona fide P2Y receptors.

$P2Y_1$	$P2Y_2$	$P2Y_4$	$P2Y_6$	$P2Y_{11}$	
100	38	40	35	33	$P2Y_1$
	100	52	44	29	$P2Y_2$
		100	42	30	$P2Y_4$
			100	28	$P2Y_6$
				100	$P2Y_{11}$

Table 3. Sequence identities among human P2Y receptors

1. $P2Y_1$ receptor

A 362 amino acid G protein-coupled receptor cloned from a chick brain cDNA library was expressed in *Xenopus* oocytes and shown to promote a Ca^{2+}-activated Cl^- current in response to adenine nucleotides (Webb et al., 1993). The pharmacological selectivity exhibited by this receptor was similar to that classically described for a P2Y receptor, and the receptor was subsequently termed the $P2Y_1$ receptor. This receptor also was cloned from turkey, stably expressed in 1321N1 human astrocytoma cells, and shown to activate phospholipase C with no effect on adenylyl cyclase activity (Filtz et al., 1994). The bovine (Henderson et al., 1995), rat (Tokuyama et al., 1995), and human (Schachter et al., 1996; Janssens et al., 1996) homologues of the $P2Y_1$ receptor were cloned subsequently. The 373 amino acid human $P2Y_1$ receptor is approximately 95% identical to the rodent and bovine $P2Y_1$ receptor and is 83% identical to the avian receptor. In addition to activation of phospholipase C, both human and turkey $P2Y_1$ receptors have been shown to activate a novel cation current after expression in *Xenopus* oocytes (O'Grady et al., 1996). Surprisingly, whereas injection of GDPβS blocked the $P2Y_1$ receptor-promoted Ca^{2+}-activated Cl^- current, it had no effect on the novel $P2Y_1$ receptor-promoted cation current. These results suggest that a G protein-independent ionotropic response may be associated with $P2Y_1$ receptor action.

Messenger RNA for the $P2Y_1$ receptor exhibits a broad tissue distribution (Webb et al., 1993) suggesting that this signaling protein may mediate the effects of adenine nucleotides, for example, in vascular smooth muscle relaxation and skeletal muscle metabolism. The human $P2Y_1$ receptor is expressed at high levels in prostate, ovary, small intestine, and colon but at non-detectable levels in heart, brain, lung, liver, and kidney (Janssens et al., 1996).

Both ADP and ATP were reported in early studies to be agonists at the $P2Y_1$ receptor with ADP often more potent than ATP (Webb et al., 1993; Filtz et al., 1994; Henderson et al., 1995; Schachter et al., 1996). The status of ATP as a $P2Y_1$ receptor agonist has been questioned by recent studies using nucleotides that were HPLC-purified just prior to assay and under conditions where conversion of ATP to ADP during the assay was minimized (Leon et al., 1997). Although ADP remained a potent agonist under these conditions, ATP was a weak antagonist rather than an agonist at the $P2Y_1$ receptor expressed in human Jurkat cells. Similarly, 2MeSADP was a potent agonist and 2MeSATP was an antagonist. Somewhat different results were obtained in studies of Ca^{2+} signaling responses to activation of the human $P2Y_1$ receptor stably expressed in 1321N1 human astrocytoma cells (Palmer et al., 1998). ATP was a full or nearly full agonist with a potency 100-fold less than ADP. Induction of agonist-induced desensitization of the $P2Y_1$ receptor prior to testing of agonist activities resulted in a decrease in ADP potency; in contrast a loss of efficacy of ATP accompanied desensitization. These results suggest that the intrinsic efficacy of ATP is less than that of ADP at the $P2Y_1$ receptor. If this proves to be the case, ATP potentially could be either an agonist or weak antagonist depending on the level of $P2Y_1$ receptor expression in a particular tissue.

In light of these uncertainties in assessing the activities of ADP and ATP, the identity of the natural agonist for the $P2Y_1$ receptor remains undefined. However, ADP clearly is more potent than ATP, and the relative significance of extracellular ATP versus ADP may depend on their relative rates of extracellular hydrolysis which in turn will depend on the tissue location of the receptor. ADP obviously will be more important in tissues where rapid hydrolysis of ATP occurs than in tissues where this is not the case. The aforementioned discussion of relative efficacies of the two molecules also means that relative $P2Y_1$ receptor expression may dictate whether ATP is an agonist at all.

The pharmacological selectivity of the $P2Y_1$ receptor has been defined in greater detail than has the selectivity of the other cloned G protein-coupled P2Y receptors. Extensive structure activity relationships have been reported for certain types of substitutions of adenine nucleotides, and a rudimentary view of the $P2Y_1$ receptor pharmacophore is established. Jacobson and coworkers have focused extensively on chain-extended 2-thioether derivatives of ATP and ADP. These molecules provide the most potent $P2Y_1$ receptor agonists available, with some compounds, e.g. 2-hexylthioATP, exhibiting potencies in the 100 pM range (Fischer et al., 1993; Boyer et al., 1995). Since several triphosphate analogues were as potent or more potent than their corresponding diphosphate analogues, it is unclear whether the issue of relative efficacy of ADP versus ATP at the $P2Y_1$ receptor also applies for this series of molecules. The selectivity of 2-thioether-substituted adenine nucleotides for the $P2Y_1$ receptor over other P2 receptors is not fully established. However, relative selectivity for a subset of P2 receptors exists since 2MeSATP is essentially inactive at the cloned receptors that are activated by uridine nucleotides, i.e. the $P2Y_2$, $P2Y_4$, and $P2Y_6$ receptors. Nonetheless, the absolute selectivity of 2-thioether-substituted adenine nucleotide analogues among adenine nucleotide-activated receptors likely is not substantial since several of the cloned P2X receptors are potently activated by 2MeSATP (Brake et al., 1994; Valera et al., 1994). Moreover, 2MeSADP is a potent agonist and 2MeSATP is a potent antagonist at the adenylyl cyclase-inhibiting P2Y receptor of platelets (Hourani et al., 1994). Although AMP itself exhibits no activity at $P2Y_1$ receptors, 2-thioether substitutions that produced ATP and ADP analogues exhibiting picomolar potency also resulted in analogues of AMP that were full agonists with potencies in some cases in the nanomolar range (Fischer et al., 1993; Boyer et al., 1996b). A single 5'-phosphate is apparently both necessary and sufficient for $P2Y_1$ receptor agonist activity since 2-thioether adenosine was not an agonist at this receptor.

Several different substitutions in the N^6-position in the adenine base had no effect and chain-extended amino- or halogen-substitution of the 8-position decreased $P2Y_1$ receptor potency (Fischer et al., 1993; Burnstock et al., 1994). 2'- or 3'-deoxyATP and 2'- or 3'-deoxyADP were inactive or exhibited low potency for the $P2Y_1$ receptor (Burnstock et al., 1994). Adenine nucleotide analogues with substitutions in the 2'- or 3'-positions generally also exhibited low potency at the $P2Y_1$ receptor, although exceptions exist. For example, 3'-NH_2 3'-deoxyATP (Burnstock et al., 1994) and 3'-benzoylbenzoyl ATP (Boyer et al., 1989) were relatively potent $P2Y_1$ receptor agonists. A number of phosphate chain modifications of ATP and ADP, e.g. ATPγS, ATPβS, ADPβS, produce agonists that are similarly potent to ATP or ADP (Fischer et al., 1993; Burnstock et al., 1994). However, the potential contribution of reduced susceptibility to hydrolysis of these analogs to their potencies relative to ADP or ATP has not been clearly established.

Progress in the study of P2 receptors has been slowed by lack of availability of selective receptor antagonists. Although suramin and reactive blue 2 interact with many different proteins, they competitively antagonize P2 receptors (Harden et al., 1995). Selectivity of antagonism among the different classes of P2 receptors has not been clearly defined but is unlikely to be extensive. Both suramin and reactive blue 2 are competitive inhibitors of the $P2Y_1$ receptor within a narrow range of concentrations (Boyer et al., 1994; Charlton et al., 1996).

Lambrecht and coworkers (Lambrecht et al., 1992) introduced pyridoxal phosphate 6-azophenyl 2',4'-disulphonic acid (PPADS) as a competitive antagonist of P2 receptors. PPADS blocks both P2X and P2Y receptors (Ziganshin et al., 1993; Boyer et al., 1994; Bultmann et al., 1994), and therefore, does not distinguish among the major P2 receptor subtypes. The similar potency of PPADS between certain of the members of the P2X and P2Y receptor classes is somewhat surprising based on the remarkably large differences in both structure and function of the ligand-gated P2X receptors and the G-protein-coupled P2Y receptors. The $P2Y_1$ receptor is competitively blocked by PPADS with a K_B of approximately 1 μM (Boyer et al., 1994; Charlton et al., 1996; Schachter et al., 1996).

Competitive antagonists of $P2Y_1$ receptors that are adenosine phosphate analogues recently have been reported. Adenosine 3'-phosphate 5'-phosphosulfate, and adenosine 3',5'-bisphosphate were shown to be partial agonists at the $P2Y_1$ receptor of turkey erythrocyte membranes (Boyer et al., 1996a). These compounds competitively antagonized the effects of full agonists and exhibited K_B values in the low μM range. The presence of a phosphate in the 2'- or 3'-position is necessary for antagonist activity. For example, both adenosine 3'-phosphate 5'-phosphate and adenosine 2'-phosphate 5'-phosphate exhibited competitive antagonist activities, whereas adenosine 5'-monophosphate did not. These molecules also were competitive antagonists without any partial agonist activity at the human $P2Y_1$ receptor stably expressed in 1321N1 human astrocytoma cells. In contrast, none of these phosphate-substituted compounds were antagonists at the $P2Y_2$, $P2Y_4$, or $P2Y_6$ receptors, or at the Gi/adenylyl cyclase-linked P2Y receptor (see below). These observations recently have been extended and a second generation antagonist, i.e. N^6-CH_3 2'-deoxy adenosine 3',5' bisphosphate, has been synthesized that competitively antagonizes the $P2Y_1$ receptor with a K_B of approximately 100 nM (Camaioni et al., 1998; Boyer et al., 1998b).

2. $P2Y_2$ receptor

The 373 amino acid mouse G protein-coupled $P2Y_2$ receptor initially was cloned by Lustig et al. (Lustig et al., 1993) utilizing expression cloning in *Xenopus* oocytes. Nucleotides promoted a Ca^{2+}-activated Cl^- current in oocytes expressing this receptor. The pharmacological selectivity of the expressed receptor was similar to that previously observed for the natively expressed receptor originally referred to as the P2U-purinergic receptor. The 376 amino acid human homologue of the $P2Y_2$ receptor subsequently was cloned (Parr et al., 1994) and proved to be 89% identical to the previously cloned murine $P2Y_2$ receptor. The human $P2Y_2$ receptor is only 40% identical to the human $P2Y_1$ receptor. Stable expression of the human receptor in 1321N1 human astrocytoma cells conferred inositol lipid signaling responses to ATP and UTP. No effect of nucleotides on cyclic AMP accumulation were observed. The human $P2Y_2$ receptor was initially cloned from a human airway epithelial cell library, and $P2Y_2$ receptor mRNA is expressed at high levels in human lung epithelial cells (Parr et al., 1994). The agonist selectivities exhibited by the cloned human $P2Y_2$ receptor are essentially identical to that of the P2Y receptor natively expressed (and called the P2U-receptor) in human airway epithelial cells (Brown et al., 1991; Lazarowski et al., 1995). Although an additional uridine nucleotide-activated receptor exists in human airway epithelial cells (Lazarowski et al., 1997b), the $P2Y_2$ receptor likely is of major importance in airway epithelial cell physiology and of considerable potential importance in drug therapy of airway diseases (Donaldson and Boucher, 1998). $P2Y_2$ receptor mRNA also is widely distributed in other human tissues, with significant levels found in heart, liver, lung, kidney, skeletal muscle, and placenta (Lustig et al., 1993; Parr et al., 1994). Very little $P2Y_2$ receptor message is found in brain.

ATP and UTP are essentially equipotent at the $P2Y_2$ receptor. Thus, the natural agonist for the $P2Y_2$ receptor could be either an adenine or uridine nucleotide. Whereas physiologically relevant release of cellular ATP has been widely observed, regulated release of UTP only has been reported for a few tissues (Saiag et al., 1995; Lazarowski et al., 1997a). The paucity of evidence for release of cellular UTP may be due to methodological rather than biological reasons, and therefore, the relative significance of ATP versus UTP as natural agonists for the $P2Y_2$ receptor remains to be established. Diadenosine tetraphosphate also potently activates the $P2Y_2$ receptor (Lazarowski et al., 1995), and since this and other diadenosine polyphosphates are apparently released from various tissues (Pintor et al., 1993), diadenosine tetraphosphate or a similar molecule also may be a physiological regulator of the $P2Y_2$ receptor. ADP and UDP originally were reported to be agonists at the $P2Y_2$ receptor (Parr et al., 1994; Lazarowski et al., 1995). However, these molecules are essentially inactive (Nicholas et al., 1996) when triphosphate contamination is eliminated and enzymatic conversion to triphosphates is reduced. Very few reports have appeared examining the effects of analogs of ATP or UTP at the $P2Y_2$ receptor. However, both ATPγS and UTPγS (Lazarowski et al., 1995; Lazarowski et al., 1996) are potent $P2Y_2$ receptor agonists. These molecules have an advantage of resistance to hydrolysis relative to their parent naturally occurring nucleotides. Most substitutions of ATP, e.g. thioether substitution in the 2-position of the adenine base, markedly decrease $P2Y_2$ receptor potency. However, substitution in the 5-position of the pyrimidine base of UTP, e.g. 5BrUTP, decreased but did not eliminate $P2Y_2$ receptor activity (Lazarowski et al., 1995). Selective antagonists of the $P2Y_2$ receptor have not yet been identified.

3. $P2Y_4$ receptor

Communi et al. (Communi et al., 1995) and Nguyen et al. (Nguyen et al., 1995) cloned a 365 amino acid G protein-coupled receptor from genomic DNA that is 51% homologous to the human $P2Y_2$ receptor, 40% homologous to the $P2Y_6$ receptor, and 35% homologous to the human $P2Y_1$ receptor. This receptor was designated the $P2Y_4$ receptor, and subsequent to stable expression in 1321N1 cells, was shown to confer UTP-promoted inositol lipid hydrolysis responses with no effect on adenylyl cyclase activity (Communi et al., 1995; Nguyen et al., 1995). The receptor originally was reported to be activated equipotently by UTP and UDP and much less potently (or not at all) by ATP. Nicholas et al (Nicholas et al., 1996) subsequently examined the selectivity of activation of the $P2Y_4$ receptor by nucleotides under conditions in which the purity and stability of agonists during assay were assessed. UTP was the most potent agonist at the $P2Y_4$ receptor, whereas UDP was essentially inactive. The variability in responses to ATP that have been observed at the $P2Y_4$ receptor may reflect its capacity to influence the metabolism of endogenous UTP that is released as a consequence of mechanical stimulation of 1321N1 human astrocytoma cells (Lazarowski et al., 1997a). Rapid measurements of Ca^{2+} responses to $P2Y_4$ receptor activation indicate that the $P2Y_4$ receptor is activated very weakly if at all by ATP. Communi and coworkers (Communi et al., 1996a) also have reported that ATP acts as a pure antagonist at short times of incubation whereas with extended incubation it behaves as a partial agonist. UTPγS is approximately equipotent to UTP as an agonist at the human $P2Y_4$ receptor (Nicholas et al., 1996). In contrast, 5BrUTP is approximately 100-fold less potent than UTP, and 5BrUDP is inactive. 2MeSATP, 2ClATP, and guanine nucleotides also are inactive at the $P2Y_4$ receptor. PPADS has been reported to be a relatively weak antagonist at the $P2Y_4$ receptor (Communi et al., 1996a). Selective antagonists of the $P2Y_4$ receptor have not yet been reported.

The principal signaling response of $P2Y_4$ receptors apparently involves Gq-mediated activation of phospholipase C. However, pertussis toxin treatment partially inhibits $P2Y_4$ receptor-mediated promotion of inositol lipid hydrolysis during short times but not long times of agonist incubation (Communi et al., 1996a), suggesting that the $P2Y_4$ receptor also may couple to members of the Gi family of G proteins.

4. $P2Y_6$ Receptor

Chang et al. (Chang et al., 1995) cloned a 328 amino acid G protein-coupled receptor from a rat aortic smooth muscle cell library that is 42%, 44%, and 35% homologous to the human $P2Y_4$, $P2Y_2$, and $P2Y_1$ receptors, respectively. Stable expression of this receptor in C6 glioma cells conferred nucleotide-stimulated promotion of inositol lipid hydrolysis. No nucleotide-promoted effects on cyclic AMP accumulation were observed. UTP was reported initially to be the most potent agonist at the $P2Y_6$ receptor (Chang et al., 1995). However, subsequent studies under conditions in which the purity and metabolism of agonists were tightly controlled have established that UDP is the most potent natural agonist at this receptor (Nicholas et al., 1996; Communi et al., 1996b). UTP was much less potent than UDP at the $P2Y_6$ receptor and only weak effects of ATP and 2MeSATP were observed. ADP and 2MeSADP also were weak agonists, although more potent than the corresponding triphosphate molecules. No selective antagonists of the $P2Y_6$ receptor have been reported. The human $P2Y_6$ receptor is 88% identical to the rat receptor and also is activated most potently by UDP and not by ATP (Communi et al., 1996b). Relatively high levels of message for the $P2Y_6$ receptor were observed in rat lung, stomach, intestine, spleen, mesentery, and aortic smooth muscle cells (Chang et al., 1995). The receptor was less abundantly expressed in heart and kidney, and was essentially absent in brain. Human $P2Y_6$ receptor message was detected in spleen, thymus, intestine, leukocytes, and placenta (Communi et al., 1996b). Evidence for $P2Y_6$ receptor expression on the apical membrane of human airway epithelial cells has been reported (Lazarowski et al., 1997b).

5. $P2Y_{11}$ receptor

A 371 amino acid P2Y receptor was cloned recently from a human placental library by Communi and coworkers (Communi et al., 1997). This receptor exhibits highest identity (33%) to the $P2Y_1$ receptor and identities of 30% or less to other cloned P2Y receptor family members. As is the case with previously cloned G protein-coupled P2Y receptors, stable expression of the $P2Y_{11}$ receptor in 1321N1 human astrocytoma cells or CHO-K1 cells resulted in conferrence of nucleotide promoted inositol lipid signaling responses. However, in contrast to the previously cloned P2Y receptors, activation of the $P2Y_{11}$ receptor also resulted in elevation of intracellular cyclic AMP levels. This result suggests that the $P2Y_{11}$ receptor also activates adenylyl cyclase, which has been previously suggested but not proven to be the case for unidentified P2Y receptors in various mammalian tissues (Griese et al., 1991; Henning et al., 1993). Studies with the $P2Y_{11}$ receptor stably expressed in 1321N1 human astrocytoma cells indicate that this receptor is much more efficiently coupled to the Gq/phospholipase C signaling pathway than the Gs/adenylyl cyclase pathway (Kennedy et al., 1998). The order of potency for activation of the $P2Y_{11}$ receptor is ATP > 2MeSATP >> ADP. The $P2Y_{11}$ receptor apparently is adenine nucleotide selective since UTP and UDP were inactive. Tissue mRNA distribution studies indicate that the $P2Y_{11}$ receptor is principally localized in cells of hematopoietic origin (Communi et al., 1997).

<u>C. Non-mammalian P2Y receptors</u>

1. Avian p2y3 receptor

A 328 amino acid G protein-coupled P2Y receptor was cloned from a chick brain library (Webb et al., 1996a) that exhibited 62%, 42%, 41%, and 38% homologies to the human $P2Y_6$, $P2Y_2$, $P2Y_4$, and $P2Y_1$ receptors, respectively (Table 4). Expression of this avian receptor in Xenopus oocytes resulted in conferrence of a nucleotide-promoted Ca^{2+}-activated Cl^- current. The order of potency of nucleotide agonists for stimulation of this response was ADP > UDP = UTP. The p2y3 receptor also was stably expressed in human Jurkat cells where UDP was shown to be the most potent agonist with ADP and UTP both approximately 10-fold less potent. Distribution of mRNA encoding this receptor has been examined in avian tissues. The highest level of expression was in the spleen with transcript also present in kidney, lung, brain, and spinal cord.

Based on the high identity between the avian p2y3 receptor and mammalian $P2Y_6$ receptor, Nicholas and coworkers (Nicholas et al., 1998) have stably expressed these receptors in 1321N1 cells and compared their pharmacological selectivities under identical conditions. Both receptors are highly UDP selective, exhibiting at least 100-fold lower potencies for UTP and ADP. ATP is essentially inactive at each receptor. A probe made from the avian p2y3 receptor identified multiple clones for $P2Y_1$, $P2Y_2$, and $P2Y_4$ receptors in two different genomic libraries but failed to identify sequence exhibiting higher homology to the p2y3 receptor. That is, no human p2y3 receptor was identified. These results strongly suggest that the avian p2y3 receptor is most likely a species homologue of the mammalian $P2Y_6$ receptor.

receptor	p2y3	Tp2Y	Xlp2Y
species	chick (turkey)	turkey	Xenopus
agonist selectivity	UDP>ADP=UTP>ATP	UTP=ATP	ATP=UTP=CTP=GTP=ITP
signaling response	Gq/PLC-β	Gq/PLC-β Gi/adenylyl cyclase	Gq/PLC-β (?)
% identity			
$P2Y_1$	38	38	38
$P2Y_2$	42	47	56
$P2Y_4$	41	56	62
$P2Y_6$	62	35	38

Table 4. G Protein-coupled P2Y receptors cloned from non-mammalian species.

2. Turkey P2Y receptor

The coding sequence of a cDNA encoding a 374 amino acid G protein-coupled receptor that exhibits identities of 56%, 47%, 38%, and 35% to the human $P2Y_1$, human $P2Y_2$, human $P2Y_4$, and rat $P2Y_6$ receptors, respectively, was cloned from a turkey blood library (Boyer et al., 1997; Boyer et al., 1998a). Stable expression of this receptor in 1321N1 human astrocytoma cells resulted in conferrence of a marked inositol phosphate response to nucleotides. Although nearest to the mammalian $P2Y_4$ receptor in overall percent identity and in identity in the predicted transmembrane spanning domains, this receptor exhibited a pharmacological selectivity that was most similar to that of the mammalian $P2Y_2$ receptor. UTP, ATP, and A_2P_4 were all potent agonists whereas ADP and UDP exhibited 100-1000-fold lower potencies. The status of this receptor as a species homologue or unique avian receptor for which a yet-to-be-identified homologue exists (or for which no mammalian homologue exists) remains to be defined.

A novel second messenger signaling response exists for this avian P2Y receptor. That is, agonist-promoted inhibition of cyclic AMP accumulation occurs over the same concentration range of agonists that promote inositol phosphate responses (Boyer et al., 1998a). The cyclic AMP inhibitory effect is blocked completely by treatment of cells with pertussis toxin whereas the inositol lipid signaling response is only partially inhibited. Similar results were observed after stable expression of the avian receptor in NIH-3T3 fibroblasts. Thus, this avian P2Y receptor appears to be dually coupled to Gq/phospholipase C-β and Gi/adenylyl cyclase.

3. Xenopus P2Y receptor

A Xenopus cDNA was isolated that encoded a 532 amino acid G protein-coupled receptor containing a long carboxy terminal domain (Bogdanov et al., 1997). Sequence contained in the first 350 amino acids of this receptor beginning from the amino terminus is 62%, 56%, 40%, and 40% identical to the human $P2Y_4$, $P2Y_2$, $P2Y_6$, and $P2Y_1$ receptors, respectively. When expressed in Xenopus oocytes this receptor promoted activation of a Cl^--activated Ca^{2+} current in response to incubation of cells with ATP and other nucleotides. ADPβS was the most effective molecule tested. However, ATP, CTP, GTP, ITP, and UTP were all equi-effective when tested at 100 μM concentrations. 2MeSATP was ineffective and diadenosine polyphosphates were only weakly effective. The relationship of this receptor to the mammalian P2Y receptors has not been established.

D. Other G Protein-Coupled P2 Receptors

The cloned mammalian G protein-coupled P2 receptors discussed thus far mediate a large percentage of the physiologically important responses to extracellular nucleotides. However, it is likely that additional mammalian P2 receptors exist that have not been cloned. For example, receptors that are specifically activated by dinucleotide polyphosphates have been proposed to exist (Pintor et al., 1991; Pintor et al., 1993). A Gi/adenylyl cyclase linked P2Y receptor(s) has been widely studied and is discussed in some detail below.

An ADP-activated aggregation-promoting P2 receptor(s) exists on platelets, i.e. the P2T receptor(s) (Haslam et al., 1981; Hourani et al., 1994). Although the existence of this receptor(s) has been realized for over three decades (Born, 1962), molecular understanding of this signaling protein(s) has evolved slowly. This lack of progress may be explained in part by the likelihood that multiple receptors comprise the "P2T receptor" response of platelets. The P2T receptor originally was distinguished from other P2 receptors by its adenine nucleotide selectivity. ADP and many analogues of ADP were shown to be agonists at the platelet P2T receptor, whereas ATP and several ATP analogues were competitive antagonists (Haslam et al., 1981; Cusack et al., 1982). ADP elevates intra-platelet Ca^{2+} levels (Hourani et al., 1994). The mechanism underlying this ADP-promoted Ca^{2+} response has remained uncertain since biochemical demonstration of P2T receptor-mediated activation of phospholipase C in platelets has been difficult to observe. However, pharmacological data suggest that the phospholipase C-activating $P2Y_1$ receptor is expressed in platelets and that this receptor mediates the Ca^{2+}-elevating effects of ADP (Leon et al., 1997). The recent results described above illustrating that ADP is a full agonist at the human $P2Y_1$ receptor while ATP is less effective or inactive also provide important insight into the action of ADP in platelets (Leon et al., 1997). Indeed, Gachet and coworkers have made a strong case that much if not all of the effects of ADP on platelet function are mediated through a $P2Y_1$ receptor.

Inhibition of adenylyl cyclase in response to ADP also occurs in platelet membranes (Cooper et al., 1979), but this biochemical effect apparently neither solely accounts for, nor is a direct sequelae of, platelet aggregation. Differential effects of ADP and some of its analogues on aggregation versus second messenger signaling responses form reasonably strong support for the idea that multiple P2T receptors exist on platelets.

Extracellular adenine nucleotides also inhibit intracellular cyclic AMP accumulation in a number of other target tissues (Okajima et al., 1987; Sato et al., 1992; Pianet et al., 1989; Yamada et al., 1992; Boyer et al., 1993). The most likely mechanism accounting for phosphodiesterase-independent decreases in cyclic AMP levels involves a Gi-promoted inhibition of adenylyl cyclase activity. The G protein-coupled P2 receptors cloned to date selectively, if not specifically, interact with components of the phospholipase C signaling cascade. Thus, adenine nucleotide-promoted inhibition (or activation) of adenylyl cyclase most likely involves a receptor or class of receptors that is uniquely different from the cloned P2Y receptors extant.

C6 rat glioma cells have provided a useful model system to study a P2 receptor that inhibits adenylyl cyclase (Valeins et al., 1992; Boyer et al., 1993; Boyer et al., 1995). Incubation of these cells with ADP, ATP, or a broad range of adenine nucleotide analogs results in marked decreases in intracellular cyclic AMP levels. In contrast, extracellular adenine nucleotides had no effect on inositol lipid hydrolysis and Ca^{2+} mobilization in C6 cells (Boyer et al., 1993; Lazarowski et al., 1994). Pertussis toxin, which ADP-ribosylates and functionally inactivates the members of the Gi class of G proteins, completely blocked the capacity of extracellular adenine nucleotides to decrease cyclic AMP levels in C6 glioma cells (Boyer et al., 1993; Boyer et al., 1995). The unique signaling properties of the P2Y receptor natively expressed in C6 glioma cells has been emphasized by stable expression of the turkey, human, or rat $P2Y_1$ receptor in these cells. Whereas no adenine nucleotide-stimulated inositol lipid signaling response is observed in wild-type C6 cells, marked effects of adenine nucleotides on inositol phosphate accumulation are observed in engineered cells (Schachter et al., 1996; Schachter et al., 1997). Thus, a receptor for extracellular adenine nucleotides exists on C6 glioma cells that couples to Gi and adenylyl cyclase without interfacing with the Gq-regulated inositol lipid signaling cascade. The agonist selectivity of this receptor is similar to that of the $P2Y_1$

receptor, and therefore, as discussed above is similar to that of the adenylyl cyclase-inhibiting P2Y receptor of platelets. Whether these are the same receptors or are unique members of a subclass of P2Y receptors that couple through Gi to inhibit adenylyl cyclase remains to be unambiguously established by molecular cloning. However, they share antagonist selectivities that distinguish them from the $P2Y_1$ receptor. For example, neither the platelet nor C6 glioma cell Gi/adenylyl cyclase-coupled P2Y receptor(s) are antagonized by adenosine 3',5' bisphosphate or PPADS (Boyer et al., 1994; Boyer et al., 1996a), which both are potent competitive antagonists of the $P2Y_1$ receptor.

<u>E. P2Y receptor structure</u>

The G protein-coupled receptors share an overall topology that predicts a seven transmembrane spanning protein with an extracellular amino terminus and a cytoplasmic carboxy terminus (Strader et al., 1994). Three dimensional structure exists for none of these proteins, and therefore, understanding of their ligand binding pockets and functional domains has been entirely deduced from a large number of mutational studies and from relatively fewer studies involving chemical modification (Strader et al., 1995). Molecular models of various G protein-coupled receptors have been generated (van Rhee and Jacobson, 1996) based on an electron density map of the G protein-coupled photoreceptor rhodopsin (Shertler et al., 1993). Van Rhee and coworkers have developed such a model for the $P2Y_1$ receptor (van Rhee et al., 1995), and site-directed mutagenesis has been carried out to confirm certain aspects of this model (Jiang et al., 1997). For example, marked effects on agonist interaction with the human $P2Y_1$ receptor resulted from mutation of residues on the extrafacial side of transmembrane domain (TM) 3, 5, 6, and 7. All of these residues were predicted by modeling to be important in agonist binding. Positively charged residues in TM3 and TM7 near the exofacial surface apparently interact with the negatively charged phosphate moieties of ATP and ADP. In contrast, positively charged residues in TM6 and TM7 of the $P2Y_2$ receptor were proposed on the basis of mutational studies to coordinate with the phosphate group(s) of UTP (Erb et al., 1995). The reader is referred to a review by van Rhee et al. (van Rhee et al., 1998) for a detailed discussion of the sequence alignment, molecular modeling, mutagenesis, and predicted binding determinants of P2Y receptors.

III. Future Directions

Remarkable progress has been made in molecular understanding of P2Y receptors in the past five years. Although biochemical studies, e.g. (Dubyak et al., 1988; Boyer et al., 1989), had strongly suggested that a group of G protein-coupled receptors existed that were activated by extracellular adenine and uridine nucleotides, it was the molecular cloning in 1993 of the $P2Y_1$ receptor by Webb et al. (Webb et al., 1993) and the $P2Y_2$ receptor by Lustig et al. (Lustig et al., 1993) that unambiguously established this to be the case. Complexities of nucleotide-regulated signaling suggested by pharmacological studies over the past three decades now can be resolved in a template of five mammalian P2Y receptors of defined structure. Several more G protein-coupled P2Y receptors remain to be cloned including a Gi/adenylyl cyclase-linked P2Y receptor that is prominent in platelets and in other tissues. Indeed, multiple receptors may exist in a P2Y receptor subclass that couple to Gi rather than Gq and phospholipase C. It will be important to carefully align tissue and cellular distribution of all of the unambiguously identified P2Y receptors with the corresponding physiological/pharmacological responses to nucleotides. The availability of cloned P2Y receptors is rapidly leading to development of selective high affinity agonists and antagonists of these receptors. This ligand development in turn will be rapidly applied to better resolve and understand the multifarious physiological responses mediated by P2Y receptors. Similarly, cloning of the P2Y receptors will lead to gene-knockout and gene-modified mice that will concurrently allow elucidation of physiological roles of each of the nucleotide-regulated receptors. Resolution of the true three-dimensional structure of P2Y receptors will follow only after the technical issues inherent in production of large amounts of pure receptor protein and production of high quality crystals from a membrane-imbedded protein are surmounted. Again, availability of DNA encoding the P2Y receptors at least places investigators at the threshold of approaches for overexpression of proteins that otherwise have proven difficult to purify from native sources. In summary, we are clearly in a new era of investigation, and the often neglected significance of nucleotides as signaling molecules, e.g. in the autonomic nervous system, no longer can be overlooked. Albeit still at a nascent stage, true molecular understanding of P2Y receptors is now conceivable.

REFERENCES

Akbar GKM, Dasari VR, Webb TE, Ayyanathan K, Pillarisetti K, Sandhu AK, Athwal RS, Daniel JL, Ashby B, Barnard EA and Kunapuli SP (1996) Molecular cloning of a novel P2 purinoceptor from human erythroleukemia cells. *J Biol Chem* **271**:18363-18367.

Berridge MJ and Irvine RF (1987) Inositol trisphosphate and diacylglycerol: two interacting second messengers. *Annu Rev Biochem* **56**:159-193.

Biddlecome GH, Bernstein G and Ross EM (1996) Regulation of phospholipase C-β1 by Gq and m1 muscarinic cholinergic receptor. Steady-state balance of receptor-mediated activation and GTPase-activating protein-promoted deactivation. *J Biol Chem* **271**:7999-8007.

Bogdanov YD, Dale L, King BF, Whittock N and Burnstock G (1997) Early expression of a novel nucleotide receptor in the neural plate of *Xenopus* embryos. *J Biol Chem* **272**:12583-12590.

Born GVR (1962) Aggregation of blood platelets by adenosine diphosphate and its reversal. *Nature* **194**:927-928.

Boyer JL, Downes CP and Harden TK (1989) Kinetics of activation of phospholipase C by P_2 purinergic receptor agonists and guanine nucleotides. *J Biol Chem* **264**:884-890.

Boyer JL, Waldo GL and Harden TK (1992) βγ-Subunit activation of G-protein-regulated phospholipase C. *J Biol Chem* **267**:25451-25456.

Boyer JL, Lazarowski ER, Chen X-H and Harden TK (1993) Identification of a P_{2Y}-purinergic receptor that inhibits adenylyl cyclase but does not activate phospholipase C. *J Pharmacol Exp Ther* **267**:1140-1146.

Boyer JL, Zohn I, Jacobson KA and Harden TK (1994) Differential effects of putative P_2-purinoceptor antagonists on phospholipase C- and adenylyl cyclase-coupled P_{2Y}-purinoceptors. *Br J Pharmacol* **113**:614-620.

Boyer JL, O'Tuel JW, Fischer B, Jacobson KA and Harden TK (1995) Potent agonist action of 2-thioether derivatives of adenine nucleotides at adenylyl cyclase-linked P2Y-purinoceptors. *Br J Pharmacol* **116**:2611-2616.

Boyer JL, Romero T, Schachter JB and Harden TK (1996a) Identification of competitive antagonists of the P2Y$_1$ receptor. *Mol Pharmacol* **50**:1323-1329.

Boyer JL, Siddiqi SM, Fischer B, Jacobson KA and Harden TK (1996b) Identification of potent P$_{2Y}$-purinoceptor agonists that are derivatives of adenosine 5'-monophosphate. *Br J Pharmacol* **118**:1959-1964.

Boyer JL, Waldo GL and Harden TK (1997) Molecular cloning and expression of an avian G protein-coupled P2Y receptor. *Mol Pharmacol* **52**:928-934.

Boyer JL, Delaney SA and Harden TK (1998a) An avian P2Y receptor that couples through Gi to inhibit adenylyl cyclase. *submitted*

Boyer JL, Mohanram A, Camaioni E, Jacobson KA and Harden TK (1998b) Competitive and selective antagonism of P2Y$_1$ receptors by N^6-methyl 2'-deoxyadenosine 3',5'-*bis*phosphate, in *Br J Pharmacol*, in press.

Boyer JL and Harden TK (1989) Irreversible activation of phospholipase C-coupled P$_{2Y}$-purinergic receptors by 3'-0-(4-benzoyl) benzoyl adenosine 5'-triphosphate. *Mol Pharmacol* **36**:831-835.

Brake AJ, Wagenbach MJ and Julius D (1994) A new structural motif for ligand-gated ion channels defined by an ionotropic ATP receptor. *Nature* **371**:519-523.

Brown HA, Lazarowski ER, Boucher RC and Harden TK (1991) Evidence that UTP and ATP regulate phospholipase C through a common extracellular 5'-nucleotide receptor in human airway epithelial cells. *Mol Pharmacol* **40**:648-655.

Bultmann R and Starke K (1994) P$_2$-purinoceptor antagonists discriminate three contraction-mediating receptors for ATP in rat vas deferens. *Naunyn Schmiedebergs Arch Pharmacol* **349**:74-80.

Burnstock G, Fischer B, Hoyle CHV, Maillard M, Ziganshin AU, Brizzolara AL, von Isakovics A, Boyer JL, Harden TK and Jacobson KA (1994) Structure activity relationships for derivatives of adenosine-5'-triphosphate as agonists at P$_2$-purinoceptors: heterogeneity within P$_{2X}$ and P$_{2Y}$ subtypes. *Drug Dev Res* **31**:206-219.

Burnstock G and Kennedy C (1985) Is there a basis for distinguishing two types of P$_2$-purinoceptors? *Gen Pharmacol* **16**:433-440.

Camaioni E, Boyer JL, Mohanram A, Harden TK and Jacobson KA (1998) Deoxyadenosine-bisphosphate derivatives as potent antagonists at P2Y$_1$ receptors. *J Med Chem* **41**:183-190.

Camps M, Carozzi A, Schnabel P, Scheer A, Parker PJ and Gierschik P (1992) Isozyme-selective stimulation of phospholipase C-β2 by G protein βγ-subunits. *Nature* **360**:684-686.

Chang K, Hanaoka K, Kumada M and Takuwa Y (1995) Molecular cloning and functional analysis of a novel P$_2$ nucleotide receptor. *J Biol Chem* **270**:26152-26158.

Charlton SJ, Brown CA, Weisman GA, Turner JT, Erb L and Boarder MR (1996) PPADS and suramin as antagonists at cloned P2Y- and P2U-purinoceptors. *Br J Pharmacol* **118**:704-710.

Clapham DE and Neer EJ (1993) New roles for G-protein βγ-dimers in transmembrane signalling. *Nature* **365**:403-406.

Clapham DE and Neer EJ (1997) G protein βγ subunits. *Annu Rev Pharmacol Toxicol* **37**:167-203.

Communi D, Pirotton S, Parmentier M and Boeynaems JM (1995) Cloning and functional expression of a human uridine nucleotide receptor. *J Biol Chem* **270**:30849-30852.

Communi D, Motte S, Boeynaems JM and Pirotton S (1996a) Pharmacological characterization of the human P2Y$_4$ receptor. *Eur J Pharmacol* **317**:383-389.

Communi D, Parmentier M and Boeynaems J-M (1996b) Cloning, functional expression and tissue distribution of the human P2Y$_6$ receptor. *Biochem Biophys Res Commun* **222**:303-308.

Communi D, Govaerts C, Parmentier M and Boeynaems J-M (1997) Cloning of a human purinergic P2Y receptor coupled to phospholipase C and adenylyl cyclase. *J Biol Chem* **272**:31969-31973.

Cooper DMF and Rodbell M (1979) ADP is a potent inhibitor of human platelet plasma membrane adenylate cyclase. *Nature* **282**:517-518.

Cusack NJ and Hourani SMO (1982) Competitive inhibition by adenosine 5'-triphosphate of the actions on human platelets of 2-chloroadenosine 5'-diphosphate, 2-azidoadenosine 5'-diphosphate and 2-methylthioadenosine 5'-diphosphate. *Br J Pharmacol* **77**:329-333.

Dixon RAF, Kobilka BK, Strader DJ, Benovic JL, Dohlman HG, Frielle T, Bolanowski MA, Bennett CD, Rands E, Diehl RE, Mumford RA, Slater EE, Sigal IS, Caron MG, Lefkowitz RJ and Strader CD (1986) Cloning of the gene and cDNA for mammalian beta-adrenergic receptor and homology with rhodopsin. *Nature* **321**:75-79.

Dohlman HG and Thorner J (1997) RGS proteins and signaling by heterotrimeric G proteins. *J Biol Chem* **272**:3871-3874.

Donaldson SH and Boucher RC (1998) Therapeutic applications for nucleotides in lung disease, in *The P2 Nucleotide Receptors* (Turner JT, Weisman GA and Fedan JS eds) pp 413-424, Humana Press, Inc. Totowa, NJ.

Dubyak GR, Cowen DS and Meuller LM (1988) Activation of inositol phospholipid breakdown in HL60 cells by P$_2$-purinergic receptors for extracellular ATP. Evidence for mediation by both pertussis toxin-sensitive and pertussis toxin-insensitive mechanisms. *J Biol Chem* **263**:18108-18117.

Erb L, Garrard R, Wang Y, Quinn T, Turner JT and Weisman GA (1995) Site-directed mutagenesis of P$_{2U}$-purinoceptors: positively charged amino acids in transmembrane helices 6 and 7 affect agonist potency and efficacy. *J Biol Chem* **270**:4185-4188.

Feolde E, Vigne P, Breittmayer JP and Frelin C (1995) ATP, a partial agonist of atypical P$_{2Y}$ purinoceptors in rat brain microvascular endothelial cells. *Br J Pharmacol* **115**:1199-1203.

Filtz TM, Li Q, Boyer JL, Nicholas RA and Harden TK (1994) Expression of a cloned P$_{2Y}$-purinergic receptor that couples to phospholipase C. *Mol Pharmacol* **46**:8-14.

Fischer B, Boyer JL, Hoyle CHV, Ziganshin AU, Brizzolara AL, Knight GE, Zimmet J, Burnstock G, Harden TK and Jacobson KA (1993) Identification of potent, selective P$_{2Y}$-purinoceptor agonists: structure activity relationships for 2-thioether derivatives of adenosine-5'-triphosphate. *J Med Chem* **36**:3937-3946.

Fredholm BB, Abbracchio MP, Burnstock G, Daly JW, Harden TK, Jacobson KA, Leff P and Williams M (1994) Nomenclature and classification of purinoceptors. *Pharmacol Rev* **46**:143-156.

Fredholm BB, Abbracchio MP, Burnstock G, Dubyak GR, Harden TK, Jacobson KA, Schwabe U and Williams M (1997a) Towards a revised nomenclature for P1 and P2 receptors. *Trends Pharmacol Sci* **18**:79-82.

Fredholm BB, Burnstock G, Harden TK and Spedding M (1997b) P2 receptor nomenclature. *Drug Dev Res* **39**:461-466.

Gilman AG (1995) G proteins and regulation of adenylate cyclase. *Angew Chem Int Ed Engl* **34**:1406-1419.

Griese M, Gobran LI and Rooney SA (1991) A$_2$ and P$_2$ purine receptor interactions and surfactant secretion in primary cultures of type II cells. *Am J Physiol* **261**:L140-L147.

Harden TK, Boyer JL and Nicholas RA (1995) P$_2$-purinergic receptors: subtype-associated signaling responses and structure. *Annu Rev Pharmacol Toxicol* **35**:541-579.

Harden TK, Filtz TM, Paterson A, Galas MC, Boyer JL and Waldo GL (1996) Regulation of phospholipase C-β isoenzymes, in *Bioactive Lipids* (Vanderhoek JY ed) pp 257-263, Plenum Press,

Haslam RJ and Cusack NJ (1981) Blood platelet receptors for ADP and for adenosine, in *Purinergic receptors* (Burnstock G ed) pp 221-286, Chapman & Hall, Ltd. London.

Henderson DJ, Elliot DG, Smith GM, Webb TE and Dainty IA (1995) Cloning and characterization of a bovine P_{2Y} receptor. *Biochem Biophys Res Commun* **212**:648-656.

Henning RH, Duin M, den Hertog A and Nelemans A (1993) Characterization of P_2-purinoceptor mediated cyclic AMP formation in mouse C2C12 myotubes. *Br J Pharmacol* **110**:133-138.

Hourani SMO and Hall DA (1994) Receptors for ADP on human blood platelets. *Trends Pharmacol Sci* **15**:103-108.

Janssens R, Communi D, Pirotton S, Samson M, Parmentier M and Boeynaems J-M (1996) Cloning and tissue distribution of the human $P2Y_1$ receptor. *Biochem Biophys Res Commun* **221**:588-593.

Janssens R, Boeynaems JM, Godart M and Communi D (1997) Cloning of a human heptahelical receptor closely related to the $P2Y_5$ receptor. *Biochem Biophys Res Commun* **236**:106-112.

Jiang Q, van Rhee AM, Nicholas RA, Schachter JB, Harden TK and Jacobson KA (1997) A mutational analysis of residues essential for ligand recognition at the human $P2Y_1$ receptor. *Mol Pharmacol* **52**:499-507.

Kaplan MH, Smith DI and Sundick RS (1993) Identification of a G protein coupled receptor induced in activated T cells. *J Immunol* **151**:628-636.

Kennedy C, Harden TK and Nicholas RA (1998) Signaling properties of the human $P2Y_{11}$ receptor. *in preparation*

Lambrecht G, Friebe T, Grimm U, Windscheif U, Bungardt E, Hildebrandt C, Baumert H, Spatz-Kumbel G and Mutschler E (1992) PPADS, a novel functionally selective antagonist of P_2 purinoceptor-mediated responses. *Eur J Pharmacol* **217**:217-219.

Lazarowski ER, Watt WC, Stutts MJ, Boucher RC and Harden TK (1995) Pharmacological selectivity of the cloned human phospholipase C-linked P_{2U}-purinergic receptor: potent activation by diadenosine tetraphosphate. *Br J Pharmacol* **116**:1619-1627.

Lazarowski ER, Stutts MJ, Watt WC, Boucher RC and Harden TK (1996) Enzymatic synthesis of UTPγS, a potent hydrolysis resistant agonist of P_{2U}-purinoceptors. *Br J Pharmacol* **117**:203-209.

Lazarowski ER, Homolya L, Boucher RC and Harden TK (1997a) Direct demonstration of mechanically induced release of cellular UTP and its implication for uridine nucleotide receptor activation. *J Biol Chem* **272**:24348-24354.

Lazarowski ER, Paradiso AM, Watt WC, Harden TK and Boucher RC (1997b) UDP activates a mucosal-restricted receptor on human nasal epithelial cells that is distinct from the $P2Y_2$ receptor. *Proc Natl Acad Sci USA* **94**:2599-2603.

Lazarowski ER and Harden TK (1994) Identification of a uridine nucleotide-selective G-protein-linked receptor that activates phospholipase C. *J Biol Chem* **269**:11830-11836.

Leon C, Hechler B, Vial C, Leray C, Cazenave J-P and Gachet C (1997) The $P2Y_1$ receptor is an ADP receptor antagonized by ATP and expressed in platelets and megakaryoblastic cells. *FEBS Lett* **403**:26-30.

Li Q, Schachter JB, Harden TK and Nicholas RA (1997) The 6H1 orphan receptor, claimed to be the p2y5 receptor, does not mediate nucleotide-promoted second messenger responses. *Biochem Biophys Res Commun* **236**:455-460.

Lustig KD, Shiau AK, Brake AJ and Julius D (1993) Expression cloning of an ATP receptor from mouse neuroblastoma cells. *Proc Natl Acad Sci USA* **90**:5113-5117.

Nathans J and Hogness DS (1983) Isolation, sequence analysis, and intron-exon arrangement of the gene encoding bovine rhodopsin. *Cell* **34**:807-814.

Newton AC (1995) Protein kinase C: structure, function, and regulation. *J Biol Chem* **270**:28495-28498.

Nguyen T, Erb L, Weisman GA, Marchese A, Heng HHQ, Garrad RC, George SR, Turner JT and O'Dowd BF (1995) Cloning, expression, and chromosomal localization of the human uridine nucleotide receptor gene. *J Biol Chem* **270**:30845-30848.

Nicholas RA, Watt WC, Lazarowski ER, Li Q and Harden TK (1996) Uridine nucleotide selectivity of three phospholipase C-activating P_2 receptors: Identification of a UDP-selective, a UTP-selective, and an ATP- and UTP-specific receptor. *Mol Pharmacol* **50**:224-229.

Nicholas RA, Li Q, Olesky M and Harden TK (1998) Evidence that the human $P2Y_6$ and avian p2Y3 receptors are species homologues. *submitted*

Okajima F, Tokumitsu Y, Kondo Y and Ui M (1987) P_2-purinergic receptors are coupled to two signal transduction systems leading to inhibition of cAMP generation and to production of inositol triphosphate in rat hepatocytes. *J Biol Chem* **262**:13483-13490.

O'Grady SM, Elmquist E, Filtz TM, Nicholas RA and Harden TK (1996) A guanine nucleotide-independent inwardly rectifying cation permeability is associated with $P2Y_1$ receptor expression in *Xenopus* oocytes. *J Biol Chem* **271**:29080-29087.

Owman C, Nilsson C and Lolait SJ (1996) Cloning of cDNA encoding a putative chemoattractant receptor. *Genomics* **37**:187-194.

Palmer RK, Harden TK and Boyer JL (1998) Apparent efficacies of ATP and ADP at the human $P2Y_1$ receptor quantitated in rapid measurements of Ca^{2+} response. *submitted*

Parr CE, Sullivan DM, Paradiso AM, Lazarowski ER, Burch LH, Olsen JC, Erb L, Weisman GA, Boucher RC and Turner JT (1994) Cloning and expression of a human P_{2U} nucleotide receptor, a target for cystic fibrosis pharmacology. *Proc Natl Acad Sci USA* **91**:3275-3279.

Pianet I, Merle M and Labouesse J (1989) ADP and, indirectly, ATP are potent inhibitors of cAMP production in intact isoproterenol-stimulated C6 glioma cells. *Biochem Biophys Res Commun* **163**:1150-1157.

Pintor J, Torres M, Castro E and Miras-Portugal MT (1991) Characterization of diadenosine tetraphosphate (Ap_4A) binding sites in cultured chromaffin cells: evidence for a P_{2Y} site. *Br J Pharmacol* **103**:1980-1984.

Pintor J and Miras-Portugal MT (1993) Diadenosine polyphosphates (Ap_xA) as new neurotransmitters. *Drug Dev Res* **28**:259-262.

Reuveny E, Slesinger PA, Inglese J, Morales JM, Iñiguez-Lluhi JA, Lefkowitz RJ, Bourne HR, Jan YN and Jan LY (1994) Activation of the cloned muscarinic potassium channel by G protein $\beta\gamma$ subunits. *Nature* **370**:143-146.

Saiag B, Bodin P, Shacoori V, Catheline M, Rault B and Burnstock G (1995) Uptake and flow-induced release of uridine nucleotides from isolated vascular endothelial cells. *Endothelium* **2**:279-285.

Sato K, Okajima F and Kondo Y (1992) Extracellular ATP stimulates three different receptor-signal transduction systems in FRTL-5 thyroid cells. *Biochem J* **283**:281-287.

Schachter JB, Li Q, Boyer JL, Nicholas RA and Harden TK (1996) Second messenger cascade specificity and pharmacological selectivity of the human $P2Y_1$ receptor. *Br J Pharmacol* **118**:167-173.

Schachter JB, Boyer JL, Li Q, Nicholas RA and Harden TK (1997) Fidelity in functional coupling of the rat P2Y$_1$ receptor to phospholipase C. *Br J Pharmacol* **122**:1021-1024.

Shertler GF, Villa C and Henderson R (1993) Projection structure of rhodopsin. *Nature* **362**:770-772.

Strader CD, Fong TM, Tota MR, Underwood D and Dixon RA (1994) Structure and function of G protein-coupled receptors. *Annu Rev Biochem* **63**:101-132.

Strader CD, Fong TM, Graziano MP and Tota MR (1995) The family of G-protein-coupled receptors. *FASEB J* **9**:745-754.

Tokuyama Y, Hara M, Jones EMC, Fan Z and Bell GI (1995) Cloning of rat and mouse P$_2$$_Y$ purinoceptors. *Biochem Biophys Res Commun* **211**:211-218.

Valeins H, Merle M and Labouesse J (1992) Pre-steady state study of the β-adrenergic and purinergic receptor interaction in C6 cell membranes: undelayed balance between positive and negative coupling to adenylyl cyclase. *Mol Pharmacol* **42**:1033-1041.

Valera S, Hussy N, Evans RJ, Adami N, North RA, Surprenant A and Buell G (1994) A new class of ligand-gated ion channel defined by P$_{2X}$ receptor for extracellular ATP. *Nature* **371**:516-519.

van Rhee AM, Fischer B, van Galen PJM and Jacobson KA (1995) Modelling the P$_{2Y}$ purinoceptor using rhodopsin as template. *Drug Des Discov* **13**:133-154.

van Rhee AM, Jacobson KA, Garrad R, Weisman GA and Erb L (1998) P2 receptor modeling and identification of ligand binding sites, in *The P2 Nucleotide Receptors* (Turner JT, Weisman GA and Fedan JS eds) pp 135-166, Humanna Press Inc. Totowa, NJ.

van Rhee AM and Jacobson KA (1996) Molecular architecture of G protein-coupled receptors. *Drug Dev Res* **37**:1-38.

Webb TE, Simon J, Krishek BJ, Bateson AN, Smart TG, King BF, Burnstock G and Barnard EA (1993) Cloning and functional expression of a brain G-protein-coupled ATP receptor. *FEBS Lett* **324**:219-225.

Webb TE, Henderson D, King BF, Wang S, Simon J, Bateson AN, Burnstock G and Barnard EA (1996a) A novel G protein-coupled P2 purinoceptor (P2Y$_3$) activated preferentially by nucleoside diphosphates. *Mol Pharmacol* **50**:258-265.

Webb TE, Kaplan MG and Barnard EA (1996b) Identification of 6H1 as a P$_{2Y}$ purinoceptor: P2Y$_5$. *Biochem Biophys Res Commun* **219**:105-110.

Yamada M, Hamamori Y, Akita H and Yokoyama M (1992) P$_2$-purinoceptor activation stimulates phosphoinositide hydrolysis and inhibits accumulation of cAMP in cultured ventricular myocytes. *Circ Res* **70**:477-485.

Yarden Y, Rodriguez H, Wong SK-F, Brandt DR, May DC, Burnier J, Harkins RN, Chen EY, Ramachandran J, Ullrich A and Ross EM (1986) The avian β-adrenergic receptor: primary structure and membrane topology. *Proc Natl Acad Sci USA* **83**:6795-6799.

Yokomizo T, Izumi T, Chang K, Takuwa Y and Shimizu T (1997) A G-protein-coupled receptor for leukotrine B$_4$ that mediates chemotaxis. *Nature* **387**:620-624.

Ziganshin AU, Hoyle CHV, Bo H, Lambrecht G, Mutschler E, Bäumert HG and Burnstock G (1993) PPADS selectively antagonizes P$_{2X}$-purinoceptor-mediated responses in the rabbit urinary bladder. *Br J Pharmacol* **110**:1491-1495.

P2 RECEPTORS IN BLOOD VESSELS

Vera Ralevic

School of Biomedical Sciences, Queen's Medical Centre, University of Nottingham, Nottingham, NG7 2UH

I. SUMMARY

Extracellular purines and pyrimidines regulate blood flow by important effects on vascular tone mediated via cell surface P2 receptors. P2 receptors are divided into two superfamilies, P2X and P2Y, according to whether they are ligand-gated cation channels or are coupled to G proteins respectively. Seven P2X and at least five P2Y receptors have been cloned to date. Members of both the P2X and the P2Y receptor families are expressed by blood vessels, being found on the smooth muscle, endothelium and perivascular nerves, where they are associated with specific effects on vascular tone. P2 receptors in blood vessels include smooth muscle $P2X_1$ receptors, which have an important role as mediators of vasoconstriction owing to ATP released as a cotransmitter from sympathetic perivascular nerves, and endothelial $P2Y_1$ (or P_{2Y}) and $P2Y_2$ (or P_{2U}) receptors which mediate vasodilatation. Vasodilator P2Y and contractile uridine nucleotide-specific receptors have been described in some vascular smooth muscles. This chapter describes the different subtypes of P2 receptors in blood vessels and examines the way in which the responses that they mediate integrate in the control of vascular tone.

II. BACKGROUND

The discovery that purine nucleotides have potent effects on the heart and vasculature was important in the early identification of these compounds as extracellular signalling molecules (Gillespie, 1934; Gaddum & Holtz, 1933; Emmelin & Feldberg, 1948; Folkow, 1949; Green & Stoner, 1950). Characterization of the different pharmacological activities of diverse purine compounds in blood vessels and other tissues contributed to the subsequent subdivision of receptors for purines into "P_1-purinoceptors" (now known as P1 receptors) at which adenosine is the principal natural ligand and "P_2-purinoceptors" (now known as P2 receptors) recognizing ATP and ADP (Burnstock, 1978), and was crucial to the further subdivision of P2 receptors into "P_{2X}-" and "P_{2Y}-purinoceptors" (Burnstock & Kennedy, 1985). Another landmark in P2 receptor research was the recognition by Seifert & Schultz in 1989 of "pyrimidinoceptors", which are specifically activated by uridine nucleotides but not by adenine nucleotides. These are distinct from the later-described "P_{2U}-purinoceptors" (now known as $P2Y_2$ receptors), which are activated equipotently by UTP and ATP (O'Connor et al., 1991), although in subsequent literature the separate identity of these two receptors for UTP has often gone unrecognised.

Following a revision of purine receptor subclassification and nomenclature P2 receptors are now divided into two superfamilies of ligand-gated inotropic P2X receptors and G protein-coupled P2Y receptors (Abbracchio & Burnstock, 1994; Fredholm et al., 1994). Further subdivision is based on structurally distinct cloned receptors, and a new system of nomenclature ($P2X_{1-7}$ and $P2Y_{1-7}$) has arisen from this (Burnstock & King, 1996) and will replace the earlier subtype nomenclature, including P_{2X}, P_{2Y}, P_{2U}, P_{2T} and P_{2Z} receptors (Burnstock & Kennedy, 1985; Gordon, 1986; O'Connor et al., 1991) as correlations between recombinant and endogenous P2 receptors are established.

It is noteworthy that whereas P2Y receptors (cloned and endogenous) are single proteins, cloned P2X proteins are receptor *subunits*, not actual receptors. As P2X receptors are formed from combinations of four or more subunits, which may be identical (homomeric receptors) or different (heteromeric receptors), it is clear that endogenous P2X receptors are not adequately represented by the seven homomeric P2X receptors cloned to date. The widespread and overlapping tissue distributions of different P2X receptor proteins suggests that many endogenous P2X receptors may be heteromers. At present it is not known how the properties of individual subunits relate to or influence the phenotype of different heteromeric P2X receptors, including ligand selectivity and susceptibility to descnsitization. For both P2X and P2Y receptors complex responses may arise from coexpression of receptors in native tissues.

Despite the handicap of a lack of potent and selective agonists and antagonists some general principles concerning subtype identities, locations and effects of P2 receptors in blood vessels can be identified. In this chapter, the new system of nomenclature based on cloned P2 receptors is used where possible in preference to the classical system to identify subtypes of P2 receptors. Thus, $P2Y_1$ replaces P_{2Y}, and $P2Y_2$ replaces P_{2U}. Where molecular evidence is currently unavailable the classical system of P2 purine receptor nomenclature is used. General principles are presented rather than a detailed examination of P2 receptors in individual types of blood vessels.

III TEXT

A. Local release of purines and pyrimidines

Purines and pyrimidines are released from a number of different sources relevant to the activation of specific P2 receptors distributed throughout the blood vessel wall. In most blood vessels ATP is released in various combinations with other neurotransmitters from perivascular nerves. ATP is released as a cotransmitter with noradrenaline (NA) and neuropeptide Y (NPY) from sympathetic perivascular nerves to mediate rapid membrane depolarization (or excitatory junction potentials (e.j.ps)), which may summate resulting in further membrane depolarization, calcium influx via voltage-dependent Ca^{2+} channels and vasoconstriction (Burnstock, 1990; Burnstock & Ralevic, 1996) (Fig. 1). Interestingly, the relative proportions of ATP and NA released as sympathetic cotransmitters varies enormously between different blood vessels and species. For example, in rat mesenteric arteries NA is the predominant sympathetic neurotransmitter and ATP has a relatively minor role, whereas in guinea-pig submucosal arterioles ATP is the main sympathetic neurotransmitter and NA acts as a prejunctional neuromodulator (see Burnstock & Ralevic, 1996).

ATP has been shown to be released from perivascular sensory-motor nerves in the rabbit ear artery (Holton & Holton, 1953, 1954; Holton, 1959), but this has not yet been demonstrated in other vessels. Neurogenic non-adrenergic non-cholinergic (NANC) relaxation mediated by ATP has been shown in guinea-pig pulmonary arteries (Liu et al., 1992), rabbit portal vein (where ATP appears to be coreleased with NO) (Brizzolara et al., 1993) (Fig. 2) and lamb small coronary arteries (Simonsen et al., 1997).

ATP and UTP are released from endothelial cells by stimuli including hypoxia and shear stress due to increased flow (Milner et al., 1990a,b; Ralevic et al., 1991; Bodin et al., 1992; Saiag et al., 1995). Rapid breakdown of ATP and UTP produces ADP and UDP respectively, which implies additional activation of the same or different P2 receptors by these nucleotides. Endothelial release of purine and pyrimidine nucleotides has most relevance for P2 receptors on the endothelial cells themselves, situated adjacent to or downstream from the site of nucleotide release. Activation of P2 receptors leads to vasodilatation (via nitric oxide (NO) and other endothelium-derived relaxing factors) and an increase in blood flow. In addition, endothelial P2 receptors have been shown to mediate further release of ATP, and presumably UTP, in an autocatalytic manner, which would serve to amplify the vasodilator response (Yang et al., 1994).

Erythrocytes are important sources of ATP (and presumably ADP following degradation of ATP) released during shear stress or hypoxia (Forrester, 1990; Ellsworth et al., 1995). Release of purines from erythrocytes into the lumen of the blood vessel suggests that their principal actions may be at endothelial cell P2 receptors, resulting in vasodilation and an increase in local blood flow.

ATP and UTP are released from activated platelets during aggregation (Goetz et al., 1971; Born et al., 1984). This is a pathophysiological process, occuring when there is damage to the function or integrity of the endothelium, and is associated with activation of P2 receptors on smooth muscle cells underlying the endothelium, and severe vasoconstriction or vasospasm and consequent ischemia.

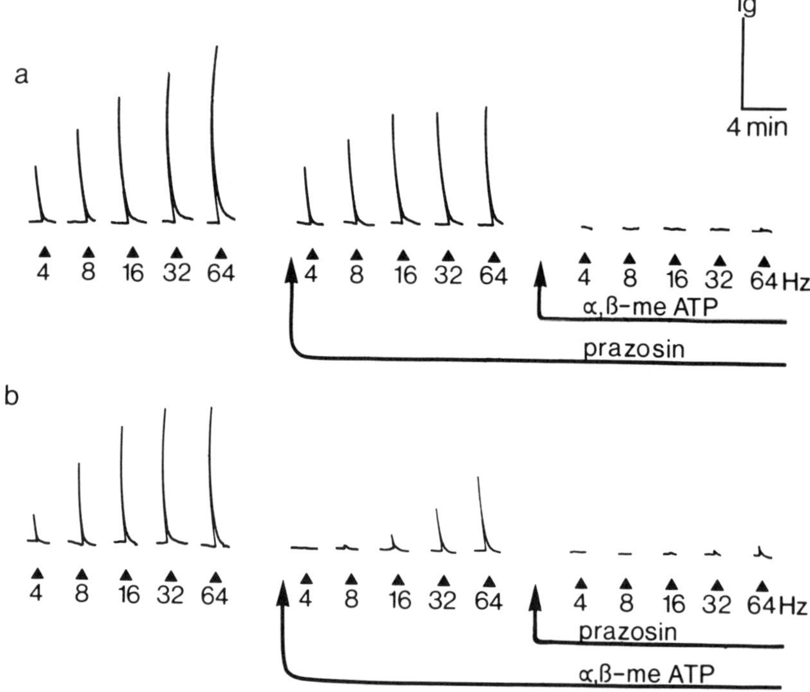

Figure 1 Contractions produced in the isolated saphenous artery of the rabbit on neurogenic transmural stimulation (0.08-0.1 msec; supramaximal voltage) for 1 sec (a,b) at the frequencies (Hz) indicated (▲). Nerve stimulations were repeated in the presence of 10 μM prazosin added before (a) or after (b) desensitization of the P$_2$-purinoceptor with α,β-methylene ATP (α,β-meATP) as indicated on the figure by the arrowed lines. The horizontal bar signifies 4 min and the vertical bar 1 g. (from Burnstock and Warland, 1987).

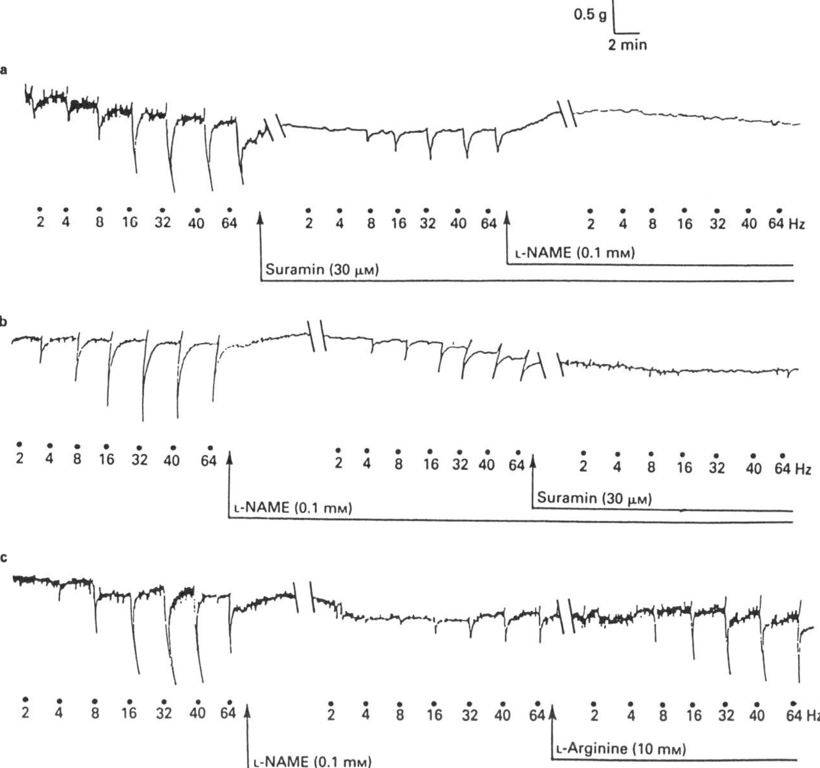

Figure 2 Relaxation of the rabbit portal vein to neurogenic transmural stimulation for 10 s. (2-64 Hz, 0.7 ms, 100 V) at 5 min intervals. Guanethidine (3.4 μM) and atropine (0.114 μM) were present throughout to block adrenergic and cholinergic neurotransmission respectively. Tone was induced with ergotamine (8.6 μM). Panel (a) shows that pre-incubation with suramin (30 μM) for 20 min reduced the nerve-mediated relaxations compared with controls and that suramin-resistant neurogenic relaxations were abolished 20 min after the addition the nitric oxide synthase inhibitor, N^G-nitro-L-arginine methyl ester (L-NAME, 0.1 mM). Panel (b) shows that neurogenic relaxations remaining after 20 min pretreatment of the tissue with L-NAME (0.1 mM) were abolished 20 min after the addition of suramin (30 μM). In (c), the effect of adding L-NAME (0.1 mM) to the tissue is shown; there was an additional rise in tone and inhibition of the response to nerve stimulation after a 20 min incubation period. The subsequent treatment of tissues with L-arginine (10 mM) for 20 min reversed this effect. Panel (d) shows the histochemical localization of reduced nicotinamide adenine dinucleotide phosphate (NADPH-)-diaphorase in the rabbit portal vein. (d1) NADPH-diaphorase reaction seen in nerve fibres between the inner circular (C) and outer longitudinal muscle coats (thin black arrows). Note also the NADPH-diaphorase positive nerves in adventitia (A) (thick black arrows). Tangentional section. (d2) NADPH-diaphorase reaction seen in a dense nerve plexus between the inner circular (C) and outer longitudinal muscle coats of the portal vein (thin black arrow). Transverse section. Calibration bar for d1 and d2 is 30 μm. Each of the traces in (a), (b) and (c) is representative of similar results in 6 separate experiments. (From Brizzolara et al., 1993).

1. P2X receptors

In smooth muscle cells ATP acts at P2X receptors, to mediate the rapid (onset within 10 ms) non-selective passage of cations (Na^+, K^+, Ca^{2+}) across the cell membrane resulting in the generation of e.j.ps, an increase in intracellular Ca^{2+} and depolarization (Bean, 1992; Dubyak & El-Moatassim, 1993). The summation of several e.j.ps results in further membrane depolarization, which leads to Ca^{2+} influx following activation of voltage-dependent Ca^{2+} channels (the primary contribution to Ca^{2+} influx) and leads to smooth muscle contraction.

The principal vascular smooth muscle P2X receptor appears to be the $P2X_1$ receptor. Although P2X receptors have not yet been cloned from vascular smooth muscle the agonist sensivity is similar to that of recombinant smooth muscle $P2X_1$ receptors cloned from vas deferens and bladder (Valera et al., 1994, 1995). For example, in isolated smooth muscle cells from rat tail artery rapid onset, rapidly desensitizing inward currents are evoked by agonists with a potency order of 2-methylthioATP (2MeSATP) = ATP $\geq$ α,β-methylene ATP (α,β-meATP) (Evans & Kennedy, 1994), which is similar to the agonist potency order of 2MeSATP > ATP > α,β-meATP mediating rapid onset, rapidly desensitizing currents at recombinant $P2X_1$ receptors (Valera et al., 1994, 1995). Immunohistochemical studies using specific antibodies have provided direct evidence for the expression of $P2X_1$ receptor protein by vascular smooth muscle. However, smooth muscle $P2X_2$ and $P2X_4$ proteins have also been identified, and it remains to be determined whether the predominant P2X receptor in the vasculature is indeed a homomeric $P2X_1$ receptor, or whether it is a heteromer that includes $P2X_1$ as one of its subunits (giving the smooth muscle P2X receptor its characteristic sensitivity to α,β-meATP).

It is noteworthy that the agonist potency order observed in isolated vascular smooth muscle preparations when agonist breakdown is not prevented, namely α,β-meATP >> 2MeSATP $\geq$ ATP, is quite different to that at the recombinant $P2X_1$ receptor, or in native tissues when ectonucleotidase activity is blocked. It is now accepted that this is primarily due to the different stabilities of purine compounds in the presence of ectonucleotidases; α,β-meATP is relatively resistant to degradation, while 2MeSATP and ATP are rapidly degraded. Thus, in the presence of ectonucleotidase inhibitors, or under conditions where ectonucleotidase activity is not involved (such as rapid concentration-clamp application of agonists on single cells), α,β-meATP is *less* potent than 2MeSATP and ATP at native and cloned $P2X_1$ receptors (Crack et al., 1994, 1995; Evans & Kennedy, 1994; Humphrey et al., 1995). An important implication is that ectonucleotidase activity critically modulates the biological response to ATP. Indeed, soluble nucleotidases have been shown to be released from sympathetic nerves along with neuronally released ATP to facilitate rapid breakdown (Todorov et al., 1997).

Rapid desensitization of responses to ATP, 2MeSATP and α,β-meATP in most vascular smooth muscle is consistent with the involvement of a $P2X_1$ receptor, since the recombinant $P2X_1$ receptor also desensitizes rapidly (in less than 100 ms) (North, 1996). However, not all vascular smooth muscle P2X receptors undergo rapid desensitization; for example there are α,β-meATP-sensitive P2X receptors in the arterial vasculature of human placenta that do not desensitize (Dobronyi et al., 1997; Ralevic et al., 1997), raising the possibility that these are heteromers formed from $P2X_1$ subunits together with a non-desensitizing P2X subunit (possibly $P2X_2$ or $P2X_4$), or that this tissue expresses a novel subtype of P2X receptor.

Of all vascular P2 receptors a physiological correlate has most clearly been established for smooth muscle P2X receptors, with an accumulation of evidence that these receptors mediate vasoconstriction in response to ATP released as a cotransmitter from sympathetic nerves (Burnstock, 1990; Burnstock & Ralevic, 1996). The biophysical and pharmacological properties of P2X receptors correspond well with a role as mediators of e.j.ps, fast postjunctional excitation and regulation of smooth muscle contractility. Both e.j.ps and the purinergic component of the vasoconstrictor response arising from stimulation of sympathetic periveascular nerves are blocked by the P2 purine receptor antagonists suramin- and pyridoxalphosphate-6-azophenyl-2',4'-disulphonic acid (PPADS), are resistant to block by adrenoceptor antagonists, and are abolished following guanethidine sympathectomy (Burnstock & Warland, 1987; Ziganshin et al., 1994 Burnstock & Ralevic, 1996) (Fig. 1).

The potency of ATP may be enhanced by postjunctional synergism involving smooth muscle P2X receptors and adrenoceptors as shown in rat femoral artery (Kennedy et al., 1985), guinea-pig and rat portal vein (Kennedy & Burnstock, 1986) and rat mesenteric arterial bed (Ralevic & Burnstock, 1990). On the other hand, rapid desensitization of most vascular smooth muscle P2X receptors implies that it may be biologically desirable to terminate the response to ATP as quickly and efficiently as possible. The combination of synergism, postjunctional desensitization and prejunctional inhibition (via P_1 receptors) following breakdown of ATP to adenosine provides tremendous potential for the fine tuning of smpathetic neurotransmission

2. P_{2U} receptors

Vascular smooth muscle P2 receptors that are activated equipotently by ATP and UTP, but not by α,β-meATP or 2MeSATP, have been described in a number of isolated blood vessels (Eltze & Ullrich, 1996; Miyagi et al., 1996; Rubino & Burnstock, 1996; Ishizaki et al., 1997; Qasabian et al., 1997) and in vascular smooth muscle cells in culture (Strøbæk et al., 1996). According to the classical system of purine receptor nomenclature this agonist potency profile classifies these receptors as P_{2U} receptors. Although current nomenclature recommends that P_{2U} receptors should be renamed $P2Y_2$ receptors this new terminology implies G protein-coupling, which although verified for endothelial P_{2U} receptors and for P_{2U} receptors on human coronary artery smooth muscle cells (Strøbæk et al., 1996) has not been confirmed for P_{2U} receptors in most other vascular smooth muscle. Thus, classification of the vascular smooth muscle P_{2U} receptor according to the new system of nomenclature awaits results from molecular and other studies.

Smooth muscle P_{2U} receptors activated equipotently by UTP and ATP mediate vasoconstriction of rat pulmonary vasculature (Rubino & Burnstock, 1996), rat renal vasculature (Eltze & Ullrich, 1996), bovine middle cerebral artery (Miyagi et al., 1996) and guinea pig portal vein (Ishizaki et al., 1997). Little is known about the physiological conditions that cause the release of UTP to activate this receptor and induce smooth muscle contraction.

Interestingly, Ca^{2+}-mobilizing P_{2U} receptors activated equipotently by ATP and UTP on cultured smooth muscle cells of rabbit pulmonary artery are not coupled to a functional response (Qasabian et al., 1997). These and other P_{2U} receptors may have a role in mitogenesis. P_{2U} receptors activated equipotently by ATP and UTP have been shown to mediate an increase in expression of both immediate-early and delayed-early cell cycle-dependent genes in cultured aortic smooth muscle cells, contrasting with the induction of only immediate-early genes by 2MeSATP in the same cells (Malam-Souley et al., 1996). P_{2U} receptors may be involved in the mitogenic effect that has been shown in response to ATP and UTP acting on vascular smooth muscle cells of rat aorta and vena cava (Erlinge et al., 1993).

3. Uridine nucleotide-specific receptors

Uridine nucleotide-specific receptors mediating contraction to UTP, but not to ATP, which show one or more of the properties of resistance to desensitization by α,β-meATP, lack of cross-tachyphylaxis with responses to ATP, and insensitivity to P2 receptor antagonists including PPADS and suramin, have been described on vascular smooth muscle in a variety of vessels and species (von Kügelgen et al., 1987, 1990; Saiag et al., 1990, 1992; Ralevic & Burnstock, 1991a; Juul et al., 1992; Windscheif et al., 1994; Lagaud et al., 1996). PPADS and suramin have, however, been shown to antagonize UTP-specific contractions in smooth muscle cells of rat tail artery (McLaren et al., 1996).

The molecular identity of the endogenous uridine nucleotide-specific receptor is currently unclear. A $P2Y_6$ receptor has been cloned from rat aortic smooth muscle and is activated by UDP > ADP > 2MeSATP, but not by UTP and ATP (Chang et al., 1995). The $P2Y_4$ receptor is a uridine-nucleotide-specific receptor, showing a strong preference for UTP over ATP (Communi et al., 1996), but cannot account for all endogenous uridine nucleotide-specific receptor-mediated responses since $P2Y_4$ mRNA appears to be expressed almost exclusively in human placenta (Communi et al., 1996), whereas uridine nucleotide-specific responses have been documented in a variety of tissues. Although current evidence suggests that they are P2Y receptors, a uridine nucleotide-specific receptor family comprising both ionotropic and G protein-coupled receptors is possible.

Not enough is currently known about the release of UTP to adequately identify a physiological role of vascular smooth muscle uridine nucleotide-specific receptors. A nonsympathetic neurogenic contractile response in guinea pig portal vein suggested to be mediated by UTP but not ATP raises the intriguing prospect of UTP neurotransmission involving uridine nucleotide-specific receptors (Ishizaki et al., 1997).

4. P_{2Y} receptors

In some mammalian blood vessels, P_{2Y} receptors have been identified on the smooth muscle, which mediate vasodilatation to ATP and 2MeSATP (Kennedy & Burnstock, 1985; Mathieson & Burnstock, 1985; Burnstock & Warland, 1987; Liu et al., 1989; Brizzolara & Burnstock, 1991; Keef et al., 1992; Corr & Burnstock, 1994; Simonsen et al., 1997) (Fig. 3). It should be noted that the classical nomenclature, based on the rank order of agonist potency, is used here to identify this receptor because of the lack of definitive information about its molecular structure or coupling. Indirect evidence that this receptor belongs to the superfamily of G protein-coupled P2Y receptors comes from the fact that the selective P2X receptor agonist α,β-meATP mediates vasoconstriction in the above vessels, implying that there are smooth muscle vasoconstrictor ionotropic P2X receptors that are distinct from the smooth muscle receptors mediating vasodilatation to ATP and 2MeSATP (P2Y). The fact that 2MeSATP is the most potent agonist at mediating smooth muscle vasodilatation via P2 receptors is most consistent with an action at $P2Y_1$ receptors. However, the subtype identity of the receptor awaits clarification.

Relaxation mediated by smooth muscle P_{2Y} receptors has been shown in rabbit hepatic artery (Brizzolara & Burnstock, 1991) (Fig. 3), rabbit mesenteric artery (Mathieson & Burnstock, 1985; Burnstock & Warland, 1987), rabbit portal vein (Kennedy & Burnstock, 1985), rabbit pulmonary artery (Qasabian et al., 1997) and human small pulmonary arteries (Liu et al., 1989). In guinea-pig and rabbit large coronary arteries (Keef et al., 1992; Corr & Burnstock, 1994) and in lamb small coronary arteries (Simonsen et al., 1997) vasodilatation is mediated by both P_{2Y} receptors on smooth muscle and their counterparts, endothelial $P2Y_1$ receptors. P_{2Y} receptors (and P_{2U} receptors) have been described in human coronary artery smooth muscle cells in culture (Strøbæk et al., 1996). The smooth muscle P_{2Y} receptor of rabbit pulmonary artery mediates relaxation independently of mobilization of intracellular Ca^{2+}, implying lack of involvement of the PLC pathway (Qasabian et al., 1997).

The mechanism of relaxation by vascular smooth muscle P_{2Y} receptors has not been established, but may involve activation of K^+ channels and hyperpolarization. In rabbit mesenteric arteries and skeletal muscle resistance arteries this appears to involve ATP-sensitive K^+ channels since glibenclamide blocks smooth muscle hyperpolarization and relaxation to ADP (Brayden, 1991).

The physiological significance of smooth muscle P_{2Y} receptors has been established for rabbit portal vein (where ATP appears to be coreleased with NO) (Kennedy & Burnstock, 1985; Brizzolara et al., 1993) (Fig. 2), guinea-pig pulmonary artery (Liu et al., 1992) and lamb small coronary arteries (Simonsen et al., 1997), where neurogenic non-adrenergic non-cholinergic relaxation mediated by ATP has been shown. It has been suggested that smooth muscle P_{2Y} receptors may mediate relaxation to ATP released from perivascular sensory-motor nerves (Burnstock, 1990).

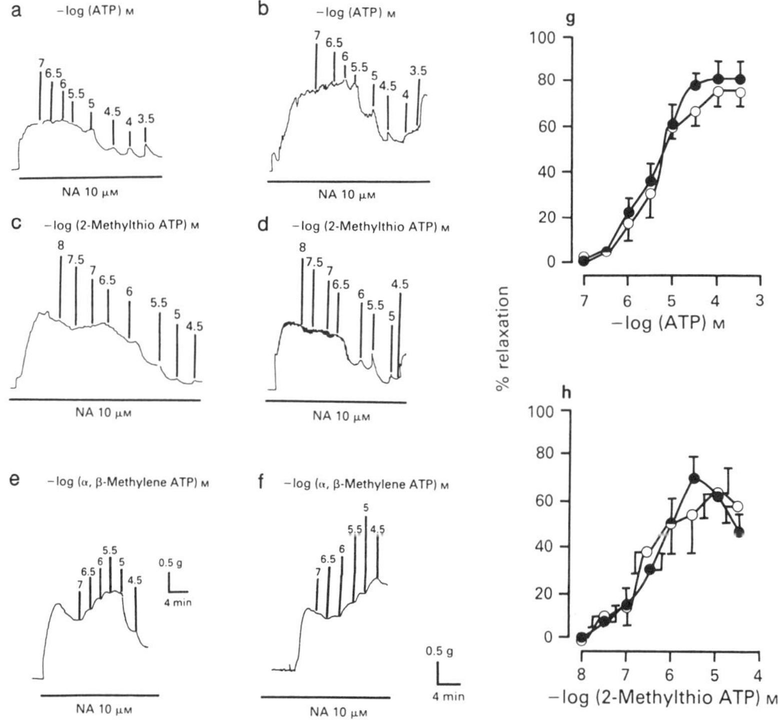

Figure 3 A. Isolated ring preparations of the rabbit hepatic artery preconstricted with noradrenaline (NA) to 75% of the maximal constriction. Response to ATP: (a) endothelium intact; (b) endothelium removed, 2-methylthio ATP; (c) endothelium intact; (d) endothelium removed, α,β-methylene ATP (e) endothelium intact; (f) endothelium removed.
Isolated ring preparations of the rabbit hepatic artery preconstricted with noradrenaline (NA) (10 µM). (g) Concentration-response curve to ATP in the presence (●) (n = 6) and in the absence (○) (n = 6) of endothelium. (h) Concentration response curve to 2-methylthio ATP in the presence (●) (n = 5) and in the absence (○) (n = 6) of endothelium. (Adapted from Brizzolara & Burnstock, 1991).

1. $P2Y_1$ receptor

The $P2Y_1$ receptor (formerly P_{2Y}) has been cloned from bovine endothelium and the recombinant receptor is activated by agonists with a potency order of 2MeSATP = ADP > ATP >> UTP (Henderson et al., 1995). The receptor is generally more sensitive to adenine nucleotide diphosphates than to triphosphates. At endogenous endothelial $P2Y_1$ receptors, as at the recombinant endothelial $P2Y_1$ receptor, ADP is a potent agonist whereas ATP is weak or inactive. The true potency of ATP at endogenous $P2Y_1$ receptors may be difficult to determine as endothelial cells coexpress $P2Y_2$ with $P2Y_1$ receptors at both of which ATP is an agonist (Communi et al., 1995; Yang et al., 1996). However, in the rat mesenteric arterial bed the disparate sensitivities of ADP and ATP at $P2Y_1$ and $P2Y_2$ receptors have clearly been shown (Ralevic & Burnstock, 1996a).In this preparation 2MeSATP and ADP are equipotent at mediating vasodilation via $P2Y_1$ receptors and are more potent than ATP and UTP acting at $P2Y_2$ receptors (Ralevic & Burnstock, 1996a). Furthermore, the P2 receptor antagonist PPADS blocks vasodilatation to 2MeSATP and ADP, but not to ATP and UTP, suggesting that ATP acts at $P2Y_2$, but not at $P2Y_1$ receptors (Ralevic & Burnstock, 1996a) (Fig. 4). The different specificities of ADP and ATP for endothelial P2 receptors may have important implications respecting the *in vivo* sources and release of these purines and is an interesting area for future research.

Coupling of $P2Y_1$ via G_q/G_{11} proteins to phospholipase C and IP_3 production is the main signal transduction pathway of this receptor, although inhibition of adenylate cyclase has been described in a clonal population of rat brain capillary endothelial cells (B10 cells) (Webb et al., 1996). Adenylate cyclase coupling appears to be a function of different G protein-coupling rather than due to different subtypes of the $P2Y_1$ receptor (Webb et al., 1996). The IP_3-mediated increase in Ca^{2+} stimulates NO synthase, leading to formation of NO, which is an important mediator of $P2Y_1$-mediated endothelium-dependent vasodilatation. Endothelium-derived hyperpolarizing factor (EDHF) and to a lesser extent prostacyclin may also be involved in $P2Y_1$ receptor-mediated endothelium-dependent relaxation. Generation of PKC (with no detectable elevations in IP_3 or cytosolic Ca^{2+}) and subsequent rapid tyrosine phosphorylation of mitogen-activated protein kinase (MAPK) appears to be the pathway by which $P2Y_1$ (and $P2Y_2$) receptors on endothelial cells mediate prostacyclin production (Bowden et al., 1995; Danby et al., 1996; Patel et al., 1996a,b). This pathway is involved in cell metabolism, secretion, gene expression and growth.

The physiological role of endothelial $P2Y_1$ receptors is currently unclear, although several possibilities are likely. $P2Y_1$ receptors may have a role as mediators of vasodilatation following ADP released from platelets during aggregation. ATP is released from endothelial cells by shear stress due to increased flow (Milner et al., 1990a,b; Ralevic et al., 1991) and may act directly at $P2Y_1$ receptors or following breakdown to ADP. Autocatalytic release of ATP (release of ATP by ATP itself) has been described in guinea-pig cardiac endothelial cells, which identifies a possible role for $P2Y_1$ receptors (Yang et al., 1994). Intriguing species differences in expression of endothelial $P2Y_1$ receptors have been noted, for example, pharmacological evidence suggests that hamster mesenteric arteries lack $P2Y_1$ receptors (but do express endothelial $P2Y_2$ receptors), although it is possible that the receptors are expressed but are not coupled to a functional response (Ralevic & Burnstock, 1996b).

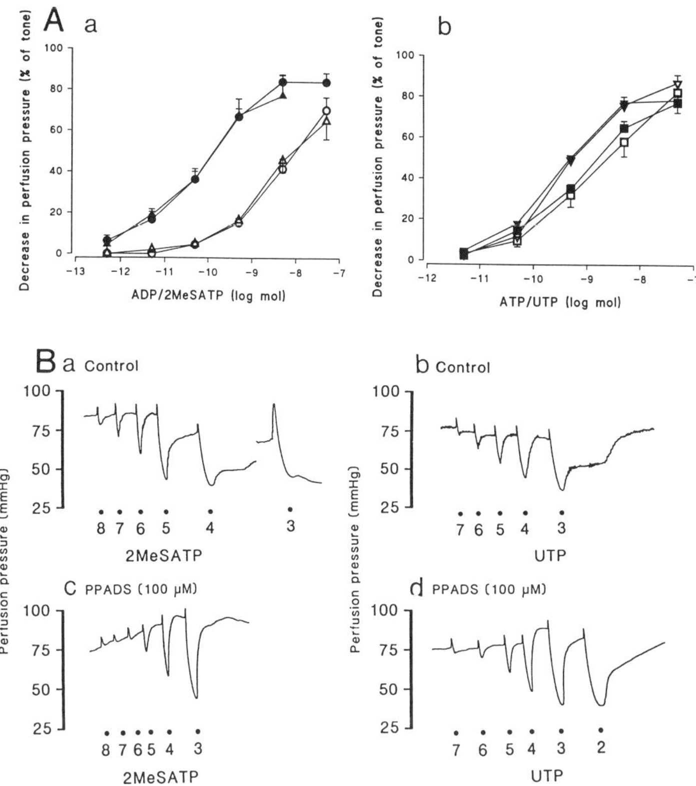

Figure 4 A. (a,b). Effect of pyridoxalphosphate-6-azophenyl-2',4'-disulphonic acid (PPADS) on endothelium-dependent vasodilator responses to a) ADP (●,O) and 2MeSATP (▲,△), b) ATP (▼,▽) and UTP (■,□) in the isolated rat mesenteric arterial bed; closed symbols are mean responses in the absence of drugs, open symbols are mean responses in the presence of drugs. PPADS is used at 10 μM against ATP and 100 μM against ADP, 2MeSATP and UTP. Note that PPADS produces an identical block of responses to ADP and 2MeSATP but does not antagonize responses to ATP and UTP. B. (a-d). Representative traces from a single rat mesenteric arterial bed preparation showing vasodilator responses to 2MeSATP and UTP in the absence (a,b) and presence (c,d) of 100 μM PPADS. Nucleotides were applied as doses of 50 μl bolus injections. Tone of the preparations was raised by 10 μM methoxamine in the absence of PPADS and 5 μM methoxamine in the presence of PPADS. PPADS had no effect on dose-dependent vasodilatation to UTP, but antagonized vasodilatation to 2MeSATP. Doses are indicated as concentration of drug (M) in the 50 μl bolus. For UTP htis corresponds to a dose-range of -11.3 to 6.3 log mol (or 0.005-500 nmol). For 2MeSATP this corresponds to a dose range of -12.3 to 7.3 log mol (or 0.0005-50 nmol). (Adapted from Ralevic & Burnstock, 1996a).

2. P2Y₂ receptor

P2Y₂ receptors (formerly P₂ᵤ) are G protein-coupled receptors activated equipotently by UTP and ATP. Although the P2Y₂ receptor has not yet been cloned from blood vessels, G protein coupling and the distinct pharmacological profile of the endogenous receptor, identifies this as a P2Y₂ receptor.

The P2Y$_2$ receptor couples to G$_{io}$ to mediate phospholipid breakdown and Ca^{2+} mobilization via PLC$_\beta$. Vasodilatation is mediated via generation primarily of NO, with a possible contribution of EDHF or prostacyclin. P2Y$_2$ receptors are colocalized with P2Y$_1$ receptors on many cells and have a common signalling pathway in PLC; however, P2Y$_2$ responses are less sensitive to manipulations of the PKC pathway (Purkiss et al., 1994; Communi et al., 1995; Gallinaro et al., 1995; Chen et al., 1996). Stimulation of phosphatidylcholine breakdown and activation of PLD (possibly via PKC) by endothelial P2Y$_2$ receptors has been reported (Purkiss & Boarder, 1992). As with the P2Y$_1$ receptor, protein tyrosine phosphorylation and MAPK activation appears to be the major route for P2Y$_2$-mediated prostacyclin production in endothelial cells (Bowden et al., 1995; Danby et al., 1996; Patel et al., 1996a,b). This occurs subsequent to activation of PKC and does not involve IP$_3$ or cytosolic Ca^{2+} (Patel et al., 1996a,b).

P2Y$_2$ receptors can be divided into two groups based on whether or not they are antagonized by suramin and PPADS, although there is currently no molecular evidence to support this subdivision (which may be due to small differences in amino acid sequence). Suramin-insensitive endothelial P2Y$_2$ receptors include those on rabbit aorta (Chinellato et al., 1994), rat mesenteric arteries (Ziyal et al., 1995), and bovine aortic endothelial cells (Wilkinson et al., 1994). Suramin-sensitive endothelial P2Y$_2$ receptors include those in hamster mesenteric arterial bed (Ziyal et al., 1995), rabbit pulmonary artery (Qasabian et al., 1997), and rat brain endothelial cells (Nobles et al., 1995). Vascular endothelial P2Y$_2$ receptors are generally resistant to blockade by PPADS, as reported in rat mesenteric arteries (Ralevic and Burnstock, 1996a) and bovine aortic endothelial cells (Brown et al., 1995). In contrast, PPADS-sensitivity of P2Y$_2$ receptors has been reported only in some non-vascular tissues (Ho et al., 1995; Reiser, 1995).

The role of a receptor equally sensitive to UTP and ATP is an enigma, particularly since ATP is coreleased with UTP, apparently from the same sources and under the same conditions. UTP (and ATP) release from platelets and endothelium (Goetz et al., 1971; Saiag et al., 1995) suggests possible roles for the P2Y$_2$ receptor in local control of vascular tone analagous to P2Y$_1$ activation by ATP released from the same sources. Enhanced leukocyte adherence to cultured pulmonary artery endothelial cells by P2Y$_2$ receptors has been shown (Dawicki et al., 1995).

3. Uridine nucleotide-specific receptor
A pertussis toxin-sensitive uridine nucleotide-specific receptor mediating an increase in intracellular calcium has been reported to coexist with P2Y$_2$ and P2Y$_1$ receptors on guinea-pig cardiac endothelial cells (Yang et al., 1996).

<u>D. P2 receptors on perivascular nerves</u>

The expression of P2X receptors on peripheral sensory nerve endings (Pelleg and Hurt, 1996; Cook & McCleskey, 1997) is consistent with a role for these receptors in nociception, and raises the possibility of a more widespread distribution of P2X receptors on sensory afferents in blood vessels. At the recombinant P2X$_3$ receptor cloned from rat dorsal root ganglion, currents evoked by ATP and α,β-meATP rapidly desensitize, whereas ATP-gated currents at endogenous P2X receptors in adult rat nodose and dorsal root ganglion neurons do not desensitize (Lewis et al., 1995). When P2X$_3$ is coexpressed in HEK293 cells with P2X$_2$ receptors (but not with other subtypes) a non-desensitizing response to ATP is observed which mimicks that seen in adult rat sensory neurons, and which can not be explained by additive effects of the two homomeric channels. This was suggested to be due to a heteromeric receptor, P2X$_2$P2X$_3$, formed from the P2X$_2$ and P2X$_3$ subunits (Lewis et al., 1995). Thus, it is possible that heteromeric P2X receptors, possibly P2X$_2$P2X$_3$ receptors, are present on the peripheral terminals of perivascular sensory-motor nerves.

A putative P3 receptor activated by nucleosides and nucleotides (2Cl-adenosine > β,γ-meATP > ATP = adenosine) and antagonized both by xanthines and α,β-meATP, is suggested to mediate inhibition of NA release from sympathetic nerves in rat tail artery and vas deferens (Shinozuka et al., 1988; Forsythe et al., 1991). Facilitation of NA release by ATP and adenosine but not by α,β-meATP, which occurs in some vessels is blocked by α,β-meATP and 8-SPT (Miyahara and Suzuki, 1987; Zhang et al., 1989; Todorov et al., 1994; Ishii et al., 1995) by a receptor suggested to be a subtype of the P3 receptor (Dalziel and Westfall, 1994). It is possible that these receptors belong to the P1 and P2 receptor superfamilies; molecular studies are needed to confirm that that they are indeed distinct from currently identified subtypes.

E. P2 receptors on platelets

Activation of P2 purine receptors on platelets may have important effects on vascular tone as a consequence of of platelet aggregation and the release of nucleotides and other vasoactive substances. Platelets express a receptor for ADP, which is currently termed the P_{2T} (or $P2Y_{ADP}$) receptor. This receptor has not yet been cloned. Recent evidence suggests that the platelet P_{2T} receptor is a $P2Y_1$ receptor (Hechler et al., 1997). The P_{2T} receptor couples to G_{i2} proteins to mediate inhibition of adenylate cyclase activity, together with activation of PLC and an increase in intracellular Ca^{2+} to promote platelet aggregation (Hourani and Hall, 1996). Platelets also express $P2X_1$ receptors, which, when activated cause the opening of non-selective cation channels and the rapid influx of Ca^{2+} (MacKenzie et al., 1996) and may contribute to the platelet aggregatory process.

F. Integrated effects of P2 receptors in the control of blood vessel tone in physiological and pathophysiological conditions

Smooth muscle and endothelial cells commonly express more than one type of P2 receptor, for example as shown for coexisting $P2Y_1$, $P2Y_2$ and uridine nucleotide-specific receptors on endothelial cells (Yang et al., 1996) and for $P2X_1$ and P_{2Y} receptors on vascular smooth muscle (Kennedy & Burnstock, 1985). Coexpression of vasoconstrictor $P2X_1$ and vasodilator P_{2Y} receptors on some vascular smooth muscle may correlate with the release of ATP from sympathetic and NANC nerves respectively. The biological significance of coexisting endothelial receptors is not clear, but may allow regulation of multiple effectors or fine tuning of agonist-evoked responses. Endothelial $P2Y_1$ and $P2Y_2$ receptors typically have a common signalling pathway in PLC, although downstream divergence of this pathway may be important. Synergism has not been reported. The different specificities of subtypes of P2Y receptors for purine and pyrimidine nucleotides suggests that source and local concentrations of ADP, ATP, UDP, UTP and adenine dinucleotides may be important determinants of the biological response, and further information on this would provide important insights into the physiological significance of P2Y receptor coexistence.

Integration of P2 purine receptors in the control of blood vessel tone is illustrated in Figure 5. Under normal conditions this is a balance between smooth muscle constrictor responses and endothelial vasodilator tone. $P2X_1$ receptors on vascular smooth muscle mediate constriction to ATP released as a neurotransmitter from sympathetic nerves; smooth muscle vasodilator P_{2Y} receptors mediate vasodilatation from NANC or sensory-motor nerves; vasoconstrictor P_{2U} and uridine nucleotide-specific receptors are also present on some vascular smooth muscle; endothelial $P2Y_1$ and $P2Y_2$ receptors mediate vasodilatation to purine nucleotides released from endothelium or erythrocytes during shear stress or hypoxia.

Normal patterns of purinergic signalling may alter dramatically under pathophysiological conditions. The net effect of purine receptors is vasodilatation if endothelial cells are intact, but vasoconstriction will predominate if the endothelium is damaged or removed (Ralevic & Burnstock, 1991b; Ralevic et al., 1995). While the first target of ATP at the intimal surface of blood vessels is generally vasodilator endothelial P2Y$_1$ receptors, this situation may change in pathophysiological conditions when the endothelium is damaged or removed and platelet aggregation is induced by platelet P$_{2T}$ (P2Y$_1$-like) receptors. Under these circumstances, collagen is exposed, to which platelets adhere, releasing ADP, ATP, UTP and adenine dinucleotides together with other substances such as 5-HT, which may promote further aggregation and vasconstriction via actions on underlying vascular smooth muscle P2 (predominantly P2X$_1$) receptors. In an inflammatory reaction, ATP may be released from sensory nerves to have effects on mast cell P$_{2Z}$ or P2X$_7$ receptors (Cockcroft & Gomperts, 1980; Tatham et al., 1988; Jaffar & Pearce, 1990; Tatham & Lindau, 1990), and may promote further ATP release following actions at P2X$_2$P2X$_3$ receptors on sensory afferents.

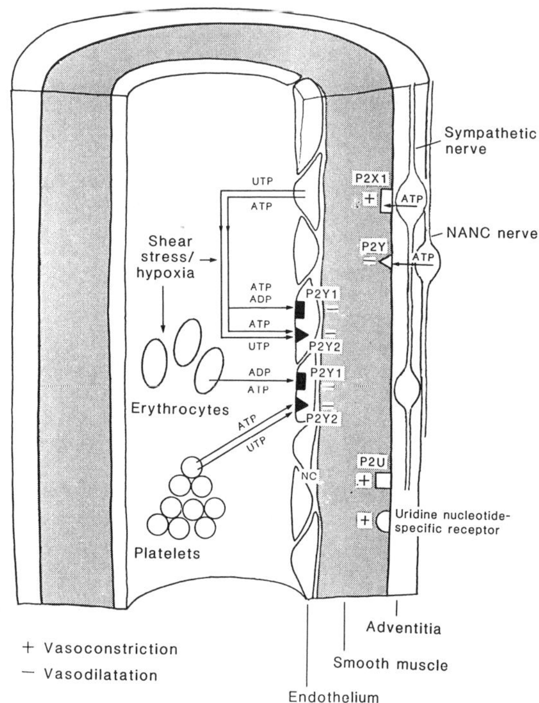

Figure 5 P2 purine receptors in the local control of blood vessel tone. ATP can be released as a cotransmitter with noradrenaline from perivascular sympathetic nerves to mediate vasoconstriction via smooth muscle P2X$_1$ receptors (P2X1). In some vessels ATP is released as a neurotransmitter from non-adrenergic non-cholinergic nerves to mediate vasodilatation via smooth muscle P$_{2Y}$ (P2Y) receptors. P$_{2U}$ (P2U) and uridine nucleotide-specific receptors on the smooth muscle mediate vasoconstriction; the physiological sources of UTP and UDP acting at these receptors are unknown. At the intimal surface of the blood vessel, endothelial P2Y$_1$ (P2Y1) and P2Y$_2$ (P2Y2) receptors mediate vasodilatation (via nitric oxide (NO)/prostacyclin (PGI$_2$)/endothelium-derived hyperpolarizing factor (EDHF)) in response to ATP and UTP released from platelets and by the actions of shear stress and hypoxia on erythrocytes and endothelial cells. Note that ADP is the principal agonist at the endothelial P2Y$_1$ receptor and at P$_{2T}$ (P2Y$_1$-like) receptors on platelets (not illustrated). NANC, non-adrenergic non-cholinergic.

IV. CONCLUSIONS AND FUTURE DIRECTIONS

The potent and diverse effects of purines and pyrimidines on blood vessel tone is consistent with the concept that these compounds have, via specific actions on P2 receptors, an important role in the control of peripheral blood flow. In common with other areas of purine and pyrimidine receptor research, studies of the vascular biology of P2 receptors urgently requires the development of potent and selective agonists and antagonists for a definitive and unambiguous characterization of P2 receptor subtypes and the actions they mediate. This is an important step towards addressing the interesting issue of the specific physiological and pathophysiological roles of vascular P2 receptors. In particular this may shed light on the enigma of the role of a receptor (P2Y$_2$ or P$_{2U}$ receptor) activated equally by ATP and UTP. Much-needed *in vivo* studies would help to characterize the short and long-term physiological roles of vascular P2 receptors.

ACKNOWLEDGEMENTS

The support of the Royal Society is gratefully acknowledged.

REFERENCES

Abbracchio MP, Burnstock G. Purinoceptors: are there familes of P$_{2X}$ and P$_{2Y}$ purinoceptors? Pharmacol Ther 1994;64:445-75

Bean BP. Pharmacology and electrophysiology of ATP-activated ion channels. Trends Pharmacol Sci 1992;13:87-90

Bodin P, Milner P, Winter R, Burnstock, G. Chronic hypoxia changes the ratio of endothelin to ATP release from rat aortic endothelial cells exposed to high flow. Proc R Soc Lond 1992;247:131-5

Born GVR, Kratzer MAA. Source and concentration of extracellular adenosine triphosphate during hemostasis in rats, rabbits and man. J Physiol 1984;354:419-29

Bowden A, Patel V, Brown C, Boarder MR. Evidence for requirement of tyrosine phosphorylation in endothelial P$_{2Y}$- and P$_{2U}$-purinoceptor stimulation of prostacyclin release. Br J Pharmacol 1995;116:2563-8

Brayden JE. Hyperpolarization and relaxation of resistance arteries in response to adenosine diphosphate. Circ Res 1991;69:1415-20

Brizzolara A, Burnstock G. Endothelium-dependent and endothelium-independent vasodilatation of the hepatic artery of the rabbit. Br J Pharmacol 1991;103:1206-12

Brizzolara AL, Crowe R, Burnstock G. Evidence for the involvement of both ATP and nitric oxide in non-adrenergic non-cholinergic inhibitory neurotransmission in the rabbit portal vein. Br J Pharmacol 1993;109:606-8

Brown C, Tanna B, Boarder MR. PPADS: an antagonist at endothelial P$_{2Y}$-purinoceptors but not P$_{2U}$-purinoceptors. Br J Pharmacol 1995;116:2413-6

Burnstock, G. "A basis for distinguishing two types of purinergic receptor." In: *Cell Membrane Receptors for Drugs and Hormones*, L. Bolis, R.W. Straub, eds., New York: Raven Press, pp. 107-118, 1978.

Burnstock G. Co-Transmission. The Fifth Heymans Memorial Lecture. Arch Int Pharmacodyn Ther 1990;304:7-33

Burnstock G, Kennedy C. Is there a basis for distinguishing two types of P$_2$-purinoceptor? Gen Pharmacol 1985;16:433-40

Burnstock G, King, BF. Numbering of cloned P2 purinoceptors. Drug Dev Res 1996;38:67-71

Burnstock G, Ralevic V. "Cotransmission." In: *The Pharmacology of Vascular Smooth Muscle*, Garland CJ & Angus J, eds. Oxford University Press, Oxford, 1996:210-32

Burnstock G, Warland JJI. P_2-Purinoceptors of two subtypes in the rabbit mesenteric artery: Reactive blue 2 selectively inhibits responses mediated via the P_{2Y}- but not the P_{2X}-purinoceptor. Br J Pharmacol 1987;90:383-91

Chang K, Hanaoka K, Kumada M, Takuwa Y. Molecular cloning and functional analysis of a novel P_2 nucleotide receptor. J Biol Chem 1995;270:26152-8

Chen BC, Lee C-M, Lee YT, Lin W-W. Characterization of signaling pathways of P_{2Y} and P_{2U} purinoceptors in bovine pulmonary artery endothelial cells. J Cardiovasc Pharmacol 1996;28:192-9

Chinellato A, Ragazzi E, Pandolfo L, Froldi G, Caparrotta L, Fassina G. Purine- and nucleotide-mediated relaxation of rabbit thoracic aorta: common and different sites of action. J Pharm Pharmacol 1994;46:337-41

Cockcroft S, Gomperts BD. The ATP^{4-} receptor of rat mast cells. Biochem J 1980;188:789-98

Communi D, Pirotton S, Parmentier M, Boeynaems J-M. Cloning and functional expression of a human uridine nucleotide receptor. J Biol Chem 1996;270:30849-52

Communi D, Raspe E, Pirotton S, Boeynaems JM. Coexpression of P_{2Y} and P_{2U} receptors on aortic endothelial cells. Comparison of cell localization and signaling pathways. Circ Res 1995;76:191-198

Cook SP, McCleskey EW. Desensitization, recovery, and Ca^{2+}-dependent modulation of ATP-gated P2X receptors in nociceptors. Neuropharmacology 1997;36:1303-1308

Corr L, Burnstock G. Analysis of P_2-purinoceptor subtypes on the smooth muscle and endothelium of rabbit coronary artery. J Cardiovasc Pharmacol 1994;23:709-15

Crack BE, Beukers MW, McKechnie KCW, Ijzerman AP, Leff P. Pharmacological analysis of ecto-ATPase inhibition: evidence for combined enzyme inhibition and receptor antagonism in P_{2X}-purinoceptor ligands. Br J Pharmacol 1994;113:1432-8

Crack BE, Pollard CE, Beukers MW, Roberts SM, Hunt SF, Ingall AH, McKechnie KCW, Ijzerman AP, Leff P. Pharmacological and biochemical analysis of FPL 67156, a novel, selective inhibitor of ecto-ATPase. Br J Pharmacol 1995;114:475-81

Danby R, Roberts J, Boarder MR. Novel second messenger pathways in P2Y receptor pharmacology: evidence that both cloned and native receptors control protein tyrosine phosphorylations. Drug Dev Res 1996;37:123

Dawicki DD, McGowan-Jordan J, Bullard S, Pond S, Rounds S. (1995) Extracellular nucleotides stimulate leukocyte adherence to cultured pulmonary artery endothelial cells. Am J Physiol 1995;268:L666-73

Dobronyi I, Hung K-S, Satchell DG, Maguire MH. Evidence for a novel P_{2X} purinoceptor in human placental chorionic surface arteries. Eur J Pharmacol 1997;320:61-4

Dubyak GR, El-Moatassim C. Signal transduction via P_2-purinergic receptors for extracellular ATP and other nucleotides. Am J Physiol 1993;265:C577-606

Ellsworth ML, Forrester T, Ellis CG, Dietrich HH. The erythrocyte as a regulator of vascular tone. Am J Physiol 1995;269:H2155-61

Eltze M, Ullrich B. Characterization of vascular P_2 purinoceptors in the rat isolated perfused kidney. Eur J Pharmacol 1996;306:139-52

Emmelin N, Feldberg W. Systemic effects of adenosine triphosphate. Br J Pharmacol 1948;3:273-84

Erlinge D, Yoo H, Edvinsson L, Reis DJ, Wahlestedt C. Mitogenic effects of ATP on vascular smooth muscle cells vs. other growth factors and sympathetic transmitters. Am J Physiol 1993;265:H1089-97

Evans RJ, Kennedy C. Characterization of P_2-purinoceptors in the smooth muscle of the rat tail artery: a comparison between contractile and electrophysiological responses. Br J Pharmacol 1994;113:853-60

Folkow B. The vasodilator action of adenosine triphosphate. Acta Physiol Scand 1949;17:311-6

Forrester, T. Release of ATP from heart: presentation of a release model using human erythrocyte. Ann New York Acad Sci 1990;603:335-352

Fredholm BB, Abbracchio MP, Burnstock G, Daly JW, Harden KT, Jacobson KA, Leff P, Williams M. Nomenclature and classification of purinoceptors. Pharmacol Rev 1994;46:143-56

Gaddum JH, Holtz P. The localization of the action of drugs on the pulmonary vessels of dogs and cats. J Physiol 1933;77:139-58

Gallinaro BJ, Reimer WJ, Dixon SJ. Activation of protein kinase C inhibits ATP-induced $[Ca^{2+}]_i$ elevation in rat osteoblastic cells: selective effects on P_{2Y} and P_{2U} signaling pathways. J Cell Physiol 1995;162:305-14

Gillespie JH. The biological significance of the linkages in adenosine triphosphoric acid. J Physiol (Lond) 1934;80:345-9

Goetz V, Prada DA, Pletscher MA. Adenine-, guanine- and uridine-5'-phosphonucleotides in blood platelets and storage organelles of various species. J Pharmacol Exp Ther 1971;178:210-5

Green HN, Stoner HB. *Biological Actions of Adenine Nucleotides*. London: Lewis, 1950

Hechler B, Léon C, Vial C, Eckly A, Cazenave J-P, Gachet C. The human $P2Y_1$
receptor is responsible for ADP-induced platelet activation. Purines and their receptors. The 5th Ann Neuropharmacol Conference 1997:Fr P.12

Henderson DJ, Elliot DG, Smith GM, Webb TE, Dainty IA. Cloning and characterisation of a bovine P_{2Y} receptor. Biochem Biophys Res Commun 1995;212:648-56

Holton FA, Holton P. The possibility that ATP is a transmitter at sensory nerve endings. J Physiol 1953;119:50-1P

Holton FA, Holton P. The capillary dilator substances in dry powders of spinal roots: a possible role of ATP in chemical transmission from nerve endings. J Physiol (Lond) 1954;126:124-40

Holton P. The liberation of ATP on antidromic stimulation of sensory nerves. J Physiol (Lond) 1959;145:494-504

Hourani SMO, Hall DA. P2T purinoceptors: ADP receptors on platelets. In: *P2 Purinoceptors: localization, function and transduction pathways*. Wiley, Chichester, 1996:53-70

Humphrey PPA, Buell G, Kennedy I, Khakh BS, Michel AD, Surprenant A, Trezise DJ. New insights on P_{2X} purinoceptors. Naunyn-Schmiedeberg's Arch Pharmacol 1995;352:585-96

Ishizaki M, Iizuka Y, Suzuki-Kusaba M, Kimura T, Satoh S. Nonadrenergic contractile response of guinea pig portal vein to electrical field stimulation mimics response to UTP but not to ATP. J Cardiovasc Pharmacol 1997;29:360-6

Jaffar ZH, Pearce FL. Histamine secretion from mast cells stimulated with ATP. Agents Actions 1990;30:64-6

Juul B, Plesner L, Aalkjaer C. Effects of ATP and UTP on $[Ca^{2+}]$i, membrane potential and force in isolated rat small arteries. J Vasc Res 1992;29:385-95

Keefe KD, Pasco JS, Eckman DM. Purinergic relaxation and hyperpolarization in guinea pig and rabbit coronary artery: role of the endothelium. J Pharmacol Exp Ther 1992;260:592-600

Kennedy C, Burnstock G. Evidence for two types of P_2-purinoceptor in the longitudinal muscle of the rabbit portal vein. Eur J Pharmacol 1985;111:49-56

Kennedy C, Burnstock G. ATP causes postjunctional potentiation of noradrenergic contractions in the portal vein of guinea pig and rat. J Pharm Pharmacol 1986;38:307-9

Kennedy C, Delbro D, Burnstock G. P_2-purinoceptors mediate both vasodilation (via the endothelium) and vasoconstriction of the isolated rat femoral artery. Eur J Pharmacol 1985;107:161-8

Lagaud GJL, Stoclet JC, Andriantsitohaina R. Calcium handling and purinoceptor subtypes involved in ATP-induced contraction in rat small mesenteric arteries. J Physiol 1996;492:689-703

Lewis C, Neidhart S, Holy C, North RA, Buell G, Surprenant A. Coexpression of P2X$_2$ and P2X$_3$ receptor subunits can account for ATP-gated currents in sensory neurones. Nature 1995;377:432-5

Liu SF, McCormack DG, Evans TW, Barnes PJ. Evidence for two P_2-purinoceptor subtypes in human small pulmonary arteries. Br J Pharmacol 1989;98:1014-20

Liu SF, Crawley DE, Evans TW, Barnes PJ. Endothelium-dependent nonadrenergic, non-cholinergic neural relaxation in guinea pig pulmonary artery. J Pharmacol Exp Ther 1992;260:541-8

MacKenzie AB, Mahaut-Smith MP, Sage SO. Activation of receptor-operated cation channels via P_{2X1} not P_{2T} receptors in human platelets. J Biol Chem 1996;271:2879-2881

Malam-Souley R, Seye C, Gadeau AP, Loirand G, Pillois X, Campan M, Pacaud P, Desgranges C. Nucleotide receptor P_{2u} partially mediates ATP-induced cell cycle progression of aortic smooth muscle cells. J Cell Physiol 1996;166:57-65

Mathieson JJI, Burnstock G. (1985) Purine-mediated relaxation and constriction of isolated rabbit mesenteric artery are not endothelium-dependent. Eur J Pharmacol 1985;118:221-9

McLaren GJ, Burke K, Sneddon P, Kennedy C. Comparison of the actions of ATP and UTP in the rat tail artery. Drug Dev Res 1996;37:135

Milner, P., Bodin, P., Loesch, A., Burnstock, G. Rapid release of endothelin and ATP from isolated aortic endothelial cells exposed to increased flow. Biochem. Biophys. Res. Commun 1990a;170:649-656

Milner P, Kirkpatrick KA, Ralevic V, Toothill VJ, Pearson JD, Burnstock G. Endothelial cells cultured from human umbilical vein release ATP, substance P and acetylcholine in response to increased flow. Proc Roy Soc Series B 1990b;241:245-8

Miyagi Y, Kobayashi S, Nishimura J, Fukui M, Kanaide H. Dual regulation of cerebrovascular tone by UTP: P_{2U} receptor-mediated contraction and endothelium-dependent relaxation. Br J Pharmacol 1996;118:847-56

Nobles M, Revest PA, Couraud P-O, Abbott NJ. Characteristics of nucleotide receptors that cause elevation of cytoplasmic calcium in immortalized rat brain endothelial cells (RBE4) and in primary cultures. Br J Pharmacol 1995;115:1245-52

North RA. P2X purinoceptor plethora. *Seminars in The Neurosciences*. 1996;8:187-94

O'Connor SE, Dainty IA, Leff P. Further subclassification of ATP receptors based on agonist studies. Trends Pharmacol Sci 1991;12:137-41

Patel V, Brown CA, Boarder MR. Novel second messenger pathways in P2Y receptor pharmacology: stimulation of MAPK by native endothelial P2Y1 and P2Y2. Drug Dev Res 1996a;37:123

Patel V, Brown C, Boarder MR. Protein kinase C isoforms in bovine aortic endothelial cells: role in regulation of P_{2Y}- and P_{2U}-purinoceptor-stimulated prostacyclin release. Br J Pharmacol 1996b;118:123-30

Pelleg A, Hurt CM. Mechanism of action of ATP on canine pulmonary vagal C fibre nerve terminals. J. Physiol. 1996;490:265-275

Purkiss JR, Boarder MR. Stimulation of phosphatidate synthesis in endothelial cells in response to P$_2$-receptor activation: evidence for phospholipase C and phospholipase D involvement, phosphatidate and diacylglycerol interconversion and the role of protein kinase C. Biochem J 1992;287:31-6

Purkiss JR, Wilkinson GF, Boarder MR. Differential regulation of inositol 1,4,5-trisphosphate by co-existing P_{2Y}-purinoceptors and nucleotide receptors on bovine aortic endothelial cells. Br J Pharmacol 1994;111:723-8

Qasabian RA, Schyvens C, Owe-Young R, Killen JP, Macdonald PS, Conigrave AD, Williamson DJ. Characterization of the P$_2$ receptors in rabbit pulmonary artery. Br J Pharmacol 1997;120:553-8

Ralevic V, Burnstock G. Postjunctional synergism of noradrenaline and adenosine 5'-triphosphate in the mesenteric arterial bed of the rat. Eur J Pharmacol 1990;175:291-9

Ralevic V, Burnstock G. Effects of purines and pyrimidines on the rat mesenteric arterial bed. Circ Res 1991a;69:1583-90

Ralevic V, Burnstock G. Roles of P$_2$-purinoceptors in the cardiovascular system. Circulation 1991b;84:1-14

Ralevic V, Burnstock G. Discrimination by PPADS between endothelial P_{2Y}- and P_{2U}-purinoceptors in the rat isolated mesenteric arterial bed. Br J Pharmacol 1996a;118:428-34

Ralevic V, Burnstock, G. Relative contribution of P_{2U}- and P_{2Y}-purinoceptors to endothelium-dependent vasodilatation in the golden hamster isolated mesenteric arterial bed. Br J Pharmacol 1996b;117:1797-802

Ralevic V, Burrell S, Kingdom J, Burnstock G. Characterization of P2 receptors for purine and pyrimidine nucleotides in human placental cotyledons. Br J Pharmacol 1997;121:1121-1126

Ralevic V, Hoyle CHV, Burnstock G. Pivotal role of phosphate chain length in vasoconstrictor *versus* vasodilator actions of adenine dinucleotides in rat mesenteric arteries. J Physiol 1995;483:703-13

Ralevic V, Milner P, Kirkpatrick KA, Burnstock G. Flow-induced release of adenosine 5'-triphosphate from endothelial cells of rat mesenteric arterial bed. Experientia 1991;48:31-4

Rubino A, Burnstock G. Evidence for a P_2-purinoceptor mediating vasoconstriction by UTP, ATP and related nucleotides in the isolated pulmonary vascular bed of the rat. Br J Pharmacol 1996;118:1415-20

Saiag B, Milon D, Allain H, Rault B, Driessche VD. Constriction of the smooth muscle of rat tail and femoral arteries and dog saphenous vein is induced by uridine triphosphate via'pyrimidinoceptors', and by adenosine triphosphate via P_{2x} purinoceptors. Blood Vess 1990;27:352-64

Saiag B, Milon D, Shacoori V, Allain H, Rault B, Van Den Driessche J. Newly evidenced pyrimidinoceptors and the P_{2x} purinoceptors are present on the vascular smooth muscle and respectively mediate the UTP- and ATP-induced contractions of the dog maxilliary internal vein. Res Commun Chem Pathol Pharmacol 1992;76:89-94

Saiag B, Bodin P, Shacoori V, Catheline M, Rault B, Burnstock G. Uptake and flow-induced release of uridine nucleotides from isolated vascular endothelial cells. Endothelium 1995;2:279-85

Seifert R, Schultz G. Involvement of pyrimidinoceptors in the regulation of cell functions by uridine and by uracil nucleotides. Trends Pharmacol Sci 1989;10:365-9

Simonsen U, García-Sacristàn A, Prieto D. Involvement of ATP in the non-adrenergic non-cholinergic inhibitory neurotransmission of lamb isolated coronary small arteries. Br J Pharmacol 1997;120:411-20

Strøbæk D, Olesen S-P, Christopersen P, Dissing S. P_2-purinoceptor-mediated formation of inositol phosphates and intracellular Ca^{2+} transients in human coronary artery smooth muscle cells. Br J Pharmacol 1996;118:1645-52

Tatham PER, Cusack NJ, Gomperts BD. Characterization of ATP^{4-} receptor that mediates permeabilisation of rat mast cells. Eur J Pharmacol 1988;147:13-21

Tatham PER, Lindau M. ATP-induced pore formation in the plasma membrane of rat peritoneal mast cells. J Gen Physiol 1990;95:459-76

Todorov L, Mihaylova-Todorova S, Westfall TD, Sneddon P, Kennedy C, Bjur RA, Westfall DP. Neuronal release of soluble nucleotidases and their role in neurotransmitter inactivation. Nature 1997;387:76-9

Valera S, Hussy N, Evans RJ, Adami N, North RA, Surprenant A, Buell G. A new class of ligand-gated ion channel defined by P_{2X} receptor for extracellular ATP. Nature 1994;371:516-9

Valera S. Characterization and chromosomal localisation of a human P2X receptor from the urinary bladder. Recept Channels 1995;3:283-9

von Kügelgen I, Bultmann R, Starke K. Interaction of adenine nucleotides, UTP and suramin in mouse vas deferens: suramin-sensitive and suramin-insensitive components in the contractile effect of ATP. Naunyn-Schmiedeberg's Arch Pharmacol 1990;342:198-205

von Kügelgen I, Häussinger D, Starke K. Evidence for a vasoconstriction-mediating receptor for UTP, distinct from the P_2 purinoceptor, in rabbit ear artery. Naunyn-Schmiedeberg's Arch Pharmacol 1987;336:556-60

Webb TE, Feolde E, Vigne P, Neary JT, Runberg A, Frelin C, Barnard EA. The P2Y purinoceptor in rat brain microvascular endothelial cells couple to inhibition of adenylate cyclase. Br J Pharmacol 1996;119:1385-92

Wilkinson GF, Purkiss JR, Boarder MR. Differential heterologous and homologous desensitization of two receptors for ATP (P2y receptors and nucleotide receptors) coexisting on endothelial cells. Mol Pharmacol. 1994;45:731-6

Windscheif U, Ralevic V, Bäumert HG, Mutschler E, Lambrecht G, Burnstock G. Vasoconstrictor and vasodilator responses to various agonists in the rat perfused mesenteric arterial bed: selective inhibition by PPADS of contractions mediated via P_{2x}-purinoceptors. Br J Pharmacol 1994;113:1015-21

Yang S, Buxton ILO, Probert CB, Talbot JN, Bradley ME. Evidence for a discrete UTP receptor in cardiac endothelial cells. Br J Pharmacol 1996;117:1572-8

Yang S, Cheek DJ, Westfall DP, Buxton IL. Purinergic axis in cardiac blood vessels. Agonist-mediated release of ATP from cardiac endothelial cells. Circ Res 1994;74:401-7

Ziganshin AU, Hoyle CHV, Lambrecht G, Mutschler E, Bäumert HG, Burnstock G. Selective antagonism by PPADS at P_{2X}-purinoceptors in rabbit isolated blood vessels. Br J Pharmacol 1994;111:923-9

Ziyal R, Ralevic V, Nickel P, Ardanuy U, Mutschler E, Lambrecht G, Burnstock G. NF023, a novel P_2-purinoceptor antagonist, selectively inhibits vasoconstrictor responses mediated via P_{2x}-purinoceptors in the rat and hamster mesenteric arterial bed. *Perspectives in Receptor Research*, Camerino, Italy 1995;P53

P2-PURINOCEPTORS AND CARDIAC FUNCTIONS

Annalisa Rubino, Autonomic Neuroscience Institute, Royal Free Hospital School of Medicine, Rowland Hill Street, London NW3 2PF, U.K.

I. SUMMARY

Since the initial observation of Drury and Szent-Gyorgyi in 1929 on the depressant effects of adenine nucleoside and nucleotides on the mammalian heart, the role of adenosine and ATP in the control of cardiovascular function has been widely investigated. Although the determination of ATP actions in many tissues, including the myocardium, is complicated by its rapid catabolism to adenosine, it was already clear in early studies that ATP and adenosine have different effects on cardiac function. Since then, a growing body of evidence has demonstrated that ATP and adenosine exert distinct, often opposing, effects on the heart via activation of specific receptor systems and mechanisms of action. As the physiological and pharmacological activity of purines was investigated, so a receptor classification developed. Techniques of molecular biology and electrophysiology have led to the cloning of recombinant receptors specific for ATP divided in two families of ionotropic ligand-gated ion channels P2X-purinoceptors and metabotropic G protein-coupled P2Y-purinoceptors. Molecular characterization and localization of several P2-purinoceptor subtypes in the heart provide functional evidence in favor of the pathophysiological role of ATP in the control of cardiac function. Following an initial discussion on the possible endogenous sources for ATP within the heart and an update on the characterization of cardiac P2-purinoceptors, this chapter will describe the direct actions of ATP on the myocardium as well as the neuromodulatory effects of ATP on the autonomic innervation of the heart. As the myocardial activity can not be divorced from the coronary blood flow, the contribution of ATP to the local control of coronary vascular tone will also be presented.

II. BACKGROUND

A. Possible origin of extracellular ATP in the heart

It is known from a number of studies that adenine nucleotides are continually released in considerable quantities in the extracellular compartments within the heart, and contribute to the homeostasis of cardiac function via activation of specific plasmamembrane receptors. Furthermore, a variety of chemical (e.g. hypoxia), mechanical (e.g. sheer stress) and pharmacological (e.g. infusion of vasoactive substances such as acethylcholine, noradrenaline) stimuli enhance ATP release from the isolated perfused heart. The precise cellular locus of ATP release in the whole perfused heart can not be easily extrapolated as nervous tissue, myocytes and coronary endothelium are all likely sources for the nucleotide. Moreover, the rapid degradation of ATP by over 95% following a single coronary passage, implies that large amounts of the nucleotide are necessary to enable ATP to be measured in the coronary effluent. However, convincing evidence detailed below demonstrates how distinct tissues contribute to an increase in the cardiac extracellular ATP under different pathophysiological conditions.

1. Autonomic innervation and intramural neurones

The role of ATP as autonomic neurotransmitter initially proposed by Burnstock (1972) is now established and a substantial body of evidence shows that ATP is co-stored and co-released with noradrenaline from sympathetic nerves in the cardiovascular system (Burnstock, 1990). In the frog heart a purinergic neurotransmission has been demonstrated in the sympathetic nerves, where ATP is co-released with adrenaline (Doyle, 1985; Hoyle and Burnstock, 1986). In the isolated guinea-pig atria, pharmacological stimulation with ouabain, forskolin and isoprenaline leads to exocytotic release of ATP of sympathetic origin. In contrast, the same stimuli fail to evoke ATP release from the ventricular myocardium which has a much sparser sympathetic innervation than atrial tissue (Katsuragi, et al., 1993; Katsuragi et al., 1995; Tokunaga et al., 1995).

The autonomic control of cardiac function includes the peripheral release of sensory neurotransmitters (namely calcitonin gene-related peptide, CGRP, and substance P, SP) from capsaicin-sensitive sensory-motor nerves (Rubino and Burnstock, 1996). The hypothesis of a role for ATP as a sensory co-transmitter derives from the seminal observation that in the ear artery the electrical stimulation of capsaicin-sensitive sensory fibres evoked vasodilatation accompanied by release of ATP (Holton, 1959). This has been substantiated by several findings, including coexistence of SP with ATP and release of ATP from sensory fibres in the spinal cord (Fyffe and Perl, 1984). Furthermore, the selective localization of P2X-purinoceptors on sensory fibres at both pre- and post-junctional sites (Vulchanova et al., 1996) strongly suggest a role for ATP as sensory neurotransmitter. The rich sensory-motor innervation of the atrial myocardium in several species including humans, must be therefore regarded as a possible neuronal source for ATP within the heart (see Rubino and Burnstock, 1996).

The contribution of intrinsic neurons to the autonomic control of cardiac function remains uncertain due to the technical difficulties to investigate their function in the absence of sympathetic and parasympathetic innervation of the myocardium. However, the localization of several neurotransmitters in subpopulations of intracardiac neurones and functional responses to autonomic neurotransmitters in isolated intracardiac neurons, strongly support the hypothesis that intrinsic neurons participate to the autonomic control of the heart (see Rubino et al., 1996). The histochemical localization of ATP in a subpopulation of intrinsic cardiac neurones by quinacrine, a fluorescent acridine dye which physically binds adenine nucleotides, provided direct evidence for neuronal ATP in the heart. Quinacrine-fluorescence, resistant to sympathetic denervation by 6-hydroxydopamine, was shown in neuronal cell bodies and fibres of guinea-pig and rabbit atrial myocardium (Crowe and Burnstock, 1982). As many of the ganglia are concentrated around the sino-atrial node and in the interatrial septum and as such are ideally placed to influence both nodal and conducting tissues, it is possible that release of ATP from intracardiac neurons may be responsible at least in part for mediating some of the direct and/or neuromodulatory effects of ATP on cardiac automaticity. Recent findings have demontrated that in rabbit sinoatrial cell node, ATP, but not its analogues or adenosine, can activate an inward rectified current via P2-purinoceptors which might contribute to the chronotropic effects seen on release of autonomic neurotransmitters (Shoda et al., 1997). Furthermore, as many intracardiac neurons project their fibres to the coronary vasculature, ATP released at external surface of blood vessels could contribute to the neuronal regulation of coronary tone via activation of vascular P2-purinoceptors (see Simonsen et al., 1997).

2. Vascular Endothelium and Smooth Muscle

The notion that endothelial cells can release ATP (Pearson and Gordon, 1979) makes the coronary endothelium an important precursor pool for adenosine and ATP involved in the moment-to-moment regulation of cardiac blood flow. Release of ATP has been shown from isolated coronary endothelial cells, following pharmacological challenge with vasoactive agonists known to be present in the blood stream (Yang et al., 1994). Moreover, ATP-evoked release of ATP from the same endothelial cells endowed with P2-purinoceptors suggest an autocrine feedback loop for the purinergic control of the coronary vascular tone (Yang et al., 1994). In the whole isolated perfused heart, studies by Paddle and Burnstock (1974) and later by Clements and Forrester (1981) showed that ATP is released from the coronary vasculature during hypoxic stimulation, thus providing early although indirect evidence for a role for ATP as a physiological modulator of coronary flow. Release of ATP from endothelial cells during shear stress and increased flow has been also shown in isolated endothelial cells, as well as in the perfused guinea-pig heart (Milner et al., 1990; Vials and Burnstock, 1996). Perfusion with radioactive adenosine at a low concentration has been used to selectively prelabel adenine nucleotide in the coronary endothelium so that increases in the release of radioactive metabolites from the whole heart can be taken as a direct measure of the release from the coronary endothelium (Schrader and Gerlach 1976). By mean of this technique Borst and Schrader (1991) have shown that under normoxic condition the endothelial release of ATP into the coronary venous effluent perfusate is at the same order of magnitude as that of adenosine. However, the coronary endothelium preferentially contributes to the ischaemia-induced release of ATP.

Although endothelial cells appear to be a major source for ATP within the coronary circulation, the contribution of vascular myocytes to release of ATP within the extracellular space in the heart can not be excluded. Release of ATP has been shown from non-vascular smooth muscle of several tissues (e.g. vas deferent and ileum) under pharmacological stimulation (Katsuragi et al., 1990;1991) Isolated smooth muscle cells of rabbit aorta in culture also release ATP under normal flow condition (Bodin et al., 1987). However, in contrast to endothelial cells, smooth muscle in culture do not show increased release of ATP under high flow condition (Bodin et al., 1987; Bodin and Burnstock, 1995). This data would suggest a minor contribution of the vascular smooth muscle to cardiac increase in extracellular ATP under pathophysiological conditions that imply coronary vasoconstriction and sheer stress.

3. Cardiac Myocytes

At the beginning of this century the concept that coronary vessels dilate in response to substances released from the hypoxic myocardium was already proposed (Barcroftt and Dixon, 1907). The release of a potent vasodilator of coronary vessels such as ATP from the heart may therefore be an important mechanism to match the blood flow supply to the myocardium with the oxygen demand of the tissue. In support of this hypothesis, release of ATP was demonstrated from isolated ventricular myocytes, depending on the presence or absence of oxygen and on the pH of the surrounding environment (Forrester and Williams, 1977; Williams and Forrester, 1983). Complexes of ATP-Ca^{2+} proteins in the sarcolemma of the isolated myocytes have been indicated as a possible source for hypoxia-evoked release of ATP from myocardial cells (Williams and Forrester, 1983).

<u>B. Receptor systems for ATP in the heart</u>

This is not intended as a comprehensive overview on the classification of P2-purinoceptors, which is largely provided in other chapters in this book. It rather gives an update on the more recent findings on the P2-purinoceptors in the heart in support of functional evidence on the cardiac effects of ATP described in the following sections.

The initial suggestion of Burnstock (1978) to distinguish P1- and P2-purinoceptors, most sensitive to adenosine and ATP respectively, has seen a number of modifications due to the growing body of information in the field. According to pharmacological evidence, mechanisms of intracellular signal transduction and molecular structure, the receptors that mediate the extracellular effects of ATP are now divided into ionotropic P2X- (ligand-gated ion channels) and metabotropic P2Y- (G protein-coupled) receptor families (Abbracchio and Burnstock, 1994). A number of distinct mammalian genes, including clones from human cell lines and tissues have been cloned within each receptor family, identified as P2X1-7 and P2Y1, P2Y2, P2Y4, P2Y6 and P2Y11 subtypes (Burnstock and King, 1996; Fredholm et al., 1997). Within the P2Y receptor family, the P2Y2, P2Y4 and P2Y6 subtypes are also activated by pyrimidine nucleotides (Communi et al., 1995; 1996).

Together with functional data, molecular evidence has been provided for the existence of cardiac receptor subtypes for ATP. In the whole rat heart as well as in isolated neonatal and adult cardiac myocytes the reverse-transcriptase polymerase chain reaction (RT-PCR) showed expression of P2Y1-, P2Y2-, P2Y4- and P2Y7-purinoceptor subtypes (Webb et al., 1996). Furthermore, a P2u receptor (or P2Y2-purinoceptor, according to the current nomenclature; Frehedolm et al., 1997) was shown in the rat heart mainly expressed in endothelial cells (Godecke et al., 1996). Within the P2Y family of ATP receptors several subtypes have been shown in the adult human heart. The human P2Y2 receptor, cloned from airway and colonic epithelial cells, has been demonstrated in the adult heart by mean of Northern blot analysis (Parr et al., 1995). Similarly, the human P2Y6 subtype, strongly expressed in the placenta, was also detected in the heart (Communi et al., 1996). Although a putative P2Y7 receptor was found to be relatively abundant in the adult human heart (Akbar et al., 1996), the receptor was later found to be that for leukotriene B4. Recently, the screening of a cDNA library from the human foetal heart provided the first characterization of P2Y4 receptors in the human heart (Bogdanov et al., 1998), previous reports showing this receptor only in the human placenta (Communi et al., 1995). Together with the P2Y4 subtype, P2Y2- and P2Y6 receptors were detected in the foetal human heart. The same study also demonstrated in the foetal heart the presence of P2X1-purinoceptors (Bugdanov et al., 1998). The human P2X1 receptor was previously cloned in the urinary bladder and also shown by RNA hybridization in several tissues such as lung, spleen, thymus, adrenal glands (Valera et al, 1995). P2X3- and P2X4-purinoceptors previously characterized in the adult human heart (Garcia-Guzman et al., 1997a; 1997b), were also shown in the foetal heart (Bugdanov et al., 1998). Binding studies have shown P2X-purinoceptors on the smooth muscle of rat arteries and arterioles (Bo et al., 1995). Furthermore, Nori et al. (1998) have

228

demonstrated that RT-PCR performed on microdissected tissues from the rat heart shows different levels of expression of P2X1-, P2X2- and P2X4-purinoceptor mRNAs in the atrial and ventricular myocardium (Fig.1). In the same study, expression of mRNAs for P2X1- P2X2- and P2X4-purinoceptors is shown by in situ hybridization in the rat coronary vessels, while no specific positivity is apparent in the myocardium (Fig. 2).

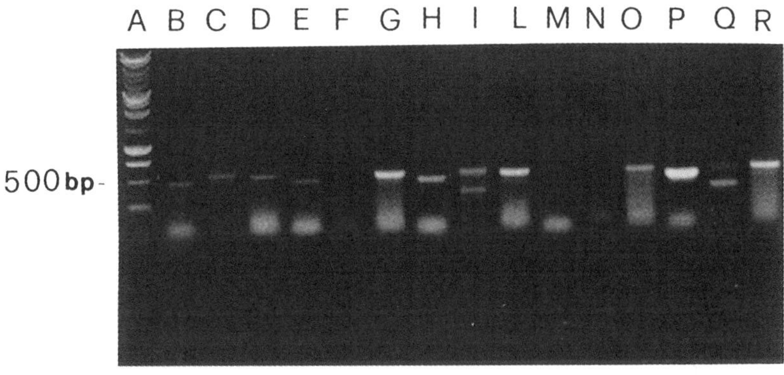

Figure 1. RT-PCR analysis of P2X-purinoceptor expression in microdissected areas of the rat heart. Lane A represents molecular weight markers. Amplification of cDNA with primers for P2X1, P2X2 and P2X4 subtypes, respectively, is shown in lanes B-D in left atrium, lanes E-G in left ventricle, lanes H-L in right atrium, lanes M-O in right ventricle. Lines P-R comparatively show amplification in aorta. Amplification of cDNA with primers specific for P2X1 and P2X4 subtypes gives a single band corresponding to about 500 bp long. Amplification of cDNA with primers specific for the P2X2 subtype gives two bands of about 500 and 300 bp long, corresponding to splice variants of P2X2-purinoceptor mRNA. Figure modified from Nori et al., 1998, with permission.

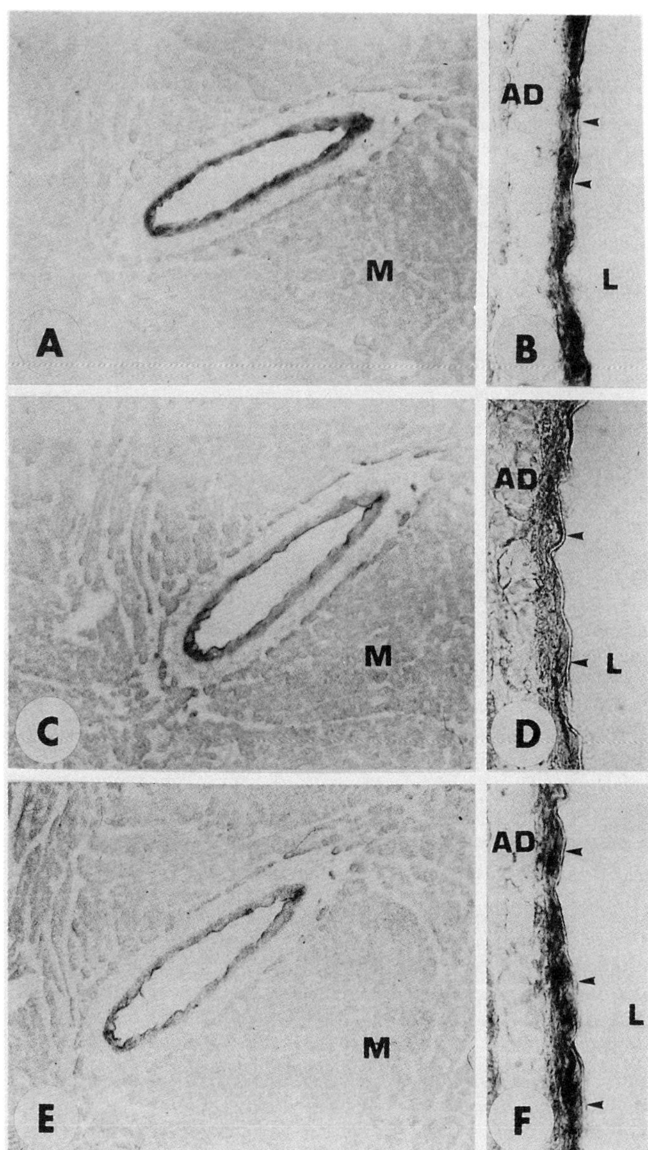

Figure 2. In sytu hybridization of three consecutive sections of rat coronary arteries, showing a large arteriole (about 150 μm diameter) in ventricular myocardium (M). mRNA transcripts for P2X1 (A), P2X2 (c) and P2X4 (E) receptors are localized at vascular level, while the surrounding myocardium is devoid of reactivity. At higher magnification (B,D,F) mRNA transcript for the receptors are detected only on the muscle layer, in close relation to the internal lamina (arrowhead). L= Lumen; AD= Adventitia. A,C,E x 84; B,D,F x 49.

The selective colocalization of mRNA transcripts for P2X1, P2X2 and P2X4 subtypes in the smooth muscle of coronary arteries is consistent with the vasoconstrictor responses of these vessels to ATP (see below). The molecular heterogeneity of P2X receptor isoforms in vascular smooth muscle substantiate the possibility of numerous heteropolymeric variants in the assembly of vascular ATP-gated ion channels. In contrast, functional evidence indicates that most of the direct and neuromodulatory effects of ATP in the heart are mediated by P2Y receptors, as detailed below.

III. TEXT

A. Direct effects of ATP on cardiac myocytes

For a long time the cardiac effects of ATP have been accounted for in large part by its rapid breakdown to adenosine and subsequent activation of P1-purinoceptors. These actions include negative chronotropism at the sinus node and dromotropism at the atrioventricular node. Furthermore, in the atrial myocardium purines have a negative inotropic effect, while in the ventricle antagonize the inotropism evoked by β-adenoceptor stimulation (reviewed in Olsson and Pearson, 1990). In isolated rat and guinea-pig atria it was shown that ATP mimics the negative inotropic effect of adenosine, although its metabolic breakdown is not a prerequisite for the activation of P1-purinoceptors (Burnstock and Meghji, 1981; 1983; Collis and Pettinger, 1982). However, several studies over the last 15 years have demonstrated a positive inotropic effect of ATP in both atrial and ventricular myocardium. A positive inotropic effect of ATP requires that the P1-purinoceptor negative inotropism is inhibited in some fashion. Mantelli et al. (1993) demonstrated that in isolated guinea-pig atria, application of increasing concentrations of ATP leads to negative inotropic responses, while in the presence of the P1-purinoceptor antagonist 1,3-dipropyl-8-cyclopentylxanthine (DPCPX), the same concentrations of ATP evoke an increase in atrial contractile force (Fig. 3). The positive inotropic effect of ATP was mimicked by its analogues α,β- and β,γ-methylene-ATP and was sensitive to antagonism by suramin and reactive blue 2, thus revealing the activation of P2Y receptors (Mantelli et al., 1993).

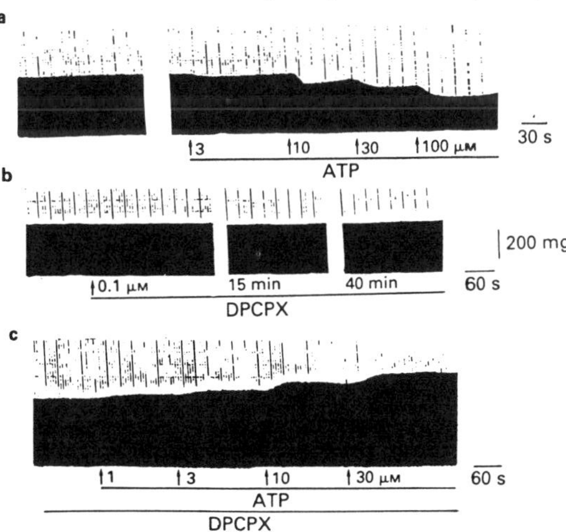

Figure 3. Inotropic effects of ATP in isolated guinea-pig atria. Typical recording of an experiment showing that application of cumulative concentrations of ATP in isolated guinea-pig left atrium electrically driven (1 Hz) increasingly reduces the basal contractile force (a); during blockade of P1-purinoceptors by the antagonist 1,3-dipropyl-8-cyclopentylxanthine (DPCPX, 0.1 μM), in the same concentration range ATP evokes an increase of atrial contratility (c). DPCPX did not show any direct effect on basal contractile force over an incubation period of 40 min (b).

In the rat ventricle blockade of coupling of P1-purinoceptors to the G-protein effector system by mean of pertussis toxin treatment resulted in positive inotropic responses to ATP, which could be additive to the maximal inotropic responses to β-adrenoceptor stimulation (Legssyer et al., 1988; Scamp et al., 1990). In line with the positive inotropism of ATP in the mammalian heart, studies on isolated rat ventricular myocytes demonstrated that ATP but not adenosine increases intracellular concentration of free Ca^{2+} and contractility (Danziger et al., 1988). The effect depended on entrance of extracellular calcium and involved also calcium-induced release of Ca^{2+} from intracellular stores (DeYoung and Scarpa, 1989; Puceat et al., 1991). ATP-evoked increase of intracellular concentration of free Ca^{2+} in isolated myocytes was potentiated by noradrenaline via β- but not α-adrenoceptor activation (DeYoung and Scarpa, 1989). An additional mechanism of ATP-mediated regulation of the intracellular free Ca^{2+} has also been suggested, dependent on extracellular concentration of inorganic phosphate, Na^+ and Ca^{2+}, which imply a coordinate transport between these ions (DeYoung and Scarpa, 1991). Furthermore, the pharmacological profile of electrical and mechanical responses indicates that in isolated myocytes activation of P2Y receptors by ATP and its analogues evoke a calcium current (I_{ca}) by an increase of both the probability of L-type calcium channel opening and the availability of functional channels (Scamp et al., 1990; 1993; Scamp and Vassort, 1994). In line with the general profile of metabotropic P2Y receptors, ATP-evoked increase of I_{ca} is mediated by a G-protein, independent from the cAMP pathway, and is associated with phosphoinositides hydrolysis and activation of protein kinase C (Legssyer et al., 1988; Scamp et al., 1990; Yamada et al., 1992). The involvement of a PLCγ in the ATP-evoked stimulation of cardiomyocytes has also been demonstrated (Puceat and Vassort, 1996). Activation of a nonspecific cation current has been described in isolated ventricular myocytes exposed to micromolar concentrations of ATP, independent from G protein activation, which does not appear to be related to occupancy of P2Y receptors (Scamp and Vassort, 1990; 1994; Zheng et al., 1993). This raises the possibility that in addition to P2Y receptors, subtype(s) of the ligand-gated ion channel of the P2X-purinoceptor family are also present on cardiomyocytes and contribute to the inotropism of ATP.

Application of ATP to isolated cardiomyocytes induces transient variations of the intracellular pH which are associated in a cause-effect relation with an increase of intracellular concentration of free $Ca^{2+.}$ It has been shown that exposure to ATP leads to acidification owing to the activation of the $HCO3^-/Cl^-$ exchanger followed by alkalinization due to the activation of the Na^+/H^+ antiport (Scamp and Vassort, 1990; Puceat et al., 1991). However, ionic mechanisms described in the cardiomyocytes following P2-purinoceptor activation also include a Cl^- conductance (Kaneda et al., 1994; Levesque and Hume, 1995), which could be of pathophysiological relevance considering the ability of chloride channels to modulate cardiac electrical activity.

The concentrations of ATP in the heart is known to vary under pathophysiological conditions such as hypoxia and ischaemia, therefore the activation of P2-purinoceptors in the myocardium is an important factor in the regulation of cardiac contractility. At high concentrations (>10 μM) of extracellular ATP, such as those found during ischaemia, maximal activation of the Cl/HCO3 exchanger would overcome the stimulation of the antiport and lead to acidification, in addition to the metabolic acidification associated with hypoxia. On this basis it has been suggested that P2-purinoceptor activation may be a contributing factor to cardiac arrhythmia following ischaemia (Vassort et al., 1994). This hypothesis is substantiated by the finding that in isolated ventricular myocytes, changes in concentration of intracellular free Ca^{2+} following activation of P2-purinoceptors are sufficient to trigger oscillatory contraction under quiescent condition. Furthermore, in electrically stimulated myocytes, activation of P2-purinoceptors potentiates the amplitude of electrically triggered contractions and induces arrhythmia (Zhang et al., 1996).

Most of the experimental evidence described above indicates that the stimulatory effects of ATP on the myocardial contractility as well as the correlated ionic and electrical events are mediated via activation of P2Y-purinoceptors. Lack of pharmacological tools to distinguish the distinct recombinant P2Y-purinoceptors so far cloned and characterized makes it difficult to identify the cardiac subtype(s). The observation that UTP and related nucleotides lack effects on Ca^{2+} transient and force of contraction of cardiac myocytes (De Young and Scarpa, 1987; Scamp et al., 1990), consistently with an absence of the P2ureceptor (recombinant P2Y2 subtype) in cardiac myocytes (Godecke et al., 1996), suggests that P2Y receptors sensitive to pyrimidines are not expressed on cardiac myocytes. Their identification in the heart should be rather correlated to the coronary vessels where UTP has potent effects. The P2Y1 and P2Y6 subtypes remains therefore the most likely candidates to mediate the complex cardiac effects of ATP.

B. Neuromodulatory effects of ATP

ATP shares with adenosine a modulatory control of the autonomic peripheral neurotransmission, by reducing the transmitter release and altering the responses of the target organ to the transmitter released in several tissues, including the myocardium. It was demonstrated that ATP co-released with noradrenaline from sympathetic nerves and its metabolic product adenosine inhibit cardiac sympathetic neurotransmission via prejunctional P1-purinoceptors in a negative neuromodulatory feed-back mechanism (Khan and Malik, 1980). However, the possible involvement of P2-purinoceptors in the neuromodulatory actions of ATP has been for long unexplored, mainly for lack of pharmacological tools. More recently, the hypothesis of peripheral release-inhibiting P2-purinoceptors that may function as autoreceptors has been investigated in several tissues (von Kugelgen et al., 1993). In the rat atria, ATP and its analogue ATPγS, but not α,β-methylene-ATP, reduced electrically evoked release of tritiated noradrenaline, the inhibitory effect being maintained in the presence of the 5'ectonucleotidase blocker α,β-methylene-ADP. The inhibitory effect of ATP was mimicked by adenosine and sensitive to the P2-purinoceptor antagonists cibacron blue 3 GA and to the P1-purinoceptor antagonist DPCPX, the antagonism by the two molecules being additive on the inhibitory effect of ATP (von Kugelgen et al., 1995). On the basis of these observations it was inferred that ATP inhibits the release of noradrenaline from cardiac sympathetic neurons via activation of both prejunctional P1- and P2-purinoceptors. The lack of any effect by the most potent agonist for P2X-purinoceptors, α,β-methylene-ATP, and the pKB value for the P2-

antagonist, cibacron blue 3GA, suggested the involvement of a prejunctional P2Y-like purinoceptor, as shown in other tissues (von Kugelgen et al., 1993). Furthermore, as cibacron blue 3 GA increased the evoked release of noradrenaline, it was proposed that prejunctional P2-purinoceptors are activated by the endogenous ligand ATP, at least under the experimental conditions described (von Kugelgen et al., 1995).

Beside reducing release of noradrenaline from cardiac sympathetic nerves, ATP modulates capsaicin-sensitive sensory-motor neurotransmission, mediated by peripheral release of CGRP within the myocardium (Rubino et al., 1992). In isolated guinea-pig atria, ATP and its analogues α,β- and β,γ-methylene-ATP mimicked the inhibitory effect of adenosine on the inotropic reponses to transmural nerve stimulation of sensory-motor nerves (see Rubino, 1993). The inhibitory effect of ATP was maintained in the presence of the P2-purinoceptor antagonist suramin, while was sensitive to the P1-purinoceptor antagonist 8-phenyltheophylline (Fig. 4A), indicating the involvement of prejunctional P1-purinoceptors in the inhibitory effect observed. However, the catabolism to adenosine did not appear to be a pre-requisite for the activation of P1-purinoceptors, as the inhibitory effect of ATP was maintained in the presence of the 5'ectonucleotidase blocker α,β-methylene-ADP (Fig. 4B) and was mimicked by ATP analogues relatively resistant to enzymatic degradation (Rubino et al., 1992). It is noteworthy that ATP shares with the sympathetic co-tranmitters noradrenaline and NPY an inhibitory control on cardiac sensory-motor neurotransmission, therefore contributing to the cross-talk between distinct branches of cardiac autonomic innervation (Rubino, 1993; Rubino and Burnstock, 1996).

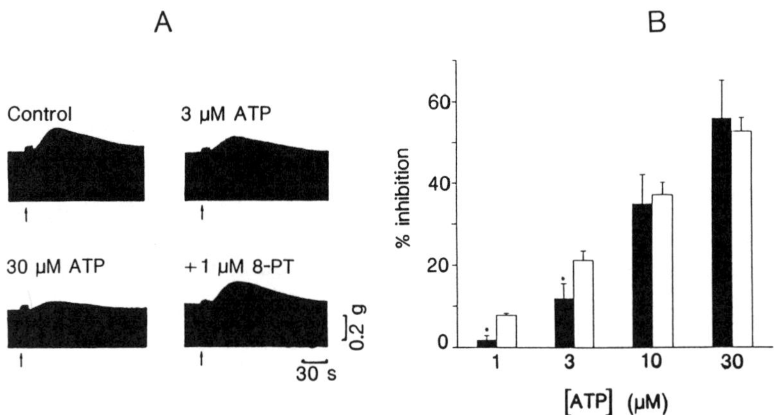

Figure 4. Concentration-dependent inhibitory effect of ATP on cardiac sensory-motor neurotransmission. (A) Recording from a typical experiment showing positive inotropic responses due to release of CGRP following transmural stimulation of sensory-motor nerves in the absence (control) and in the presence of ATP. The addition to the bathing solution of the P1-purinoceptor antagonist 8-phenyltheophylline (8-PT, 1 µM) reversed the inhibitory effect of 30 µM ATP. The arrows indicate the beginning of transmural nerve stimulation. (B) The concentration-dependent inhibitory effect of ATP on sensory-motor neurotransmission (open columns) was still obtained in the presence of 10 µM α,β-methylene-ADP (solid columns), a significant reduction of the inhibitory effect being

shown only at the lowest concentrations of ATP. Figure modified from Rubino et al., 1992, with permission.

Extrinsic sympathetic, parasympathetic and sensory-motor fibres terminate in the intrinsic plexuses and influence the activity of intracardiac neurons via different mechanisms and receptors. Intracardiac neurons may also mediate some reflex-like actions and exert a certain degree of independent local control of the cardiac function. Studies in cultured guinea-pig and rat intracardiac neurons demonstrated that a large population of these cells respond to exogenous application of ATP (Saffrey et al.,1992; Allen and Burnstock, 1990). In a study on cultured guinea-pig intracardiac neurons two distinct cellular populations were identified (Allen and Burnstock, 1990). In some cells ATP produced a transient increase in membrane conductance, resulting in a large inward current. In the other cell population a three-component response to ATP was shown. The P2X agonist α,β-methylene-ATP only weakly mimicked electrical responses to ATP, while the P2Y agonist 2-methylthioATP was more active than ATP. Furthermore, functional responses to ATP and its analogues were attenuated by reactive blue 2, shown to display a degree of selectivity in antagonizing P2Y-purinoceptors. On the basis of these lines of evidence, the presence of P2Y-purinoceptors on intracardiac neurons has been suggested (Allen and Burnstock, 1990). This raises the possibility that in the heart they may act at pre- or post-junctional level to modulate transmitter release from intracardiac neurons and thereby modulate the activity of the effector tissue. Studies on adult guinea-pig cardiomyocytes in long term cultures and co-cultures with intrinsic cardiac neurons have confirmed this hypothesis, showing that intrinsic cardiac neurons which possess P2-purinoceptors can greatly enhance myocyte contractile rate following activation by ATP and its analogues (Horackova et al., 1994).

C. Vascular actions of ATP on the coronary circulation

As the coronary flow directly relates to the cardiac contractility, the direct and neuromodulatory effects of ATP on the myocardium can not be divorced from its vascular actions on the coronary circulation. In the isolated rat heart ATP produces a biphasic response via vasoconstrictor P2X- and vasodilator P2Y-purinoceptors (Hopwood and Burnstock, 1987). However, in contrast with other blood vessels where ATP synergizes with noradrenaline to evoke vasoconstriction via P2X-purinoceptors and α-adrenoceptors, respectively, in the coronary circulation the predominant effect of ATP is vasodilatation via P2Y-purinoceptors (Corr and Burnstock, 1991, Matsumoto et al., 1997).

A neurogenic vasodilatation mediated by ATP has been shown in lamb small coronary arteries (Simonsen et al., 1997). Several studies have shown that coronary vasodilatation to ATP requires an intact endothelium, although in the left anterior descending coronary arteries of the rabbit vasodilatation to ATP occurs in the absence of endothelium, via smooth muscle P2Y-purinoceptors (Corr and Burnstock, 1991; Keef et al., 1992). The contribution of smooth muscle P1-purinoceptors to vasodilator responses to ATP has also been suggested in the rabbit coronary artery (Keef et al., 1992). ATP evokes relaxation of large canine coronary arteries by an endothelium-dependent mechanism whereas adenosine appears to be more effective than ATP in relaxing small coronary arteries in an endothelium-independent fashion (White and Angus, 1987). In the guinea-pig and rat coronary vascular bed it has been shown that vasodilatation to ATP is sensitive to treatment with nitric oxide synthase inhibitors and indomethacin, thus implying that stimulation of endothelial P2Y-purinoceptors evokes release of nitric oxide and prostanoids (Hopwood et al., 1989; Vials and Burnstock, 1992; 1994). In contrast, vasodilatation to the P2Y agonist 2-methylthioATP is sensitive to treatment with L-NAME (an inhibitor of nitric oxide synthasis), but is resistant to prevention of prostanoid formation by indomethacin (Fig. 5).

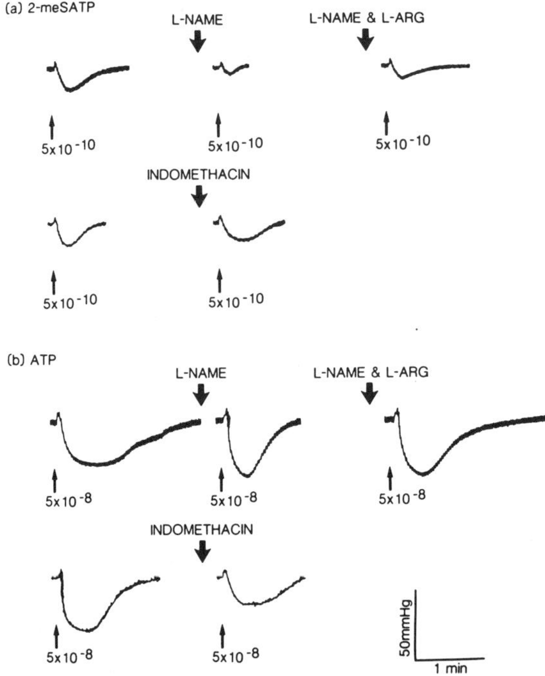

Figure 5. Inhibition of nitric oxide and prostanoid production during vasodilator responses to P2Y-purinoceptor stimulation. Typical perfusion traces from in vitro guinea-pig perfused heart showing vasodilator responses to 2-methylthioATP (a) and ATP (b) in the absence and in the presence of L-nitroarginine methyl ester (L-NAME, 30 μM) and indomethacin 1 μM. L-arginine (L-ARG, 1.5 mM) partially reversed the inhibitory effect of L-NAME. Figure from Vials and Burnstock, 1994, with permission.

The distinction between the vasodilator effect of ATP and its analogue could be explained by the fact that in several blood vessels, including the coronary circulation, ATP shares with UTP an additional endothelial receptor of the P2Y family where ATP and UTP are both more potent agonists than 2-methylthioATP. The involvement of an endothelial P2u receptor in the ATP/UTP-evoked vasodilatation has been shown in the isolated rat heart (Godecke et al., 1996). In contrast, in the canine epicardial coronary artery, the localization of a UTP-sensitive P2Y receptor on the smooth muscle leads to a potent vasoconstrictor effect of pyrimidine nucleotides (UTP, UDP), opposing to the potent endothelium-dependent vasodilator effect of ATP (Matsumoto et al., 1997).

In the mammalian heart immediate vasodilatation of the coronary arteries occurs in response to hypoxia, quickly satisfying the oxygen requirement when imposition of an increased workload increases oxygen usage and demand of the myocardium and a state of hypoxia is likely to be brought about. Although adenosine has been for long time regarded as the major agent responsible for vasodilatation during hypoxic perfusion of the heart, pharmacological evidence has been provided indicating that ATP is involved in hypoxia-evoked coronary dilatation. In a study in the isolated rat heart, the P2Y antagonist reactive blue 2 reduced by about 40% vasodilator responses to hypoxic perfusion. Furthermore, hydroquinone treatment reduced the hypoxic vasodilatation of the rat coronary vascular bed by over 60%, indicating the contribution of endothelial nitric oxide (Hopwood et al., 1989). The functional data together with evidence for release of ATP in the coronary effluent from the isolated rat heart following hypoxia (Clemens and Forrester, 1981; Vial et al., 1987) support the view that ATP plays a major role in the local control of coronary blood flow.

IV. CONCLUSION AND FUTURE DIRECTIONS

The molecular characterization of receptor subtypes for ATP in the heart of several species, including humans, supports functional data showing that ATP plays many different and synergistic roles in the maintenance of cardiac homeostasis via activation of specific P2-purinoceptors. Furthermore, the presence of a set of specific P2-purinoceptor subtypes in the foetal heart and neonatal myocytes, many of them not differing from those in the adult heart, strongly suggest the contribution of ATP to differentiation processes and to the control of cardiovascular function at the foetal stage of development.

P2Y receptors are involved in the direct effects of ATP on cardiomyocytes and mediate its neuromodulatory actions on cardiac autonomic innervation and intracardiac neurons. P2Y receptors on cardiomyocytes share intracellular mechanisms of activation involved in the positive inotropism following α-adrenoceptor stimulation. Both receptor systems increase the turnover of phosphatidylinositol which leads to an increase in formation of inositol trisphosphate, and diacylglicerol and activation of protein kinase C. Although subsequent pathways seem to differ, both types of stimulation influence intracellular pH. They modulate Na-H antiport activity and facilitate the recovery of pH following an intracellular acid challenge. Evidence of a complementary action of β- but not α-adrenoceptor and P2-purinoceptor activation on calcium homeostasis and inotropism suggests that noradrenaline and ATP released from sympathetic innervation may act in concert to stimulate cardiac contractility. This could be an important mechanism to develop maximal force of contraction in healthy cells to compensate for the failing heart from which ATP could be released. Interestingly, as ATP and noradrenaline may synergize in increasing cardiac contractility via P2Y-purinoceptors and β-adrenoceptors, respectively, the vasodilator effect of ATP on the coronary vasculature following stimulation of vascular P2Y-purinoceptors complements β-adrenoceptor stimulation and consequent vasodilatation to noradrenaline.

Several subtypes of the P2X-purinoceptor family have been characterized in the myocardium. Functional data and their localization on vascular structures indicate that their major role is the control of coronary vascular tone, although the presence of P2X-purinoceptors on ventricular myocytes could not be excluded.

Furthermore, the existence of P2X sensory receptors for ATP should be considered within the heart. The development of selective agonists and antagonists for the numerous subtypes of recombinant P2-purinoceptors characterized is awaited for a better understanding of the role that endogenous ATP plays in the regulation of cardiac performance. Given the suggested contribution of ATP to the regulation of coronary tone and cardiac contractility during pathophysiological events such as ischaemia and arrythymia, the possibility of targeting specific P2-purinoceptor subtypes may hold exciting therapeutic potentials.

Acknowledgments
This work was supported by the British Heart Foundation. Mr. Roy Jordan is gratefully thanked for his excellent assistance during the preparation of the manuscript.

REFERENCES.

Abbracchio M, Burnstock G. Purinoceptors: Are there families of P2X and P2Y purinoceptors. Pharmacol Ther 1994; 64:445-475

Akbar GKM, Dasari VR, Webb TE, Ayyanathan K, Pillarisetti K, Sandhu AK, Athwal RS, Daniel JL, Ashby B, Barnard EA, Kunapuli SP. Molecular cloning of a novel P2 purinoceptor from human erythroleukemia cells. J Biol Chem 1996;271:18363-18367

Allen TGJ, Burnstock G. The action of adenosine 5'-triphosphate on guinea-pig intracardiac neurones in culture. Br J Pharmacol 1990;100:269-276

Barcroft J, Dixon WE. The gaseous metabolism of the mammalian heart. Part 1. J Physiol (Lond) 1907;35:182-204

Bo X, Burnsyock, G. Heterogeneous distribution of [3H](,(-methylene ATP binding sites in blood vessels. J Vasc Res 1993;30:87-101

Bodin P, Bailey D, Burnstock G. Increased flow-induced ATP release from isolated vascular endothelial cells but not smooth muscle cells. Br J Pharmacol 1987;103:1203-1205

Bodin P, Burnstock G. Synergistic effect of acute hypoxia on flow-induced release of ATP from cultured endothelial cells. Experientia 1995;51:256-259

Bogdanov Y, Rubino A, Burnstock, G Characterization of subtypes of the P2X and P2Y families of ATP receptors in the foetal human heart. Life Sci 1998;62:697-703

Borst MM, Schrader J. adenine nucleotide release from isolated perfused guinea pig hearts and extracellular formation of adenosine. Circ Res 1991;68:797-806

Burnstock G. Purinergic nerves. Pharmacol Rev 1972;24:509-581

Burnstock Geoffrey "A basis for distinguishing two types of purinergic receptors." In Cell Membrane Receptors for Drug and Hormones: A multidisciplinary approach, Straub RW, Bolis L ed. New York Raven Press, 1978

Burnstock G. Noradrenaline and ATP as cotransmitters in sympathetic nerves. Neuroche Int 1990;17:357-368

Burnstock G, Meghji P. Distribution of P1-and P2-purinoceptors in guinea-pig and frog heart. Br J Pharmacol 1981;73:879-885

Burnstock G, Meghji P. The effect of adenyl compounds on the rat heart. Br J Pharnmacol 1983;79:211-218

Burnstoc G, King BF. Numbering of cloned P2-purinoceptors. Drug Devel Res 1996;38:67-71

Clemens MG, Forrester T. Appearance of adenosine triphosphate in the coronary sinus effluent from isolated working rat heart in response to hypoxia. J Physiol (Lond) 1981;312:143-158

Collis MG, Pettinger SJ. Can ATP stimulate P1-receptors in guinea-pig atrium without conversion to adenosine. Eur J Pharmacol 1982; 81:521-529

Communi D, Pirotton S, Parmentier M, Boeynaems JM. Cloning and functional expression of a human uridine nucleotide receptor. J Biol Chem 1995;270:30849-30852

Communi D, Parmentier M, Boeynaems JM. Cloning, functional expression and tissue distribution of the human P2Y6 receptor. Biochem Byophys Res Commun 1996;222:303-308

Corr L, Burnstock, G. Vasodilator response of coronary smooth muscle to the sympathetic co-transmitters noradrenaline and adenosine 5'-triphosphate. Br J Pharmacol 1991;104:337-342

Crowe R, Burnstock G. Fluorescent histochemical localization of quinacrine-positive neurones in the guinea-pig and rabbit atrium. Cardiovasc Res 1982;16:384-390

Danziger RS, Raffaelli S, Moreno-Sanchez R, Sakai M, Capogrossi MC, Spurgeon HA, Lakatta EG. Extracellular ATP has a potent effect to enhance cytosolic calcium and contractility in single ventricular myocytes. Cell Calcium 1988;9:193-199

DeYoung MB, Scarpa A. extracellular ATP induces Ca2+ transient in cardiac myocytes which are potentiated by norepinephrine. FEBS Lett 1987;223,53-58

DeYoung MB, Scarpa A. ATP receptor-induced Ca2+ transients in cardiac myocytes: sources of mobilized Ca2+. Am J Physiol 1989;260:C1182-1190

DeYoung MB, Scarpa A. Extracellular ATP activates Na+, Pi, and Ca2+ transport in cardiac myocytes. Am J Physiol 1991;260:C1182-C1190

Doyle FMS. A non-adrenergic non-cholinergic (NANC) excitatory response in frog atria. Br J Pharmacol 1985;85:211P

Drury AN, Szent-Gyorgyi A. The physiological activity of adenine compounds with special reference to their action upon the mammalian heart. J Physiol (Lond) 1929;68:213-237

Forrester T, Williams CA. Release of adenosine triphosphate from isolated adult heart cells in response to hypoxia. J Physiol (Lond) 1977;268:371-390

Fredholm BB, Abbracchio MP, Burnstock, Dubyak GR, Harden K, Jacobson KA, Schwabe U, Williams M. Towards a revised nomenclature for P1 and P2 receptors
TiPS 1997;18:79-82.

Fyffe REW, Perl ER. Is ATP a central synaptic mediator for certain primary afferent fibres from mammalian skin? Proc Natl Acad Sci USA 1984;81:6890-6893

Garcia-Guzman M, Sthumer W, Soto F. Molecular characterization and pharmacological properties of the hyman P2X3 purinoceptor. Mol Brain Res 1997a;47:59-66

Garcia-Guzman M, Soto F, Gomez-Hernandez JM, Lund P, Stuhmer W. Characterization of recombinant human P2X4 receptor reveals pharmacological differences to the rat homologous. Mol Pharmacol 1997b;51:109-118

Godecke S, Decking UKM, Godecke A, Schrader J. Cloning of the rat P2u receptor and its potential role in coronary vasodilatation. Am J Physiol 1996;270:C570-C577

Holton P. The liberation of adenosine triphosphate on antidromic stimulation of sensory nerves. J Physiol (Lond) 1959;145:494-504

Hopwood AM, Burnstock G. ATP mediates coronary vasoconstriction via P2x-purinoceptors and coronary vasodilataion via P2y-purinoceptors in the isolated perfused rat heart. Eur J Pharmacol 1987;136:49-54

Hopwood AM, Lincoln J, Kirkpatrick, KA, Burnstock, G. Adenosine 5'-triphosphate, adenosine and endothelium-derived-relaxing factor in hypoxic vasodilatation of the heart. Eur J Pharmacol 1989;165:323-326

Horackova M, Huang MH, Armour JA. Purinergic modulation of adult guinea pig cardiomyocytes in long ter cultures and co-coltures with extracardiac and intrinsic cardiac neurones. Cardiovasc Res 1994;28:673-679

Hoyle CHV, Burnstock G. Evidence that ATP is a neurotransmitter in the frog heart. Eur J Pharmacol 1986;124:285-289

Kaneda M, Fukui K, Doi K. Activation of chloride current by P2-purinoceptors in rat ventricular myocytes. Br J Pharmacol 1994;111:1355-1360

Katsuragi T, Tokunaga T, Usune S,, Furukawa T. A possible coupling of postjunctional ATP release and transmitters' receptor stimulation in smooth muscle. Life Sci 1990;46:1301-1307

Katsuragi T, Tokunaga T, Ogawa S, Soejima O, Furukawa T. Existence of ATP-evoked ATP release system in smooth muscle. J Pharmacol Exp Ther 1991;259:513-518

Katsuragi T, Tokunaga T, Ohba M, Sato C, Furukawa T. Implication of ATP released from atrial but not papillary muscle segments of guinea pig by isoproterenol and forskolin. Life Sci 1993;53:961-967.

Katsuragi T, Tokunaga T, Sato C, Furukawa T. Possible neuronal origin of ATP release evoked by forskolin and ouabain from guinea-pig atrial segments. Eur J Pharmacol 1995;282:213-217

Keff KD, Pasco JS, Eckman DM. Purinergic relaxation and hyperpolarization in guinea-pig and rabbit coronary artery: role of the endothelium. J Pharmacol Exp Ther 1992;260:592-600

Khan MT, Malik KU. Inhibitory effects of adenosine and adenine nucleotides on potassium-evoked efflux of [3H]-noradrenaline from the rat isolated heart:lack of relationship to prostaglandins. Br J Pharmacol 1980;68:551-561

Legssyer A, Poggioli J, Renard D, Vassort G. ATP and other adenosine compounds increase mechanical activity and inositol triphosphate production in rat heart. J Physiol (Lond) 1988; 401:185-199

Levesque PC, Hume JR. ATPo but not cAMP activates a chloride conductance in mouse ventricular myocytes. Cardiovasc Res 1995;29:336-343

Mantelli L, Amerini S, Filippi S, Ledda F. Blockade of adenosine receptors unmasks a stimulatory effect of ATP on cardiac contractility. Br J Pharmacol 1993;109;1268-1271

Matsumoto T, Nakane T, Chiba S. UTP induces vascular responses in the isolated and perfused canine epicardial coronary artery via UTP-preferring P2Y receptors. Br J Pharmacol 1997;122:1625-1632

Milner P, Bodin P, Loesch A, Burnstock G. Rapid release of endothelin and ATP from isolated aortic endothelial cells exposed to increased flow. Byochem Byophys Res Commun 1990;170:649-656.

Nori S, Fumagalli L, Bo X, Bogdanov Y, Burnstock G. Coexpression of mRNAs for P2X1, P2X2 and P2X4 receptors in rat vascular smooth muscle: an in situ hybridization and RT-PCR study. J Vasc Res 1998;916:179-185.

Olsson RA, Pearson JD. Cardiovascular purinoceptors. Physiol Rev 1990;70:761-845

Paddle BM, Burnstock, G. Release of ATP from perfused heart during coronary vasodilatation. Blood Vessels 1974;11:110-119

Parr CE, Sullivan DM, Paradiso AM, Lazarowski ER, Burch LH, Olsen JC, Erb L, Weisman GA, Boucher RC, Turner JT. Cloning and expression of a human P2U nucluotide receptor, a target for cystic fibrosis pharmacotherapy. Proc Natl Acad Sci USA 1995;91:3275-3279

Pearson JD, Gordon JL, vascular endothelial and smooth muscle cells in culture selectively release adenine nucleotides. Nature 1979;28:384-386

Puceat M, Clement O, Scamp F, Vassort G. Extracellular ATP-induced acidification leads to cytosolic calcium transient rise in single rat cardiac myocytes. Biochem J 1991:274:55-62

Puceat m, Vassort G. Purinergic stimulation of rat cardiomyocytes induces tyrosine phosphorylation and membrane association of phospholipase C(: a major mechanism for InsP3 generation. Biochem J 1996;318: 723-728.

Rubino A, Amerini S, Ledda F, Mantelli L. ATP modulates the efferent function of capsaicin-sensitive neurones in guinea-pig isolated atria. Br J pharmacol 1992;105:516-520

Rubino A. Non-adrenergic non-cholinergic (NANC) neuronal control of the atrial myocardium. Gen Pharmacol 1993;24:539-545

Rubino A, Burnstock G. Capsaicin-sensitive sensory-motor neurotransmission in the peripheral control of cardiovascular function. Cardiovasc Res 1996;31:467-479

Rubino Annalisa, Hassal Candice JS, Burnstock Geoffrey "Autonomic Control of the Myocardium: Non-adrenergic Non-cholinergic (NANC) Mechanisms." In Nervous Control of the Heart, Shepherd JT, Vatner SF ed. Harwood Academic Publishers, 1996

Saffrey MJ, Hassal CJS, Allen TG, Burnstock G. Ganglia within the gut, heart, urinary bladder, and airways:studies in tissue culture. Int Rev Cytol 1992;136:93-144

Scamp F, Vassort G. Mechanism of extracellular ATP-induced depolarization in rat isolated ventricula myocytes. Pfluger Arch 1990;417:309-316

Scamp F, Legssyer E, Mayoux E, Vassort G. The mechanism of positive inotropy induced by adenosine triphosphate in rat heart. Circ Res 1990;7:1007-1016

Scamp F, Nilius B, Alvarez J, Vassort G. Modulation of L-type Ca channel activity by P2-purinergic agonist in cardiac cells. Pflugers Arch 1993;422:465-471

Scamp F, Vassort G. Pharmacological profile of the ATP-mediated increase in L-type calcium current amplitude and activation of a non-specific cation current in rat ventricular cells. Br J Pharmacol 1994;113:982-986.

Schrader J, Gerlach E. Compartmentation of cardiac adenine nucleotides and formation of adenosine. Pfluger Arch 1976;367:129-235

Shoda M, Hagiwara N, Kasanuki H, Hosoda S. ATP-activated cation current in rabbit sino-atrial node cells. J Mol Cell Cardiol 1997;29:689-695

Simonsen U, Garcia-Sacristan A, Prieto D. Involvement of ATP in the non-adrenergic non-cholinergic inhibitory neurotransmission of lamb isolated coronary small arteries. Br J Pharmacol 1997;120:411-420

Tokunaga T, Katsuragi T, Sato C, Furukawa T. ATP release evoked by isoprenaline from adrenergic nerves of guinea-pig atrium. Neurosci Lett 1995;186:95-98

Valera S, Talabot F, Evans RJ, Gos A, Antonarakis SE, Morris MA, Buell GN. Characterization and chromosomal localization of a human P2x receptor from the urinary bladder. Recept Channel 1995;3:283-2899

Vassort G, Puceat M, Scamps F. Modulation of myocardial activity by extracellular ATP. Trends Ccardiovasc Med 1994;4:236-240

Vials A, Burnstock, G. Effects of nitric oxide synthase inhibitors L-NG -nitroarginine and L-NG-nitroarginine methyl ester on responses to vasodilators of the guinea-pig coronary vasculature. Br J Pharmacol 1992;107:604-609

Vials AJ, Burnstock, G. differential effects of ATP- and 2-methylthioATP-induced relaxation in guinea pig coronary vasculature. J Cardiovasc Pharmacol 1994;23:757-764

Vial A, Burnstock G. ATP release from the isolated perfused guinea-pig heart in response to increased flow. J Vasc Res 1996;33:1-4

Vial C, Owen P. Opie LH, Posel D. Significance of release of adenosine triphosphate and adenosine induced by hypoxia or adrenaline in perfused rat heart. J Mol Cell Cardiol 1987;19:187-197

von Kugelgen I, Kurtz K, Starke K. Axon terminal P2-purinoceptors in feedback control of sympathetic transmitter release. Neuroscience 1993;56:263-267

von Kugelgen I, Stoffel D, Starke K. P2-purinoceptor-mediated inhibition of noradrenaline release in rat atria. Br J Pharmacol 1995;115:247-254

Vulchanova L, arvidsson U, Riedl M, Wang J, Buell G, Surprenant A, North RA, Elde R. Differential distribution of two ATP-gated ion channel (P2X receptors) determined by immunocytochemistry. Proc Natl Acad Sci USA 1996;93-8063-8067

Yamada M, Hamammori Y, Akita H, Yokoyama M. P2-purinoceptor activation stimulates phosphoinositide hydrolysis and inhibits accumulation of cAMP in cultured ventricular myocytes. Cir Res 1992;70:477-485.

Yang S, Cheek DJ, Westfall DP, Buxton ILO. Purinergic axis in cardiac blood vessels. Agonist-mediated release of ATP from cardiac endothelial cells. Circ Res 1994;74:401-407

Webb TE, Boluyt MO, Barnard EA Molecular biology of P2Y purinoceptors: expression in the rat heart. J Autoom Pharmacol 1996;16:303-307

White TD, Angus JA. Relaxant effects of ATP and adenosine on canine large and small coronary arteries in vitro. Eur J Pharmacol 1987;143:119-126

Williams CA, Forrester T. Possible source of adenosine triphosphate released from rat myocytes in response to hypoxia and acidosis. Cardiovasc Res. 1983;17:301-312

Zhang BX, Ma X, McConnel BK, Damron DS, Bond M. Activation of purinergic receptors triggers oscillatory contractions in adult rat ventricular myocytes. Circ Res 1996;79:94-102.

Zheng JS, Christie A, Levy MN, Scarpa A. Modulation by extracellular ATP of two distinct currents in rat myocytes. Am J Physiol 1993;264:C1411-C1417

VASCULAR P_2 - RECEPTORS AND THEIR POSSIBLE ROLE IN HYPERTENSION

Charles Kennedy, Department of Physiology and Pharmacology, University of Strathclyde, Glasgow, UK

I. SUMMARY

Multiple subtypes of P_2-receptor are present in blood vessels. Most arteries and veins show a high density of P_{2X1}-receptor mRNA and prominent P_{2X1}-receptor expression. When activated, P_{2X1}-receptors carry a depolarizing cation current, leading to calcium ion influx and vasoconstriction. Extracellular calcium ions enter the smooth muscle cells through the P_{2X1}-receptor ion channel and via L-type calcium channels opened by cell depolarisation. These receptors mediate the cotransmitter function of ATP when released from sympathetic nerves. A number of studies have reported an increased purinergic component of vascular sympathetic neurotransmission in hypertensive animal models, but more studies are required before a firm conclusion can be reached. P_{2Y}-receptors are present on both vascular endothelial cells and smooth muscle cells. In many tissues endothelial P_{2Y1}- and P_{2Y2}-receptors mediate release of nitric oxide and so induce vasodilation. Removal of the endothelium abolishes these effects. Excitatory P_{2Y}-receptors are also present in the smooth muscle cells of some vessels. These mediate release of intracellular calcium ion stores and activation of depolarizing chloride and cation currents. These receptors may play a role in the regulation of vascular tone by locally-released ATP and UTP.

II. INTRODUCTION

Adenosine 5'-triphosphate (ATP) was discovered in 1929 (Fiske and SubbaRow, 1929; Lohmann, 1929). In the same year the first report of the effects of purine nucleosides and nucleotides on the cardiovascular system was published (Drury and Szent-Györgi, 1929). Since then, hundreds of studies have shown that ATP and its breakdown products adenosine 5'-diphosphate (ADP), adenosine 5'-monophosphate (AMP) and adenosine, are active on single vascular smooth muscle and endothelial cells, isolated arteries and veins, perfused vascular beds and in the whole animal (see Burnstock, 1980; Burnstock and Kennedy, 1986, for reviews). Consequently, a number of roles have been proposed for these compounds in the control of vascular tone and blood pressure under physiological and pathophysiological conditions (see Burnstock, 1980; Burnstock and Kennedy, 1986; Boarder, et al., 1995, and references therein).

In this chapter I will discuss the distribution, properties and functions of the various receptors through which ATP acts to evoke vasoconstriction and vasodilation. I will also review the evidence that the neurotransmitter role of ATP from perivascular sympathetic nerves is altered in hypertension.

III. TEXT

P_2-Receptor Classification

In 1978, Burnstock suggested that purines could act through two types of receptors which he named P_1- and P_2-purinoceptors. At the P_1-purinoceptor adenosine and AMP are agonists, whereas ATP and ADP have no effect. At the P_2-purinoceptor the opposite is seen, i.e., ATP and ADP are potent agonists and adenosine and AMP have no effect. In 1985, Burnstock and Kennedy started the subdivision of the P_2-purinoceptor, with the identification of the P_{2X}-and P_{2Y}-subtypes. This was based on a comparison of the potency order of several agonists in a number of preparations. In one group of tissues, including vascular smooth muscle, the ATP analogue α,β-methyleneATP (α,β-meATP) was much more potent than ATP or 2-methylthioATP (2-meSATP) at evoking smooth muscle contraction The receptor mediating this action was named the P_{2X}-purinoceptor. In contrast, in another group of tissues, again including vascular preparations, 2-meSATP was much more potent than ATP, which in turn was more potent than α,β-meATP, at producing relaxation or inhibition. This receptor was named the P_{2Y}-purinoceptor.

In 1986, Gordon proposed a further two P_2-purinoceptor subtypes. The P_{2T}-purinoceptor mediates aggregation of platelets and the P_{2Z}-purinoceptor mediates pore formation in mast cells. The final subtype to be proposed and widely accepted was the P_{2U}-purinoceptor, so called because uridine 5'-triphosphate (UTP) was a potent agonist (O'Connor, et al., 1991).

These P_2-subtypes were all based on pharmacological studies on native receptors in a wide range of isolated tissues and cells. The next major step came with the cloning of most of these subtypes. Since 1994, at least 12 P_2-purinoceptors have been cloned and functionally expressed. Also, they have been renamed P_2-receptors to take into account the activity of pyrimidine as well as purine nucleotides. These form two families (Abbracchio and Burnstock, 1994). P_{2X}-receptors are a new structural family of ligand-gated cation channels (see Buell, et al., 1996; North, 1996). These channels are permeable to sodium, potassium and calcium ions and when activated cause depolarization and excite cells. In contrast, P_{2Y}-receptors are all members of the superfamily of G protein-coupled receptors and, in most cases, activation leads to generation of inositol 1,4,5-trisphosphate (IP_3) and the release of internal calcium stores (Boarder, et al., 1995; Harden, et al., 1995).

To date, seven subtypes of P$_{2X}$-receptor have been cloned. Most arteries and veins show a high density of P$_{2X1}$-receptors in ligand-binding studies (Bo and Burnstock, 1993) and high levels of P$_{2X1}$-mRNA (Valera, et al., 1994; Collo, et al., 1996). However, P$_{2X1}$-mRNA has not been found in rat cerebral arteries (Collo, et al., 1996) and functional P$_{2X1}$-receptors are absent in human extrarenal arteries and veins (Von Kügelgen, et al., 1995). When functionally expressed, the cloned P$_{2X1}$-receptor carries an inward, depolarizing cationic current (I$_{cat}$) (Valera, et al., 1994). ATP activates similar currents via native P$_{2X1}$-receptors in isolated smooth muscle cells from numerous arteries, including the rabbit ear artery (Benham and Tsien, 1987; Benham, et al., 1987) and portal vein (Xiong, et al., 1991), rat tail artery (Evans and Kennedy, 1994; McLaren, et al., 1997), aorta (Pacaud, et al., 1995), portal vein (Pacaud, et al., 1994) and pulmonary artery (Hartley and Kozlowski, 1997) and human saphenous vein (Loirand and Pacaud, 1995). I$_{cat}$ activates with a latency of a few milliseconds, consistent with the action of a ligand-gated cation channel. The current rapidly reaches a peak, but is not maintained in the continued presence of agonist. Rather, fast desensitisation is seen (Figure 1). Activation of cloned (Valera, et al., 1994) and native (Evans and Kennedy, 1994; McLaren, et al., 1997) P$_{2X1}$-receptors is inhibited by the antagonists suramin and pyridoxalphosphate-6-azophenyl-2',4'-disulphonic acid (PPADS).

Vasoconstriction mediated via P$_{2X1}$-receptors is dependent upon the influx of extracellular calcium ions into the smooth muscle cells by two routes. The first is the P$_{2X1}$-receptor ion channel itself, as it permeable to calcium ions and P$_{2X1}$-receptor activation in the absence of extracellular calcium does not lead to a rise in intracellular calcium ion levels (Benham and Tsien, 1987; Benham, et al., 1987; Kitajima, et al., 1993, 1994; Pacaud, et al., 1994, 1995; Loirand and Pacaud, 1995; Lagaud, et al., 1996). In some arteries, the calcium influx can then cause the release of intracellular calcium ion stores (Loirand and Pacaud, 1995; Pacaud, et al., 1994; Pacaud and Loirand, 1995). The second route of calcium ion influx is via L-type calcium channels opened by cell depolarization (Kitajima, et al., 1993; McMillan, et al., 1994; Andriantsitohaina, et al., 1995). Consequently, contractions mediated by P$_{2X1}$-receptors are abolished in the absence of extracellular calcium ions (Kitajima, et al., 1993; Galligan, et al., 1995).

One point to note is that the P$_{2X}$-receptor was defined originally by an agonist potency profile of α,β-meATP >> 2-meSATP $\geq$ ATP. However, it is now clear that the potency of ATP and 2-meSATP in intact tissues is greatly decreased (100-1000-fold) by breakdown by ecto-ATPase (Figure 1). When agonist action is studied in the absence of breakdown, ATP and 2-meSATP are, in fact, slightly more potent than α,β-meATP at the P$_{2X1}$-receptor (Kennedy and Leff, 1995, and references therein).

P$_{2Y}$-Receptors in Vascular Smooth Muscle

Although P$_{2X1}$-receptors are the most widely studied and best characterised site of action of ATP in arterial smooth muscle, other P$_2$-receptors are also present. However, in most cases the properties and functions of these receptors are relatively poorly characterised, though they are most likely to be various subtypes of P_{2Y}-receptor. ATP causes generation of IP_3 and release of intracellular calcium ion stores in the aorta of rat (Phaneuf, et al., 1987; Tawada, et al., 1987), cow (Tada, et al., 1992) and pig (Kalthof, et al., 1993), in the rat mesenteric artery (Lagaud, et al., 1996) and in human coronary artery (Strøbæk, et al., 1996a). The generation of IP_3 is not mimicked by α,β-meATP, indicating that P_{2X1}-receptors were not involved (Tawada, et al., 1987; Guild, et al., 1992; Kalthof, et al., 1993). In these studies, no direct link was made between activation of second messenger systems and vascular smooth muscle tone.

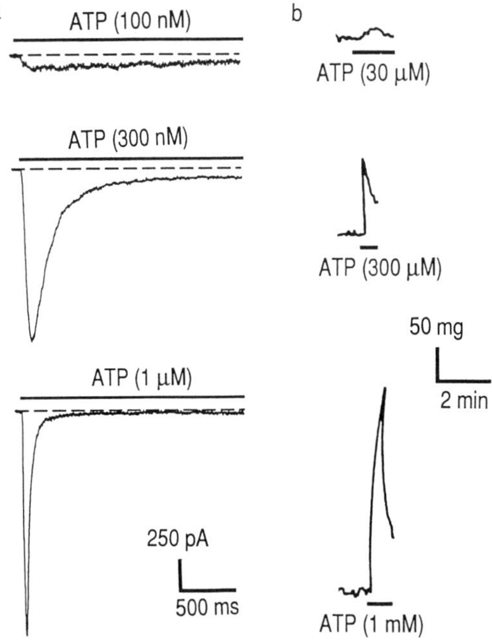

Figure 1. Excitatory effects of ATP in the rat tail artery. a) Traces show the inward currents evoked by ATP (100 nM-1 µM) in single smooth muscle cells when applied rapidly for 2 seconds using a U-tube superfusion system, as indicated by the solid bars. The dashed lines represent zero current levels. The holding potential was -60 mV. Each record is from a different cell. Figure reproduced from Evans and Kennedy, 1994. b) Contractions evoked by ATP (30 µM-1 mM) in isolated rat tail artery ring segments.

Excitatory P_{2Y}-receptors

In electrophysiological studies, ATP acts independently of P_{2X1}-receptors to depolarise smooth muscle cells of rabbit aorta (Pavenstädt, et al., 1991) and to activate a small, maintained cationic current in rabbit portal vein (Xiong, et al., 1991) and slow, calcium-dependent chloride currents in rabbit portal vein (Xiong, et al., 1991) and pulmonary artery (Halliwell, et al., 1994; Hartley and Kozlowski, 1997), the aorta of pig (Droogmans, et al., 1991) and rat (Von der Weid, et al., 1993) and human coronary artery (Strøbæk, et al., 1996b). In most vessels the chloride current was activated by calcium released from intracellular stores. In contrast, in the rabbit portal vein the chloride current was triggered by calcium influx via the slow cationic current.

UTP, a potent agonist at the P_{2Y2}-, P_{2Y4}- and P_{2Y6}-receptor subtypes (Nicholas, et al., 1996; Communi and Boeynaems, 1997), also causes synthesis of IP_3 and release of internal calcium stores in the same vascular smooth muscle cells as ATP (Tawada, et al., 1987; Kalthof, et al., 1993; Pediani, et al., 1995), as well as depolarisation (Pavenstädt, et al., 1991). We have shown that although UTP can activate P_{2X1}-receptors in smooth muscle cells of the rat tail artery, this is only seen at high concentrations, 100 times higher than those at which ATP is active (McLaren, et al., 1997). However, UTP is equipotent with ATP at evoking contraction (Saïag, et al., 1990; Evans and Kennedy, 1994; McLaren, et al., 1996), suggesting that UTP causes contraction of this artery via a site distinct from P_{2X1}-receptors.

UTP also evokes vasoconstriction in many central and peripheral blood vessels (see Seifert and Schultz, 1989; Abbracchio and Burnstock, 1994). Cerebral arteries have been particularly well studied in several species, including humans (Urquilla, 1978; Urquilla, et al., 1978; Shirisawa, et al., 1983; Hardebo, et al., 1987a,b; Von Kügelgen, et al., 1990). For example, in human and canine middle cerebral arteries, UTP causes a vasoconstriction which is very long-lasting. When the effects of UTP and ATP have been compared in the same study, they were found to act via a common non-P_{2X1}-receptor in some arteries (Eltze and Ullrich, 1996; Miyage, et al., 1996; McLaren, et al., 1996), whereas in others, UTP acts at a receptor at which ATP is not an agonist (Von Kügelgen, et al., 1987, 1990; Ralevic and Burnstock, 1991; Juul, et al., 1992; Windscheif, et al., 1994; Lagaud, et al., 1996; Rubino and Burnstock, 1996).

Of the cloned P_{2Y}-receptor subtypes at which UTP is active, the P_{2Y2}-receptor is the best characterised. ATP and UTP have a similar potency as agonists, and suramin and PPADS have weak or no antagonist action (Charlton, et al., 1996a,b; Nicholas, et al., 1996). A lack of effect of suramin and PPADS has also been reported for native P_{2Y}-receptors in intact arteries, for example in the rat mesenteric (Windscheif, et al., 1994), pulmonary (Rubino and Burnstock, 1996) and renal (Eltze and Ullrich, 1996) vascular beds. However, the other UTP-sensitive P_{2Y}-receptor subtypes are not as well characterised and further studies are necessary before their contribution to the responses to UTP and ATP can be determined.

Inhibitory P_{2Y}-Receptors

In many blood vessels, ATP and UTP cause vasodilation via P_{2Y1}- and P_{2Y2}-receptors located on vascular endothelial cells (see Burnstock and Kennedy, 1986; O'Connor, et al., 1991; Boarder, et al., 1995). In most cases, removal of the endothelium abolishes these actions. However, in a few vessels, ATP can produce vasodilation in the absence of an intact endothelial layer. This effect was first seen in the rabbit mesenteric artery (Mathieson and Burnstock, 1985). In the rabbit aorta, ATP, but not 2-meSATP, caused endothelium-independent vasodilation, indicating that P_{2Y1}-receptors were not involved (Chinellato, et al., 1992; Ragazzi, et al., 1993). Also, relaxation was not due to the breakdown of ATP to adenosine, as the action of these agents was unaffected by adenosine receptor antagonists.

Vasodilatory P_{2Y}-receptors have most often been reported in pulmonary and coronary arteries. P_{2Y1}-receptors are present in the rabbit coronary artery (Corr and Burnstock, 1991, 1994; Keef, et al., 1992), rabbit pulmonary artery (Qasabian, et al., 1997) and human small pulmonary artery (Liu, et al., 1989), since 2-meSATP, as well as ATP, produces endothelium-independent vasodilation in this tissue. In single smooth muscle cells of human coronary artery, 2-meSATP, ATP and UTP, but not α,β-meATP, evoke the generation of IP_3, a release of intracellular calcium ion stores and the activation of a hyperpolarizing potassium current and a depolarizing chloride current (Strøbæk, et al., 1996a,b).

In contrast, P_{2Y1}-receptors are absent from smooth muscle cells of the rat pulmonary artery (Bakhramov, et al., 1996; Hartley and Kozlowski, 1997). ATP and UTP, but not 2-meSATP or α,β-meATP, evoked a depolarizing chloride current and a hyperpolarizing potassium current. Presumably, in this and other vessels, the net effect of activation of these two opposing currents will depend upon the relative calcium sensitivity of the chloride and potassium channels and the resting membrane potential of the smooth muscle cells.

Unidentified P_2-Receptors

Several studies have described receptors which are not easily classified as the various subtypes described above. In the arterial bed of human placental cotyledons, vasoconstriction evoked by α,β-meATP is not inhibited by PPADS (Ralevic, et al., 1997). In human placental chorionic arteries, ATP and α,β-meATP, but not UTP, evoke suramin-sensitive vasoconstriction (Dobronyi, et al., 1997). Desensitisation of the response to α,β-meATP does not inhibit ATP, indicating separate sites of action. α,β-meATP also has an unusual action in the rabbit saphenous artery as it stimulates metabolism of inositol phosphates, implying an action at P_{2Y}-receptors (Nally, et al., 1992). At present it is unclear if these results represent novel P_2-receptor subtypes.

Vascular Functions of P_2-Receptors

P_{2X}-Receptors

The best characterised function of ATP in blood vessels is as an excitatory cotransmitter with noradrenaline and neuropeptide Y from perivascular sympathetic nerves (Figure 2). Sympathetic neurotransmission is not the simple single neurotransmitter process previously envisaged. Rather, multiple transmitters act and interact, in varying degrees depending upon tissue, to produce an integrated control of smooth muscle tone (see Burnstock, 1996; Sneddon, et al., 1996 for reviews).

Tetrodotoxin- and guanethidine-sensitive, calcium-dependent release of ATP has been demonstrated in a number of blood vessels following sympathetic nerve stimulation (see: Burnstock and Kennedy, 1986; Burnstock, 1996). Initially, ATP and noradrenaline were thought to be stored in the same vesicles in symapathetic nerves. However, it is now clear that there are at least two populations of vesicles. One contains mainly ATP, and, perhaps, some noradrenaline, whereas this is reversed in the second population (Todorov, et al., 1996).

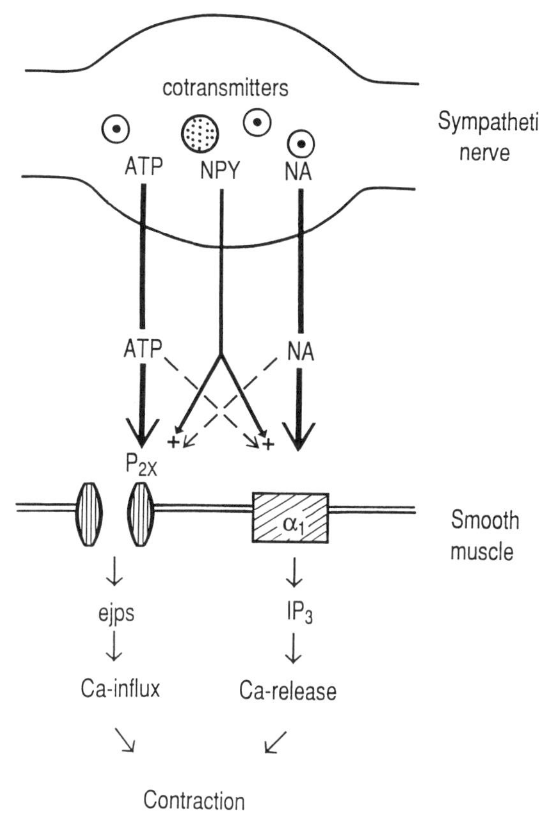

Figure 2. Schematic representation of cotransmission by ATP and noradrenaline in sympathetic nerves. ATP = adenosine 5'-triphosphate, NPY = neuropeptide Y, NA = noradrenaline, $P_{2X} = P_{2X}$-receptor, $\alpha_1 = \alpha_1$-adrenoceptor, e.j.p. = excitatory junction potential, IP_3 = inositol 1,4,5-trisphosphate. + indicates that ATP and noradrenaline can interact synergistically and that neuropeptide Y appears to act mainly by potentiating contractions to ATP and noradrenaline. The receptors for NPY have been omitted for clarity.

When ATP is released from perivascular sympathetic nerves, it acts at postjunctional P_{2X1}-receptors to evoke a rapid, transient excitatory junction potential (e.j.p.). With sufficient stimulation the e.j.p. summate and the membrane can become sufficiently depolarized to open voltage-dependent calcium channels and so initiate a calcium action potential and contraction. In contrast, noradrenaline acts via G protein-coupled receptors, mainly α_1-adrenoceptors, to generate IP_3 and release of intracellular calcium stores. Neuropeptide Y appears to act mainly by potentiating contractions to ATP and noradrenaline.

The relative contribution of ATP and noradrenaline induced activation to the total contraction varies greatly between vessels. For example, in the rat tail artery activation by ATP only contributes to approximately 10% of the peak contraction (Bao, et al., 1993). In the rabbit ear artery this increases to 20-60% (Kennedy, et al., 1986), whereas in the rabbit mesenteric artery ATP is the sole excitatory neurotransmitter, although noradrenaline is released and contributes to feedback inhibition (Ramme, et al., 1987).

The functions of vascular P$_{2Y}$-receptors are not well defined. P$_{2Y}$-receptors do not appear to be involved in the excitatory vascular neurotransmitter actions of ATP as purinergic e.j.p. and neurogenic contractions are abolished by desensitisation of the P$_{2X1}$-receptor (see above). This implies that P$_{2Y}$-receptors are not activated by neuronally-released ATP. Instead, ATP and noradrenaline were thought, initially, to be stored in the same vesicles in sympathetic nerves. However, it is now clear that there are at least two populations of vesicles. One contains mainly ATP, and perhaps some noradrenaline, whereas this is reversed in the mediate actions of ATP released from non-neuronal sources (see: Burnstock and Kennedy, 1986).

In view of the widespread presence of endothelial P$_{2Y}$-receptors, it has been suggested that they may also play a role in local regulation of vascular tone by ATP (Burnstock and Kennedy, 1986). Under normal conditions the level of purines in the blood is low. However, intracellular stores of ATP can be released from smooth muscle cells, endothelial cells and from blood cells during platelet aggregation. UTP can also be released from endothelial cells (Saïag, et al., 1995). Thus, P$_{2Y}$-receptors on endothelial and smooth muscle cells may be activated by locally-released ATP and UTP.

An example of local control of blood flow by ATP was reported in isolated perfused guinea-pig heart (Hopwood, et al., 1989). Induction of hypoxia in this preparation induced coronary vasodilation which was inhibited by Reactive Blue 2, an antagonist at P$_{2Y}$-receptors, or by an inhibitor of nitric oxide. This suggested that hypoxia induced the release of ATP, which acted at endothelial P$_{2Y}$-receptors to generate nitric oxide, which, in turn, evoked vasodilation.

Recently, ATP has been proposed to function as an inhibitory neurotransmitter in the lamb coronary artery (Simonsen, et al., 1997). In this tissue, ATP, 2-meSATP and electrical field stimulation evoked vasodilation, but UTP and α,β-meATP had little effect. In each case, vasodilation was inhibited by the P$_{2Y}$-receptor antagonists suramin and Basilene Blue E-3G. This inhibitory neurogenic action of ATP may play an important role in maintaining blood flow to the heart.

Changes in Vascular Functions of P$_2$-Receptors in Hypertension

Hypertension is recognised as a major health problem, which if uncorrected leads to cardiovascular complications such as an increased probability of stroke, myocardial infarction and heart failure. However, apart from a few cases, the underlying cause of hypertension is unknown. Many factors have been suggested to contribute to the genesis of the disease, but, in general, it is difficult to determine if these factors are caused by, rather than the cause of, hypertension. In most cases of essential hypertension, elevated arterial blood pressure is associated with an increase in peripheral vascular resistance (Head, 1989) and it has been suggested that an enhanced activity of sympathetic nerves and adaptive structural changes in the vessel wall are important components in this change.

250

If there is greater activity of sympathetic nerves in hypertension then it would be reasonable to expect that ATP would play a greater role as a neurotransmitter. In a small number of studies the contribution of ATP to neurogenic contractions was investigated using the spontaneously hypertensive rat (SHR). Vidal, et al., (1986) found in the tail artery an increased contribution by neuronally-released ATP to neurogenic contractions of the SHR compared with control animals, suggesting that an increased purinergic response could be involved in hypertension. However, subsequent studies found no evidence for a greater purinergic component to neurotransmission in SHR tail artery compared with normotensive controls (Dalziel, et al., 1989; Muir and Wardle, 1989).

Evidence for a greater purinergic component to neurotransmission has also been found in vessels from a hypertensive rabbit model (Bulloch and McGrath, 1992). In the distal saphenous artery of normotensive animals ATP had little or no effect as an excitatory cotransmitter, but in preparations from hypertensive animals ATP now played a significant role. Likewise, in two vessels, the ileocolic and proximal saphenous arteries, in which ATP was a transmitter in the normotensive state, an increase in its function was seen in the hypertensive state. Furthermore, there was less negative feedback via presynaptic α_2-adrenoceptors in the hypertensive, compared with the normotensive vessels. This suggested that the increase in ATP release could be due to the reduction in ongoing autoinhibition.

Clearly more studies are required before a firm conclusion can be reached regarding the role of ATP in hypertension. Nonetheless, since ATP has an important role as an excitatory cotransmitter with noradrenaline and neuropeptide Y in perivascular sympathetic nerves in the normotensive state, there is still scope for targeting the purinergic component of neurotransmission as a therapy for hypertension.

IV. CONCLUSIONS AND FUTURE DIRECTIONS

P_2-receptors are widely distributed in the cardiovascular system. P_{2X1}-receptors are expressed in most arteries and veins and they mediate the sympathetic cotransmitter function of ATP in these vessels. The relative contribution of ATP and noradrenaline as neurotransmitters is vessel-dependent. P_{2Y}-receptors present in endothelial and smooth muscle cells can mediate both vasodilation and vasoconstriction. These receptors may play a role in the regulation of vascular tone by locally-released ATP and UTP.

The P_{2X1}-receptors present in vascular smooth muscle and the P_{2Y}-receptors present in vascular endothelial cells have been intensively characterised. However, much less is known about other vascular P_2-receptors and this may be a fruitful area for future research. Multiple recombinant P_{2Y}-subtypes have been identified, each with distinct pharmacological properties, but there has been little attempt to correlate these with individual receptors present in blood vessels. The physiological functions of the native P_2-receptors is also unclear. Finally, although a number of studies have reported an increased purinergic component of vascular sympathetic neurotransmission in hypertensive animal models, more studies are required before a firm conclusion can be reached.

V. REFERENCES

Abbracchio MP, Burnstock G. Purinoceptors: are there families of P2X and P2Y purinoceptors? *Pharmac Ther*, 64: 445-475, 1994.

Andriantisitohaina R, Lagaud GJ-L, Andre A, Muller B, Stoclet J-C. Effects of cGMP on calcium handling in ATP-stimulated rat resistance arteries. *Am J Physiol*, 268: H1223-H1231, 1995.

Bakhramov A, Hartley SA, Salter KJ, Kozlowski, RZ. Contractile agonists preferentially activate Cl over K currents in arterial myocytes. *Biochem Biophys Res Com*, 227: 168-175, 1996.

Bao J-X, Stjärne L. Dual contractile effects of ATP released by field stimulation revealed by effects of α,β-methyleneATP and suramin in rat tail artery. *Brit J Pharmacol*, 110: 1421-1428, 1993.

Benham CD, Bolton, TB, Byrne NG, Large WA. Action of externally applied adenosine triphosphate on single smooth muscle cells dispersed from rabbit ear artery. *J Physiol*, 387: 473-488, 1987.

Benham CD, Tsien, RW. A novel receptor-operated Ca^{2+}-permeable channel activated by ATP in smooth muscle. *Nature*, 328: 275-278, 1987.

Bo X, Burnstock G. Heterogeneous distribution of [^{3}H] α,β-methyleneATP binding sites in blood vessels. *J Vasc Res*, 30: 87-101, 1993.

Boarder MR, Weisman GA, Turner JT, Wilkinson GF. G-protein-coupled P_2-purinoceptors: from molecular biology to functional responses. *Trends Pharmacol Sci*, 16: 133-139, 1995.

Buell G, Collo G, Rassendren F. P2X Receptors: An emerging channel family. *Eur J Neurosci*, 8: 2221-2228, 1996.

Bulloch JM, McGrath JC. Evidence for increased purinergic contribution in hypertensive blood vessels exhibiting cotransmission. *Brit J Pharmacol*, 107: 145P, 1992 (abstract).

Burnstock G. A basis for distinguishing two types of purinergic receptor. In: Straub RW (ed), *Cell Membrane Receptors for Drugs and Hormones: A Multidisciplinary Approach*, New York: Raven Press, 1987, pp107-118.

Burnstock G. Cholinergic and purinergic regulation of blood vessels. In: Bohr DF, Somlyo AP, Sparks HV, Geiger SR (eds), *Handbook of Physiology*, sec 2, The Cardiovascular System, Vol II, Vascular Smooth Muscle. Washington DC: American Physiological Society, 1980, pp 567-612.

Burnstock G. Introduction: Purinergic Neurotransmission. *Seminars Neurosci*, 8: 171-174, 1996.

Burnstock G, Kennedy C. Is there a basis for distinguishing two types of P_2-purinoceptor? *Gen Pharmacol* 5: 433-440, 1985.

Burnstock G, Kennedy C. A dual function for adenosine 5'-triphosphate in the regulation of vascular tone. Excitatory cotransmitter with noradrenaline from perivascular nerves and locally released inhibitory intravascular agent. *Circ Res*, 58: 319-330, 1986.

Charlton SJ, Brown CA, Weisman GA, Turner JT, Erb L, Boarder MR. PPADS and suramin as antagonists at cloned P2Y- and P2U-purinoceptors. *Brit J Pharmacol*, 118: 704-710, 1996a.

Charlton SJ, Brown CA, Weisman GA, Turner JT, Erb L, Boarder MR. Cloned and transfected P2Y4 receptors: characterisation of a suramin and PPADS-insensitive response to UTP. *Brit J Pharmacol*, 119: 1301-1303, 1996b.

Chinellato A, Ragazzi E, Pandolfo L, Froldi G, Caparrotta L, Fassina G. Pharmacological characterisation of a new purinergic receptor site in rabbit aorta. *Gen Pharmacol*, 23: 1067-1071, 1992.

Collo G, North AN, Kawashima E, Merlo-Pich E, Neidhart S, Surprenant A, Guell G. Cloning of P2X$_5$ and P2X$_6$ receptors and the distribution and properties of an extended family of ATP-gated ion channels. *J Neurosci*, 16: 2495-2507, 1996.

Communi D, Boeynaems JM. Receptors responsive to extracellular pyrimidine nucleotides. *Trends Pharmacol Sci*, 18: 83-86, 1997.

Corr L, Burnstock G. Vasodilator response of coronary smooth muscle to the sympathetic cotransmitters noradrenaline and adenosine 5'-triphosphate. *Brit J Pharmacol*, 104: 337-342, 1991.

Corr L, Burnstock G. An analysis of P$_2$ -purinoceptor subtypes on the smooth muscle and endothelium of rabbit coronary artery. *J Cardiovasc Pharmacol*, 23: 709-715, 1994.

Dalziel HH, Machaly M, Sneddon P. Comparison of the purinergic contribution to sympathetic vascular responses in SHR and WKY rats in vitro and in vivo. *Eur J Pharmacol*, 173: 19-26, 1989.

Dobronyi I, Hung KS, Satchell DG, Maguire, MH. Evidence for a novel P$_{2X}$ purinoceptor in human placental chorionic surface arteries. *Eur J Pharmacol*, 320: 61-64, 1997.

Droogmans G, Callewaert G, Declerck I, Casteels R. ATP-induced Ca^{2+}-release and Cl-current in cultured smooth muscle cells from pig aorta. *J Physiol*, 440, 623-634, 1991.

Drury AN, Szent-Györgi A. The physiological activity of adenine compounds with special reference to their action upon the mammalian heart. *J Physiol*, 68: 213-237, 1929.

Eltze M, Ullrich B. Characterisation of vascular P$_2$-purinoceptors in the rat isolated perfused kidney. *Eur J Pharmacol*, 306: 139-152, 1996.

Evans RJ, Kennedy C. Characterisation of P$_2$-purinoceptors in the smooth muscle of the rat tail artery; a comparison between contractile and electrophysiological responses. *Brit J Pharmacol*, 113: 853-860, 1994.

Fiske CH, SubbaRow Y. Phosphorus compounds of muscle and liver. *Science*, 70: 381-382, 1929.

Galligan JJ, Herring A, Harpstead T. Pharmacological characterisation of purinoceptor-mediated constriction of submucosal arterioles in guinea-pig ileum. *J Pharmacol Exp Ther*, 1425-1430, 1995.

Gordon JL. Extracellular ATP: effects, sources and fate. *Biochem J*, 233: 309-319, 1986.

Guild SB, Jenkinson S, Muir TC. The interaction between noradrenaline and ATP upon polyphospho-inositide metabolism and contraction in tail arteries from normo- and hypertensive rats. *J Pharm Pharmacol*, 44: 836-840, 1992.

Halliwell RM, Wang Q, Hogg RC, Large WA. Synergistic action of histamine and adenosine triphosphate on the response to noradrenaline in rabbit pulmonary artery smooth muscle cells. *Pflügers Arch*, 426: 433-439, 1994.

Hardebo JE, Kahrstrom J, Owman C, Salford LG. Endothelium-dependent relaxation by uridine tri- and diphosphate in isolated human pial vessels. *Blood Vessels*, 24: 150-155, 1987.

Hardebo JE, Kahrström J, Owman C. P$_1$- and P$_2$ -purine receptors in brain circulation. *Eur J Pharmacol*, 144: 343-352, 1987.

Harden TK, Boyer JL, Nicholas RA. P$_2$-purinergic receptors: subtype-associated signalling responses and structure. *Ann Rev Pharmacol Toxicol*, 35: 541-579, 1995.

Hartley SA, Kozlowski, RZ. Electrophysiological consequences of purinergic receptor stimulation in isolated rat pulmonary arterial myocytes. *Circ Res*, 80: 170-178, 1997.

Head RJ. Hypernoradrenergic innervation: its relationship to functional and hyperplastic changes in the vasculature of the spontaneously hypertensive rat. *Blood Vessels*, 26: 1-20, 1989.

Hopwood AM, Lincoln J, Kirkpatrick KA, Burnstock G. Adenosine 5'-triphosphate, adenosine and endothelium-derived relaxing factor in hypoxic vasodilation of the heart. *Eur J Pharmacol*, 165: 323-326, 1989.

Juul B, Plesner L, Aalkjær C. Effects of ATP and UTP on [Ca^{2+}], membrane potential and force in isolated rat small arteries. *J Vasc Res*, 29: 385-395, 1992.

Kalthof B, Bechem M, Flocke K, Pott L, Schramm M. Kinetics of ATP-induced Ca^{2+} transients in cultures of pig aortic smooth muscle cells depend on ATP concentration and stored Ca^{2+}. *J Physiol*, 466: 245-262, 1993.

Keef KD, Pasco JS, Eckman DM. Purinergic relaxation and hyperpolarisation in guinea-pig and rabbit coronary artery: role of the endothelium. *J Pharmacol Exp Ther*, 260: 592-600, 1992.

Kennedy C, Leff P. How should P$_{2X}$-purinoceptors be characterised pharmacologically? *Trends Pharmacol Sci*, 16: 168-174, 1995.

Kennedy C, Saville V, Burnstock G. The contributions of noradrenaline and ATP to the responses of the rabbit central ear artery to sympathetic nerve stimulation depend on the parameters of stimulation. *Eur J Pharmacol*, 122: 291-300, 1986.

Kitajima S, Ozaki H, Karaki H. The effects of ATP and α,β-methylene-ATP cytosolic Ca^{2+} level and force in rat isolated aorta. *Brit J Pharmacol*, 110: 263-268, 1993.

Kitajima S, Ozaki H, Karaki, H. Role of different subtype of P$_2$-purinoceptor on cytosolic Ca^{2+} levels in rat aortic smooth muscle. *Eur J Pharmacol*, 266: 263-267, 1994.

Lagaud GJL, Stoclet JC, Andriantsitohaina R. Calcium handling and purinoceptor subtypes involved in ATP-induced contraction in rat small mesenteric arteries. *J. Physiol*, 492: 689-703, 1996.

Liu SF, McCormack DG, Evans TW, Barnes PJ. Evidence for two P$_2$-purinoceptor subtypes in human small pulmonary arteries. *Brit J Pharmacol*, 98: 1014-1020, 1989.

Lohmann K. Über die pyrophosphatfraktion im muskel. *Naturwiss*, 17: 624-625, 1929.

Loirand G, Pacaud P. Mechanism of ATP-induced rise in cytosolic Ca^{2+} in freshly isolated smooth muscle cells from human saphenous vein. *Pflugers Archiv*, 430: 429-436, 1995.

Mathieson JJI, Burnstock G. Purine-mediated relaxation and constriction of isolated rabbit mesenteric artery are not endothelium-dependent. *Eur J Pharmacol*, 118: 221-229, 1985.

McLaren GJ, Burke K, Sneddon P, Kennedy C. Characterisation of the sites at which ATP acts to evoke contraction of the rat tail artery. *Brit J Pharmacol*, 119: 370P, 1996 (abstract).

McLaren GJ, Sneddon P, Kennedy C. Comparison of the actions of ATP and UTP at P$_{2X1}$-receptors in smooth muscle of the rat tail artery. *Eur J Pharmacol*, 1997 (in press).

McMillan MR, Snook J, Garland CJ, Murphy TV. Sources of Ca^{2+} in noradrenaline and adenosine 5'-triphosphate (ATP)-evoked contraction of rabbit proximal saphenous artery. *Brit J Pharmacol*, 112: 417P, 1994 (abstract).

Miyagi Y, Kobayashi S, Nishimura J, Fukui M, Kanaide H. Dual regulation of cerebrovascular tone by UTP: P$_{2U}$-receptor-mediated contraction and endothelium-dependent relaxation. *Brit J Pharmacol*, 118: 847-856, 1996.

Muir TC, Wardle KA. Vascular smooth muscle responses in normo- and hypertensive rats to sympathetic nerve stimulation and putative transmitters. *J Autonomic Pharmacol*, 9: 23-34, 1989.

Nally JE, Muir TC, Guild SB. The effects of noradrenaline and adenosine 5'-triphosphate on polyphos-phoinositide and phosphatidylcholine hydrolysis in arterial smooth muscle. *Brit J Pharmacol*, 106: 865-870, 1992.

Nicholas RA, Watt WC, Lazarowski ER, Li Q, Harden TK. Uridine nucleotide selectivity of three phospholipase C-activating P_2-receptors: identification of a UDP-selective, a UTP-selective and an ATP- and UTP-specific receptor. *Mol Pharmacol*, 50: 224-229, 1996.

North RA. P_{2X} purinoceptor plethora. *Seminars Neurosci*, 8: 187-194, 1996.

O'Connor SE, Dainty IA, Leff P. Further subclasification of ATP receptors based on agonist studies. *Trends Pharmacol Sci*, 12: 137-141, 1991.

Pacaud P, Grégoire G, Loirand G. Release of Ca^{2+} from intracellular store in smooth muscle cells of rat portal vein by ATP-induced Ca^{2+} entry. *Brit J Pharmacol*, 113: 457-462, 1994.

Pacaud P, Loirand G. Release of Ca^{2+} by noradrenaline and ATP from the same Ca^{2+} store sensitive to both $InsP_3$ and Ca^{2+} in rat portal vein myocytes. *J Physiol*, 484: 549-555, 1995.

Pacaud P, Malam-Souley R, Loirand G, Desgranges C. ATP raises $[Ca^{2+}]_i$ via different P2-receptor subtypes in freshly isolated and cultured aortic myocytes. *Am J Physiol*, 269: H30-H36, 1995.

Pavenstädt H, Lindeman V, Lindeman S, Kunzelmann K, Spath M, Greger R. Effect of depolarising and hyperpolarising agents on the membrane potential difference of primary cultures of rabbit aorta vascular smooth muscle cells. *Pflugers Arch*, 419: 69-75, 1991.

Pediani JD, Wilson SM, Gordon J, Dominiczak, AF, McGrath JC. External ATP mobilises Ca^{2+} from a thapsigargin-sensitive internal store in cultured, rat aortic smooth muscle cells. *J Physiol*, 483: 106P, 1995 (abstract).

Phaneuf S, Berta P, Casanova J, Cavadore J-C. ATP stimulates inositol phosphates accumulation and calcium mobilisation in a primary culture of rat aortic myocytes. *Biochem Biophys Res Commun*, 143: 454-460, 1987.

Qasabian RA, Schyvens C, Owe-Young R, Killen JP, Macdonald PS, Conigrave AD, Williamson DJ. Characterisation of P_2 receptors in rabbit pulmonary artery. *Brit J Pharmacol*, 120: 553-558, 1997.

Ragazzi E, Chinellato A, Pandolfo L, Froldi G, Caparrotta L, Prosdocimi M, Aliev G, Fassina G. Atherosclerosis-related remodelling of aortic relaxation to purines in the Watanabe Heritable Hyperlipidemic rabbit. *J Pharmacol Exp Ther*, 266: 1091-1096, 1993.

Ralevic V, Burnstock G. Effects of purines and pyrimidines on the rat mesenteric arterial bed. *Circ Res*, 69: 1583-1590, 1991.

Ralevic V, Burrell S, Kingdom J, Burnstock G. Characterisation of P_2 receptors for purine and pyrimidine nucleotides in human placental cotyledons. *Brit J Pharmacol*, 121: 1121-1126, 1997.

Ramme D, Regenold JT, Starke K, Busse R, Illes P. Identification of the neuroeffector transmitter in jejunal branches of the rabbit mesenteric artery. *Naunyn-Schmiedeberg's Arch Pharmacol*, 336: 267-273, 1987.

Rubino A, Burnstock G. Evidence for P_2-purinoceptor mediating vasoconstriction by UTP, ATP and related nucleotides in the isolated pulmonary vascular bed of the rat. *Brit J Pharmacol*, 118: 1415-1420, 1996.

Saïag B, Milon D, Herve A, Rault B, Van Den Driessche J. Constriction of the smooth muscle of rat tail and femoral arteries and dog saphenous vein is induced by uridine triphosphate via pyrimido-ceptors, and by adenosine triphosphate via P_{2X}-purinoceptors. *Blood Vessels*, 27: 352-364, 1990.

Saïag B, Bodin P, Shacoori V, Catheline M, Rault B, Burnstock G. Uptake and flow-induced release of uridine nucleotides from isolated vascular endothelial cells. *Endothelium*, 2: 279-285, 1995.

Seifert R, Schultz G. Involvement of pyrimidoceptors in the regulation of cell functions by uridine and by uracil nucleotides. *Trends Pharmacol Sci*, 10: 365-369, 1989.

Shirasawa Y, White RP, Robertson JT. Mechanisms of the contractile effect induced by uridine 5-triphosphate in canine cerebral arteries. *Stroke*, 14: 347-355, 1983.

Simonsen U, Garcia-Sacristán A, Prieto D. Involvement of ATP in the non-adrenergic non-cholinergic inhibitory neurotransmission of lamb isolated coronary small arteries. *Brit J Pharmacol*, 120: 411-420, 1997.

Sneddon P, McLaren GJ, Kennedy C. Purinergic cotransmission: sympathetic nerves. *Seminars Neurosci*, 8:201-206, 1996.

Strøbæk D, Olesen SP, Christopheresen P, Dissing S. P_2 -purinoceptor-mediated formation of inositol phosphates and intracellular Ca^{2+} transients in human coronary artery smooth muscle. *Brit J Pharmacol* , 118: 1645-1652, 1996a.

Strøbæk D, Christopheresen P, Dissing S, Olesen SP. ATP activates K and Cl channels via purinoceptor mediated release of Ca^{2+} in human coronary artery smooth muscle. *Am J Physiol*, 271: C1463-C1471, 1996b.

Tada S, Okajima F, Mitsui Y, Kondo Y, Ui M. P_2 purinoceptor-mediated cyclic AMP accumulation in bovine vascular smooth muscle cells. *Eur J Pharmacol*, 227: 25-31, 1992.

Tawada Y, Furukawa K-I, Shigekawa M. ATP-induced calcium transient in cultured rat aortic smooth muscle cells. *J Biochem*, 102: 1499-1509, 1987.

Todorov LD, Mihaylova-Todorova SM, Craviso GL, Bjur RA, Westfall DP. Evidence for the differential release of the cotransmitters ATP and noradrenaline from sympathetic nerves of the guinea-pig vas deferens. *J Physiol*, 496: 731-748, 1996.

Urquilla PR. Prolonged contraction of isolated human and canine cerebral arteries induced by uridine 5'-triphosphate. *Stroke*, 9: 133-136, 1978.

Urquilla PR, Van Dyke K, Trush M. Structure-activity relation of pyrimidine nucleotides and nucleoside in canine isolated cerebral vessels. *J Pharm Pharmac*, 30: 189-190, 1978.

Valera S, Hussy N, Evans RJ, Adami N, North RA, Surprenant A, Buell G. A new class of ligand-gated ion channel defined by P_{2X} receptor for extracellular ATP. *Nature*, 371: 519-523, 1994.

Vidal M, Hicks PE, Langer SZ. Differential effects of α,β-methyleneATP on responses to nerve stimulation in SHR and WKY tail arteries. *Naunyn-Schmiedeberg's Arch Pharmacol*, 332: 384-390, 1986.

Von der Weid P-Y, Serebryakov VN, Orallo F, Bergmann C, Snetkov VA, Takeda K. Effects on ATP cultured smooth muscle cells from rat aorta. *Brit J Pharmacol*, 108: 638-645, 1993.

Von Kügelgen I, Häussinger D, Starke K. Evidence for a vasoconstriction-mediating receptor for UTP, distinct from the P_2-purinoceptor, in rabbit ear artery. *Naunyn-Schmiedeberg's Arch Pharmacol*, 336: 556-560, 1987.

Von Kügelgen I, Krumme B, Schaible U, Schollmeyer PJ, Rump LC. Vasoconstrictor responses to the P_{2X}-purinoceptor agonist β,γ-methylene-L-ATP in human cutaneous and renal blood vessels. *Brit J Pharmacol*, 116: 1932-1936, 1995.

Von Kügelgen I, Starke K. Evidence for two separate vasoconstriction-mediating nucleotide receptors, both distinct from the P_{2X}-receptor, in rabbit basilar artery: a receptor for pyrimidine nucleotides and a receptor for purine nucleotides. *Naunyn-Schmiedeberg's Arch. Pharmacol*, 341: 538-546, 1990.

Windscheif U, Ralevic V, Bäumert HG, Mutschler E, Lambrecht G, Burnstock G. Vasoconstrictor and vasodilator responses to various agonists in the rat perfused mesenteric arterial bed: selective inhibition by PPADS of contractions mediated via P_{2X}-purinoceptors. *Brit J Pharmacol*, 113: 1015-1021, 1994.

Xiong Z, Kitamura K, Kuriyama H. ATP activates cationic currents and modulates the calcium current through GTP-binding protein in rabbit portal vein. *J Physiol*, 440: 143-165, 1991.

NEUROMODULATION IN THE CARDIOVASCULAR SYSTEM

Clemens Allgaier and Peter Illes; Institut für Pharmakologie und Toxikologie
Universität Leipzig, LeipzigGermany

I. SUMMARY

Purines are substantially involved in postjunctional and prejunctional neuromodulation in the cardiovascular system. Co-stored with noradrenaline and acetylcholine in sympathetic and parasympathetic nerves, ATP may be differentially released in response to nerve stimulation. ATP released from the nerve terminals of sympathetic neurones may contribute to neurogenic smooth muscle contraction through activation of $P2X_1$ receptors, or alternatively induce blood vessel relaxation through endothelial $P2Y_1$ receptors. ATP also induces ATP liberation from smooth muscle and endothelial cells, possibly reflecting a feedforward mechanism. Presynaptically, ATP may modulate transmitter release via P2 autoreceptors. Adenosine generated by the breakdown of ATP or arising from other sources has direct vasodilator effects in all vascular beds except in the kidney and the placenta. Adenosine is able to inhibit transmitter release via prejunctional A_1 receptors thereby diminishing α_1 adrenoceptor-mediated vasoconstriction. Presumably, not only adenosine but also ATP inhibits transmitter release through activation of presynaptic A_1 receptors which might explain various observations previously ascribed to the involvement of a hybrid P_3 receptor.

II. BACKGROUND

Nucleosides, in particular adenosine, and nucleotides, in particular ATP, produce various cardiovascular effects that are not directly related to their role in energy metabolism. In 1929, Drury and Szent-Györgyi demonstrated that adenosine injected into anaesthetised animals has potent actions on the heart and circulation, causing bradycardia, coronary vasodilation, and a decrease in systemic blood pressure. Using crystallised adenosine from heart muscle extracts Lindner and Rigler (1931) corroborated the coronary vasoactivity of adenosine for a number of species and suggested that adenosine is a physiological modulator of coronary blood flow. The involvement of adenosine in the metabolic regulation of coronary blood flow was also suggested by the increase in adenosine formation during hypoxia (Berne 1963; Gerlach et al. 1963). Sattin and Rall (1970) demonstrated adenosine-induced cAMP formation in rodent brain providing the first evidence for the existence of adenosine receptors. Subsequently, it became clear that there was a bi-directional modulation of adenylate cyclase activity by different adenosine receptors either inhibiting or stimulating the enzyme. Work to date has provided the existence of additional adenosine receptors and the involvement of a variety of post-receptor mechanisms (see below). The direct vasodilator effect of adenosine occurring in all vascular beds except in the kidney and the placenta is mediated by A_2 receptor activation, while cardiac A_1 receptors improve oxygen supply (Olsson and Pearson 1990).

ATP plays important physiological and pathophysiological roles in a variety of systems, including that of a neurotransmitter in peripheral and central neurons (Burnstock 1990; Edwards and Gibb 1993; Evans 1993; Illes and Nörenberg 1993; Illes et al. 1996). Co-stored in sympathetic and parasympathetic nerves, it accounts for many actions caused by autonomic nerve stimulation not due to the primary neurotransmitters noradrenaline and acetylcholine (Burnstock 1990; Hoyle 1996; Sneddon et al. 1996; von Kügelgen and Starke 1991a). However, the biological responses to ATP are often complicated by the presence of functionally distinct ATP receptor subtypes on the different cells that comprise tissues or organ systems, and the rapid catabolism of ATP to adenosine.

ATP has very early been shown to produce important cardiovascular effects (Green and Stoner 1950). ATP affects cardiovascular function and regional blood flow by its ability to modulate most vascular smooth muscle cells, cardiac myocytes, and various types of blood vessels (for review see e.g.: Olsson and Pearson 1990; von Kügelgen and Starke 1991a). Vasoconstriction is primarily associated with stimulation of smooth muscle P2X receptors (corresponding to the cloned $P2X_1$ receptor) that act as ligand-gated cation channels. ATP-dependent relaxation of intact blood vessels is primarily associated with endothelial P2 receptors (corresponding to the $P2Y_1$ receptor) that mobilise Ca^{2+} from intracellular stores followed by the rapid release of vasorelaxant agents, such as PGI_2 and NO (for review see: Boarder et al. 1995; Dubyak and El-Moatassim 1993).

This review highlights purinoceptor-mediated neuromodulation in the cardiovascular system. Studies from other tissue preparations, such as the vas deferens, are additionally included if the results may be of principal significance. Since multiple receptors mediate the effects of adenosine, ATP, and related nucleosides and nucleotides, the present state of view of purinoceptor nomenclature is briefly summarised in the following. For a comprehensive discussion see e.g. Abbrachio and Burnstock (1994), Abbrachio et al. (1993), Burnstock and King (1996), Fredholm et al. (1994, 1997), and Dalziel and Westfall (1994).

III. TEXT

A. <u>Purinoceptor nomenclature and post receptor mechanisms</u>

The P1 adenosine receptor family comprises 4 separate adenosine receptors. All of them appear G protein-coupled. The A_1 receptor couples with Gi and Go but not with Gs or Gz. Receptor activation can induce various cellular responses, such as inhibition of adenylate cyclase (Londos et al. 1980), stimulation of K^+ conductance (Trussel and Jackson 1985), inhibition of a Ca^{2+} conductance (Scholz and Miller 1991), and stimulation of phospholipase C (Gerwins and Fredholm 1995). Via A_1 receptors adenosine is able to decrease neurotransmiter release (Fredholm and Dunwiddie, 1988). A_2 receptors (A_{2A}, A_{2B}) have been defined by their ability to stimulate adenylate cyclase, presumably via the Gs protein. However, the possibility that A_2 receptors can interact with effectors other than adenylate cyclase cannot be ruled out. A_{2A} receptors are found on neutrophils, platelets, blood vessels, and very abundantly in the striatum, particularly in the GABAergic striatopallidal neurones (Ongini and Fredholm 1996). The A_{2B} receptor is widely distributed in the organism, but in low abundance. Less is known about the A_3 receptor inhibiting forskolin-stimulated cAMP accumulation in CHO cells (Linden 1994) presumably via Gi or Go. In RBL-2H3 cells, A_3 receptor activation stimulates inositol IP_3 production and accumulation of intracellular Ca^{2+} (Ramkumar et al. 1993). This receptor may be implicated in the 8-(p-sulfophenyl)-theophylline-resistant hypotensive response to adenosine analogs in the pithed rat (Fozzard and Carruthers 1993).

Many P2 receptors have now been cloned. They are subdivided into 2PX and 2PY receptor families, each represented by 7 members. P2X receptors are intrinsic cation channels permeable to Na^+, K^+, and Ca^{2+}. Neuronal P2X receptors mediate fast synaptic responses. P2Y receptors are G protein-coupled; they comprise the previously pharmacologically classified P2Y, P2U, and P2T receptors. P2Y receptors mediate slow synaptic signals or release of Ca^{2+} from intracellular pools in response to ATP or UTP. They are often linked to phospholipase C (O'Conner 1991), but their transduction mechanisms also include inhibition of adenylate cyclase (Boyer et al. 1993) and mobilisation of arachidonic acid (Bruner and Murphy 1993).

B. Purinergic co-transmission

Co-transmission of noradrenaline, ATP, and other agents [e.g. neuropeptide Y (NPY)] initially proposed by Burnstock in 1976 seems to be rather the rule than the exception in sympathetic nerves (Figure 1). Early evidence was mainly provided by experiments performed on the isolated vas deferens and arteries of various species (for review see: Burnstock 1990, 1996; Sneddon et al. 1996; von Kügelgen and Starke 1991a).

In the vas deferens, ATP is co-released with noradrenaline producing a biphasic contraction of the smooth muscle (von Kügelgen and Starke 1991a). The early phase of the response was prevented by application of α,β-methylene ATP, a metabolically stable ATP analogue that first activates and then desensitises P2X receptors, arylazidoaminopropionyl ATP (ANAPP3), a photo-affinity label which covalently binds to P2X receptors after irradiation, and the P2 purinoceptor antagonists suramin and pyridoxalphosphate-6-azophenyl-2`,4`-disulphonic acid (PPADS) (McLaren et al. 1994) indicating that it was mediated by ATP. The late phase of the neurogenic response or the contraction produced by exogenous noradrenaline were not affected under these conditions but selectively abolished by prazosin, an α_1 adrenoceptor antagonist. The actions of ATP and noradrenaline are brought about by different transduction mechanisms. P2X purinoceptor activation produces rapid muscle contraction due to membrane depolarisation followed by Ca^{2+} influx through voltage-dependent calcium channels. α_1 Adrenoceptor stimulation mediates the slow phase of the neurogenic contraction mainly by IP_3 formation followed by Ca^{2+} release from the sarcoplasmatic reticulum.

Sympathetic innervation of arteries is more complex than that of the vas deferens and considerably differs between vessels. The degree to which ATP and noradrenaline contribute to vasoconstriction varies from one artery to another (Hirst and Edwards 1989; Sneddon et al. 1996; Starke et al. 1992; von Kügelgen et al. 1991a). In dog and rat mesentery arteries, as well as in rabbit ileocolonic arteries the neurogenic response is mediated by ATP and noradrenaline. However, in rabbit saphenous artery (MacDonald et al. 1992) and small jejunal arteries (Ramme et al. 1987; Evans and Cunnane 1992) sympathetic vasoconstriction is largely mediated by ATP, whereas in rat tail artery (Bao 1993) or rabbit pulmonary artery (Weitzell et al. 1979), endogenous ATP exhibits little or no constrictor effect.

Transmitter release strongly depends on the stimulation conditions and can be differentially modulated by prejunctional receptors (von Kügelgen and Starke 1991a, and below). This suggests the occurrence of sub-populations of vesicles containing different proportions of the various transmitters (Burnstock

1990; Todorov et al. 1994a; von Kügelgen and Starke 1991a). In guinea-pig vas deferens, the ATP/[³H]noradrenaline overflow ratio decreased with increasing pulse number (20 - 540 pulses) but increased with frequency (1 - 10 Hz) (von Kügelgen and Starke 1991b). At higher frequencies of stimulation, NPY is additionally released. NPY has little direct postjunctional action in the vas deferens and many vessels but seems to potentiate the action of ATP and noradrenaline. Prejunctionally, NPY reduces the release of noradrenaline (Burnstock 1990; Westfall et al. 1995).

Only a minor proportion of the endogenous ATP measured by HPLC or the luciferin-luciferase assay comes from neuronal sources (von Kügelgen and Starke 1991a). In the rabbit aorta and the guinea-pig vas deferens, endogenous ATP contributed to only about 3% and 17% of the stimulation-evoked overflow of ATP, respectively (Sedaa et al. 1990; Vizi and Burnstock 1988). The majority comes from smooth muscle and endothelial cells subsequent to α_1 adrenoceptor stimulation by released noradrenaline (Figure 1). In addition, ATP itself releases ATP from smooth muscle cells by activation of P2X receptors (von Kügelgen and Starke 1991a; Katsuragi et al. 1991). Accordingly, studies on ATP release in intact tissues have to be performed in the presence of α_1 and P2 purinoceptor blockers such as prazosin and suramin to prevent non-neuronal liberation of ATP. In contrast, in primary cultures of sympathetic neurones (Allgaier et al. 1994b) evoked overflow of ATP is exclusively of neuronal origin (von Kügelgen et al. 1994). In this preparation, the molar ratio of evoked ATP/[³H]-noradrenaline overflow was estimated to 1:6 subsequent to stimulation by bursts of 100 pulses/10 Hz.

Co-release of ATP with acetylcholine from postganglionic parasympathetic nerves occurs, e.g. in the rabbit portal vein (Brizzolara et al. 1993). Neurogenic relaxation in the presence of guanethidine and atropine, which was independent of the endothelium, was diminished by the P2 purinoceptor antagonist suramin.

<u>C. Modulation of sympathetic transmitter release by purinoceptors and the a2-autoreceptor</u>

1. Purinoceptors that induce transmitter release
ATP is known to mediate fast excitatory synaptic transmission in mammalian neurones via P2X receptors (Evans et al. 1992; Illes et al. 1996; Nieber et al. 1997; Silinsky et al. 1992). In primary cultures of rat (Boehm 1994; Boehm et al. 1995; von Kügelgen et al. 1997) and chick sympathetic neurones (Allgaier et al. 1995b), ATP was able to induce noradrenaline release. The P2 receptor antagonists reactive blue 2 and suramin diminished the response to ATP. 2-Methylthio-ATP, but not α,β-methylene-ATP or adenosine analogues, mimicked the action of ATP, presumably by activation of a rapidly desensitising P2X receptor.

In chick sympathetic neurones, the ATP-induced noradrenaline release was also reduced by the nicotine receptor antagonists hexamethonium and mecamylamine (Allgaier et al. 1995b). Vice versa, noradrenaline release elicited by the nicotine receptor agonist 1,1-dimethyl-4-piperazinium (DMPP) was reduced by suramin and reactive blue 2. Moreover, the effects of DMPP and ATP were not additive at high nucleotide concentrations suggesting that ATP may activate a subpopulation of cholinergic nicotine receptors to trigger transmitter release from this preparation. Similarly, in cultured Xenopus myotonal muscle cells (Igusa 1988) and *membranes* of isolated rat superior cervical ganglion cells

(Nakazawa and Inoue 1993; Nakazawa 1994), ATP activated nicotinc receptors, while in studies with *cultured* rat superior cervical ganglion neurones hexamethonium or mecamylamine failed to affect the ATP-induced cation conductance (Cloues at al. 1993) or noradrenaline release (Boehm 1994). Differences between species or the preparations used may account for the discrepancies observed.

2. *Prejunctional α_2-autoreceptors and purinoceptors*

The observation that a neurotransmitter can act presynaptically (prejunctionally) to modulate its own release was first made for noradrenergic neurones. Noradrenaline was shown to inhibit its own release via α_2-adrenoceptors located at the nerve terminal (Figure 1)

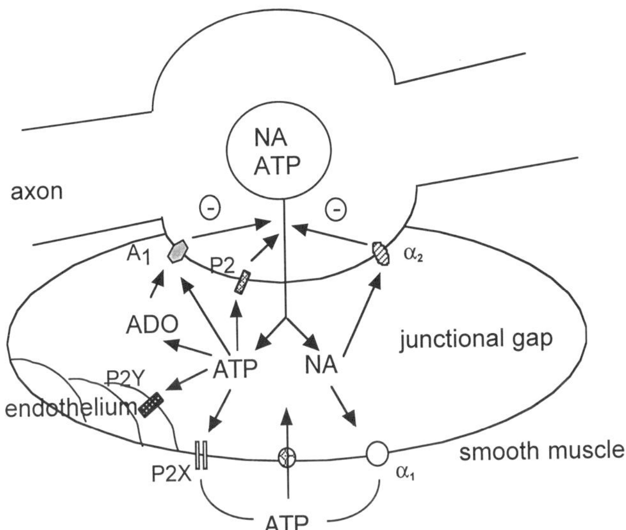

Figure 1. A generalised scheme for noradrenergic-purinergic co-transmission. Noradrenaline (NA) and ATP released from nerve terminals may contribute to neurogenic smooth muscle contraction through activation of α_1-adrenoceptors and/or P2X receptors. In intact blood vessels, however, ATP may induce relaxation through endothelial P2Y receptors. Both NA and ATP may additionally elicit postjunctional ATP liberation. Presynaptically, NA and ATP may modulate transmitter release via both α_2 and P2 autoreceptors. Inhibitory prejunctional A_1 adenosine (ADO) receptors may be activated not only by ADO generated by the cleavage of ATP, but also directly by the nucleotide.

(for review see e.g.: Starke et al. 1989). The proportion of transmitter and co-transmitter(s) depends on the frequency and duration of stimulation, and similarly also the modulation of transmitter release by presynaptic receptors depends on the stimulation frequency. In some tissues such as the rat tail artery and rat kidney, the release of noradrenaline and ATP was affected by presynaptic α_2-adrenoceptors in a similar manner (Bohmann et al. 1995; Msghina et al. 1992), while in the guinea-pig vas deferens, the α_2-adrenoceptor mediated inhibition of noradrenaline release exceeded that of ATP release (Sperlagh and

Vizi 1992; Driessen et al. 1993). In synaptosomes prepared from the myenteric plexus of the guinea-pig ileum, ATP release evoked by high K^+, acetylcholine, or nicotine was even insensitive to the α_2-adrenoceptor agonist clonidine, while noradrenaline release was significantly reduced (Hammond et al. 1988). Different populations of vesicles containing different proportions of noradrenaline and ATP, or axons differentially controlled by presynaptic receptors may account for the discrepancies observed (von Kügelgen et al. 1996).

3. P1 adenosine receptors

There is a large body of evidence that the release of various transmitters including noradrenaline, acetylcholine, dopamine, serotonin, and glutamate is inhibited through stimulation of presynaptic A_1 adenosine receptors (Figure 1) (for review see: Fredholm and Dunwiddie 1988; Olsson and Pearson 1990). Much of what is known comes from studies on the central nervous system, however, also autonomic transmitter release is inhibited by adenosine. Prejunctional A_1 adenosine receptors are present in various sympathetically innervated smooth muscle organs including blood vessels, such as the portal vein and the mesenteric artery of rabbits (Brown and Collis 1983; Illes et al. 1988). Both pertussis toxin (Dolphin and Prestwich 1985) and N-ethylmaleimide (Allgaier et al. 1987; Wurster et al. 1990) impair adenosine neuromodulation indicating the coupling to Gi/Go; however, the final transduction mechanism of presynaptic A_1 receptors is as yet unclear.

Table 1. Purinoceptor-mediated modulation of noradrenaline and ATP release from sympathetically innervated tissues and cultured sympathetic neurons

Species	Tissue/ preparation	Drug (μM)	Maximal effect on noradrenaline release	Maximal effect on ATP release	References
adenosine A_1 receptor					
rat	tail artery	CPA (0.1)	$\downarrow$ by 70%	n.d.	Gonçalvez and Queiroz 1996
guinea-pig	vas deferens	adenosine (300)	n.d.	$\downarrow$ by 90%	Kirkpatrick and Burnstock 1992
guinea-pig	vas deferens	adenosine (100)	$\downarrow$ by 40%	$\downarrow$ by 70%	Driessen et al. 1994
adenosine			A_{2A}		*receptor*
rat	tail artery	CGS-21680 (0.03)	$\uparrow$ by 58%	n.d.	Gonçalvez and Queiroz 1996
P2 purinoceptor					
rat	tail artery	2-methylthio-ATP (100)	$\downarrow$ by 49%	n.d.	Gonçalvez and Queiroz 1996
rat	atria	cibacron blue 3GA (30)	$\uparrow$ by 70%	n.d.	von Kügelgen et al. 1995
mouse	vas deferens	suramin (300)	$\uparrow$ by 65%	n.d.	von Kügelgen et al. 1993
		reactive blue 2 (10)	$\uparrow$ by 45%	n.d.	
chick	cultured sympathetic neurons	2-methylthio-ATP (30)	$\uparrow$ by 65%	n.d.	Allgaier et al. 1995

Stimulation-evoked overflow of endogenous ATP or previously incorporated [^{3}H]-noradrenaline was measured. CPA, N^6-cyclopentyladenosine; CGS-21680, 2-*p*-(2-carboxyethyl) phenylethylamino-5'-*N*-ethylcarboxamidoadenosine. $\uparrow$, increase; $\downarrow$, decrease: n.d., transmitter release was not determined

The prejunctional A_1 receptors may respond to adenosine arising from the breakdown of neuronally released ATP, or adenosine from other sources, such as the effector muscle cell (Illes et al. 1988). Accordingly, also release of ATP from sympathetic neurones is presumably subject to A_1 adenosine receptor inhibition. In an early experiment, KCl-evoked [^{3}H]-purine release in the pulmonary artery of the rabbit was attenuated by adenosine and enhanced by the adenosine receptor antagonist theophylline (Katsuragi and Su 1982). In the guinea-pig vas deferens, adenosine and the A_1 receptor-selective agonist 2-chlorocyclopentyladenosine diminished the electrically evoked overflow of ATP even to a greater extent than the release of noradrenaline (Table 1) (Driessen et al. 1994; Kirkpatrick and Burnstock 1992). These data again suggest differential modulation of sympathetic transmitter release by prejunctional receptors.

Nucleotides can also directly stimulate prejunctional A_1 adenosine receptors without a previous breakdown to adenosine (Figure 1). This has been demonstrated for various tissues and species including rat atria (von Kügelgen et al. 1995), rat and mouse vas deferens (Kurz et al. 1993), and guinea-pig ileum (Wiklund and Gustafsson 1987). In rat atria, ATP and ATPγS, a metabolically stable analogue of ATP inhibited noradrenaline release, and these effects were diminished by 8-cyclopentyl-1,3-dipropylxanthine (DPCPX), a selective A_1 adenosine receptor antagonist (von Kügelgen et al. 1995). These data may suggest that ATP in those instances in which it is co-released with noradrenaline can inhibit its own release either via A_1 adenosine or P2 receptors (see below) each located presynaptically at the axon terminals.

In addition, A_{2A} adenosine receptors may contribute to sympathetic transmitter release in some tissues. In rat tail artery, CGS-21680, a selective A_{2A} receptor agonist, increased electrically evoked noradrenaline release (Table 1) (Gonçalves and Queiroz 1996). Thus, the existence of A_{2A} receptors may also explain other facilitatory effects of adenosine observed e.g. in the saphenous artery of the rabbit (Todorov et al. 1994). To date, there is no convincing evidence that sympathetic transmitter release is additionally regulated by hybrid P_3 adenosine receptors (Shinozuka et al. 1990) exhibiting responses to both nucleosides and nucleotides which are blocked by methylxantines as a discriminating criterion.

4. P2 receptors

It seems feasible that ATP can directly modulate transmitter release by acting on prejunctional P2 receptors (Figure 1). In the vas deferens and some arteries (e.g. rat atria and tail artery), ATP-induced activation of prejunctional P2 autoreceptors inhibited noradrenaline release (Table 1) (Gonçalves and Queiroz 1996; Kurz et al. 1993; von Kügelgen et al. 1989, 1993, 1995). Cibacron blue 3GA, a P2 receptor antagonist, reduced the inhibition caused by ATP and ATPγS without significantly affecting the inhibition caused by N^6-cyclopentyladenosine (CPA), an A_1 receptor agonist. The antagonism caused by concurrent application of cibacron blue 3GA and DPCPX was additive indicating a dual, A_1 and P2 receptor-mediated regulation of noradrenaline release (von Kügelgen et al. 1995). In rabbit artery (Miyahara and Suzuki 1987), guinea-pig ileum (Sperlagh and Vizi 1991), and primary cultures of sympathetic neurones (Allgaier et al. 1994a, 1995a) P2 purinoceptor activation facilitated transmitter release (Table 1). In cultured sympathetic neurones, the facilitation caused by 2-methylthio ATP and ATP, but not by α,ß-methylene-ATP was exclusively P2 receptor-mediated, since i) it was antagonised by suramin and reactive blue 2, and ii) P1 adenosine receptors were not detectable by pharmacological approaches (Allgaier et al. 1994b, 1995a). Thus, this preparation may offer the opportunity to investigate the P2 purinoceptor transduction mechanism independent of P1 adenosine receptor-mediated effects. 2-Methylthio-ATP-induced facilitation of noradrenaline release from chick sympathetic neurones remained virtually unaffected subsequent to treatment with pertussis toxin suggesting the involvement of a P2X receptor (Allgaier et al.; unpublished observation).

IV. CONCLUSIONS

The data summarised above indicate that autonomic transmitter release in the cardiovascular system is regulated by purinoceptors in a complex manner. Considerable differences have been detected between various species and tissues investigated. However, 3 observations may be of more general significance: 1) the co-transmitters noradrenaline and ATP appear differentially released in response to nerve stimulation, and this release is differentially modulated by various presynaptic receptors, 2) ATP co-released with noradrenaline is presumably subject to P2 purinoceptor-mediated autoregulation, and 3) released ATP activates not only prejunctional P2 receptors but also adenosine receptors (mainly A_1) which may explain various observations previously ascribed to the involvement of a hybrid P_3 receptor. Whether changes in the purinoceptor-regulated modulation of autonomic transmitter release contribute to cardiovascular disorders is an interesting aspect of additional studies.

REFERENCES

Abbrachio MP, Burnstock G (1994) Purinoceptors: are there families of P2X and P2Y purinoceptors. Pharmacol Ther 64:445-475

Abbrachio MP, Cattabeni F, Fredholm BB, Williams M (1993) Purinoceptor nomenclature: a status report. Drug Dev Res 28:207-213

Allgaier C, Hertting G, von Kügelgen O (1987) The adenosine receptor-mediated inhibition of noradrenaline release possibly involves a N-protein and is increased by α_2-autoreceptor blockade. Br J Pharmacol 90:403-412

Allgaier C, Pullmann F, Schobert A, von Kügelgen I, Hertting G (1994a) P_2 Purinoceptors modulating noradrenaline release from sympathetic neurons in culture. Eur J Pharmacol 252:R7-R8

Allgaier C, Schobert A, Belledin M, Jackisch R, Hertting G (1994b) Modulation of electrically evoked [^{3}H]-noradrenaline release from cultured chick sympathetic neurons. Naunyn-Schmiedeberg's Arch Pharmacol 350:258-266

Allgaier C, Wellmann H, Schobert A, von Kügelgen I (1995a) Cultured chick sympathetic neurons: modulation of electrically evoked noradrenaline release by P_2-purinoceptors. Naunyn-Schmiedeberg's Arch Pharmacol 352:17-24

Allgaier C, Wellmann H, Schobert A, Kurz G, von Kügelgen I (1995b) Cultured chick sympathetic neurons: ATP-induced noradrenaline release and its blockade by nicotinic receptor antagonists. Naunyn-Schmiedeberg's Arch Pharmacol 352:25-30

Bao JX(1993) Sympathetic neuromuscular transmission in rat tail artery: a study based on electrochemical, electrophysiological and mechanical recording. Acta Physiol Scand Suppl 148, 610:1-58

Berne RM (1963) Cardiac nucleotides in hypoxia: possible role in regulation of coronary blood flow. Am J Physiol 204:317-322

Boarder MR, Weisman GA, Turner JT, Wilkinson GF (1995) G protein-coupled P_2 purinoceptors: from molecular biology to functional responses. Trends Pharmcol Sci 16:133-139

Boehm S (1994) Noradrenaline release from rat sympathetic neurons evoked by P_2-purinoceptor activation. Naunyn-Schmiedeberg's Arch Pharmacol 350:454-458

Boehm S, Huck S, Illes P (1995) UTP- and ATP-triggered transmitter release from rat sympathetic neurones via separate receptors. Br J Pharmacol 116:2341-2343

Bohmann C, Rump LA, Schaible U, von Kügelgen I (1995) α-Adrenoceptor modulation of norepinephrine and ATP release in isolated kidneys of spontaneously hypertensive rats. Hypertension 25:1224-1231

Boyer JL, Lazarowski ER, Chen X-H, Harden TK (1993) Identification of a P_{2Y}-purinergic receptor that inhibits adenylyl cyclase. J Pharmacol Exp Ther 267:1140-1146

Brizzolara AL, Crowe R, Burnstock G (1993) Evidence for the involvement of both ATP and nitric oxide in non-adrenergic, non-cholinergic inhibitory neurotransmission in the rabbit portal vein. Br J Pharmacol 109:606-608

Brown CM, Collis MG (1983) Adenosin A_1 receptor mediated inhibition of nerve stimulation-induced contractions of the rabbit portal vein. Eur J Pharmacol 93:277-282

Bruner G, Murphy S (1993) Purinergic P_{2Y} receptors on astrocytes are directly coupled to phospholipase A_2. Glia 7:219-224

Burnstock G (1972) Purinergic nerves. Pharmacol Rev 24:509-581

Burnstock G (1976) Do some nerve cells release more than one transmitter? Neuroscience 1:239-248

Burnstock G (1990) Co-transmission. Arch Int Pharmacodyn 304:7-33

Burnstock G (1996) Introduction: purinergic transmission. Sem Neurosci 8:171-174

Burnstock G, King BF (1996) Numbering of cloned P2 purinoceptors. Drug Dev Res 38:67-71

Cloues R, Jones S, Brown DA (1993) Zn^{2+} potentiates ATP-activated currents in rat sympathetic neurons. Pflügers Arch 424:152-158

Dalziel HH, Westfall DP (1994) Receptors for adenine nucleotides and nucleosides: subclassification, distribution, and molecular characterization. Pharmacol Rev 46:449-466

Driessen B, von Kügelgen I, Starke K (1993) Neural ATP release and its α_2-adrenoceptor-mediated modulation in guinea-pig vas deferens. Naunyn Schmiedeberg's Arch Pharmacol 348:358-366

Driessen B, von Kügelgen I, Starke K (1994). P_1-purinoceptor-mediated modulation of neural noradrenaline and ATP release in guinea-pig vas deferens. Naunyn Schmiedeberg's Arch Pharmacol 350:42-48

Drury AN, Szent-Györgyi A (1929) The physiological activity of adenine compounds with especial reference to their action upon the mammalian heart. J Physiol (Lond) 68:213-237

Dubyak GR, El-Moatassim (1993) Signal transduction via P_2-purinergic receptors for extracellular ATP and other nucleotides Am J Physiol 265:C577-C606

Edwards FA, Gibb AJ (1993) ATP - a fast neurotransmitter. FEBS Lett 325:86-89

Evans RJ, Derkach V, Surprenant A (1992) ATP mediates fast synaptic transmission in mammalian neurons. Nature 357:503-505

Fozard JR, Carruthers AM (1993) Adenosine A_3 receptors mediate hypotension in the angiotensin II-supported circulation of the pithed rat. Br J Pharmacol 109:3-5

Fredholm BB, Dunwiddie TV (1988) How does adenosine inhibit transmitter release? Trends Pharmacol Sci 9:130-134

Fredholm BB, Abbracchio MP, Burnstock G, Daly JW, Harden TK, Jacobson KA, Leff P, Williams M (1994) Nomenclature and classification of purinoceptors. Pharmacol Rev 46:143-156

Fredholm BB, Abbracchio MP, Burnstock G, Dubyak GR, Harden TK, Jacobson KA, Schwabe U, Williams M (1994) Nomenclature and classification of purinoceptors. Trends Pharmacol Sci 18:79-82

Gerlach E, Deuticke B, Dreisbach RH (1963) Der Nukleotid-Abbau im Herzmuskel bei Sauerstoffmangel und seine mögliche Bedeutung für die Coronardurchblutung. Naturwissenschaften 50:228-229

Gerwins P, Fredholm BB (1995) Activation of phospholipase C and phospholipase D by stimulation of adenosine A_1, bradykinin or P_{2U} receptors does not correlate well with protein kinase C activation. Naunyn Schmiedeberg's Arch Pharmacol 351:194-201

Gonçalves J, Queiroz G (1996) Purinoceptor modulation of noradrenaline release in rat tail artery: tonic modulation mediated by inhibitory P_{2Y} and facilitatory A_{2A}-purinoceptors. Br J Pharmacol 117:156-160

Green HN, Stoner HB (1950) Biological actions of the adenine nucleotides, pp 1-221, HK Lewis & Co Ltd, London, UK

Hammond JR, MacDonald WF, White TD (1988) Evoked secretion of [^{3}H]noradrenaline and ATP from nerve varicosities isolated from the myenteric plexus of the guinea-pig ileum. Can J Physiol Pharmacol 66:369-375

Hirst GDS, Edwards FR (1989) Sympathetic neuroeffector transmission in arteries and arterioles. Physiol Rev 69:546-604

Hoyle CHV (1996) Purinergic co-transmission: parasympathetic and enteric nerves. Sem Neurosci 8:207-215

Igusa Y (1988) Adenosine 5'-triphosphate activates acetylcholine receptor channels in cultured Xenopus myotomal muscle cells. J Physiol 405:169-185

Illes P, Nörenberg W (1993) Neuronal ATP receptors and their mechanism of action. Trends Pharmacol Sci 14:50-54

Illes P, Jackisch R, Regenold JT (1988) Presynaptic P1-purinoceptors in jejunal branches of the mesenteric artery and their possible function. J Physiol 397:13-29

Illes P, Nieber K, Nörenberg W (1996) Electrophysiological effects of ATP on brain neurones. J Autonom Pharmacol 16:407-411

Katsuragi T, Su C (1982) Augmentation by theophylline of [^{3}H]purine release from vascular adrenergic nerves: evidence for presynaptic autoinhibition. J Pharmacol Exp Ther 220:152-156

Katsuragi T, Tokunaga T, Ogawa S, Soejima O, Sato C, Furukawa T (1991) Existence of ATP-evoked ATP release system in smooth muscles. J Pharmacol Exp Ther 259:513-518

Kirkpatrick K, Burnstock G (1992) Evidence that the inhibition of ATP release from sympathetic nerves by adenosine is a physiological mechanism. Gen Pharmacol 23:1045-1050

Kurz K, von Kügelgen I, Starke K (1993) Prejunctional modulation of noradrenaline release in mouse and rat vas deferens: contribution of P$_1$- and P$_2$-purinoceptors. Br J Pharmacol 110:1465-1472

Linden J (1994) Cloned adenosine A$_3$ receptors: pharmacological properties, species differences and receptor functions. Trends Pharmcol Sci 15:298-306

Lindner F, Rigler R (1931) Über die Beeinflussung der Weite der Herzkranzgefäße durch Produkte des Zellkernstoffwechsels. Pfluegers Arch 226:697-708

Londos C, Cooper DMF, Wolff J (1980) Subclasses of external adenosine receptors. Proc Natl Acad Sci 77:2551-2554

MacDonald A, Daly CJ, Bulloch JM, McGrath JC (1992) Contributions of alpha1-adrenoceptors, alpha 2-adrenoceptors and P$_{2X}$-purinoceptors to neurotransmission in several rabbit isolated blood vessels: role of neuronal uptake and autofeedback. Br J Pharmacol 105:347-354

McLaren GJ, Lambrecht G, Mutschler E, Bäumert HG, Sneddon P, Kennedy C (1994) Investigation of the actions of PPADS, a novel P$_{2X}$-purinoceptor antagonist, in the guinea-pig isolated vas deferens. Br J Pharmacol 111:913-917

Miyahara H, Suzuki H (1987) Pre- and postjunctional effects of adenosine triphosphate on noradrenergic transmission in the rabbit ear artery. J Physiol 389:423-440

Msghina M, Mermet C, Gonon F, Stjärne L (1992) Electrophysiological and electrochemical analysis of the secretion of ATP and noradrenaline from the sympathetic nerves in rat tail artery: effects of α_2-adrenoceptor agonists and antagonists and noradrenaline reuptake blockers. Naunyn Schmiedeberg's Arch Pharmacol 346:173-186

Nakazawa K (1994) ATP-activated current and its interaction with acetylcholine-activated current in rat sympathetic neurons. J Neurosci 14:740-750

Nakazawa K, Inoue K (1993) ATP- and acetylcholine-activated channels co-existing in cell-free membrane patches from rat sympathetic neuron. Neurosci Lett 163:97-100

Nieber K, Poelchen W, Illes P (1997) Role of ATP in fast excitatory synaptic potentials in locus coeruleus neurones of the rat. Br J Pharmacol 122:423-430

O'Conner SE, Dainty IA, Leff P (1991) Further subclassification of ATP receptors based on agonist studies. Trends Pharmacol Sci 12:137-141

Olsson RA, Pearson JD (1990) Cardiovascular purinoceptors. Physiol Rev 70:761-845

Ongini E, Fredholm BB (1996) Pharmacology of adenosine A$_{2A}$ receptors. Trends Pharmacol Sci 17:364-371

Ramme D, Regenold JT, Starke K, Busse R, Illes P (1987) Identification of the neuroeffector transmitter in jejunal branches of the rabbit mesenteric artery. Naunyn Schmiedeberg's Arch Pharmacol 336:267-273

Ramkumar V, Stiles GL, Beaven MA, Ali H (1993) The A_3 adenosine receptor is the unique adenosine receptor which facilitates release of allergic mediators in mast cells. J Biol Chem 268:16887-16890

Sattin A, Rall TW (1970) The effect of adenosine and adenine nucleotides on the cyclic adenosine 3',5'-monophosphate content of guinea-pig cerebral cortex slices. Mol Pharmacol 6:13-23

Scholz KP, Miller RJ (1991) Analysis of adenosine actions on Ca^{2+} currents and synaptic transmission in cultured rat hippocampal pyramidal neurones. J Physiol (Lond) 435:373-393

Sedaa KO, Bjur RA, Shinozuka K, Westfall DP (1990) Nerve and drug-induced release of adenine nucleosides and nucleotides from rabbit aorta. J Pharmacol Exp Ther 252:1060-1067

Shinozuka K, Bjur RA, Westfall DP (1990) Effects of α,ß-methylene ATP on the prejunctional purinoceptors of the sympathetic nerves of the rat caudal artery. J Pharmacol Exp Ther 254:900-904

Silinsky EM, Gerzanich V, Vanner SM (1992) ATP mediates excitatory synaptic transmission in mammalian neurones. Br J Pharmacol 106:762-763

Sneddon P, McLaren GJ, Kennedy C (1996) Purinergic co-transmission: sympathetic nerves. Sem Neurosci 8:201-205

Sperlagh B, Vizi ES (1991) Effect of presynaptic P_2 receptor stimulation on transmitter release. J Neurochem 56: 1466-1470

Starke K, Göthert M, Kilbinger H (1989) Modulation of neurotransmitter release by presynaptic autoreceptors. Rev Physiol Biochem Pharmacol 77:1-124

Starke K, von Kügelgen I, Bültmann R (1992) Noradrenaline-ATP co-transmission: operation in blood vessels and co-transmitter release ratios. Jap J Pharmacol 58:166P-173P

Todorov LD, Bjur RA, Westfall DP (1994a) Temporal dissociation of the release of the sympathetic co-transmitters ATP and noradrenaline. Clin Exp Pharmacol Physiol 21:931-932

Todorov LD, Bjur RA, Westfall DP (1994b) Inhibitory and facilitatory effects of purines on transmitter release from sympathetic nerves. J Pharmacol Exp Ther 268:985-989

Trussel LO, Jackson MB (1985) Adenosine-activated potassium conductance in cultured striatal neurons. Proc Natl Acad Sci USA 82:4857-4861

Vizi ES, Burnstock G (1988) Origin of ATP release in the rat vas deferens: concomitant measurement of [^{3}H]noradrenaline and [^{14}C]ATP. Eur J Pharmacol 158:69-77

von Kügelgen I (1996) Modulation of neural ATP release through presynaptic receptors. Sem Neurosci 8:247-257

von Kügelgen I, Starke K (1991a) Noradrenaline-ATP co-transmission in the sympathetic nervous system. Trends Pharmacol Sci 12:319-324

von Kügelgen I, Kurz K, Starke K (1993) Axon terminal P_2-purinoceptors in feedback control of sympathetic transmitter release. Neuroscience 56:263-267

von Kügelgen I, Schöffel E, Starke K (1989) Inhibition by nucleotides acting at presynaptic P_2-receptors of sympathetic neuro-effector transmission in the mouse isolated vas deferens. Naunyn-Schmiedeberg's Arch Pharmacol 340:522-532

von Kügelgen I, Stoffel D, Starke K (1995) P_2-purinoceptor-mediated inhibition of noradrenaline release in rat atria. Br J Pharmacol 115:247-254 von Kügelgen I, Allgaier C, Schobert A, Starke K (1994) Co-release of noradrenaline and ATP from cultured sympathetic neurons. Neuroscience 61:199-202

von Kügelgen I, Nörenberg W, Illes P, Schobert A, Starke K (1997) Differences in the mode of stimulation of cultured rat sympathetic neurons between ATP and UDP. Neuroscience 78:935-941

Westfall TC, Yang C-L, CurfmanFalvey M (1995) Neuropeptide-Y-ATP interactions at the vascular sympathetic neuroeffector junction. J Cardiol Pharmacol 26:682-687

Weitzell R, Tanaka T, Starke K (1979) Pre- and postsynaptic effects of yohimbine stereoisomers on noradrenergic transmission in the pulmonary artery of the rabbit. Naunyn Schmiedeberg's Arch Pharmacol 308:127-136

Wiklund NP, Gustafsson LE (1987) On the nature of endogenous purines modulating cholinergic neurotransmission in the guinea-pig ileum. Acta Physiol Scand 131:11-18

Wurster S, Nakov R, Allgaier C, Hertting G (1990) Involvement of N-ethylmaleimide-sensitive G proteins in the modulation of evoked [³H]noradrenaline release from rabbit hippocampus synaptosomes. Neurochem Int 17:149-155

ADENOSINE RECEPTORS, BLOOD VESSEL AND INFLAMMATION

Gail W. Sullivan, University of Virginia, Charlottesville, VA, USA.

I. SUMMARY

Adenosine and adenosine analogs, by binding to neutrophil (PMN) and monocyte/macrophage adenosine receptors, can affect leukocyte viability, adherence, phagocytic capacity, motility and release of oxidative and non-oxidative products. Adenosine and selective A_1 adenosine receptor agonists enhance PMN adherence to endothelium and surfaces coated with matrix proteins, and increase PMN chemotaxis and immune complex-stimulated superoxide release. In addition, agonist binding to monocyte A_1 receptors results in increased $Fc\gamma R$-mediated phagocytosis. In contrast, adenosine and selective agonists of A_{2A} adenosine receptors diminish stimulated PMN adherence, degranulation, and oxidative activity. Adenosine binding to A_2 receptors on monocytes results in decreased $Fc\gamma R$-mediated phagocytosis, diminished viability, reduced tumor necrosis factor-α (TNF) release but increased production of interleukin-10. These adenosine-mediated changes in PMN and moncyte/macrophage function have the potential to alter the course of an inflammatory response in cardiovascular diseases such as ischemia/reperfusion injury, atherosclerotic plaque development and inflammatory aneurysm formation.

II. TEXT

The inflammatory response serves the purpose of eliminating harmful agents from the body. There is a wide range of pathogenic insults that can initiate an inflammatory response including infection, allergens, noxious chemicals, and toxins. Other stimuli also can stimulate inflammation. These include ischemia/reperfusion, hypoxia, mechanical and thermal trauma. Inflammation normally is a very localized event which serves in expulsion, attenuation by dilution, and isolation of the damaging agent and injured tissue. The body's response becomes an agent of disease when it results in inappropriate injury to host tissues in the process of eliminating a pathogen or responding to another inflammatory stimulus.

Inflammation is a component of pathogenesis in several cardiovascular conditions. Examples include: ischemia/reperfusion injury (Hartmann, Robinson *et al.* 1977; Romson, Hook *et al.* 1983; Mullane, Read *et al.* 1984; Chatelain, Latour *et al.* 1987; Engler 1987) (as reviewed in [Frangogiannis, Youker *et al.* 1996; Sharma and Das 1997]), atherosclerosis (reviewed in [Ross 1993]), inflammatory aortic aneurysms (Walker, Bloor *et al.* 1972; Pennell, Hollier *et al.* 1985; Girardi and Coselli 1997), and restenosis following balloon angioplasty (reviewed in [Ross 1993]). The cells involved with inflammation include blood cells, the vascular endothelium, vascular smooth muscle, fibroblasts, and myocytes. The triggers for inflammation in cardiovascular diseases range from mechanical trauma to the blood vessel in angioplasty to hypercholesterolaemia in atherosclerosis.

Following acute stimulation, resident inflammatory cells (*e.g.* macrophages and mast cells) and vascular changes promote an inflammatory cascade by recruiting PMN to the site of injury. Although monocytes collect slowly at inflammatory foci, given favorable conditions the monocytes evolve into long term resident accessory cells, macrophages, and lipid-rich foam cells. Macrophages are multi-functional cells which play a key role in the slower processes of atherosclerotic plaque and inflammatory aortic aneurysm formation. Macrophages produce and secrete an array of cytokines, complement, lipids, reactive oxygen species, proteases and growth factors that modulate surrounding vascular tissue function. In addition, macrophages are phagocytes that consume dying apoptotic tissue, pathogens and debris and act as accessory cells processing antigen for presentation to lymphocytes.

Platelets and damaged vascular tissues release through exocytosis and cell leakage adenosine and adenine nucleotides that become converted to adenosine through the action of ectonucleotidases. The local concentration of adenosine is controlled by rates of blood flow, cellular adenosine uptake, and the activities of enzymes that metabolize adenosine (Pearson and Gordon 1979; Girardi and Coselli 1997). There is evidence from *in vivo* studies that adenosine and adenosine analogs can affect the inflammatory response in vascular disease. Adenosine acting as an agonist at adenosine A_{2A} receptors reduces reperfusion injury in part by decreasing the inflammatory response (Olafsson, Forman *et al.* 1987; Pitarys II, Virmani *et al.* 1991; Nolte 1992; Schlack, Schäfer *et al.* 1993; Zhao, Sato *et al.* 1996; Jordan, Zhao *et al.* 1997). In addition, blockage of A_1 adenosine receptors with antagonists decreases ischemia/reperfusion injury (Neely and Keith 1995). Hence, adenosine produced at sites of vascular injury can alter the course of an inflammatory response.

From radioligand binding assays, there is evidence that human PMN possess both A_1 and A_{2A} adenosine receptors (Cronstein, Rosenstein *et al.* 1985; Schrier and Imre 1986; Martini, Di Sacco *et al.* 1991). In addition, reverse transcriptase-polymerase chain reaction analysis indicates that human PMN contain message for adenosine A_3 receptors (Bouma, Trudi *et al.* 1997). Experiments using monoclonal antibodies to the adenosine A_1 receptor suggest that cultured human monocytes (4-6 hr) express A_1 receptors (Salmon, Brogle *et al.* 1993). Radioligand binding assays indicate that rabbit alveolar macrophages contain A_2 receptors (Hasday and Sitrin 1987). From functional assays there is evidence that uncultured human monocytes contain low levels of A_{2A} receptors (Eigler, Greten *et al.* 1997) and monocytes on long-term culture (4-6 days) and isolated macrophages from rheumatoid synovial fluid express high levels of A_{2A} receptors (Eppell, Newell *et al.* 1989). Polymerase chain reaction studies report that a human macrophage cell line contains message for adenosine A_3 receptors (Sajjadi, Takabayashi *et al.* 1996). The present chapter reviews the role of adenosine binding to its receptors on the regulation of the inflammatory response in cardiovascular diseases.

A. Neutrophil and monocyte viability and differentiation

Circulating PMN are short-lived cells with a life span measured in hours. Following migration from the blood, PMN soon become apoptotic expressing surface factors that promote their own elimination by phagocytic macrophages with minimal rupture and release of toxic products. Although adenosine itself does not alter human PMN apoptosis *in vitro*, adenosine analogs with a potency order suggesting A_{2A} receptor function delay PMN apoptosis. Further enhancement of PMN survival occurs when an adenosine A_{2A} agonist is combined with a type IV phosphodiesterase (PDE) inhibitor (Walker, Rocchini *et al.* 1997). Dense platelet granules contain ATP, ADP and the dinucleoside polyphosphates Ap_3A and Ap_4A that are released upon platelet activation. Delay of PMN apoptosis *in vitro* occurs when ATP or one of the dinucleoside polyphosphates are present. It has not been determined if the delay in apoptosis is from adenosine formed from nucleotide degradation in culture (Gasmi, McLennan *et al.* 1996).

Adenosine can influence the viability of macrophages and the development of monocytes into macrophages. With *in vitro* culture human monocytes develop into antigen presenting accessory cells within 1-2 days and then go on to develop into macrophages by 8 days (Najar, Ruhl *et al.* 1990). If endogenous adenosine deaminase (ADA) is inhibited with erythro-9-(2 hydroxy-3-nonyl) adenine (EHNA) adenosine accumulates and most cells die within 3 days (Fischer, B. *et al.* 1976). Exogenous adenosine, ATP, ADP, AMP arrest monocyte to macrophage development at the accessory cell stage (Najar, Ruhl *et al.* 1990). The adenosine analog 2-chloroadenosine (2-CA) also decreases the viability of mouse macrophages (Ohtani, Kumazawa *et al.* 1982).

The expression of adenosine receptors changes during monocyte/macro-phage development. Fresh human monocytes have low numbers of A_1 receptors. On culture (4-6hr), A_1 adenosine receptors appear on the surface (Salmon, Brogle *et al.* 1993). As monocytes develop into macrophages in culture they express more A_2 adenosine receptors that become more functionally apparent after 2 days of culture. At this point, both IgG Fc mediated and complement-mediated phagocytosis are inhibited by adenosine and adenosine analogs (μM) with an order of potency suggesting binding of A_2 adenosine receptors (Eppell, Newell *et al.* 1989; Salmon, Brogle *et al.* 1993). In contrast, binding of A_1 selective agonists to cultured monocytes at much lower concentrations (pM) results in increased FcγR-mediated phagocytosis (Salmon, Brogle *et al.* 1993). Hence, monocytes developing into macrophages go through a series of changes (some that are sensitive to adenosine) that affect their functional state.

<u>B. Macrophage complement and cytokine production</u>
Macrophages are both phagocytic effectors and secreting directors in the inflammatory response. Through the production of a changing repertoire of cytokines and other products macrophages coordinate the initiation, maintenance and resolution of an inflammatory response. It has been observed that agonist activity at human adenosine A_2 receptors inhibits complement C2 production (Lappin and Whaley 1984). In addition, adenosine and analogs of adenosine influence macrophage cytokine production and release. Adenosine (μM) inhibits TNF release from mouse macrophages by a post-transciptional effect, and exogenous adenosine (100mg/kg) *in vivo* decreases serum concentrations of TNF in endotoxin challenged mice (Parmely, Zhou *et al.* 1993). TNF has many actions when produced and released at inflammatory sites that can alter vascular function. TNF promotes angiogenesis, inhibits endothelial proliferation, stimulates the expression of adherence factors on leukocytes and endothelial cells, primes PMN and enhances adherent PMN oxidative activity, and depresses cardiac function in heart failure (as reviewed by [Sharma and Das 1997]).

McWhinney *et al.* (McWhinney, Dudley *et al.* 1996) stably transfected a TNF-luciferase gene construct into a mouse macrophage cell line (J774.1). When the transfected cells are stimulated with endotoxin the luciferase gene is expressed. Adenosine counteracts this induction and the order of potency for inhibiting luciferase activity suggests that adenosine A_3 receptors could play a role in the inhibition of TNF gene expression. The inhibition by the agonist N^6-2-(4-amino-3-phenyl)ethyladenosine (APNEA) is reversed by treatment of the cells with the non-selective adenosine receptor antagonist 1,3-dipropyl-8-*p*-acrylate)-xanthine (BW-1433) (Salvatore, Jacobson *et al.* 1993). Using a human macrophage cell line (U937) Sajjadi *et al.* (Sajjadi, Takabayashi *et al.* 1996) observed a suppressive effect of A_3 adenosine receptor agonists at the level of TNF gene transcription.

We and others have observed that endogenous adenosine and adenosine analogs reduce human monocyte (Bouma, Stad *et al.* 1994; Eigler, Greten *et al.* 1997) and rat macrophage (Reinstein, Lichtman *et al.* 1994) TNF release by binding to adenosine A_{2A} receptors. The activity of adenosine in human monocyte cultures is blocked by the A_{2A} receptor antagonist 8-(3-chlorostyryl) caffeine (Eigler, Greten *et al.* 1997). Thus, there is evidence for transcriptional and post-transcriptional suppressive effects of adenosine on TNF production *in vitro* and *in vivo* via A_{2A} and/or A_3 adenosine receptor binding. In addition, endotoxin-stimulated release of interleukin-6 (IL-6) and interleukin-8 (IL-8) are decreased by adenosine analogs with an order of potency suggesting A_2 adenosine receptor activity (Bouma, Stad *et al.* 1994).

Interleukin-1 (IL-1) is an endothelial cell growth inhibitor that promotes endothelial senescence. Conversely, IL-1 stimulates proliferation of fibroblasts and smooth muscle cells (as reviewed by [Greisler 1997]). In mouse macrophages, adenosine does not affect IL-1 mRNA or IL-1 release (Parmely, Zhou *et al.* 1993), but the adenosine receptor agonist $R(-)-N^6$-(2-phenylisopropyl)adenosine (R-PIA;10μM) does decrease (30%) IL-1ß release from endotoxin-stimulated human monocytes (Vraux, Chen *et al.* 1993). IL-1ß stimulates the production of other cytokines from monocytes/macrophages including TNF, IL-6 and IL-8. Adenosine and the adenosine analogs 5'N-ethylcarboxamidoadenosine (NECA), 2-CA and N^6-cyclopentyladenosine (CPA) inhibit IL-1ß-induced human monocyte release of TNF. Only 2-CA inhibits IL-1ß-induced IL-6 and IL-8 release, and adenosine (100μM) increases IL-6 and IL-8 release from human monocytes (Bouma, Stad *et al.* 1994).

Interleukin-10 (IL-10) has antiinflammatory activity. It decreases endotoxin-stimulated TNF release from monocytes, inhibits oxidative activity and lowers the expression of leukocyte adhesion molecules. TNF, H_2O_2 and endotoxin stimulate the release of IL-10 from monocytes. Adenosine (10-100μM) enhances stimulated human monocyte production of IL-10 (Le Moine, Stordeur *et al.* 1996). *In vivo* in mice, the selective A_{2A} adenosine receptor agonist 2-[p-(2-carboxyethyl)-phenethylamino]-5'-N-ethyl-carboxamido-adenosine (CGS21680; 0.2-2mg/kg) decreases endotoxin-stimulated [TNF] and increases [IL-10] in plasma (Hasko, Szabo *et al.* 1996). By enhancing IL-10 production adenosine may promote resolution of an inflammatory response.

<u>C. Neutrophil and monocyte Recruitment and Adherence</u>
It is homo- and heterotypic interactions of vascular and blood elements that orchestrate an inflammatory response. Homotypic aggregations of PMN have the potential to obstruct blood flow. Although some have not observed an effect of adenosine on PMN aggregation (Cronstein, Kramer *et al.* 1983; Grinstein and Furuya 1986), Skubitz *et al* .(Skubitz, Wickham *et al.* 1988) report that adenosine (IC_{50}=~1μM) reduces human PMN aggregation stimulated by the chemoattractant formyl-methionyl-leucyl-phenylalanine (fMLP) and C_{5a}-rich zymosan-activated plasma. Conversely, aggregation is augmented by adenosine deaminase. By decreasing homotypic aggregation, adenosine may decrease clot formation in small blood vessels.

PMN adhesion to extracellular matrix proteins affects their activation within interstitial tissue. Adenosine and adenosine analogs do not affect adherence of unstimulated human PMN to matrix protein coated surfaces (Cronstein, Levin *et al.* 1992), and there is little effect of adenosine itself on TNF-stimulated PMN adherence to matrix protein coated surfaces (De La Harpe and Nathan 1989), but the selective A_{2A} receptor agonist CGS 21680, especially with the type IV PDE inhibitor rolipram, does decrease TNF-induced PMN adherence to a fibrinogen coated surface (Sullivan, Carper *et al.* 1995) (**Figure 1A**).

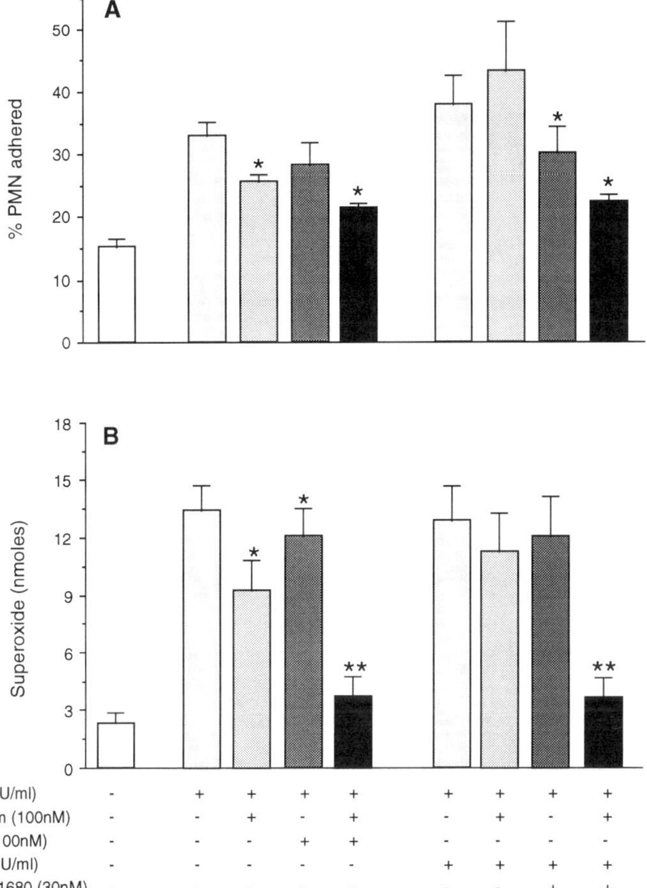

Figure 1: Rolipram with adenosine decreases TNF-stimulated adherence and synergistically decreases the oxidative activity of TNF-stimulated PMN adhering to fibrinogen. PMN (1 X 10^6/ml) were incubated 90min at 37°C ±rolipram (100nM), ± adenosine (ADO; 100nM), ± CGS21680 (30nM), and ±TNF (1U/ml) and then PMN adherence (**A**), and the oxidative burst of the adhering PMN (**B**) assayed. Adenosine, CGS 21680, and/or rolipram decrease activity compared to TNF-stimulated activity (*p<0.05). The decrease in the PMN oxidative burst with rolipram and adenosine or CGS 21680 is synergistic (**p<0.05). Mean ± SEM of 5 experiments. Reprinted from *Int. J. Immunopharmac.* 17(10), Sullivan, GW, Carper, HT and GL Mandell, The specific type IV phosphodiesterase inhibitor rolipram combined with adenosine reduces tumor necrosis factor-α-primed neutrophil oxidative activity, p. 793-803 Copyright (1995), with kind permission from Elsevier Science, Ltd, The Boulevard, Langford Lane, Kidlington 0X5 1GB, UK

Divergent effects of the adenosine analogs NECA and CPA were observed on human PMN adherence to matrix protein coated surfaces. The A_1 adenosine receptor selective agonist CPA (pM), increases PMN adherence to gelatin but not to fibrinogen coated surfaces. In contrast, NECA (nM; non-selective) decreases PMN adherence to fibrinogen coated surfaces, but NECA has a lesser effect on stimulated PMN adherence to gelatin coated surfaces (Cronstein, Levin *et al.* 1992). This suggests that adenosine could have a biphasic role. By binding to A_1 receptors, very low concentrations of adenosine increase adherence and thus may promote an inflammatory response; whereas, higher concentrations of adenosine (*e.g.* as found in injured tissues) by binding to PMN A_{2A} adenosine receptors can decrease adherence and thus may promote resolution of an inflammatory response (Cronstein, Daguma *et al.* 1990).

Heterotypic adherence of leukocytes with vascular endothelium is the first step in inflammatory cell transmigration from the blood to the surrounding tissue. Unstimulated PMN adherence to endothelium is not affected by adenosine. Treatment of human PMN with fMLP doubles their adherence to human umbilical cord endothelial cells (HUVECs). This stimulated adherence is decreased 60% by the adenosine agonist 2-CA (10μM) (Cronstein, Levin *et al.* 1986). As with adherence to matrix proteins, NECA and CPA display opposite effects on PMN adherence to endothelium. NECA decreases fMLP-stimulated PMN adherence; whereas; the selective A_1 agonist CPA increases PMN adherence (Cronstein, Levin *et al.* 1992).

Surface molecules that affect adhesion between cells include selectins, integrins, immunoglobulin type molecules and leucine-rich glycoproteins. Selectins expressed on both PMN and endothelial cells are key for initial PMN rolling. Following rolling stimulated PMN can become more adherent and spread on endothelium via integrin binding. Both the amount of PMN integrin expression and the activation state of the integrin molecules affects binding (Vedder and Harlan 1988; Schleiffenbaum, Moser *et al.* 1989). Adenosine acting via A_2 receptors decreases fMLP-stimulated human PMN Mac-1 (CD11b/CD18) expression (Wollner, Wollner *et al.* 1993) and thus may diminish adherence. L-selectin on PMN appears to be the target of adenosine action, and the effect is from a change in function of the selectins rather than their expression. During flow, 2-CA inhibits rolling and increases the number of stationary PMN. In an *in vivo* model, adenosine inhibits PMN rolling in rat mesentery (Asako, Wolf *et al.* 1993). 2-CA does not affect the integrin-dependent act of PMN migrating through an *in vitro* endothelial monolayer (Firestein, Bullough *et al.* 1995).

In the early phases of myocardial infarction there is heightened sympathetic activity. This results in increased catecholamine levels in the injured tissues along with increased adenosine concentrations. Bazzoni *et al.* observed that epinephrine plus adenosine have an additive effect on decreasing fMLP-stimulated human PMN adherence to endothelium. Thus, the activities of adenosine may be supported by the action of other endogenous inhibitors of PMN adherence (Bazzoni, Dejana *et al.* 1991).

The release of chemoattractants and cytokines regulate leukocyte-endothelial interaction and recruitment of leukocytes to areas of inflammation. Adenosine affects endothelial cytokine production. Adenosine (100μM) reduces IL-6 and IL-8 production from HUVEC cells maximally stimulated with IL-1, TNF or endotoxin (Bouma, Vandenwildenberg *et al.* 1996). Macrophages produce cytokines such as TNF and IL-1 that promote the recruitment of PMN by stimulating endothelial cells to express adherence factors, and inducing PMN to express the ligands for endothelial adhesion factors. Bouma *et al.* (Bouma, Vandenwildenberg *et al.* 1996) have observed that adenosine (250μM) decreases endothelial E-selectin expression when the endothelial cells are stimulated with TNF or IL-1ß. Intercellular adhesion molecule-1 [ICAM-1] and vascular cell adhesive molecule-1 [VCAM-1]) on the surface of endothelial cells act as ligands for integrin binding. Adenosine inhibits expression of VCAM-1 in TNF-stimulated HUVECs, but not expression of ICAM-1. Hence, adenosine by modulating endothelial function can affect leukocyte recruitment.

<u>D. Neutrophil migration</u>

After activated PMN adhere to endothelium, they migrate out of the vascular lumen between endothelial cells into the underlying tissue following gradients of chemoattractants. The first studies showing that adenosine could affect PMN migration came from experiments with rabbit cells (Garcia-Castro, Mato *et al.* 1983). They showed that although adenosine does not affect PMN chemotaxis, it does counteract migration inhibition by 3-deazaadenosine. Rose *et al.* (Rose, Hirshhorn *et al.* 1988) observed that adenosine (1nM-10μM) and the adenosine receptor agonists NECA and PIA (pM) increase human PMN chemotaxis without effect on PMN non-directed migration. This phenomenon is mediated by adenosine binding to PMN A_1 adenosine receptors (Cronstein, Daguma *et al.* 1990). TNF decreases PMN non-directed and directed migration (Shalaby, Palladino Jr. *et al.* 1987). We observed (Sullivan, Linden *et al.* 1990) that adenosine and adenosine receptor agonists (A_2> A_1 selective receptor agonists) restore TNF-inhibited PMN non-directed migration and chemotaxis (**Figure 2**). Although dibutryl cAMP (0.1-10μM; dbcAMP) restores TNF-inhibited chemotaxis, we were unable to show a clear correlation between adenosine activity and cAMP in PMN (Sullivan, Linden *et al.* 1990). A confounding problem in these experiments was the formation of adenosine from the breakdown of dbcAMP. In summary, adenosine (pM) stimulates PMN migration via binding to A_1 adenosine receptors, and at higher concentrations can counteract the inhibitory effects of 3-deazaadenosine and TNF on PMN migration.

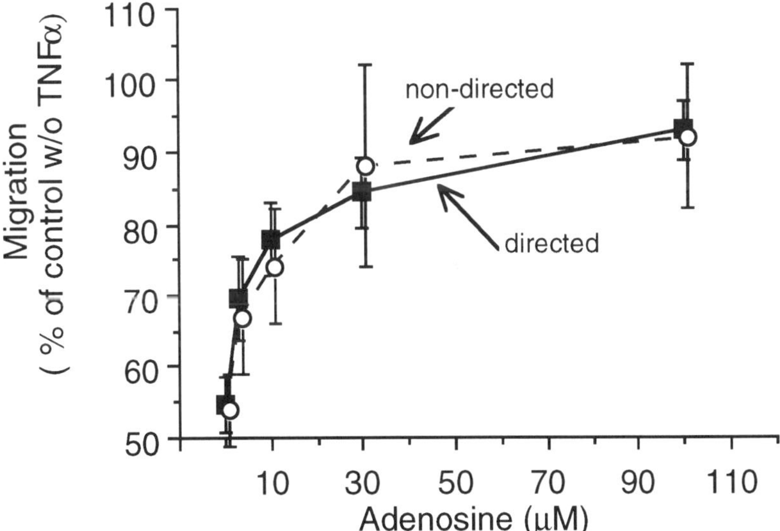

Figure 2: Adenosine restores TNF-inhibited non-directed and directed migration to fMLP (p<0.050). PMN (5 X 10^6/ml) were incubated ±adenosine for 30min at 37°C and ±TNF (10U/ml) and then non-directed migration and directed migration to fMLP (100nM) under agarose were determined.   Control non-directed migration=0.78±0.09mm and control directed migration=2.76±0.03mm. Mean ±

SEM of 7 experiments. Reprinted from *Journal of Immunology* 145 (5) Sullivan, GW, Linden, J, Hewlett, EL, Carper, HT, Hylton, JB and GL Mandell, Adenosine and related compounds counteract tumor necrosis factor-α inhibition of neutrophil migration: Implication of a novel cyclic AMP-independent action on the cell surface, p. 1537-1544, Copyright (1990), with kind permission from *The American Association of Immunologists.*

E. Exocytosis of lysosomal granule products

PMN contain lysosomal granules which are released by exocytosis into phagosomes and to the outside of the cell upon stimulation with a range of factors that include activated complement, bacteria and microbial products, immune complexes, and TNF. Release of proteolytic enzymes from adherent PMN granules can dislodge endothelial cells from an underlying biological surface in the absence of detectable cytotoxicity (Harlan, Killen *et al.* 1981). There are at least three types of lysosomal granules within human PMN, each containing a distinct repertoire of protein products including Mac-1, cytochrome b (a component of the NADPH oxidase), proteolytic enzymes, and microbicidal compounds that not only serve to kill phagocytized pathogens but also can injure host tissue when released to the outside of the cell through exocytosis and upon cell death. Degranulation through exocytosis occurs in stages. The more labile secondary (2°; specific) and tertiary granules are released prior to the primary (1°; azurophilic) granules. Cronstein *et al.* (Cronstein, Kramer *et al.* 1983) observed that adenosine (100μM) in the presence of the ADA inhibitor EHNA partially decreases (27%) fMLP-stimulated 1° granule release from human PMN. There is a similar effect on PMN degranulation with the A_1/A_2 adenosine receptor agonist NECA (50nM-10μM) (Cronstein, Kramer *et al.* 1988). Grinstein and Furuya (Grinstein and Furuya 1986) failed to see a significant effect of adenosine (10μM) on PMN granule exocytosis in response to fMLP. Using the 1° granule marker elastase, Zhang and Fredholm (Zhang and Fredholm 1994) observed a decrease in fMLP-stimulated release of 1° granules with NECA and adenosine (IC_{50}=14 nM and 64nM, respectively). Iannone *et al.* (Iannone, Zimmerman *et al.* 1987) observed that adenosine and NECA decrease 1° granule release in response to fMLP (IC_{50} =1-10nM). In whole blood, the adenosine receptor agonists 2-CA and N^6-(3-iodobenzyl)-adenosine-5'-N-methyluronamide (IB-MECA; IC_{50} =1-10μM) decrease endotoxin- and TNF -stimulated 1° granule release (Bouma, Trudi *et al.* 1997).

A_2 selective receptor agonists are more active than A_1 agonists at inhibiting stimulated degranulation. When adenosine or an A_2 adenosine receptor agonist is combined with a PDE inhibitor at concentrations that inhibit stimulated degranulation, PMN [cAMP] is increased (Iannone, Zimmerman *et al.* 1987; Schmeichel and Thomas 1987; Cronstein, Kramer *et al.* 1988). The effects of methylxanthine PDE inhibitors are biphasic. At <1μM they enhance 1° granule release stimulated by fMLP, but at higher concentrations they inhibit degranulation. A possible explanation for this is that at lower concentrations, methylxanthines are non-selective adenosine receptor antagonists that block endogenous adenosine binding to A_2 adenosine receptors and thus promote degranulation, but at higher concentrations methylxanthines inhibit PDE activity increasing [cAMP] resulting in a net inhibition of degranulation. Collectively, these studies suggest that adenosine and adenosine A_2 receptor agonists decrease human PMN 1° granule release, possibly by altering PMN [cAMP].

There are more recent data (Richter 1992) indicating that adenosine can also affect human PMN 2° granule release. Adenosine and 2-CA decrease adherent PMN degranulation of lactoferrin (2° granule marker) when the cells are stimulated with fMLP (IC_{50}=1μM) or TNF (IC_{50}=100μM). Thus, 2° granule exocytosis is sensitive to adenosine regulation in adherent PMN.

Nucleotides are released from stimulated platelets and from injured tissue. In contrast to adenosine's effect on PMN degranulation, ATP (10-60μM) stimulates human PMN exocytosis of vitamin B_{12} binding protein (2° granule marker) and to a lesser degree lysozyme (1° and 2° granule marker) but not ß-glucuronidase (1° granule marker) (Balazovich and Boxer 1990). In addition, ATP stimulates expression of Mac-1 which is stored within gelatinase positive granules (Freyer, Boxer *et al.* 1988). Hence, released nucleotides have a pro-inflammatory action stimulating degranulation of the more labile 2° and tertiary PMN granules. Conversely, adenosine formed by ectonucleotidase breakdown of adenine nucleotides inhibits degranulation of the less labile 1° granules.

Mouse peritoneal macrophages contain ß-galactosidase within granules that is released by zymosan particle ingestion. Zymosan-induced secretion can be decreased by treatment with adenosine, ATP, ADP, AMP and 3-deazadenosine. Adenosine activity is not blocked by the competitive adenosine receptor antagonist theophylline indicating that the site of activity may be intracellular or that the activity is mediated by a theophylline insensitive receptor such as A_3 in rodents (Riches, Watkins *et al.* 1985).

F. Production and scavenging of reactive oxidative species

In 1982, it was discovered that adenosine can inhibit mouse macrophage oxidative activity (Pike and Snyderman 1982). In guinea-pig macrophages adenosine (μM) in the presence of ADA-inhibition and especially in the presence of L-homocysteine decreases superoxide release in response to fMLP (Pike and Snyderman 1982). Adenosine also inhibits the production of reactive oxygen species (ROS) from human PMN (Cronstein, Kramer *et al.* 1983). PMN in suspension display a rapid oxidative burst in response to both soluble (*e.g.* fMLP, complement components, and cytokines), and particulate stimuli (*e.g.* immune complex, bacteria, and yeast). ROS are clearly essential for host defense from infection, but can also cause damage to host tissues when released from PMN.

Of the ROS, hydrogen peroxide appears particularly key in the destruction of vascular endothelium (Sacks, Moldow *et al.* 1978; Weiss, Young *et al.* 1981; Ager and Gordon 1984). Most reports agree that adenosine readily inhibits PMN release of ROS stimulated by fMLP (IC_{50} =~100nM) (Cronstein, Rosenstein *et al.* 1985; Roberts, Newby *et al.* 1985; Fredholm, Zhang *et al.* 1996). The effects of adenosine are not blocked by the adenosine uptake inhibitor dipyridamole indicating that adenosine doesn't need to enter the cell to have its effect (Cronstein, Kramer *et al.* 1983; Roberts, Newby *et al.* 1985), and adenosine activity is abolished by added ADA (Cronstein, Kramer *et al.* 1983). Inhibition of endogenous ADA enhances the oxidative suppressive activity of adenosine (Cronstein, Kramer *et al.* 1983). Adenosine inhibition of oxidative activity is blocked by non-selective adenosine receptor competitive antagonists (Cronstein, Rosenstein *et al.* 1985; Roberts, Newby *et al.* 1985; Schrier and Imre 1986; Thiel and Bardenheuer 1992), and the activity is readily reversible if ADA is added just prior to stimulation (Cronstein, Kramer *et al.* 1983). Adenosine is not a scavenger of oxidative products (Cronstein, Rosenstein *et al.* 1985). Order of potency data (Cronstein, Rosenstein *et al.* 1985; Iannone, Zimmerman *et al.* 1987; Dianzani, Brunelleschi *et al.* 1994; Fredholm, Zhang *et al.* 1996) with analogs of adenosine indicate that adenosine decreases PMN-oxidative activity by binding to A_{2A} receptors. Adenosine activity is not antagonized by selective A_1 adenosine receptor antagonists, but is antagonized by adenosine A_{2A} selective receptor antagonists (Fredholm, Zhang *et al.* 1996). In summary, adenosine in a reversible manner decreases stimulated PMN oxidative activity by specific binding to adenosine A_{2A} receptors on the cell surface.

Vasculitis can be initiated by immune complex stimulation of PMN. Adenosine, NECA and 2-CA do not decrease human PMN hydrogen peroxide release stimulated by immune complexes (Cronstein, Kubersky *et al.* 1987), and an adenosine A_1 receptor agonist (CPA) actually increases superoxide release in response to stimulation with immune complexes (Salmon and Cronstein 1990). Thus, A_1 and A_{2A} adenosine receptors have different actions on PMN function, and adenosine has divergent effects on the response with different physiological stimuli (Cronstein, Daguma *et al.* 1990; Salmon and Cronstein 1990).

Although TNF is not a strong stimulus for suspended PMN, it will prime suspended PMN for an increased oxidative burst to a second stimulus (Shalaby, Palladino Jr. *et al.* 1987), and will cause a slow marked prolonged activation of PMN adhering to a biological surface (Nathan 1987). It has been observed that adenosine decreases TNF-primed fMLP-stimulated suspended human PMN oxidative activity. The effect of adenosine is not on the priming step (Stewart and Harris 1993) but rather on the augmented response to the second stimulus (Barnes, Mandell *et al.* 1995). Thus, in suspended PMN, adenosine does not affect priming by TNF, but can inhibit the enhanced oxidative response to a second stimulus following TNF-priming.

Adenosine completely inhibits TNF-stimulated (IC_{50}=~500nM) or fMLP-stimulated (IC_{50}=~25nM) adherent human PMN oxidative activity when the PMN are adhered to a serum coated surface. Adenosine acts in the middle of a 45 min lag period prior to the appearance of oxidative activity with little effect before or after this period. The effect of adenosine is less on cell adherence, than on spreading of the adhered cells, and is dependent on the cytoskeleton (De La Harpe and Nathan 1989; Nathan and Sanchez 1990). Although adenosine receptor agonists have minimal effects on f-actin formation in response to stimulation (Sullivan, Linden *et al.* 1990; Cronstein and Haines 1992), adenosine receptor occupancy prevents the reorganization of the cytoskeleton stimulated by TNF (Nathan and Sanchez 1990).

Occupancy of adenosine receptors can affect intracellular [cAMP], and increasing cyclic AMP with adenylyl cyclase agonists and PDE inhibitors results in depressed PMN oxidative activity (Smolen, Korchak *et al.* 1980; Iannone, Zimmerman *et al.* 1987; Schmeichel and Thomas 1987; Bengis-Garber and Gruener 1996; Carletto, Biasi *et al.* 1997). Some studies fail to indicate a connection between adenosine-stimulated cAMP production and PMN oxidative activity (Cronstein, Kramer *et al.* 1988; Cronstein, Haines *et al.* 1992; Spisani, Pareschi *et al.* 1996), but others support a role for cAMP in adenosine-reduced PMN oxidative activity (Nathan 1987; Nielson and Vestal 1989; Mueller, Montoya *et al.* 1992; Sullivan, Carper *et al.* 1995; Sullivan, Luong *et al.* 1995). It has been observed that the PKA inhibitor *N*-(2-aminoethyl)-5-isoquinolinesulfonamide (H-9) and a peptide PKA inhibitor significantly reverse adenosine inhibition of PMN superoxide production (Mueller, Montoya *et al.* 1992). Nielson and Vestal observed that adenosine at physiologically achievable concentrations significantly increases PMN [cAMP] in resting cells and in PMN stimulated with fMLP (Nielson and Vestal 1989). In suspended PMN, we observed a synergistic effect in decreasing TNF-primed fMLP-stimulated superoxide production with endogenous concentrations of adenosine or selective A_{2A} adenosine receptor agonists in the presence of PDE inhibitors (Sullivan, Carper *et al.* 1995) **(Figure 3)**

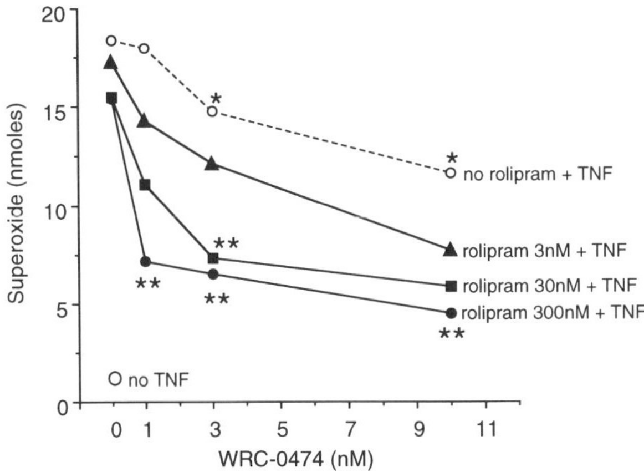

Figure 3: Synergistic effect of rolipram and the adenosine A_{2A} agonist 2-(N'-[2-isopropylethylidene]hydrazino)adenosine (WRC-0474) on TNF-primed fMLP-stimulated PMN superoxide release. PMN (1 X 10^6/ml) were incubated 30min at 37°C with ADA (1U/ml), ±rolipram (100nM), ±WRC-0474 (1nM-10nM), and ±TNF (10U/ml) and then stimulated 10min with fMLP (100 nM). WRC-0474 decreases the PMN oxidative burst (*$p<0.05$). The decrease in the PMN oxidative burst with rolipram and WRC-0474 is synergistic (**$p<0.05$). Mean ± SEM of 3 experiments. Reprinted from *Int. J. Immunopharmac.* 17(10), Sullivan, GW, Carper, HT and GL Mandell, The specific type IV phosphodiesterase inhibitor rolipram combined with adenosine reduces tumor necrosis factor-α-primed neutrophil oxidative activity, p. 793-803 Copyright(1995), with kind permission from Elsevier Science, Ltd, The Boulevard, Langford Lane, Kidlington 0X5 1GB, UK

In adherent PMN, TNF halves PMN [cAMP]. Cyclic AMP decreases TNF-induced PMN ROS release, spreading and microfilament reorganization in a reversible and non-toxic manner (Nathan and Sanchez 1990). We observed a synergistic effect between the A_{2A} adenosine receptor agonist CGS 21680 (30nM) and the type IV PDE inhibitor rolipram to decrease adherent PMN superoxide release stimulated by TNF (Sullivan, Carper *et al.* 1995) (**Figure 1B**). Thus, the data in primed and stimulated adherent PMN argue for a role of cAMP in adenosine diminished oxidative activity.

In vivo, adenosine decreases PMN superoxide release. Infusion of adenosine into endotoxin-exposed pigs has no effect on intracellular production of ROS. In contrast, adenosine markedly reduces extracellular release of superoxide (Thiel, Holzer *et al.* 1997) Thus, adenosine may differentially affect intracellular ROS which are key to killing ingested organisms and released ROS that have the potential to harm tissues.

In addition, adenosine alters ROS scavenger activity within vascular tissue. Adenosine by a pertussis toxin sensitive activity that is inhibitable with theophylline induces an increase in ROS scavenger enzyme activity (superoxide dismutase, catalase and glutathione peroxidase) in endothelium, smooth muscle, and myocytes. Conversely, ADA reduces scavenger activity (Maggirwar, Dhanraj *et al.* 1994). Adenosine by decreasing ROS production and stimulating scavengers could lessen vascular injury.

As with degranulation, adenine nucleotides have opposite effects on ROS release compared to adenosine. Kuhns *et al.* (Kuhns, Wright *et al.* 1988) observed that ATP at concentrations equal to those found at sites of thrombus formation prime human PMN oxidative activity to a second stimulus, and with immune complex stimulation, human PMN display enhanced oxidative activity with ATP, ADP, AMP and adenosine (Ward, Cunningham *et al.* 1988; Ward, Cunningham *et al.* 1988; Walker, Cunningham *et al.* 1989; Axtell, Sanborg *et al.* 1990; Moon, van Der Zee *et al.* 1990). The observation that ADP counteracts adenosine-reduced adherent PMN oxidative activity, indicates that the balance between ADP and adenosine may control adherent PMN oxidative activity (De La Harpe and Nathan 1989). It is possible that activated platelets could cause either a decrease or an increase in PMN ROS release depending on the stimulus and the relative proportions of adenosine and nucleotides present. Platelets decrease PMN ROS release and this activity is eliminated with ADA suggesting that adenosine is the active agent (Moon, van Der Zee *et al.* 1990). McGarrity *et al.* extracted a fraction from platelets and collected a supernatant from thrombin activated platelets that decrease by an ADA sensitive mechanism PMN oxidative activity (McGarrity, Stephenson *et al.* 1988).

Decreasing PMN superoxide release by adenosine can also indirectly affect vascular integrity and function by affecting nitric oxide (NO) metabolism. NO is produced both by constitutive nitric oxide synthase (cNOS) in vascular endothelium and through induction of nitric oxide synthase (iNOS) within phagocytes and vascular tissues. NO stimulates vasorelaxation, and inhibits smooth muscle growth. Superoxide produced by phagocytes acts as a NO scavenger by reacting with NO to produce peroxynitrite. Peroxynitrite and its decomposition product hydroxyl radical are toxic to vascular tissues. Endotoxin and ATP act synergistically in the murine macrophage line Raw 264.7 to stimulate iNOS activity (Tonetti, Sturla *et al.* 1994). Peroxynitrite could be a contributing factor in development of the inflammatory response in atherosclerosis and ischemia/reperfusion injury where the presence of activated phagocytes release excessive superoxide and NO (Muijsers, Folkerts *et al.* 1997). By decreasing superoxide production adenosine both increases the availability of the vasorelaxing agent NO to the tissues and also prevents tissue damage by reducing peroxynitrite formation.

III. CONCLUSIONS AND FUTURE DIRECTIONS

Recent research has been directed at utilizing the endogenous increase in adenosine that occurs in injured tissues to quell the inflammatory response (Gruber, Hoffer *et al.* 1989; Cronstein, Eberle *et al.* 1991; Cottam, Wasson *et al.* 1993; Cronstein, Naime *et al.* 1993; Cronstein, Naime *et al.* 1994; Firestein, Boyle *et al.* 1994; Cronstein 1995; Cronstein, Naime *et al.* 1995; Rosengren, Bong *et al.* 1995; Gadangi, Longaker *et al.* 1996). The metabolite 5-amino-4-imidazole carboxamide (AICA) riboside enhances adenosine accumulation during ATP catabolism. Gruber *et al.* (Gruber, Hoffer *et al.* 1989) observed that treatment of ischemic canine myocardium with AICA riboside results in less recruitment of indium-labeled PMN. It is also possible to increase [adenosine] in injured tissues by inhibiting 5-aminoimidazole-4-carboxamide ribonucleotide (AICAR) transformylase with methotrexate or sulfasalazine. In a murine air pouch model, methotrexate decreases PMN accumulation stimulated by carrageenan. This effect is reversed by the adenosine A_{2A} receptor antagonist DMPX, but not by the A_1 receptor antagonist DPCPX (Cronstein, Naime *et al.* 1993). Similar results are observed with sulfasalazine (Gadangi, Longaker *et al.* 1996). Salicylates by promoting ATP hydrolysis increases tissue concentrations of adenosine. Salicylate treatment decreases fMLP-stimulated PMN adherence to HUVECs by a mechanism that is reversed by ADA (Cronstein, Van de Stouwe *et al.* 1994).

Inhibition of adenosine kinase results in a decrease in nucleotide synthesis and increased [adenosine] (Firestein, Boyle *et al.* 1994). The adenosine kinase inhibitor 4-amino-1-(5-amino-5-deoxy-1-ß-D-ribofuranosyl)-3-bromo-pyrazolo[3,4-D]pyrimidine (GP-1-515) decreases PMN adhesion to endothelium (Rosengren, Bong *et al.* 1995) and diminishes fMLP, and IL-8-stimulated PMN adhesion to human aortic endothelial cells (Firestein, Bullough *et al.* 1995). In addition, treatment with GP-1-515 results in decreased PMN adherence to myocytes. The adenosine receptors that affect adherence appear to be on the PMN (Bullough, Magill *et al.* 1995). Hence, by increasing the accumulation of adenosine at inflammatory foci it may be possible to control the course cardiovascular diseases mediated by inflammation.

By altering the activity at specific adenosine receptors with selective analogs, it may be possible to modify the inflammatory cascade in particular tissues. Because of divergent actions at different adenosine receptor types, receptor specificity should be considered when developing adenosine analogs for therapeutic use. Since adenosine receptors are ubiquitous, alteration of leukocyte function has to be studied in the context of potential systemic responses. In ischemia/reperfusion models manipulation of specific adenosine receptors affects the progression of injury (Olafsson, Forman *et al.* 1987; Pitarys II, Virmani *et al.* 1991; Nolte 1992; Schlack, Schäfer *et al.* 1993; Neely and Keith 1995; Zhao, Sato *et al.* 1996; Jordan, Zhao *et al.* 1997). Further research should concentrate on developing selective adenosine receptor agonists and antagonists that localize and act within targeted tissues.

IV. REFERENCES

Ager A and Gordon JL. Differential effects of hydrogen peroxide on indices of endothelial cell function. J. Exp. Med.; 1984; 159: 592-603.

Asako H, Wolf RE and Granger DN. Leukocyte adherence in rat mesenteric venules: Effects of adenosine and methotrexate. Gastroenterology; 1993; 104: 31-37.

Axtell RA, Sanborg RR, Smolen JE, Ward PA and Boxer LA. Exposure of human neutrophils to exogenous nucleotides causes elevation in intracellular calcium, transmembrane calcium fluxes, and an alteration of a cytosolic factor resulting in enhanced superoxide production in response to FMLP and arachidonic acid. Blood; 1990; 75: 1324-1332.

Balazovich KJ and Boxer LA. Extracellular adenosine nucleotides stimulate protein kinase C activity and human neutrophil activation. J. Immunol.; 1990; 144: 631-637.

Barnes CR, Mandell GL, Carper HT, Luong S and Sullivan GW. Adenosine modulation of tumor necrosis factor-α-induced neutrophil activation. Biochem. Pharmacol.; 1995; 50: 1851-1857.

Bazzoni G, Dejana E and Del Maschio A. Adrenergic modulation of human polymorphonuclear leukocyte activation: Potentiating effect of adenosine. Blood; 1991; 77: 2042-2048.

Bengis-Garber C and Gruener N. Protein kinase A downregulates the phosphorylation of p47 phox in human neutrophils: A possible pathway for inhibition of the respiratory burst. Cell. Signalling; 1996; 8: 291-296.

Bouma MG, Stad RK, Van den Wildenberg FAJM and Buurman WA. Differential regulatory effects of adenosine on cytokine release by activated human monocytes. J. Immunol.; 1994; 153: 4159-4168.

Bouma MG, Trudi MMA, van den Wildenberg FAJM and Buurman WA. Adenosine inhibits neutrophil degranulation in activated human whole blood: Involvement of adenosine A2 and A3 receptors. J. Immunol.; 1997; 158: 5400-5408.

Bouma MG, Vandenwildenberg F and Buurman WA. Adenosine inhibits cytokine release and expression of adhesion molecules by activated human endothelial cells. Amer. J. Physiol.; 1996; 39: C 522-C 529.

Bullough DA, Magill MJ, Firestein GS and Mullane KM. Adenosine activates A2 receptors to inhibit neutrophil adhesion and injury to isolated cardiac myocytes. J. Immunol.; 1995; 155: 2579-2586.

Carletto A, Biasi D, Bambara LM, Caramaschi P, Bonazzi ML, Lussignoli S, Andrioli G and Bellavite P. Studies of skin-window exudate human neutrophils - increased resistance to pentoxifylline of the respiratory burst in primed cells. Inflammation; 1997; 21: 191-203.

Chatelain P, Latour L-G, Tran D, De Lorgeril M, Dupras G and Bourassa M. Neutrophil accumulation in experimental myocardial infarcts: relation with extent of injury and effect of reperfusion. Lab. Invest.; 1987; 75: 1083-1090.

Cottam HB, Wasson DB, Shih HC, Raychaudhuri A, Di Pasquale G and Carson DA. New adenosine kinase inhibitors with oral antiinflammatory activity: synthesis and biological evaluation. J. Med. Chem.; 1993; 36: 3424-3430.

Cronstein BN. The antirheumatic agents sulphasalazine and methotrexate share an anti-inflammatory mechanism. Brit. J. Rheumatol.; 1995; 34: 30-32.

Cronstein BN, Daguma L, Nichols D, Hutchison AJ and Williams M. The adenosine/neutrophil paradox resolved: Human neutrophils possess both A1 and A2 receptors that promote chemotaxis and inhibit O_2^- generation, respectively. J. Clin. Invest.; 1990; 85: 1150-1157.

Cronstein BN, Eberle MA, Gruber HE and Levin RI. Methotrexate inhibits neutrophil function by stimulating adenosine release from connective tissue cells. Proc. Natl. Acad. Sci. USA; 1991; 88: 2441-2445.

Cronstein BN and Haines KA. Stimulus-response uncoupling in the neutrophil: Adenosine A2-receptor occupancy inhibits the sustained, but not the early, events of stimulus transduction in human neutrophils by a mechanism independent of actin-filament formation. Biochem. J.; 1992; 281: 631-635.

Cronstein BN, Haines KA, Kolasinski S and Reibman J. Occupancy of Gas-linked receptors uncouples chemoattractant receptors from their stimulus-transduction mechanisms in the neutrophil. Blood; 1992; 80: 1052-1057.

Cronstein BN, Kramer SB, Rosenstein ED, Korchak HM, Weissmann G and Hirshhorn R. Occupancy of adenosine receptors raises cyclic AMP alone and in synergy with occupancy of chemoattractant receptors and inhibits membrane depolarization. Biochem. J.; 1988; 252: 709-715.

Cronstein BN, Kramer SB, Weissmann G and Hirschhorn R. Adenosine: A physiological modulator of superoxide anion generation by human neutrophils. J. Exp. Med.; 1983; 158: 1160-1177.

Cronstein BN, Kubersky SM, Weissmann G and Hirschhorn R. Engagement of adenosine receptors inhibits hydrogen peroxide (H_2O_2) release by activated human neutrophils. Clin. Immunol. Immunopathol.; 1987; 42: 76-85.

Cronstein BN, Levin RI, Belanoff J, Weissmann G and Hirschhorn R. Adenosine: An endogenous inhibitor of neutrophil-mediated injury to endothelial cells. J. Clin. Invest.; 1986; 78: 760-770.

Cronstein BN, Levin RI, Philips M, Hirschhorn R, Abramson SB and Weissmann G. Neutrophil adherence to endothelium is enhanced via adenosine A1 receptors and inhibited via adenosine A2 receptors. J. Immunol.; 1992; 148: 2201-2206.

Cronstein BN, Naime D and Firestein G. The antiinflammatory effects of an adenosine kinase inhibitor are mediated by adenosine. Arthritis Rheumat.; 1995; 38: 1040-1045.

Cronstein BN, Naime D and Ostad E. The antiinflammatory mechanism of methotrexate. Increased adenosine release at inflamed sites diminishes leukocyte accumulation in an in vivo model of inflammation. J. Clin. Invest.; 1993; 92: 2675-2682.

Cronstein BN, Naime D and Ostad E. The antiinflammatory effects of methotrexate are mediated by adenosine. Adv. Exp. Med. Biol.; 1994; 370: 411-416.

Cronstein BN, Rosenstein ED, Kramer SB, Weissmann G and Hirschhorn R. Adenosine; A physiologic modulator of superoxide anion generation by human neutrophils, adenosine acts via an A2 receptor on human neutrophils. J. Immunol.; 1985; 135: 1366-1371.

Cronstein BN, Van de Stouwe M, Druska L, Levin RI and Weissmann G. Nonsteroidal antiinflammatory agents inhibit stimulated neutrophil adhesion to endothelium: Adenosine dependent and independent mechanisms. Inflammation; 1994; 18: 323-335.

De La Harpe J and Nathan CF. Adenosine regulates the respiratory burst of cytokine-triggered human neutrophils adherent to biological surfaces. J. Immunol.; 1989; 143: 596-602.

Dianzani C, Brunelleschi S, Viano I and Fantozzi R. Adenosine modulation of primed human neutrophils. Eur. J. Pharmacol.; 1994; 263: 223-226.

Eigler A, Greten TF, Sinha B, Haslberger C, Sullivan GW and Endres S. Endogenous adenosine curtails lipopolysaccharide-stimulated tumour necrosis factor synthesis. Scand. J. Immunol.; 1997; 45: 132-139.

Engler R. Consequences of activation and adenosine-mediated inhibition of granulocytes during myocardial ischemia. Fed. Proc.; 1987; 46: 2407-2412.

Eppell BA, Newell AM and Brown EJ. Adenosine receptors are expressed during differentiation of monocytes to macrophages in vitro: Implications for regulation of phagocytosis. J. Immunol.; 1989; 143: 4141-4145.

Firestein GS, Boyle D, Bullough DA, Gruber HE, Sajjadi FG, Montag A, Sambol B and Mullane KM. Protective effect of an adenosine kinase inhibitor in septic shock. J. Immunol.; 1994; 152: 5853-5859.

Firestein GS, Bullough DA, Erion MD, Jimenez R, Ramirez-Weinhouse M, Barankiewicz J, Smith CW, Gruber HE and Mullane KM. Inhibition of neutrophil adhesion by adenosine and an adenosine kinase inhibitor. The role of selectins. J. Immunol.; 1995; 154: 326-334.

Fischer D, B. VDWM, Snyderman R and Kelley WN. A role for adenosine deaminase in human monocyte maturation. J. Clin. Invest.; 1976; 58: 399-407.

Frangogiannis NG, Youker KA and Entman ML. The role of the neutrophil in myocardial ischemia and reperfusion. Myocardial Ischemia: Mechanisms. Reperfusion. Protection. Basal, Birkhäuser Verlag. 263-284. 1996.

Fredholm BB, Zhang Y and van der Ploeg I. Adenosine A2A receptors mediate the inhibitory effect of adenosine on formyl-met-leu-phe-stimulated respiratory burst in neutrophil leucocytes. Naunyn-Schmiedeberg's Arch. Pharmacol.; 1996; 354: 262-267.

Freyer DR, Boxer LA, Axtell RA and Todd III RF. Stimulation of human neutrophil adhesive properties by adenine nucleotides. J. Immunol.; 1988; 141: 580-586.

Gadangi P, Longaker M, Naime D, Levin RI, Recht PA, Montesinos MC, Buckley MT, Carlin G and Cronstein BN. The anti-inflammatory mechanism of sulfasalazine is related to adenosine release at inflamed sites. J. Immunol.; 1996; 156: 1937-1941.

Garcia-Castro I, Mato JM, Vasanthakumar G, Wiesmann WP, Schiffmann E and Chiang PK. Paradoxical effects of adenosine on neutrophil chemotaxis. J. Biol. Chem.; 1983; 258: 4345-4349.

Gasmi L, McLennan AG and Edwards SW. The diadenosine polyphosphates Ap3A and Ap4A and adenosine triphosphate interact with granulocyte-macrophage colony-stimulating factor to delay neutrophil apoptosis: implications for neutrophil: platelet interactions during inflammation. Blood; 1996; 87: 3442-3449.

Girardi LN and Coselli JS. Inflammatory aneurysm of the ascending aorta and aortic arch. Ann. Thorac. Surg.; 1997; 64: 251-253.

Greisler HP. Macrophages. The Basic Science of Vascular Disease. Armonk, Futura Publishing Co. 227-244. 1997.

Grinstein S and Furuya W. Cytoplasmic pH regulation in activated human neutrophils: Effects of adenosine and *pertussis toxin* on Na^+/H^+ exchange and metabolic acidification. Biochim. Biophys. Acta; 1986; 889: 301-309.

Gruber HE, Hoffer ME, McAllister DR, Laikind PK, Lane TA, Schmid-Schoenbein GW and Engler RL. Increased adenosine concentration in blood from ischemic myocardium by AICA riboside: Effects on flow, granulocytes, and injury. Circulation; 1989; 80: 1400-1411.

Harlan JM, Killen PD, Harker LA, Striker GE and Wright DG. Neutrophil-mediated endothelial injury in vitro. J. Clin. Invest.; 1981; 68: 1394-1403.

Hartmann JR, Robinson JA and Gunnar RM. Chemotactic activity in the coronary sinus after experimental myocardial infarction: Effects of pharmacologic interventions on ischemic injury. Am. J. Cardiol.; 1977; 40: 550-555.

Hasday JD and Sitrin RG. Adenosine receptors on rabbit alveolar macrophages: Binding characteristics and effects on cellular function. J. Lab. Clin. Med.; 1987; 110: 264-272.

Hasko G, Szabo C, Nemeth Z, Kvetan V, Pastores SM and Vizi ES. Adenosine receptor agonists differentially regulate IL-10, TNF-a, and nitric oxide production in RAW 264.7 macrophages and in endotoxemic mice. J. Immunol.; 1996; 157: 4634-4640.

Iannone MA, Zimmerman TP, Reynolds-Vaughn R and Wolberg G. Effects of adenosine on human neutrophil function and cyclic AMP content. Topics and Perspectives in Adenosine Research. Berlin Heidelberg, Springer-Verlag. 286-298. 1987.

Jordan JE, Zhao Z-Q, Sato H, Taft S and Vinten-Johansen J. Adenosine A2 receptor activation attenuates reperfusion injury by inhibiting neutrophil accumulation, superoxide generation and coronary endothelial adherence. J. Pharmacol. Exp. Ther.; 1997; 280: 301-309.

Kuhns DB, Wright DG, Nath J, Kaplan SS and Basford RE. ATP induces transient elevations of $[Ca^{2+}]$ in human neutrophils and primes these cells for enhanced O_2^- generation. Lab. Invest.; 1988; 58: 448-453.

Lappin D and Whaley K. Adenosine A2 receptors on human monocytes modulate C2 production. Clin. Exp. Immunol.; 1984; 57: 454-460.

Le Moine O, Stordeur P, Schandene L, Marchant A, Degroote D, Goldman M and Deviere J. Adenosine enhances Il-10 secretion by human monocytes. J. Immunol.; 1996; 156: 4408-4414.

Maggirwar SB, Dhanraj DN, Somani SM and Ramkumar V. Adenosine acts as an endogenous activator of the cellular antioxidant defense system. Biochem. Biophys. Res. Commun.; 1994; 201: 508-515.

Martini C, Di Sacco S, Tacchi P, Bazzichi L, Soletti A, Bondi F, Ciompi ML and Lucacchini A. A2 adenosine receptors in neutrophils from health volunteers and patients with rheumatic disease. Purine and pyrimidine Metabolism in Man. New York, Plenum Press. 459-462. 1991.

McGarrity ST, Stephenson AH, Hyers TM and Webster RO. Inhibition of neutrophil superoxide anion generation by platelet products: Role of adenine nucleotides. J. Leukocyte Biol.; 1988; 44: 411-421.

McWhinney CD, Dudley MW, Bowlin TL, Peet NP, Schook L, Bradshaw M, De M, Borcherding DR and Edwards III CK. Activation of adenosine A3 receptors on macrophages inhibits tumor necrosis factor-α. Eur. J. Pharmacol.; 1996; 310: 209-216.

Moon DG, van Der Zee H, Weston LK, Gudewicz PW, Fenton II JW and Kaplan JE. Platelet modulation of neutrophil superoxide anion production. Thrombosis Haemostasis; 1990; 63: 91-96.

Mueller H, Montoya B and Sklar LA. Reversal of inhibitory pathways in neutrophils by protein kinase antagonists: A rational approach to the restoration of depressed cell function. J. Leukocyte Biol.; 1992; 52: 400-406.

Muijsers RBR, Folkerts G, Henricks PAJ, Sadeghi-Hashjin G and Nijkamp FP. Peroxynitrite: A two-faced metabolite of nitric oxide. Life Sci.; 1997; 60: 1833-1845.

Mullane KM, Read N, Salmon JA and Moncada S. Role of leukocytes in acute myocardial infarction in anesthctizcd dogs: Rclationship to myocardial salvage by anti-inflammatory drugs. J. Pharmacol. Exp. Ther.; 1984; 228: 510-522.

Najar HM, Ruhl S, Bru-Capdeville AC and Peters JH. Adenosine and its derivatives control human monocyte differeniation into highly accessory cells versus macrophages. J. Leukocyte Biol.; 1990; 47: 429-439.

Nathan CF. Neutrophil activation on biological surfaces. Massive secretion of hydrogen peroxide in response to products of macrophages and lymphocytes. J. Clin. Invest.; 1987; 80: 1550-1560.

Nathan CF and Sanchez E. Tumor necrosis factor and CD11/CD18 (β2) integrins act synergistically to lower cAMP in human neutrophils. J. Cell Biol.; 1990; 111: 2171-2181.

Neely CF and Keith IM. A1 adenosine receptor antagonists block ischemia-reperfusion injury of the lung. Am. J. Physiol.; 1995; 268: L1036-L1046.

Nielson CP and Vestal RE. Effects of adenosine on polymorphonuclear leucocyte function, cyclic 3' : 5'-adenosine monophosphate and intracellular calcium. Brit. J. Pharmacol.; 1989; 97: 882-888.

Nolte D. Reduction of postischemic leukocyte-endothelium interaction by adenosine via A2 receptor. Naunyn-Schmiedeberg's Arch. Pharmacol.; 1992; 346: 234-237.

Ohtani A, Kumazawa Y, Fujisawa H and Nishimura C. Inhibition of macrophage function by 2-chloroadenosine. J. Reticuloendothel. Soc.; 1982; 32: 189-200.

Olafsson B, Forman MB, Puett DW, Pou A, Cates CU, Friesinger GC and Virmani R. Reduction of reperfusion injury in the canine preparation by intraconary adenosine: importance of the endothelium and the no-reflow phenomenon. Lab. Invest; 1987; 76: 1135-1145.

Parmely MJ, Zhou W-W, Edwards III CK, Borcherding DR, Silverstein R and Morrison DC. Adenosine and a related carbocyclic nucleoside analogue selectively inhibit tumor necrosis factor-α production and protect mice against endotoxin challenge. J. Immunol.; 1993; 151: 389-396.

Pearson JD and Gordon JL. Vascular endothelial and smooth muscle cells in culture selectively release adenine nucleotides. Nature; 1979; 281: 384-386.

Pennell RC, Hollier LH, Lie JT, Bernatz PE, Joyce JW, Pairolero PC, Cherry KJ and Hallett JW. Inflammatory abdominal aortic aneurysms: a thirty-year review. J. Vasc. Surg.; 1985; 2: 859-869.

Pike MC and Snyderman R. Transmethylation reactions regulate affinity and functional activity of chemotactic factor receptors on macrophages. Cell; 1982; 28: 107-114.

Pitarys II CJ, Virmani R, Vildibill Jr HD, Jackson EK and Forman MB. Reduction of myocardial reperfusion injury by intravenous adenosine administered during the early reperfusion period. Circulation; 1991; 83: 237-247.

Reinstein LJ, Lichtman SN, Currin RT, Wang J, Thurman RG and Lemasters JJ. Suppression of lipopolysaccharide-stimulated release of tumor necrosis factor by adenosine: Evidence for A2 receptors on rat kupffer cells. Hepatology; 1994; 19: 1445-1452.

Riches DWH, Watkins JL, Henson PM and Stanworth DR. Regulation of macrophage lysosomal secretion by adenosine, adenosine phosphate esters, and related structural analogues of adenosine. J. Leukocyte Biol.; 1985; 37: 545-557.

Richter J. Effect of adenosine analogues and cAMP-raising agents on TNF-, GM-CSF, and chemotactic peptide-induced degranulation in single adherent neutrophils. J. Leukocyte Biol.; 1992; 51: 270-275.

Roberts PA, Newby AC, Hallett MB and Campbell AK. Inhibition by adenosine of reactive oxygen metabolite production by human polymorphonuclear leucocytes. Biochem. J.; 1985; 227: 669-674.

Romson JL, Hook BG, Kunkel SL, Abrams GD, Schork MA and Lucchesi BR. Reduction of the extent of ischemic myocardial injury by neutrophil depletion in the dog. Circulation; 1983; 67: 1016-1023.

Rose FR, Hirshhorn R, Weissmann G and Cronstein BN. Adenosine promotes neutrophil chemotaxis. J. Exp. Med.; 1988; 167: 1186-1194.

Rosengren S, Bong GW and Firestein GS. Anti-inflammatory effects of an adenosine kinase inhibitor. Decreased neutrophil accumulation and vascular leakage. J. Immunol.; 1995; 154: 5444-5451.

Ross R. The pathogenesis of atherosclerosis: a perspective for the 1990s. Nature; 1993; 362: 801-809.

Sacks T, Moldow CF, Craddock PR and Bowers TK. Oxygen radicals mediate endothelial cell damage by complement-stimulated granulocytes: An in vitro model of immune vascular damage. J. Clin. Invest.; 1978; 61: 1161-1167.

Sajjadi FG, Takabayashi K, Foster AC, Domingo RC and Firestein GS. Inhibition of TNF-α expression by adenosine: Role of A3 adenosine receptors. J. Immunol.; 1996; 156: 3435-3442.

Salmon JE, Brogle N, Brownlie C, Edberg JC, Kimberly RP, Chen B-X and Erlanger BF. Human mononuclear phagocytes express adenosine A1 receptors: A novel mechanism for differential regulation of Fcγ receptor function. J. Immunol.; 1993; 151: 2775-2785.

Salmon JE and Cronstein BN. Fcg receptor-mediated functions in neutrophils are modulated by adenosine receptor occupancy: A1 receptors are stimulatory and A2 receptors are inhibitory. J. Immunol.; 1990; 145: 2235-2240.

Salvatore CA, Jacobson MA, Taylor HE, Linden J and Johnson RG. Molecular cloning and characterization of the human A3 adenosine receptor. Proc. Natl. Acad. Sci. USA; 1993; 90: 10365-10369.

Schlack W, Schäfer S, Borchard U and Thämer V. Adenosine A2-receptor activation at reprfusion reduces infarct size and improves myocardial wall function in dog heart. J. Cardiovascular Pharmacol.; 1993; 22: 89-96.

Schleiffenbaum B, Moser R, Patarroyo M and Fehr J. The cell surface glycoprotein Mac-1 (CD11b/CD18) mediates neutrophil adhesion and modulates degranulation independently of its quantitative cell surface expression. J. Immunol.; 1989; 142: 3537-3545.

Schmeichel CJ and Thomas LL. Methylxanthine bronchodilators potentiate multiple human neutrophil functions. J. Immunol.; 1987; 138: 1896-1903.

Schrier DJ and Imre KM. The effects of adenosine agonists on human neutrophil function. J. Immunol.; 1986; 137: 3284-3289.

Shalaby MR, Palladino Jr. MA, Hirabayashi SE, Eessalu TE, Lewis GD, Shepard HM and Aggarwal BB. Receptor binding and activation of polymorphonuclear neutrophils by tumor necrosis factor-alpha. J. Leukocyte Biol.; 1987; 41: 196-204.

Sharma HS and Das DK. Role of cytokines in myocardial ischemia and reperfusion. Mediators of Inflammation; 1997; 6: 175-183.

Skubitz KM, Wickham NW and Hammerschmidt DE. Endogenous and exogenous adenosine inhibit granulocyte aggregation without altering the associated rise in intracellular calcium concentration. Blood; 1988; 72: 29-33.

Smolen JE, Korchak HM and Weissmann G. Increased levels of cyclic adenosine-3',5'-monophosphate in human polymorphonuclear leukocytes after surface stimulation. J. Clin. Invest.; 1980; 65: 1077-1085.

Spisani S, Pareschi MC, Buzzi M, Colamussi ML, Biondi C, Traniello S, Pagani Zecchini G, Paglialunga Paradisi M, Torrini I and Ferretti ME. Effect of cyclic AMP level reduction on human neutrophil responses to formylated peptides. Cell. Signalling; 1996; 8: 269-277.

Stewart AG and Harris T. Adenosine inhibits platelet-activating factor, but not tumour necrosis factor-α-induced priming of human neutrophils. Immunology; 1993; 78: 152-158.

Sullivan GW, Carper HT and Mandell GL. The specific type IV phosphodiesterase inhibitor rolipram combined with adenosine reduces tumor necrosis factor-α-primed neutrophil oxidative activity. Int. J. Immunopharmac.; 1995; 17: 793-803.

Sullivan GW, Linden J, Hewlett EL, Carper HT, Hylton JB and Mandell GL. Adenosine and related compounds counteract tumor necrosis factor-α inhibition of neutrophil migration: Implication of a novel cyclic AMP-independent action on the cell surface. J. Immunol.; 1990; 145: 1537-1544.

Sullivan GW, Luong LS, Carper HT, Barnes CR and Mandell GL. Methylxanthines with adenosine alter TNFα-primed PMN activation. Immunopharmacol.; 1995; 31: 19-29.

Thiel M and Bardenheuer H. Regulation of oxygen radical production of human polymorphonuclear leukocytes by adenosine: the role of calcium. Pflügers Arch.; 1992; 420: 522-528.

Thiel M, Holzer K, Kreimeier U, Moritz S, Peter K and Messmer K. Effects of adenosine on the functions of circulating polymorphonuclear leukocytes during hyperdynamic endotoxemia. Infect. Immun.; 1997; 65: 2136-2144.

Tonetti M, Sturla L, Bistolfi T, Benatti U and De Flora A. Extracellular ATP potentiates nitric oxide synthase expression induced by lipopolysaccharide in RAW 264.7 murine macrophages. Biochem. Biophys Res Commun.; 1994; 203: 430-435.

Vedder NB and Harlan JM. Increased surface expression of CD11b/CD18 (Mac-1) is not required for stimulated neutrophil adherence to cultured endothelium. J. Clin. Invest.; 1988; 81: 676-682.

Vraux VL, Chen YL, Masson I, DeSousa M, Giroud JP, Florentin I and Chauvelot-Moachon L. Inhibition of human monocyte TNF production by adenosine receptor agonists. Life Sci.; 1993; 52: 1917-1924.

Walker BAM, Cunningham TW, Freyer DR, Todd III RF, Johnson KJ and Ward PA. Regulation of superoxide responses of human neutrophils by adenine compounds. Lab. Invest.; 1989; 61: 515-521.

Walker BAM, Rocchini C, Boone RH, Ip S and Jacobson MA. Adenosine A2a receptor activation delays apoptosis in human neutrophils. J. Immunol.; 1997; 158: 2926-2931.

Walker DI, Bloor K, Williams G and Gillie I. Inflammatory aneurysms of the abdominal aorta. Brit. J. Surg.; 1972; 59: 609-614.

Ward PA, Cunningham TW, McCulloch KK and Johnson KJ. Regulatory effects of adenosine and adenine nucleotides on oxygen radical responses of neutrophils. Lab. Invest.; 1988; 58: 438-447.

Ward PA, Cunningham TW, Walker AM and Johnson KJ. Differing calcium requirements for regulatory effects of ATP, ATPγS and adenosine on 0_2^- responses of human neutrophils. Biochem. Biophys. Res. Commun.; 1988; 154: 746-751.

Weiss SJ, Young J, LoBuglio AF, Slivka A and Nimeh NF. Role of hydrogen peroxide in neutrophil-mediated destruction of cultured endothelial cells. J. Clin. Invest.; 1981; 68: 714-721.

Wollner A, Wollner S and Smith JB. Acting via A2 receptors, adenosine inhibits the upregulation of Mac-1 (Cd11b/CD18) expression on FMLP-stimulated neutrophils. Amer J Resp Cell Mol Biol 1993; 9:179-185.

Zhang Y and Fredholm BB. Propentofylline enhancement of the actions of adenosine on neutrophil leukocytes. Biochem. Pharmacol.; 1994; 48: 2025-2032.

Zhao Z-Q, Sato H, Williams MW, Fernandez AZ and Vinten-Johansen J. Adenosine A2-receptor activation inhibits neutrophil-mediated injury to coronary endothelium. Am. J. Physiol.; 1996; 271: H1456-H1464.

P2 PURINOCEPTORS AND REGULATION OF THE FUNCTION OF PLATELETS, ERYTHROCYTES AND MAST CELLS

Francesco Di Virgilio and Simonetta Falzoni, Department of Experimental and Diagnostic Medicine, Section of General Pathology, University of Ferrara, Ferrara, Italy.

I. SUMMARY

Platelets, erythrocytes and mast cells have been a focus of interest in purinergic receptor biology for many years. These cells have been invaluable for the development of new pharmacological agonists and antagonists (notably platelets), the molecular cloning of members of the P2Y family, the dissection of the signal transduction pathways (especially erythrocytes), and the identification recently of the elusive permeabilizing ATP receptor (mast cells). Platelets, erythrocytes and mast cells not only express P2 receptors but also release ATP, thus setting the stage for an autocrine/paracrine loop that might have a key role in the control of local blood flow and tissue responses to noxious agents.

II. INTRODUCTION

The number of different cell types whose functions are affected by extracellular nucleotides has increased enormously during recent years. Amidst this wealth of reports of new and exciting actions of ATP and other purinergic or pyrimidinergic effectors, platelets, erythocytes and mast cells, three cell types that had been instrumental in the early studies, have lost some of their original attractiveness, even though ATP analogues with inhibitory activity at the the platelet purinergic receptor have entered clinical trials as antithrombotic agents (Humphries, et al., 1995). This relative oblivion might be a side effect of the explosive development of our knowledge on the structure and function of purinergic P2 receptors from other tissues, compared to the comparatively slow progress of our understanding of the molecular structure of the receptor subtypes expressed in platelets, erythrocytes and mast cells.

However, this situation will change rapidly with the increased awareness of the crucial role that extracellular ATP and its metabolites play in the regulation of regional blood flow and in the modulation of the inflammatory reaction. ATP is released from endothelial cells as well as from erythrocytes during hypoxia and, thus, might trigger a vasodilatory response mediated by release of nitric oxide from endothelial cells, a response known to be triggered by P2Y receptors. Furthermore, at sites of endothelial cell injury, whcre collagcn is exposcd and platclet aggregation occurs, ATP (and ADP) secreted during the platelet release reaction may again act at P2X receptors of smooth muscle cells or P2Y receptors of endothelial cells to trigger vasoconstriction or vasodilation, respectively.

III. TEXT

Platelets

Early evidence for the presence of a purinergic receptor on platelet plasma membrane goes back to 1962, when Born reported that exogenous ADP could trigger platelet aggregation (Born, 1962). Rather surprisingly, ATP acted as a competitive antagonist, thus identifying a unique pharmacological profile among P2 receptors (Hourani and Hall, 1994). This receptor was then named P2T (Gordon, 1986). The issue of whether ADP stimulation of platelets is mediated by just one or more P2 receptor subtypes has not been resolved yet. The platelet receptor, P2T, is the only P2 receptor included in the original classification by Gordon that has not been identified at the molecular level, despite numerous attempts. In addition, the array of second messengers (or of transducing events) set in motion by ADP is very complex, suggesting the participation of a serpentine (seven membrane-spanning) receptor (P2Y-like) as well as a ligand-gated channel (P2X-like) (Hourani and Hall, 1994).

Very recently, Leon, et al. (1997) reported the cloning from a human placental cDNA library of a P2 receptor identical to the previously isolated human $P2Y_1$ but for the insertion of a serine residue in transmembrane domain 3. The pharmacological profile of this newly cloned P2Y receptor is superimposable to that of the human P2T receptor (see Table I): ADP and 2MeSATP (2-methylthioadenosine 5'-triphosphate) are agonists, while ATP and its analogues are either ineffective or antagonistic. Furthermore, PCR analysis yielded amplification products of the expected size from several megacariocytic cell lines and blood platelets. Thus, it is suggested that the elusive platelet P2T receptor is in fact a $P2Y_1$ receptor. An obvious implication of these studies is that the $P2Y_1$ receptor should be an ADP rather than an ATP receptor.

In the study by Leon, et al. (1997), the purity of commercially available ATP was checked by HPLC. When ATP solutions were carefully purified by HPLC, ADP caused an increase in the cytosolic Ca^{2+} concentration, $[Ca^{2+}]_i$, by inducing both release from intracellular stores and influx across the plasma membrane. This second mechanism for $[Ca^{2+}]_i$ increase does not seem to depend entirely on the well known activation of a delayed influx through a Ca^{2+} channel opened by store depletion (Fasolato, et al., 1994), but rather on the opening of a true ADP-gated plasma membrane cation channel.

A patch-clamp study performed in human platelets by MacKenzie and co-workers (1996) has revealed the presence of a cation channel activated by ATP, ADP and $\alpha\beta$-methylene-ATP (α,β-MeATP) , while UTP and AMP were ineffective. As also reported in other cell types (Baricordi, et al., 1996), in order to be able to reveal this current, a pretreatment with apyrase, a nucleotide hydrolysing enzyme, was necessary, suggesting that

TABLE I: Pharmacological profile of heterologously-expressed human $P2Y_1$ and native human P2T ($[Ca^{2+}]_i$ increases).

Agonist	'P2T' receptor	$P2Y_1$ receptor
2MeSADP	+(full) EC_{50}=41 nM	+(full) EC_{50}=49 nM
ADP	+(full) EC_{50}=800 nM	+(full) EC_{50}=204 nM
ADPαS	+(partial)	+(nd)
ADPβS	+(partial)	+(nd)
Antagonists		
Sp-ATPαS	+(competitive)	+(competitive)
ATP	+(competitive)	+(competitive)
βγMeATP	+(competitive)	+(competitive)
2ClATP	+(competitive)	?
2MeSATP	+(non-competitive)	?
Suramin	+	+
PPADS	+	+
Without effect		
αβMeATP	+	+
UTP	+	+
Adenosine	+	+

Abbreviations used only in this Table: 2MeSADP, 2-methylthio ADP; Sp-ATPαS, adenosine 5'-*O*-(1-thiotriphosphate) (Sp isomer); βγMeATP, β,γ-methylene-ATP; PPADS, pyridoxalphosphate-6-azophenyl-2',4'-disulfonic acid. For experimental details, see Leon et al., 1997.

---

the platelet P2X purinoceptor is chronically desensitized by release of endogenous ATP and ADP. The nucleotide-activated channel was permeable to both Na^+ and Ba^{2+}, thus presumably also to Ca^{2+}, and showed inactivation and desensitization, features compatible with an ATP-activated channel of the $P2X_1$ subtype. Altogether, these reports would reconcile platelets with the current paradigm common to most other cell types, i.e., expression of both P2Y and P2X receptors.

The well documented ability of ADP to to mobilize Ca^{2+} from intracellular stores (Hallam and Rink, 1995) fits well with the presence of a P2Y receptor, but rather intriguingly, literature reports on the ability of this nucleotide to trigger phospholipase C (PLC) activation and inositol 1,4,5-trisphosphate generation (IP3), are contradictory, leading some authors to conclude that ADP is at best a weak activator of PLC (Gachet and Cazenave, 1991; Hourani and Hall, 1994). This could imply that some other pathway independent of PLC activation is responsible for mobilization of the intracellular Ca^{2+}

pool. Alternative mechanisms yet to be investigated in platelets could be Ca^{2+}-induced Ca^{2+} release or cyclicADP ribose generation, both acting at the ryanodine receptor expressed by both excitable and nonexcitable cells.

Like most other aggregating agents, ADP causes inhibition of stimulated adenylate cyclase (Haslam and Rosson, 1975), an effect that might be relevant for the *in vivo* action of ADP as many anti-aggregating agonists, such as adenosine and prostacycline, stimulate adenylate cyclase (Hourani and Cusack, 1991). However, the correlation between ADP-induced inhibition of adenylate cyclase and aggregation is far from clear. Some ADP analogues (e.g., 2-MeSADP) are more potent inhibitors of adenylate cyclase than are aggregating agents, leading some investigators to postulate that these effects are mediated by separate platelet receptors (see for discussion Hourani and Hall, 1996). The affinity reagent 5'-fluorosulphonylbenzoyladenosine (FSBA) inhibits ADP-induced aggregation, but has no effect on ADP inhibition of adenylate cyclase. A protein of about 100 kDa labelled by this reagent was claimed to be the ADP receptor responsible for aggregation ("aggregin"; Colman, 1992). Later studies by Colman and co-workers lead to the development of a new affinity reagent [8-(4-bromo-2,3-dioxobuthylthio)ADP, 8-BDB-TADP] that labelled aggregin and inhibited all effects of ADP (Puri, et al., 1995). Based on the known antagonist activity of ATP at the platelet ADP receptor, high affinity stable analogues are actively being investigated.

The increase in $[Ca^{2+}]_i$ and, to a lesser extent, inhibition of adenylate cyclase are powerful platelet-aggregating events. Thus, it is reasonable to hypothesize that one of the main physiological roles of the platelet ADP receptor(s) is in hemostasis. During this process, platelets aggregate, and release pro-coagulant mediators providing a physical substrate for blood coagulation. ADP is normally present in a very low concentration in non-coagulated blood; on the contrary, it is stored, together with ATP, in high concentrations within platelet dense granules. Thus, ADP during hemostasis acts as an autocrine/paracrine activators that acts in synergy with other factors in stimulation and spreading of platelet thrombus formation.

Extracellular ADP originates both from ADP that is released as such during the process of platelet aggregation, and as a consequence of the action of ecto-nucleotidases on ATP. The role of ecto-nucleotidases is complex, as these enzymes, on one hand generate the potent pro-aggregating agent ADP, but, on the other, also produce, as an end product of their activity, adenosine, a potent inhibitor of platelet aggregation (Marcus and Seifer, 1993). Since ecto-nucleotidases are strongly expressed on the plasma membrane of endothelial cells, the final result of the initial activation is dependent on the integrity of the endothelial lining of the vessel (Marcus, et al., 1991). In the presence of an intact endothelium, pro-thrombotic agents released by activated platelets will be counteracted by the generation of anti-thrombotic factors from the endothelium. Some of these factors either are ATP metabolites (adenosine), or are released as a consequence of endothelial stimulation by ATP (nitric oxide). On the contrary, in the absence of an intact endothelium, ectonucleotidase activity will be reduced, thus also reducing adenosine formation. Furthermore, under these conditions ATP will easily access the subendothelial smooth muscle layer, where the main effect will be activation of P2X channels, leading to contraction and, therefore, vasoconstriction (Burnstock, 1987). The source of extracellular

ATP need not to be platelets themselves in the first instance. An intravascular (embolus, atherosclerotic plaque) or extravascular (trauma) event that injures the vessel wall may easily release large amounts of cellular ATP, thus initiating the whole cascade leading to ADP generation.

Erythrocytes

Erythrocytes were one of the first blood cells on which the effects of extracellular ATP were tested (Parker and Snow, 1972); then, for many years interest for purinergic mechanisms in these cells faded. Interest has been rekindled in the nineties with a series of elegant studies on turkey erythrocytes, mainly performed in the laboratory of Kendall Harden, that lead to a thorough description of the signalling pathway associated to P2Y receptor activation (Boyer, et al., 1996b).

Red blood cells express a typical $P2Y_1$ receptor coupled to phospholipase $C\beta$ via a G protein of the G_q family. The predicted structure is that of a typical serpentine receptor showing little homology to receptors for adenosine, and high homology to receptors for interleukin-8, thrombin, vasoactive intestinal peptide, platelet-activating factor, angiotensin II and neuropeptide Y. Within the P2Y subfamily, homology is highest for $P2Y_2$, $P2Y_4$ and $P2Y_6$.

The agonist order of potency fits nicely with that initially proposed by Burnstock and Kennedy for a P2Y receptor (Burnstock and Kennedy, 1985) (see also Table II). Stimulation of phospholipase C then leads to IP3 and diacylglicerol formation. Turkey erythrocyte

---

TABLE II: Pharmacological profile of turkey erythrocyte P2Y receptor.

Agonist	EC_{50} (nM)
2MeSATP	8
2MeSADP	6
ADPβS	96
ATPγS	1260
ATP	6350
ADP	5550
App(NH)p	4450
UTP	143000
α,β-MeATP	>100000
β,γ-MeATP	>100000
AMP	inactive
Adenosine	inactive

Abbreviations used only in this Table: App(NH)p, adenosine 5'-(β,γ-imido)triphosphonate. Adapted from Boyer, et al., 1996b.

---

plasma membranes have also provided a model for binding assays and photoaffinity labelling of P2 purinergic receptors, but with results that have been difficult to reconcile with studies performed with intact cells.   On the other hand, the turkey erythrocyte membrane model proved to be a reliable *in vitro* system for pharmacological analysis. This has lead to the synthesis of AMP-analogues, i.e., 2-(5-hexenylthio)AMP, that are hundreds-fold more potent than ATP or ADP (Boyer, et al., 1996c). Quite interestingly, these compounds are inactive at the $P2Y_2$ receptor, thus providing the basis for the development of highly needed purinergic agonists selective for the various members of the P2Y subfamily.

The erythrocyte plasma membrane model has also been useful for the development of hydrolysis-resistant P2Y receptor antagonists based on sulfate derivatives of adenine nucleotides, e.g., adenosine 3'-phosphate 5'-phosphosulfate (Boyer et al. 1996a). Among other more common P2 antagonists, pyridoxal phosphate 6-azophenyl 2', 4'- disulphonic acid (PPADS) were found to be a competitve antagonist of the turkey P2Y receptor.

It is not known whether erythrocytes also express receptors of the P2X subtype. However, red blood cells were among the first cell types shown to undergo an increase in transmembrane ion fluxes in response to extracellular ATP (Parker and Snow, 1972). These authors reported that concentrations of extracellular ATP in excess of 0.1 mM caused a net gain of Na^+ and loss of K^+ paralleled by an increase in water content. These effects were prevented by Mg^{2+} and hexokinase plus glucose, and were potentiated by EDTA. Other nucleotides were ineffective.

Increases in permeability of erythrocyte plasma membrane caused by extracellular ATP were also reported later by Trams, et al. (1980). According to these authors, in human blood a dramatic accumulation of extracellular adenylates was observed when ATP was accumulated to a concentration in excess of 200 µM. It was concluded that exogenously-added ATP had caused a permeability change in erythrocyte plasma membrane that allowed for leakage of cytoplasmic ATP (a sort of "ATP-induced ATP release"). No further characterization of the pathway of efflux of cytoplasmic ATP followed, thus we cannot assume this as evidence for the expression in red blood cells of an ATP-permeabilizing receptor of the P2Z ($P2X_7$) subtype. However, taken together the early reports of Parker and Snow and those of Trams and colleagues may suggest the presence of an ATP-gated channel that recognizes as a preferred ligand ATP in its tetraanionic form and may also, under some experimental conditions, allow efflux of larger molecules.

Thus, while it is well ascertained that erythrocytes express P2Y receptors, it is less clear whether they also express P2X receptors and of what subtype. On the contrary, it is well documented that ATP can be released from erythrocytes in response to ATP itself or other factors such as hypoxia (Bergfeld and Forrester, 1992). The pathway of hypoxia-induced ATP release has not been elucidated, but it has been proposed to involve the membrane protein known as band 4.5. ATP release from erythrocytes could contribute to regulation of local blood flow as this nucleotide has a well recognized vasodilatory activity (Boeynaems and Pearson, 1990). ATP acts on P2Y receptors of the vascular endothelium to promote release of nitric oxide and, thus, vessel wall relaxation. Under ischemic conditions ATP release from blood cells could be one of the local responses aimed at increasing blood flow. Presence of an intact endothelial lining is crucial for this effect since endothelial cell removal or damage converts vasodilation into vasoconstriction. In case there is a massive release of ATP in the blood, the process of ATP release from red blood cells may become self-sustaining and lead to generalised vasodilation and potentially to circulatory shock. This model for the pathogenesis of traumatic shock has been described in detail by Trams and colleagues (1980), but as of now remains highly hypothetical.

The ability of red blood cells to release ATP has also been invoked as a mechanism for the claimed anti-tumoral activity of exogenously administered ATP (Rapoport and Fontaine, 1989). It has been known for some time that intraperitoneal injection of ATP causes inhibition of the growth of tumors implanted in the foot pad of $CB6F_1$ mice, however the mechanism of this antitumor effect has remained unknown (Rapoport and Fontaine, 1989). Suggestions have been put forward that the antitumor activity of ATP could be mediated by the expansion of an intra-erythrocyte ATP pool of the host. This, in turn, would lead to an increase in the blood plasma levels of this nucleotide. At this stage it is fair to say that this interpretation is very speculative, and other hypothesis have been proposed (see Estrela, et al., 1995).

Mast Cells

Extracellular ATP is long known to be a powerful agonist for mast cells. Early studies by Sugyama (1971) and Dahlquist and Diamant (1974) showed that ATP triggered a fast and massive release of histamine and other mediators of immediate hypersensitivity from mast cells. Gomperts and co-workers extensively characterized this effect and discovered that secretion was due to activation of an ATP^{4-} receptor that caused plasma membrane lesions leading to loss of inorganic phosphate, sugar phosphates and nucleotides, and uptake of several low molecular weight markers (such as fluorescein, HEDTA- ^{57}Co chelate, stable analogues of nucleotides) from the external medium (Bennet, et al., 1981). The filtration properties of the ATP lesions were not exactly measured, but during a typical ATP stimulation lasting 20-45 min there was no uptake of [^{3}H]inulin (MW 5200), suggesting that the ATP-dependent pore of mast cells could have a molecular cut-off similar to that later determined in macrophages by Steinberg et al. (1990). This was further confirmed by Tatham and Lindau who reported that the largest molecules that can be loaded into mast cells by ATP permeabilization fall in the 600-1000 D molecular weight range (Thatham and Lindau, 1990).

Studies on the efflux of phosphorylated metabolites also suggested that pore size was dependent on the ATP^{4-} concentration, as release of higher MW molecules (up to the maximum of 1000 D) could be obtained by raising extracellular ATP or lowering the divalent cation concentration of the suspending medium. This conclusion has been confirmed by patch clamp analysis of ATP lesions in mast cells which sugggests that the mast cell pores in the presence of ATP undergo dynamic fluctuations depending on the ATP concentration (Tatham and Lindau, 1990). The Hill coefficient of 2 calculated for the mast cell and thymocyte pores would suggest that two ATP^{4-} molecules are required to recruit an additional element (subunit) into an existing pore (see also Pizzo, et al., 1991, for the thymocyte ATP pore).

The optimal ATP^{4-} concentration that causes maximal permeabilization is above 8 μM, but under these conditions histamine secretion is severely inhibited, a likely consequence of the extensive plasma membrane damage and/or depletion of intracellular metabolites. The optimal ATP^{4-} dose for secretion is 3 μM, a concentration at which there is release of inorganic phosphate but only a minor leakage of higher molecular weight metabolites (Cockcroft and Gomperts, 1979). All these observations point to the presence of a typical ATP-permeabilizing receptor in mast cells, although some of the most recently characterized $P2Z/P2X_7$ agonists, such as benzoylbenzoylATP, have not been investigated in this cell type (for a pharmacological study in mast cells of other currently used ATP analogues see Tatham, et al., 1988).

A few studies have also been performed in mast-cell like cells, such as the P-815, Mc9 or RBL cells. P-815 mouse mastocytoma cells appear to posses an, as yet, poorly characterized ATP-gated cation channel permeable to both Ca^{2+} and Na^+ (Zanovello, et al., 1990); likewise, these cells are found positive for $P2X_7$ expression by Northern blotting (Collo, et al., 1997). Activation of Ca^{2+} uptake by P815 cells requires ATP concentrations >100 μM and is not mimicked by other nucleotides. Rather intriguingly, Ca^{2+} uptake is greatly enhanced by pre-treatment with a stable PGI_2 analogue (Nagishi, et al., 1991), suggesting that the ATP receptor of mast cells might be up-regulated by inflammatory mediators, as already shown for the macrophage $P2Z/P2X_7$ receptor (Falzoni, et al., 1995; Humphrey and Dubyak, 1996). Sustained stimulation by extracellular ATP of P815 cells eventually leads to cell death by apoptosis (Zanovello, et al., 1990). DNA fragmentation starts after a lag of about 2-3 h and is nearly complete within 6-7 h. Although a thorough study has not been performed, this effect likely depends on the drastic upsetting of intracellular ion homeostasis caused by the opening of the ATP-ligated channel. In particular, it appears that the crucial event is Na^+ uptake and K^+ loss, while Ca^{2+} appears to be unnecessary.

Mast cells also express P2 receptors of the P2Y subtype, although some mast cell-derived cell lines may lack them (Osipchuk and Cahalan, 1996; Sudo et al, 1996). In mast cell monolayers applications of ATP at a concentration of 0.5 to 10 μM cause large transient increases in intracellular Ca^{2+} suggestive of release from intracellular stores. ATP triggers hydrolysis of plasma membrane phosphoinositides and, thus, generation of inositol triphosphate. However this effect is biphasic with stimulation at low, and inhibition at high, concentrations (Cockcroft and Gomperts, 1980). It is not clear whether activation of phosphoinositide turnover is only due to the P2Y receptor, or may also be consequent to the Ca^{2+} influx triggered by opening of the ATP-gated pore. In the early experiments reported by Cockcroft and Gomperts (1980), activation of phosphoinositide turnover overlaps closely with stimulation of histamine secretion and leakage of intracellular phosphorylated metabolites, thus pointing to involvement of a single receptors in all these responses, presumably $P2Z/P2X_7$.

Mast cells possess large granules that contain different inflammatory mediators or enzymes as well as ATP. This could generate an autocrine/paracrine mechanism capable

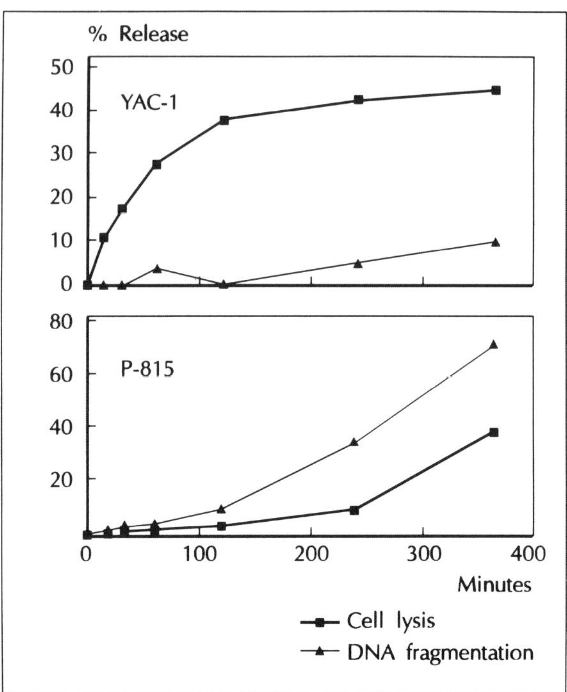

Figure 1: DNA fragmentation caused by P2X receptor activation in the P815 masto-cytoma cell line. Release of low MW DNA fragments after stimulation with ATP from P-815 cells was compared to that observed in YAC-1 lymphoma cells, a cell type that upon ATP stimulation undergoes fast necrosis rather than apoptosis. (Reprinted with permission from Steinberg and Di Virgilio, 1991.)

of spreading an activation signal as nicely shown by Osipchuk and Cahalan (1992). These authors provided convincing evidence that mechanical perturbation of a single mast cell in a cluster triggered a wave-like spread of cytoplasmic Ca^{2+} ($[Ca^{2+}]_i$) increase in the neighbouring cells. The spread of $[Ca^{2+}]_i$ increase was inhibited by pretreatment with a desensitizing concentration of ATP or suramin. Very interestingly, also spreading of antigen-induced $[Ca^{2+}]_i$ responses appeared to be, at least in part, mediated by ATP release. Shortly after antigen stimulation, some mast cells showed repetitive $[Ca^{2+}]_i$ spikes that were blocked by suramin. Antigen stimulation also lead to accumulation of ATP into the extracellular space.

While it is clear that mast cells can release ATP and, thus, spread an inflammatory signal through the interstitial tissue, they are not very likely the only cell type normally resident in the interstitium that is capable of actively generating an extracellular ATP signal, as lymphocytes and macrophages also can secrete this nucleotide in response to inflammatory agents (Filippini, et al., 1990; Ferrari, et al., 1997). Altogether, these findings suggest an putative new scenario for the activation of the inflammatory reaction in which accumulation of ATP and ADP into an inflamed tissue might be a central event that triggers activation of resident cells, such as tissue mast cells, and platelet activation. These early events in turn cause release of other inflammatory mediators (i.e., histamine and serotonine), thus leading to spreading of the inflammatory response and recruitment of additional immune and inflammatory cells.

IV. CONCLUSIONS AND FUTURE DIRECTIONS

It is clear that P2 purinoceptors play an important role in the regulation of platelet, erythrocyte and mast cell physiology. The P2Y and P2X subtypes are likely to mediate different responses and thus are presumably activated in different conditions. A key factor is obviously the concentration of ATP in the pericellular milieu, as EC_{50} for the two subtypes differs by at least one order of magnitude. At lower extracellular ATP concentrations P2Y receptors are more likely to be turned on, while at higher (micromolar or low millimolar) concentrations P2X receptors will be also triggered. This may suggest that the P2Y subtype plays a major role under normal physiological conditions (i.e. low extracellular ATP), while the P2X subtype comes into play when the ATP concentration in the blood or in the interstitial fluid raises as a consequence of tissue damage or stimulated release.

Identification of sources and mechanisms of ATP release and development of animal models in which the genes coding for different P2 receptors have been deleted (knockout mouse) will be clearly one of the most important goals for the next few years. All data so far available lead one to postulate an important role for purinergic P2 receptors (and extracellular ATP) in the modulation of the inflammatory response. Thus, there is little doubt that this field will become more attractive to both academic research and the pharmaceutical industry.

V. REFERENCES

Baricordi OR, Ferrari D, Melchiorri L, Chiozzi P, Hanau S, Chiari E, Rubini M, Di Virgilio F. An ATP-activated channel is involved in mitogenic stimulation of T lymphocytes. *Blood, 87*: 682-690, 1996.

Bennett JP, Cockcroft S, Gomperts BD. Rat mast cells permeabilized with ATP secrete histamine in response to calcium ions buffered in the micromolar range. *J Physiol, 317*: 335-343, 1981.

Bergfeld GR, Forrester T. Release of ATP from human erythrocytes in response to a brief period of hypoxia and hypercapnia. *Cardiovasc Res, 26*: 40-47, 1992.

Boeynaems J-M, Pearson JD. P_2 purinoceptors on vascular endothelial cells: physiological significance and transduction mechanisms. *Trends Pharmacol Sci, 11*: 34-37, 1990.

Born GVR. Aggregation of blood platelets by adenosine diphosphate and its reversal. *Nature, 194*: 927-929, 1962.

Boyer JL, Schachter JL, Romero T, Harden TK. Identification of competitive antagonists of the $P2Y_1$-purinergic receptor. *Mol Pharmacol, 50*: 1323-1329, 1996a.

Boyer JL, Schachter JB, Sromek SM, Palmer RK, Jacobson KA, Nicholas RA, Harden TK. Avian and human homologues of the P2Y receptor: pharmacological, signalling and molecular properties. *Drug Dev Res, 39*: 253-261, 1996b.

Boyer JL, Siddiqi SM, Fischer B, Jacobson KA, Harden TK. Analogs of adenosine monophosphate that are potent P2Y purinergic receptor agonists. *Brit J Pharmacol, 118*: 1959-1964, 1996c.

Burnstock G. Local control of blood pressure by purines. *Blood Vessels, 24*: 156-160, 1987.

Burnstock G, Kennedy C. Is there a basis for distinguishing two types of P2 purinoceptor? *Gen Pharmacol, 16*: 433-440, 1985.

Burnstock G, Wood JN. Purinergic receptors: their role in nociception and primary afferent neurotransmission. *Curr Opin Neurobiol, 6*: 526-532, 1996.

Cockcroft S, Gomperts BD. ATP induces nucleotide permeability in rat mast cells. *Nature, 279*: 541-542, 1979.

Cockcroft S, Gomperts BD. The ATP^{4-} receptor of rat mast cells. *Biochem J, 188*: 789-798, 1980.

Collo G, Neidhart S, Kawashima E, Kosco-Vilbois M, North RA, Buell G. Tissue distribution of the P2X$_7$ receptor. *Neuropharm, 36*: 1277-1283, 1997.

Colman RW. Platelet ADP receptors stimulating shape change and inhibiting adenylate cyclase. *News Physiol Sci, 7*: 274-278, 1992.

Dahlquist R, Diamant B. Interaction of ATP and calcium on the rat mast cell: effect on histamine release. *Acta Pharmacol Toxicol, 34*: 368-384, 1974.

Estrela JM, Obrador E, Navarro J, Lasso de la Vega MC, Pellicer JA. Elimination of Ehrlich tumours by ATP-induced growth inhibition, glutathion depletion and X rays. *Nature Med, 1*: 84-88, 1995.

Falzoni S, Munerati M, Ferrari D, Spisani S, Moretti S, Di Virgilio F. The purinergic P2Z receptor of human macrophage cells. Characterization and possible physiological role. *J Clin Invest, 95*: 1207-1216, 1995.

Fasolato C, Innocenti B, Pozzan T. Receptor-activated Ca^{2+} influx: how many mechanisms for how many channels? *Trends Pharmacol Sci, 15*: 77-83, 1994.

Ferrari D, Chiozzi P, Falzoni S, Hanau S, Di Virgilio F. Purinergic modulation of Interleukin 1-β release from microglial cells stimulated with bacterial endotoxin. *J Exp Med, 185*: 579-582, 1997.

Filippini A, Taffs RE, Sitkovsky MV. Extracellular ATP in T-lymphocyte activation: possible role in effector functions. *Proc Natl Acad Sci USA, 87*: 8267-8271, 1990.

Gachet C, Cazenave JP. ADP-induced blood platelet activation: a review. *Nouv Rev Fr Hematol, 33*:347-358, 1991.

Gordon JL. Extracellular ATP: effects, sources and fate. *Biochem J 233*: 309-319, 1986.

Hallam TJ, Rink TJ. Responses to adenosine diphosphate in human platelets loaded with the fluorescent calcium indicator quin2. *J Physiol 368*: 131-146, 1985.

Haslam RJ, Rosson GM. Effects of adenosine on levels of adenosine cyclic 3',5'-monophosphate in relation to adenosine incorporation and platelet aggregation. *Mol Pharmacol 11*: 528-544, 1975.

Hourani SMO, Cusack NJ. Pharmacological receptors on blood platelets. *Pharmacol Rev 43*: 243-298, 1991.

Hourani SMO, Hall DA. ADP receptors on human blood platelets. *Trends Pharmacol Sci 15*: 103-108, 1994.

Hourani SMO, Hall DA. P2T purinoceptors: ADP receptors on platelets. *Ciba F Symp 198*: 53-70, 1996.

Humphreys BD, Dubyak GR. Induction of the P2Z/P2X$_7$ nucleotide receptor and associated phospholipase D activity ny lipopolysaccharide and IFN-γ in the human THP-1 monocytic cell line. *J Imunol 157*: 5627-5637, 1996.

Humphries RG, Robertson MJ, Leff P. A novel series of P2T purinoceptor antagonists: definition of the role of ADP in arterial thrombosis. *Trends Pharmacol Sci 16*: 179-181, 1995.

Leon C, Hechler B, Vial C, Leray C, Cazenave J-P, Gachet C. *FEBS Lett 403*: 26-30, 1997.

MacKenzie AB, Mahaut-Smith MP, Sage SO. Activation of receptor-operated cation channels via $P2X_1$ not P2T purinoceptors in human platelets. *J Biol Chem 271*: 2879-2881, 1996.

Marcus AJ, Safier LB. Thromboregulation: multicellular modulation of platelets reactivity in hemostasis and thrombosis. *FASEB J 7*: 516-522, 1993.

Marcus AJ, Safier LB, Hajjar KA, Ullman HL, Islam N, Broekman MJ, Eiroa AM. Inhibition of platelet function by an aspirin-insensitive endothelial cell ADPase. Thromboregulation by endothelial cells. *J Clin Invest 88*: 1690-1696, 1991.

Negishi M, Hashimoto H, Ichikawa A. Differential regulation of thrombin- or ATP-induced mobilization of intracellular Ca^{2+} by prostacyclin receptor in mouse mastocytoma cells. *Biochem Biophys Res Commun 176*: 102-107, 1991.

Osipchuk Y, Cahalan M. Cell-to-cell spread of calcium signals mediated by ATP receptors in mast cells. *Nature 359*: 241-244, 1992.

Parker JC, Snow RL. Influence of external ATP on permeability and metabolism of dog red blood cells. *Am J Physiol 223*: 888-893, 1972.

Pizzo P, Zanovello P, Bronte V, Di Virgilio F. Extracellular ATP causes lysis of mouse lymphocytes and activates a plasma membrane ion channel. *Biochem J 274*: 139-144, 1991.

Puri RN, Kumar A, Chen H, Colman RF, Colman RW. Inhibition of ADP-induced platelet responses by co-valent modification of agrregin, a putative ADP receptor, by 8-(4-bromo-2,3-dioxothio)ADP. *J Biol Chem 270*: 24482-24488, 1995.

Rapoport E, Fontaine J. Anticancer activities of adenine nucleotides in mice are mediated through expansion of erythrocyte ATP pools. *Proc Natl Acad Sci USA 86*: 1662-1666, 1989.

Steinberg TH, Buisman HP, Greenberg S, Di Virgilio F, Silverstein SC. Effects of extracellular ATP on mononuclear phagocytes. *Ann NY Acad Sci 603*: 120-129, 1990.

Steinberg TH, Di Virgilio F. Cell-mediated cytotoxicity: ATP as an effector and the role of the target cells. *Curr Opin Immunol 3*: 71-75, 1991.

Sudo N, Tanaka K, Koga Y, Okumura Y, Kubo C, Nomoto K. Extracellular ATP activates mast cells via a mechanism that is different from the activation induced by the cross-linking of Fc receptors. *J Immunol 156*: 3970-3979, 1996.

Sugiyama K. Calcium-dependent histamine release with degranulation from isolated mast cells. *Jpn J Pharmacol 21*: 209-226, 1971.

Tatham PER, Cusack NJ, Gomperts BD. Characterisation of the ATP4- receptor that mediates permeabilisation or rat mast cells. *Eur J Pharmacol 147*: 13-21, 1988.

Tatham PER, Lindau M. ATP-induced pore formation in the plasma membrane of rat peritoneal mast cells. *J Gen Physiol 95*:459-476, 1990.

Trams EG, Kaufmann H, Burnstock G. A proposal for the role of ecto-enzymes and adenylates in traumatic shock. *J Theor Biol 87*: 609-621, 1980.

Zanovello P, Bronte V, Rosato A, Pizzo P, Di Virgilio F. Responses of mouse lymphocytes to extracellular ATP. II. Extracellular ATP causes cell-type dependent lysis and DNA fragmentation. *J Immunol 145*: 1545-1550, 1990.

PURINORECEPTORS AND T CELL FUNCTION: REGULATION OF LYMPHOCYTE ACTIVATION BY EXTRACELLULAR ATP AND ADENOSINE

Michail Sitkovsky, John Armstrong, Masahiro Koshiba and Sergey Apasov
Laboratory of Immunology, National Institute of Allergy and Infectious
Diseases, National Institutes of Health, Bethesda, Maryland 20892 USA

I. SUMMARY

Purinergic receptors are implicated in different important processes in the
neuronal system and in regulation of the cardiovascular system. They provide
novel molecular targets and offer exciting pharmacologic possibilities. However,
the development of new drugs which target different purinergic receptors must
be done with full understanding of the role of these receptors in the immune
system thereby necessitating the studies of the effects of extracellular ATP and
adenosine and of the role of purinergic receptors in T-cell-mediated immunity.
 Our studies were guided by the hypothesis that signaling through purinergic
receptors by extracellular ATP (extATP) and/or adenosine (Ado) may interfere
with antigen presentation and with T cell antigen receptor (TCR)-triggered
activation pathways in differentiating thymocytes as well as with expansion and
effector functions of peripheral cytotoxic T cells (CTL) and T-helper cells.
Evidence to support this model is suggested by studies of the contribution of
purinergic receptor expression to the *in vitro* functional responses of T cells.
It has been shown that some purinergic receptors in T cells are expressed in a
differentiation-dependent manner (e.g. $P2X_7/P2Z$) or follow the pattern of
immediate early genes (e.g. $p2Y_2$), thereby providing the basis for the belief that
these receptors may function in feed back regulatory mechanisms of T cell
activation.   The A_{2A} adenosine receptor is the predominant Gs-coupled
adenosine receptor in murine thymocytes and T cells, and the activation of A_{2A}
receptors results in the accumulation of cAMP.   Subsequent activation of
cAMP-dependent protein kinase (PKA) accounts for the observed antigen
receptor (TCR)-antagonistic properties of extAdo.   While *in vitro* studies
suggest a possible role of purinergic receptors, the use of purinergic receptor
gene deficient mice may provide conclusive evidence of their contribution to
normal processes and the pathogenesis of diseases such as severe combined
immunodeficiency associated with adenosine deaminase deficiency, or ADA
SCID.

II. BACKGROUND

Purinergic receptors are implicated in different important processes in the
neuronal system and in regulation of the cardiovascular system (Linden, 1994;
Olah & Stiles, 1995; Jacobson et al, 1996; Svenningsson et al, 1997; Harden
et al, 1997).   These novel molecular targets offer exciting pharmacologic

possibilities, but the development of new drugs which target different purinergic receptors must be done with full understanding of the role of these receptors in the immune system. This necessitates studies of the effects of extracellular ATP and adenosine and of the role of purinergic receptors in T-cell-mediated immunity.

Antigen-specific T cell-mediated cellular responses represent an important mechanism of anti-viral, anti-bacterial, and anti-tumor immunity.   T lymphocytes express cell surface antigen receptors (TCR) which enable them to recognize MHC class I and II -presented peptides from e.g. infectious agents on the surface of antigen-presenting cells (Davis et al, 1997). The recognition of an antigen by TCR is followed by triggering of complex transmembrane signaling pathways (Berrideg, 1997; Alberola et al, 1997; Qian & Weiss, 1997).

Both the maturation of thymocytes into mature peripheral T cells and their effector functions are antigen-receptor (TCR)- driven (Robey & Fowlkes, 1994; Anderson et al, 1996; Malissen & Malissen, 1996; Hogquist et al, 1996; Bevan, 1997). Thus, the recognition of an antigen by molecules of TCR and co-receptors results in transmembrane signaling to T cells and triggering of different effector functions including expression of the lethal hit-delivering Fas Ligand (Rouvier et al, 1993; Nagata & Suda, 1995), exocytosis of cytolytic granules by cytotoxic T cells (CTL) (Henkart & Sitkovsky, 1994; Sitkovsky & Henkart, 1996), and lymphokine secretion by T-helper cells.

Mature T lymphocytes are produced by the processes of positive and negative selection in thymus. Only thymocytes which express TCR molecules with moderate affinity and moderate TCR-signaling potential survive (positive selection) in the thymus, while thymocytes with inadequate TCR affinity and/or signaling potential die from "neglect"; those with too high an affinity are culled by negative selection. Processes of positive and negative selection ensure that effective T-cells will mature into peripheral T cells, but those with potential autoimmune reactivity will not (Surch & Sprent, 1994, Robey & Fowlkes, 1994; Anderson et al, 1996; Malissen & Malissen, 1996; Hogquist et al, 1996; Bevan, 1997).

Thus, TCR and co-receptor-driven transmembrane signaling plays a decisive role in the selection, generation, expansion, and effector functions of T-lymphocytes. Our studies of purinergic receptors in T lymphocytes are based on the underlying assumption that TCR signaling can be enhanced and/or antagonized by signaling from physiologically abundant molecules in the extracellular environment.   Such signaling could be mediated by G-protein-coupled receptors and/or by ligand-gated channels (Apasov et al, 1995).

It was during our studies of the molecular mechanisms of T cell effector functions (Berrebi et al, 1987; Kincaid et al; Trenn et al, Takayama et al, 1987) that we started to consider the role of extracellular ATP (Filippini & Sitkovsky, 1990; Filippini et al, 1990; Redegeld et al, 1993) and of its catabolite, adenosine (Apasov et al, 1995) in the regulation of TCR-driven processes in lymphocytes. Accordingly, the differentiation, expansion, and effector functions of T lymphocytes are expected to be regulated not only by antigen/TCR interactions but also by signaling through P1 (adenosine) and P2 (ATP) purinergic receptors (Apasov et al, 1995).

In this review we will describe: i) the effects of extracellular ATP and adenosine on T-cells; ii) the identification of predominant receptors for ATP and adenosine on T cells, and iii) implications of purinergic receptor function in the mechanism of the human immunological disease, ADA SCID.

III. TEXT

A. Effects of Extracellular ATP and Adenosine on Different T-Cell Subsets

To understand the regulation of the T-cell antigen receptor (TCR) repertoire-forming processes of thymocyte selection and the mechanism of lymphotoxic effects of purines due to Adenosine Deaminase (ADA) deficiency, we investigated whether extracellular adenosine triphosphate (ATP_{ext}) and the ATP catabolite, adenosine may play roles in the process of negative selection of thymocytes by signaling through purinergic receptors. It is believed that negative selection is a process which ensures that potentially autoreactive (autoimmune) T cells will not emerge from the thymus and that those thymocytes with too high TCR affinity for self-antigen and/or too strong signaling potential [Surch & Sprent, 1994; Vacchio et al, 1994) will be eliminated by apoptosis. It is for this reason that TCR-triggered thymocyte apoptosis is considered to model negative selection of thymocytes.

The effects of extracellular ATP and adenosine were studied after incubation of different concentrations of these compounds and/or their analogs with lymphocytes followed by measuring the number of intact, early apoptotic and dead cells using flow cytometry methods (Apasov et al, 1994). Measurements of forward *vs* side scatter and of DNA degradation allowed the use of a thymocyte apoptosis assay as the long-term read-out of ATP- and adenosine-mediated signaling. It was shown that signaling through the P2 and P1 receptors causes different apoptotic processes, as evidenced by different patterns of apoptotic thymocyte distribution in FACS analysis using propidium iodide (PI) incorporation and forward scatter parameters. Importantly, we found [29] that different subpopulations of thymocytes responded differently:

    1. Double positive (DP), immature $CD4^+8^+$ thymocytes are possibly killed through both P2 (ATP_o) and P1 (Ado) class purinergic receptors, with P1 being more potent in apoptosis induction. In this respect, the data demonstrating the impaired ability of slow-hydrolyzable AMP-PCP to kill double positive thymocytes supports the interpretation that adenosine formation from ATP_o is responsible for apoptosis triggering.

    2. Single positive (SP) $CD4^+8^-$ thymocytes are killed mostly through ATP (P2 class purinergic receptors), with adenosine having no detectable effect.

    3. Single positive $CD48^+$ thymocytes also were not killed by adenosine and were found much less susceptible to P2 signaling.

4. Double negative $CD4^-8^-$ are susceptible to both ATP and adenosine equally, but to a lesser degree than double positive thymocytes.

Thus, signaling through purinergic receptors may have dramatic consequences in thymocyte development as evidenced by a differential susceptibility of immature DP and SP thymocytes to apoptosis-inducing signals through P1 (adenosine) and P2 (ATP) purinergic receptors. Moreover, signaling through P1 and P2 receptors has been shown to be able to interfere (antagonize or synergize) with TCR- and steroid-triggered signaling, and ATP_{ext} has been shown to have an additive effect with TCR crosslinking in induction of thymocyte death (Apasov et al, 1994; McConkey et al, 1994) further supporting a possible role for purines in the regulation of thymocyte selection.

These properties of ATP_{ext} and adenosine are consistent with the possibility of their involvement in the regulation of T-cell generation due to a differential expression of purinergic receptors in different thymocyte subsets. Specifically, different thymocyte subpopulations have different susceptibility to the effects of ATP and Ado, with P1 receptors being mostly expressed in immature (DP and DN) thymocytes, and P2 receptors being expressed in more mature (SP and $TCR^{hi}CD69^{hi}$) subpopulations, i.e. those identified as cells that undergo positive selection in the thymus.

Especially interesting was the finding that of all tested apoptotic stimuli, only ATP_{ext} was efficient in triggering rapid, protein synthesis-independent lysis of $CD4^+8^-$ thymocytes ($CD4^+SP$) and peripheral $CD4^+$ T cells, while in addition it induced a slower DNA fragmentation and protein synthesis-dependent apoptosis in all thymocyte subsets. In contrast, extracellular adenosine preferentially targeted $CD4^+8^+$ (double positive) thymocytes for apoptosis.

The rapid ATP-induced death of $CD4^+$ cells is consistent with the opening of ATP-gated membrane pores ($P2X_7$ receptor, formally known as p2z), and this was confirmed in direct studies of $[Ca^{++}]_i$ and cell permeability increases.

<u>B. Identification of predominant receptors for ATP and adenosine on T cells</u>

Biochemical and pharmacological studies of extATP-induced changes in intracellular $[Ca^{++}]$ and cell permeabilization revealed that the $P2X_7$ receptor (Buell et al, 1996; Suprenant et al, 1996) is the predominant receptor in murine T cells (Chused et al 1996).

Flow cytometry of Ca^{++}-sensitive, dye-loaded thymocytes showed that ATP_o can be utilized to trigger extracellular Ca^{++} translocation into thymocytes after signaling through purinergic receptors. In these studies (Chused et al 1996), extracellular ATP was found to increase $[Ca^{++}]$ in thymocytes, although ATP did not release calcium from intracellular stores. Furthermore, the $[Ca^{++}]_i$ response was not rapidly reversible and was not mediated by the phosphoinositol phosphate pathway. ATP was found to depolarize thymocytes and permeabilize the membrane to small molecules such that Ca^{2+} and ethidium (314 Da) ions were allowed entry, but propidium (415 Da) ions were not. These findings indicated that thymocytes constitutively express only $P2X_7$-type purinoceptors.

T cell subsets were found to differ in their response to ATP, with $CD4^+8^+$ double-positive thymocytes being the least responsive, and $CD4^-8^+$ single-positive thymocytes, $CD8^+$ splenic T cells, $CD4^+8^-$ single-positive thymocytes, and $CD4^+$ splenocytes showing increasing reactivity.

Even within peripheral T cell populations the $P2X_7$ receptor was found to be differentially expressed. In this regard, $P2X_7$ expression correlates with three subsets of splenic $CD4^+$ T cells defined by CD44 and CD45RB differentiation antigens, such that P2x7 activity was found to be highest in the memory T cell population; i.e. $P2X_7^{lo}$ cells are $CD44^{bright}CD45RB^{bright}$, $p2x7^{int}$ are $CD44^{dull}CD45RB^{int}$; and $P2X_7^{hi}$ are $CD44^{bright}CD45RB^{dull}$. It is suggested that $p2X_7$ receptor-mediated signaling could be involved both in the regulation of differentiation and cell death in the thymus and in peripheral T lymphocytes.

Thus, the pattern of distribution of $P2X_7$ receptors as determined by cell-permeabilization and Ca^{++} measurement studies (Chused et al, 1996) is in agreement with results of experiments where the read-out of effects of ext ATP was the cell death assay (Apasov et al, 1995).

C. Expression of Purinergic Receptors in T Cells as Immediate Early Genes

The differentiation-related expression of the $P2X_7$ receptor (Chused et al, 1996) suggested that this receptor could be differentially involved at different stages of T cell development and in different effector functions. This, in turn, suggested that the expression of these receptors could be regulated by T cell activation and differentiation-driving stimuli. Therefore, we sought to determine if the expression of purinergic receptors changes during the early stages of T cell activation in short term assays. Since no appropriate mAb to purinergic receptors were available to test this hypothesis, changes in mRNA expression were followed after activation of T cells with such different activating stimuli as steroids and mAb to TCR/CD3 complex. The first cloned purinergic receptor was the $P2Y_2$ receptor (Lustig et al 1993; Webb et al, 1993), and we used cDNA probes for rat $P2Y_2$ (formerly a p2u receptor) to clone and sequence the mouse $P2Y_2$ receptor (Koshiba et al, 1997). Quantitative RT-PCR was then utilized in studies of expression of $P2Y_2$ mRNA after activation of mouse thymocytes by anti-TCR mAb or steroid hormones.

In these studies we observed a transient and protein synthesis-independent enhancement of $P2Y_2$ mRNA expression in mouse thymocytes after the addition of steroid hormone or monoclonal antibody (mAb) to crosslink the T-cell receptor (TCR).

Flow cytometry of thymocyte subsets excluded the possibility that the observed increases in $p2y_2$ receptor mRNA expression were due to the enrichment of steroid-treated cells with a $P2Y_2$ mRNA-rich thymocyte subset. We proposed therefore that the rapid increase of $P2Y_2$ receptor mRNA expression could be a common early event in responses of T-cells to different activating stimuli.

As these experiments progressed the $P2X_1$ cDNA was cloned (Brake ta al, 1994; Valera et al, 1994), and we extended our studies to include an examination of TCR- and steroid-triggered $P2X_1$ mRNA expression. Surprisingly, no dexamethasone (DEX)-induced increase in mRNA expression for the ligand-gated ion channel $P2X_1$ receptor was detected in mouse thymocytes, raising questions about the validity of the previously suggested role of p2x1 receptors in thymocyte apoptosis (Owens et al, 1991). Triggering of TCR-mediated intracellular signaling pathways through crosslinking of TCR or by addition of phorbol ester and Ca^{2+} ionophore also resulted in the up-regulation of $P2Y_2$, but not $P2X_1$, receptor mRNA. Thus, our data call into question the validity of experiments (Chvatchko et al, 1996) in which conclusion were based on the controversial finding that TCR up-regulates $P2X_1$ expression. Taken together with the recently discovered ability of nucleotide receptor-initiated signaling to antagonize or enhance the effects of TCR-crosslinking or steroids on thymocytes, the observed rapid up-regulation of $p2Y_2$ receptor mRNA expression may reflect an immediate early gene response where newly expressed cell surface nucleotide receptors provide regulatory feed-back signaling from extracellular ATP in the T-cell differentiation process.

The definitive implication of extracellullar ATP receptors in T cell differentiation and effector functions must await the development of new tools and of P2 receptor gene deficient animals, but studies of the roles of adenosine receptors have been facilitated by the use of selective agonists and antagonists of well characterized adenosine receptors.

D. Extracellular Adenosine A_{2A} Receptor-Mediated Signaling Inhibits Expansion of T Cells

Adenosine has been the subject of intensive and prolonged investigation due to its clear role in the regulation of the cardiovascular and nervous systems (Linden, 1994; Olah & Stiles, 1995; Jacobson et al, 1996; Svenningsson et al, 1997). It was demonstrated in our studies that the A_{2A} receptor is the predominant Gs-coupled adenosine receptor in murine thymocytes and T cells and that the activation of these A_{2A} receptors on cells with low Adenosine Deaminase (ADA) activity results in the accumulation of cAMP. Subsequent activation of cAMP-dependent protein kinase (PKA) accounts for most of the observed effects of adenosine on T cell functions, including antagonism of TCR-triggered IL-2 receptor alpha chain upregulation and cell proliferation (Huang et al, 1997).

The analysis of the expression of adenosine receptors in lymphocytes was facilitated by the availability of specific agonists and antagonists and of specific cDNA probes [2]. This allowed us to demonstrate that it is the A2A and not the A2B or A1 receptor that is predominantly expressed in mouse T cells

(Huang et al, 1997, Hirschhorn, 1995).

It was found that the exposure of T cells to extracellular adenosine causes increased levels of cAMP. Agonists of the A_{2A} receptor (e.g. CGS21680), but not of other receptors increase levels of cAMP, while antagonists of this receptor decreased the adenosine-induced cAMP accumulation (Huang et al, 1997, Koshiba et al, 1997). Results of Northern blot studies also confirmed that the A_{2A} mRNA is the predominantly expressed adenosine receptor transcript in mouse T cells (Huang et al, 1997, Koshiba et al, 1997).

The exposure of T cells to extAdo may result in blocking the expansion of T cells (Huang et al, 1997) due to a strong inhibition of TCR-triggered proliferation and upregulation of the IL-2 receptor a chain (CD25). These effects of adenosine were observed even at low concentrations of extracellular adenosine. The inhibition of up-regulation of IL-2 receptor by extAdo is sufficient to explain the inhibition of expansion of T cells. These effects of extracellular Ado are likely to be mediated by A_{2A} receptor-mediated signaling rather than by intracellular toxicity of adenosine catabolites, since: i) poorly metabolized adenosine analogs cause the accumulation of cAMP and strong inhibition of TCR-triggered CD25 upregulation; ii) the A_{2A}, but not the A_1 or A_3, receptors are the predominantly expressed and functionally coupled adenosine receptors in mouse peripheral T and B lymphocytes, and the adenosine-induced cAMP accumulation in lymphocytes correlates with the expression of A_{2A} receptors; iii) the specific agonist of the A_{2A} receptor, CGS21680, induces increases in $[cAMP]_i$ in lymphocytes, while the specific antagonist of the A_{2A} receptor, chlorostyrylcaffeine (CSC), inhibits the effects of Ado and CGS21680; and iv) the agonist-induced increases in $[cAMP]_i$ mimic the adenosine-induced inhibition of TCR-triggered CD25 upregulation and splenocyte proliferation. These studies suggest a possible role of adenosine receptors in the regulation of lymphocyte expansion and point to the A_{2A} purinergic receptor on T cells as a potentially attractive pharmacological target.

E. Implications of Purinergic Receptor Function in the Mechanism of ADA SCID

Normal development and functions of immune cells require adenosine deaminase (ADA) activity. The absence or low level of ADA in humans results in severe combined immunodeficiency (SCID) which is characterized by hypoplastic thymus, T-lymphocyte depletion, and autoimmunity. ADA SCID is currently explained by direct intracellular lymphotoxicity of accumulated adenosine and/or its metabolites (reviewed in Hirschhorn, 1995). Our experiments showed that T cell depletion could be due in part to extracellular adenosine (extAdo)-induced signaling which inhibits TCR-driven activation and expansion of peripheral T cells.

Accumulation of extracellular and intracellular adenosine (Ado) can also result from hypoxic conditions (Hoskin et al, 1994) and these conditions *in vivo* should be analyzed since we expect the impairment of T-cell dependent immune response (Koshiba et al, 1997) in hypoxic tissues such as large, solid tumors.

In support of a "signaling" mechanism of ADA SCID we found that extracellular Ado (extAdo) suppresses all tested T cell receptor (TCR)-triggered effector functions of T lymphocytes including TCR-triggered FasL mRNA upregulation in cytotoxic T lymphocytes (CTLs), perforin-mediated cytotoxicity, and cytotoxic granule exocytosis. Strong evidence in support of the signaling model_is that T cells which would normally apoptose from TCR signaling are rescued when those signals are sent in the presence of extracellular adenosine. Moreover, a brief exposure to Ado is sufficient to observe inhibition of TCR-triggered effector functions, and this observation is important since it may explain the prolonged effects of extracellular adenosine even in the presence of normal ADA activities.

The "memory" of T cells to exposure to extAdo is best explained by sustained increases in cAMP. The demonstration of cross-talk between A_{2A} receptors and the TCR in both directions supports the possible role of A_{2A} receptors in mechanisms of extAdo-mediated immunosuppression *in vivo* under ADA deficiency and hypoxic conditions in, *e.g.*, solid tumors.

IV. FUTURE DIRECTIONS

Significant insights into the expression and function of purinergic receptors have been accomplished using biochemical and pharmacological studies in short- term assays (seconds to minutes). The most important goal, however, is to determine whether signaling through purinergic receptors is important in initiating and/or modulating physiological responses. Such responses in many cases are longer (hours and days), and traditional tools often are not able to conclusively implicate purinergic receptors in cellular responses. Furthermore, studies of the physiological role of purinergic receptors using pharmacological tools in long term assays as in (Chvatchko, 1996; Ferrari et al, 1997a; Ferrari et al, 1997b) should be viewed with caution since misinterpretation is possible because we have found that some probes for purinergic receptors, previously reported to be specific (Ferrari et al, 1997a; Ferrari et al, 1997b) show pleiotropic effects [Smith and Sitkovsky, manuscript in preparation]. Moreover, similarities in the time course of expression of purinergic receptors and of *in vivo* responses (Chvatchko, 1996) should not be overinterpreted.

The direct evidence to implicate purinergic receptors in cellular functions remains to be provided using a new generation of specific tools including mAb to different epitopes of these receptors and gene targeting techniques. The first steps in that direction have now been taken with the development of monoclonal Ab to recombinant A_{2A} adenosine purinergic receptors (Robeva et al, 1996) and of animals with deficiency in expression of individual genes for purinergic receptors (Ledent et al, 1997).

The breeding of different TCR-transgenic mice with purinergic receptor gene- deficient animals will provide definitive models to study the roles of these receptors in the functioning of the immune system.

REFERENCE

Alberola-Ila J, Takaki, Kerner JD, Perlmutter RM. Differential signaling by lymphocyte antigen receptors. Annu Rev Immunol 1997;15:125-54

Anderson G, Moore NC, Owen JJ, Jenkinson EJ. Cellular interactions in thymocyte development. Annu Rev Immunol 1996;14:73-79

Apasov SG, Koshiba M, Chused TM, Sitkovsky MV. Effects of extracellular ATP and adenosine on different thymocyte subsets: possible role of ATP-gated channels and G protein-coupled purinergic receptors. J Immunol 1997;158:5095-5105

Apasov S, Koshiba M, Redegeld F, Sitkovsky M. Role of extracellular ATP and P1 and P2 classes of purinergic receptors in T-cell development and cytotoxic T lymphocyte effector functions. Immunol Rev 1995;146:5-19

Berrebi G, Takayama H, Sitkovsky MV. Antigen-receptor interaction requirement for conjugate formation and lethal-hit triggering by cytotoxic T lymphocytes can be bypassed by protein kinase C activators and Ca++ ionophores. Proc Natl Acad Sci USA 1987;84:1364-1368
Berridge MJ. Lymphocyte activation in health and disease. Crit Rev Immunol 1997;17:155-78

Bevan MJ. In thymic selection, peptide diversity gives and takes away. Immunity 1997;7:175-8

Brake A, Wagenbach MJ, Julius D. New structural motif for ligand-gated ion channels defined by an ionotropic ATP receptor. Nature 1994;371:519-524

Buell G, Collo G, Rassendren F. P2X receptors: an emerging channel family. Eur J Neurosci 1996;8:2221-8

Chused TM, Apasov S, Sitkovsky MV. Murine T lymphocytes modulate activity of an ATP-activated p2z-type purinoreceptor during differentiation. J Immunol 1996;157:1371-1380

Chvatchko Y, Valera S, Aubry JP, Renno T, Buell G, Bonnefoy JY. The involvement of an ATP-gated ion channel, P(2X1), in thymocyte apoptosis. Immunity 1996;5:275-83

Davis MM, Lyons DS, Altman JD, McHeyzer-Williams M, Hampl J, Boniface JJ, Chien Y. T cell receptor biohcemistry, repertoire selection and general features of TCR and Ig structure. CIBA Found Symp 1997;204:94-104

Ferrari D, Chiozzi P, Falzoni S, DalSusino M, Melchiorri L, Baricordi OR, DiVirgilio F. Extracellular ATP triggers IL-1 beta release by activating the purinergic P2Z receptor of human macrophages. J Immunol 1997;159:1451-8

Ferrari D, Chiozzi P, Falzoni S, Hanau S, DiVirgilio F. Purinergic modulation of interleukin-1 beta release from microglial cells stimulated with bacterial endotoxin. J Exp Med 1997;185:579-82

Filippini A, Sitkovsky MV. "Alternative Molecular Mechanisms of T-Cell Receptor Regulated Effector Functions of Cytolytic T-Lymphocytes." In *Proceedings of the International Conference "Molecular Aspects of the Immune Response", vol. X4.* J Immunol Immunopharmacol, p 149, 1990.

Filippini A, Taffs RE, Agui T, Sitkovsky MV. Ecto-ATPase activity in cytolytic T-lymphocytes. Protection from the cytolytic effects of extracellular ATP. J Biol Chem 1990;265:334-340

Harden TK, Lazarowski ER, Boucher RC. Release, metabolism and interconversion of adenine and uridine nucleotides: implications for G protein-coupled P2 receptor agonist selectivity. Trends Pharmacol Sci 1997;18:43-6

Henkart PA, Sitkovsky MV. Cytotoxic lymphocytes. Two ways to kill target cells. Curr Biol 1994;4:923-925

Hirschhorn R. Adenosine deaminase deficiency: molecular basis and recent developments. Clin

Immunol Immunopathol 1995;76:S219-S227

Hogquist KA, Bevan MJ. The nature of the peptide/MHC ligand involved in positive selection. Semin Immunol 1996;8:63-8

Hoskin DW, Reynolds T, Blay J. 2-chloroadenosine inhibits the MHC-unrestricted cytolytic activity of anti-CD3-activated killer cells: evidence for the involvement of a non-A1/A2 cell-surface adeonsine receptor. Cell Immunol 1994;159:85-93

Huang S, Koshiba M, Apasov, S, Sitkovsky M. Role of A_{2A} adenosine receptor- mediated signaling in inhibition of T cell activation and expansion. Blood 1997;90:1600-1610

Jacobson KA, vonLubitz DK, Daly JW, Fredholm BB. Adenosine receptor ligands: differences with acute versus chronic treatment. Trends Pharmacol Sci 1996; 17:109-13

Kincaid R, Takayama H, Billingsley M, Sitkovsky M. Differential expression of calmodulin-binding proteins in B-, T-lymphocytes and thymocytes. Nature 1987;330:176-178

Koshiba M, Apasov S, Sverdlov V, Chen P, Erb L, Turner JT, Weisman, GA, Sitkovsky MV. Transient up-regulation of P2Y2 nucleotide receptor mRNA expression is an immediate early gene response in activated thymocytes. Proc Natl Acad Sci USA 1997;94:831-836

Koshiba M, Kojima H, Huang S, Apasov S, Sitkovsky MV. Memory of extracellular adenosine/A_{2A} purinergic receptor-mediated signalling in murine T cells. J Biol Chem 1997;272:25881

Ledent C, Vaugeois JM, Schiffmann SN, Pedrazzini T, El Yacoubi M, Vanderhaeghen JJ, Costentin J, Heath JK, Vassart G, Parmentier M. Aggressiveness, hypoalgesia and high blood pressure in mice lacking the adenosine A_{2A} receptor. Nature 1997;388:674-8

Linden J. Cloned adenosine A3 receptors: pharmacological properties, species differences and receptor functions. Trends Pharmacol Sci 1994;15:298-306

Lustig KD, Shiau AK, Brake AJ, Julius D. Expression cloning of an ATP receptor from mouse neuroblastoma cells. Proc Natl Acad Sci USA

Malissen B, Malissen M. Functions of TCR and pre-TCR subunits: lessons from gene ablation. Curr Opin Immunol 1996;8:383-93

McConkey DJ, Nicotera P, Orrenius S. Signalling and chromatin fragmentation in thymocyte apoptosis. Immunol Rev 1994;142:343-63

Nagata S, Suda T. Fas and Fas ligand:lpr and gld mutations. Immunol Today 1995;16:35-43

RECENT ADVANCES IN CARDIAC ADENOSINE METABOLISM

Jürgen Schrader, Ulrich Decking and Thomas Stumpe, Department of Cardiovascular Physiology, Heinrich-Heine-University, Düsseldorf, Germany

I. SUMMARY

Adenosine can be formed intracellularly (cytosolic 5′-nucleotidase and SAH-hydrolase) and extracellularly (ecto-5′-nucleotidase). The major source for cardiac adenosine formation is the intracellular dephosphorylation of AMP. More than 90% of the adenosine formed is rephosphorylated to AMP via adenosine kinase (1.95 nmol/min/g) and only a small fraction (0.06 nmol/min/g) escapes the metabolic cycle between AMP and adenosine and is released into the vascular space. Therefore, intracellular adenosine formation exceeds by far adenosine release into the extracellular space. However, since the intracellular adenosine concentration is normally very low and adenosine is continuously formed extracellularly from released adenine nucleotides, the concentration gradient for adenosine in the normoxic heart is from extracellular to intracellular. This gradient is rapidly reversed during hypoxia or pharmacological inhibition of adenosine kinase.

Inadequate supply of oxygen is the most important physiological trigger for the formation of adenosine. The critical PO_2 below which adenosine accumulates found to be about 3 mmHg. In the range between 5 - 11 mmHg, cardiomyocytes have the remarkable ability to downregulate their oxygen consumption by about 40 % (hibernation) before metabolic signs of hypoxia occur. Therefore a reduction in oxygen supply as such is not a sufficient cause for the formation of adenosine but the imbalance between ATP formation and consumption.

AMP is both a precursor of adenosine but also a product when rephosphorylated by adenosine kinase. The functional significance of this metabolic cycle was only recently discovered. It was shown that cardiac hypoxia causes inhibition of adenosine kinase thereby shunting adenosine from the salvage pathway (adenosine Ù AMP) to extracellular release. Inhibition of adenosine kinase by hypoxia therefore amplifies changes of free cytosolic adenosine and by this means translates a minor decrease in cardiac energetic into a substantial rise of adenosine.

Future studies must resolve which factors govern the *in vivo* activity of adenosine kinase and how this is linked to the known physiological effects of adenosine. Whether a disturbance of the active metabolic cycle between AMP and adenosine can result in enhanced uric acid production in certain disease states remains to be investigated.

II. INTRODUCTION

The history of adenosine research started with the isolation of AMP from cardiac muscle and the examination of the cardiovascular effects of adenine nucleotides and nucleosides by Drury and Szent-Györgyi (1929). Albert Szent-Györgyi received the Nobel Prize for his demonstration that the hydrolysis of ATP provides the energy for muscular contraction. The other two key papers which greatly influenced modern adenosine research came from Berne's (1963) and Gerlach's (Gerlach, et al., 1963) laboratory in 1963. Both groups provided evidence suggesting that adenosine may be an important metabolite linking regulation of coronary blood flow to cardiac energy metabolism through work-induced hydrolysis of ATP.

Aside from the functional role adenosine may play in various organs it was from the beginning crucial to know by which metabolic pathways adenosine is formed, which cells serve as sinks and sources for adenosine and which factors govern the relationship between energy metabolism and adenosine formation. The initial view that the main pathway for adenosine formation is by dephosphorylation of AMP linked to ATP turnover was modified in later years in many respects. We now know that aside from AMP, adenosine is formed from S-adenosylhomocysteine (SAH) and once formed adenosine is rapidly rephosphorylated to AMP or deaminated to inosine. Many laboratories have contributed to our present understanding of adenosine metabolism and major questions appear now to be much better understood. Most of the new concepts have been developed in the cardiovascular system. However, the general conclusions and principal insights are applicable to other organs and tissues as well.

Various aspects of cardiac adenosine metabolism have been reviewed by Schrader (1990), Sparks and Bardenheuer (1986), Belardinelli et al.(1989) and Olsson and Pearson (1990). There have been specialized reviews on the history of adenosine research (Olsson, 1992), the role of adenosine in pain during angina pectoris (Sylven, 1993), circadian variations (Chagoya de Sanchez, 1995), adenosine receptor subtypes (Linden, 1994; Olah and Stiles, 1995), electrophysiology of adenosine in the heart (Belardinelli, et al., 1995; Shen and Kurachi, 1995), myocardial protection (Lasley and Mentzer, 1995) and cardiovascular pharmacology of purines (Rogen, et al., 1997). The following review will summarize the major recent advances in our understanding of the metabolism of adenosine.

III. TEXT

Compartmentation of cardiac adenosine metabolism

First evidence for compartmentation of adenosine metabolism within an intact organ came from studies in which isolated perfused hearts were prelabeled with radioactive adenine or adenosine (Schrader and Gerlach, 1976). Measurement of the specific activity of adenosine released into the coronary effluent perfusate revealed that the specific radioactivity of precursor adenine nucleotides was significantly lower compared with that of released adenosine. Since precursor and product must have the same specific activity in a homogeneous system this suggested that released adenosine was derived from a highly labeled adenine nucleotide pool within the heart. Using autoradiography under similar experimental conditions this pool was later identified to be the vascular endothelium (Nees, et al., 1985). This led to the new concept, that the endothelium forms a tight metabolic barrier for intravascularly applied adenosine.

The molecular basis of this barrier function resides in a highly active adenosine transport system combined with rapid intracellular conversion to AMP via adenosine kinase. Using human umbilical vein endothelial cells, adenosine transport was shown to be mediated by a single facilitated diffusion system (K_M: 70 µM) with high maximum velocity (V_{max}: 600 pmol/10^6 cells/s) that could be inhibited by nitrobenzylthioinosine (Sobrevia, et al., 1994). Adenosine taken up from extracellular production sites or formed intracellularly is rapidly phosphorylated by low K_M (< 2.5 µM) adenosine kinase, maintaining a low intracellular concentration of adenosine. Since this enzyme exhibits a sufficiently high V_{max} in endothelium (Deussen, et al., 1993; Mistry and Drummond, 1986), it is very likely that under normal conditions there is a concentration gradient of adenosine from the extracellular to the intracellular space (Mattig, et al., 1996).

The demonstration of a high affinity uptake system for adenosine by endothelial cells was thereafter used by several laboratories to selectively prelabel endothelial adenine nucleotides within an intact organ. By measuring e.g. the specific activity of cAMP released from prelabeled hearts as an index of the specific activity of intracellular AMP, the precursor of adenosine, the contribution of the coronary endothelial adenine nucleotides to coronary venous adenosine release was estimated to be 14% in the guinea pig heart (Kroll, et al., 1987). Using a similar approach Becker and Gerlach arrived at an estimate of 25% (Becker and Gerlach, 1987). Selective endothelial prelabeling was also used to determine changes of the relative contribution of the endothelium versus cardiomyocytes to venous adenosine release. It was shown that during hypoxia (Deussen, et al., 1986), hypercapnic perfusion (Deussen, et al., 1986), acidosis (Bardenheuer, et al., 1987), catecholamines (Bardenheuer, et al., 1987) and acetylcholine (Bardenheuer, et al., 1987) the contribution of the endothelium to global adenosine release always decreased. This suggests that adenosine release from cardiomyocytes dominates over vascular adenosine release.

Selective uptake of intracoronary applied adenosine by endothelial cells has important functional implications. Provided the endothelial barrier to be tight, this would imply that endothelial cells could mediate vasodilation by arterially applied adenosine. Adenosine receptors have been identified on endothelial cells (Mustafa, et al., 1995) and endothelial cells have been shown to release NO when stimulated with adenosine (Li, et al., 1995). Whether or not endothelial derived vasodilators mediate the coronary vascular effects of adenosine under *in vivo* conditions has not been explored. Another implication of selective endothelial adenosine uptake is that the dose-response curve for adenosine obtained by conventional intracoronary application may not be identical when adenosine is formed intracellularly by cardiomyocytes and then diffuses to its site of action at vascular resistance vessels. The threshold for coronary vasodilation with intracoronary mode of application is 10^{-8} M, and maximal vasodilation is reached at 10^{-6} M (Schrader, et al., 1977). Elevating interstitial adenosine concentration with specific inhibitors of adenosine kinase and adenosine deaminase it was found, that the dose response curve in the dog heart is surprisingly steep: threshold concentration was 7×10^{-7} M and half maximal effect was 1.4×10^{-6} M (Kroll and Stepp, 1996; Stepp, et al., 1996). Future studies must incorporate this knowledge when exploring the functional role of adenosine in setting vascular tone.

Adenosine and cardiac energetics

A key feature in the regulation of cardiac performance is that myocardial O_2 consumption and coronary blood flow are closely matched. This implies at the organ level that mechanisms are available by which O_2 delivery through the coronary circulation can adequately balance the O_2 needs of the myocardium. In addition, due to the high rate of ATP turnover in the heart, important regulatory mechanisms exist at the cellular level by which changes in energy requirements (hydrolysis of ATP) are finely balanced by the rate of energy conversion (production of ATP), thus resulting in a stable steady state of cardiac performance.

There has been much effort in recent years to further define the underlying mechanisms of these two processes. The classical model of respiratory control by free cytosolic ADP has been seriously questioned and a multifactorial control of oxidative phosphorylation has been proposed (for review see Chance, et al., 1986; Heineman and Balaban, 1990). In reviewing the recent literature it is striking to note that predominantly in saline perfused heart preparations there was a correlation between free ADP and cardiac performance, whereas in more intact models e.g. the blood perfused dog heart *in situ* there were no measurable changes in free ADP despite substantial increases in cardiac work (Balaban, et al., 1986).

In a recent study, Decking et al. investigated the relationship between free ADP and cardiac performance using a newly developed isolated guinea pig heart preparation capable of performing external work and suitable for NMR spectroscopic investigations (Decking, et al., 1997). It was found that cardiac energy status and free ADP remained unaltered when myocardial O_2 consumption was increased by changes in afterload. This finding clearly indicates that a rise in free ADP is not obligatory for respiratory control. Most likely the increased afterload simultaneously increased coronary perfusion pressure and thereby assured adequate tissue oxygenation in this model. Along with this interpretation is the finding, that in the same experimental model adrenergic stimulation with dobutamine caused free ADP to increase and this effect was attenuated by a fluorocarbon emulsion which doubled the oxygen carrying capacity of the perfusion medium. Similar findings were reported for Langendorff hearts perfused with erythrocytes (Ning, et al., 1992). It is highly likely that the apparent correlation between free ADP and myocardial O_2 consumption observed in several studies does not reflect a regulatory role of ADP but is indicative of inadequate tissue oxygenation.

The complexity of respiratory control in the heart is highlighted by the finding that addition of pyruvate to glucose containing medium increased cardiac performance and lowered free ADP (Decking, et al., 1997). Provision of pyruvate is known to elevate mitochondrial NADH and thereby increases the driving force for oxidative phosphorylation. The decreased free ADP might subsequently act as a feedback signal, which limits oxidative phosphorylation.

That tissue oxygenation is in fact crucial for changes in free ADP was recently demonstrated by Stumpe and Schrader (Stumpe and Schrader, 1997) using isolated cardiomyocytes exposed to a constant ambient PO_2. It was found that a threefold increase in respiratory rate by electrical stimulation was associated with an increase in free ADP only when ambient PO_2 was below 4.6 mmHg. Above this value there were no changes in cellular energetics, again suggesting that tissue oxygenation may have been a crucial but uncontrolled variable in previous *in vivo* studies.

Free ADP is linked to AMP by action of myokinase and the substrate concentration of AMP is an important determinant for the formation of adenosine. Assuming myokinase to be in equilibrium one would assume that changes in ADP translate by mass action into changes in AMP and consequently in adenosine. The experimental evidence supports this assumption. In both models, the working heart preparation (Decking, et al., 1997) or isolated cardiomyocytes (Stumpe and Schrader, 1997) a parallelism between free ADP, AMP and adenosine was observed suggesting that the mechanisms governing the changes in ADP may be similar to those which are important in the formation and release of adenosine. Surprisingly, in the well oxygenated heart addition of pyruvate to glucose containing medium decreased free AMP but had no impact on adenosine.

Using a mathematical model of adenosine metabolism it was predicted that a doubling of free AMP results in approximately a doubling of adenosine release into the coronary effluent perfusate (Kroll, et al., 1993). Contrary to this prediction a variety of interventions had only a small impact on the free AMP concentration, despite pronounced changes in adenosine release (Decking, et al., 1997): Reducing arterial PO_2, elevating preload or decreasing afterload as well as adrenergic stimulation with dobutamine or isoproterenol augmented the release of adenosine several-fold but increased free AMP only little if at all. These findings suggests that an amplification mechanism must exist in the heart, whereby under conditions of limited oxygen availability small changes in AMP are translated into large changes in adenosine. Theoretically this could be activation of cytosolic 5′-nucleotidase or inhibition of adenosine kinase.

Tissue PO_2 as a trigger for enhanced adenosine

It has long been known that a considerable decrease in oxygen supply to the heart induces a rise in adenosine release (e.g., Schrader, et al., 1977). However, the precise relationship between cytosolic PO_2, cardiac energy status and free cytosolic adenosine has only recently been explored on the cellular level (Smolenski, et al., 1991; Stumpe and Schrader, 1997). As shown in Figure 1, oxygen consumption in stimulated rat cardiomyocytes (9 Hz) remained unchanged as long as ambient PO_2 was above 10.7 mmHg. Oxygen consumption (and thus ATP formation) only declined when PO_2 was decreased below this level, the critical PO_2.

The concept of a critical PO_2 was originally introduced to define the threshold at which oxygen consumption decreases in whole animals or isolated cells when oxygen delivery is reduced (Schumacker, et al., 1993). This concept can be more generally applied and defines the PO_2 below which tissue metabolites are changed. While oxygen consumption started to decline below a PO_2 of 10.7 mmHg, a decrease in phosphocreatine

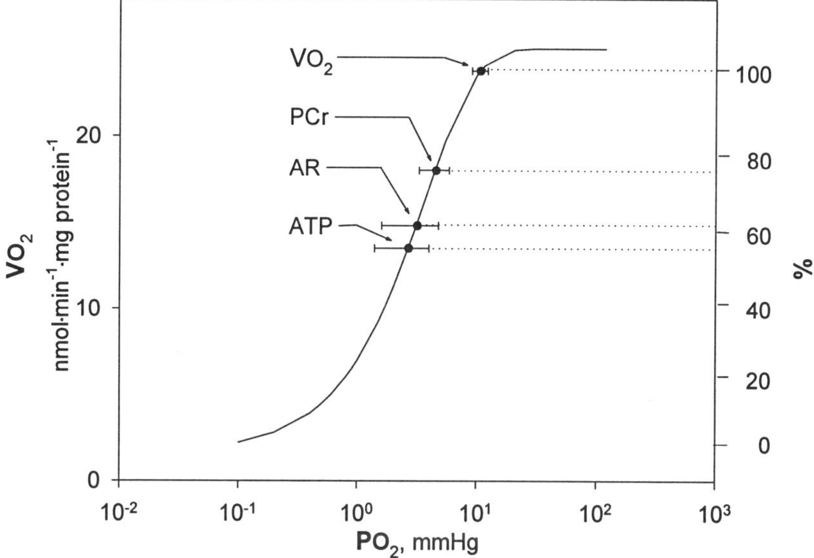

Figure 1: Oxygen consumption (VO_2) of stimulated rat cardiomyocytes at different levels of ambient PO_2. The critical PO_2 values for VO_2 were calculated from the original data. Also depicted are the critical PO_2 for phosphocreatine (PCr), cytosolic adenosine concentration (AR), and ATP, which were determined in parallel measurements. Data from Stumpe and Schrader (1997).

was only observed when PO_2 fell below 4.6 mmHg (Fig. 1). The criticial PO_2 for a rise in adenosine (3.2 mmHg) and a decline in ATP (2.7 mmHg) was even lower. Since adenosine only increased at a PO_2 level where energy status was already compromized, this is in line with the suggestion that an increase in free AMP is a prerequisite for enhanced adenosine formation but not its sufficient cause.

From these data it is obvious that stimulated cardiomyocytes have the ability to decrease their oxygen consumption by about 40 % before metabolic signs of hypoxia (increase in adenosine and decrease in ATP) occur. It is therefore neither a decrease in tissue PO_2 as such, nor the reduction in oxygen consumption but most likely an imbalance between ATP formation and consumption, resulting in a rise in cytosolic AMP which governs cytosolic adenosine. This interpretation is in line with the work of Bardenheuer and Schrader (1986) who demonstrated that adenosine release from the heart is not controlled by the metabolic rate but by an imbalance between oxygen supply and demand.

A coordinated suppression of energy demand as a means to balance ATP production and ATP utilization constitutes a fundamental defense strategy of cells against hypoxia (Hochachka, 1986). By reducing energy expenditure cells avoid early energy failure. This phenomenon of cell protection is not restricted to isolated cells but can also be observed in the intact heart where it is termed short term hibernation (Arai, et al., 1991; Heusch and Schulz, 1996). The molecular mechanisms underlying metabolic downregulation in the heart are presently not understood and may include changes in the efficiency of mitochondrial function, changes in cellular calcium handling or a more direct PO_2 dependent effect on contractile proteins (Marban, 1991). The important implication of this finding is that hypoxia as such is not a sufficient cause for the formation of adenosine in the steady state as long as compensatory mechanisms are operative and cardiac energy status remains unaltered. Only when these adaptive, energy sparing mechanisms are exhausted there is a true mismatch between ATP supply and ATP demand. As a consequence adenosine is formed and released at an accelerated rate.

What are the implications of these findings as to the role adenosine may play in the regulation of coronary blood flow? In the light of the findings discussed above it is highly unlikely that in well oxygenated cardiomyocytes a rise in energy turnover results in an increase in adenosine formation. A rise in cytosolic adenosine requires a low PO_2 (microhypoxia, $PO_2 < 3$ mmHg) which may not be present in blood perfused dog hearts even during increases in cardiac work (Kroll and Martin, 1994). Nevertheless, there are numerous reports in the literature over the past 20 years demonstrating a correlation between coronary blood flow and adenosine (Berne, 1980). It is conceivable that this adenosine was not entirely derived from cardiomyocytes, but was also formed by vascular smooth muscle or coronary endothelium, both of which are capable of producing adenosine (Skladanowski, et al., 1996). However, there is no known signal which links cardiac energy requirements to adenosine release of vascular cells. As far as adenosine formation by cardiomyocytes is concerned, the near perfect match between oxygen delivery via the coronary circulation and cardiac energy requirements makes a global impairment of tissue oxygenation unlikely. There might be small perfusion domains of low PO_2 triggering enhanced formation of adenosine. However, although coronary blood flow displays a high degree of local flow heterogeneity, there is no evidence for tissue hypoxia in low flow areas (Sonntag, et al., 1996). In conclusion, as long as there is an adequate oxygen supply to the heart, vasodilator substances other than adenosine must be responsible for the observed increase in coronary blood flow. A similar conclusion was reached by Kroll and Feigl, demonstrating that in the intact dog heart adenosine is unlikely to control basal coronary blood flow (Kroll and Feigl, 1985).

Role of the metabolic cycle between adenosine and AMP

AMP is both a precursor of adenosine which is catalyzed by 5′-nucleotidase and a product of adenosine when this nucleoside is phosphorylated by adenosine kinase. The potential importance of this metabolic cycle between AMP and adenosine was first addressed by Arch and Newsholme (1978). These authors assumed adenosine kinase to be saturated under *in vivo* conditions and suggested that the cycle might serve to enhance the formation of adenosine at a given substrate concentration. At this time, however, it was not known that the free cytosolic adenosine concentration is only very low due to protein binding so that adenosine kinase is normally not working under saturating conditions. Furthermore, the input from S-adenosylhomocystein-hydrolase to the formation of adenosine was not considered.

First data on the turnover rate of the AMP-adenosine cycle came from use of potent inhibitors of adenosine kinase. Blocking the enzyme with iodotubercidin caused a dose dependent increase in the adenosine concentration of the coronary effluent perfusate of isolated hearts, reaching levels which were comparable to those observed with severe ischemia. Coronary blood flow increased in parallel and this effect could be abolished by enzymatically converting vasoactive adenosine into inactive inosine by coinfusion of adenosine deaminase. These data demonstrated that more than 90% of the adenosine produced within the heart is normally rephosphorylated to AMP without escaping into the venous effluent. The magnitude of this cycle is such that inhibition of adenosine rephosphorylation causes maximal coronary vasodilation under otherwise normoxic conditions. Kroll, et al. (1993), suggested that the rapid cycling may serve to amplify the relative importance of AMP hydrolysis over transmethylation in controlling cytosolic adenosine concentrations. In addition, it was postulated that endogenous inhibition of adenosine kinase might turn this cycle into an amplification system transforming small changes in adenosine formation into major changes in cytosolic adenosine and, ultimately, adenosine release.

More recently Decking, et al. (1997), addressed the physiological function of the AMP-adenosine metabolic cycle by combining measurements of the free cytosolic concentration of AMP using NMR-spectroscopy with the formation and release of adenosine in the heart. Selective blockade of adenosine deaminase and adenosine kinase indicated that flux through adenosine kinase decreased from 85% to 35% during hypoxia. Mathematical model analysis further indicated that the observed decrease in flux rate corresponds to an over 90% inhibition of adenosine kinase under this condition. Together these data clearly indicate that hypoxia potently inhibits adenosine kinase *in vivo*, thereby shunting adenosine from the salvage pathway (adenosine to AMP) to venous adenosine release. The AMP-adenosine metabolic cycle therefore appears to play an important physiological role: it amplifies small changes in free adenosine into major changes in adenosine. In doing so inhibition of adenosine kinase translates a minor decrease in cardiac energy status into a rise of cytosolic adenosine. Most likely this mechanism forms the molecular basis for the known high sensitivity of the cardiac adenosine system to changes in the O_2 supply-to-demand ratio.

Very little is known at present which factors mediate the hypoxia-induced inhibition of adenosine kinase. *In vitro* work suggests that ADP, AMP (Palella, et al., 1980), Pi (Gorman, et al., 1997), a deficiency of ATP (DeJong, 1977; Palella, et al., 1980) and an increase in pH and Mg^{2+} (Palella, et al., 1980) can inhibit adenosine kinase. Under the experimental conditions cited above (Decking, et al., 1997) which were carried out in the isolated perfused guinea pig heart, there was no evidence for changes in pH and intracellular Mg^{2+} as determined by ^{31}P-NMR. The observed hypoxia-induced increase in free ADP (40 to maximally 90 μmol/l) was only slightly inhibitory in purified adenosine kinase from human placenta (Palella, et al., 1980) while AMP only acted as competitive inhibitor at much higher concentration compared with those found *in vivo*. It is also unlikely that the small decrease in ATP observed with hypoxia can influence the activity of adenosine kinase because of the very low K_M-value of this enzyme. More recently it was found that adenosine kinase can be potently inhibited by P_i (Gorman, et al., 1997). Since hypoxia is associated with a rapid fall in PCr and a concomitant rise in P_i, this mechanism may be important *in vivo*. However, it is presently unclear whether the hypoxia-induced inhibition of adenosine kinase is the sum of the individual metabolite changes listed above or whether other unknown factors are involved.

Figure 2 summarizes our current knowledge of fluxes and cytosolic concentrations in cardiac adenosine metabolism. In the normoxic heart AMP and not SAH-hydrolase is the major substrate for adenosine formation. The majority of adenosine formed is rephosphorylated and only a minor portion is released into the coronary venous effluent. When switching to hypoxia, the increase in cytosolic AMP and adenosine production is not sufficient to explain the substantial rise of cytosolic adenosine and adenosine release observed under *in vivo* conditions. Hypoxia-induced inhibition of the key regulatory enzyme, adenosine kinase, limits rephosphorylation of cytosolic adenosine. Thereby adenosine is shunted from the salvage pathway to venous release, amplifiying the relatively small change in free AMP.

IV. CONCLUSIONS AND FUTURE DIRECTIONS

Rapid signal amplification for adenosine appears to be the functional role of the AMP-adenosine metabolic cycle during impaired tissue oxygenation. It is equally possible that adenosine kinase may be modulated by factors independent of tissue PO_2. Very little, however, is known at present on the individual factors which can control the activity of this enzyme and which are biologically relevant. Important questions to be answered concern the influence of ions known to change under various physiological conditions. It is also pertinent to explore whether there is posttranslational modification of adenosine kinase, e.g., by phosphorylation, and if so, which signaling pathway may be involved.

There is already some indirect evidence that besides the oxygen dependent regulation of adenosine kinase there may be oxygen independent mechanisms operative. A characteristic feature, e.g., of the cerebral circulation is the fact, that whenever metabolic rate of the brain increases there is an increase in cerebral blood flow which is in excess to the metabolic needs (Raichle, 1987). As a consequence, an increase in tissue PO_2 was observed by several laboratories (Ogawa, et al., 1993). Metabolic activation was also shown to be associated with enhanced formation of vasoactive adenosine (Ko, et al., 1990). This situation is opposite to that found in the heart and several other tissues, where an increase in metabolic rate is most often associated with a slight decrease in tissue PO_2, so that the match between oxygen supply and oxygen demand is not perfect. Since a decreased tissue oxygenation is not associated with functional activation of the brain it is rather unlikely that brain energetic parameters change sufficiently to cause an increase in free AMP. On the other hand the AMP-adenosine metabolic cycle is known to be highly active in the brain tissue and pharmacological inhibition was reported to depress neuronal activity in hippocampal slices (Pak, et al., 1994). It is therefore quite conceivable that inhibition of adenosine kinase by energy independent mechanisms might link functional activation to adenosine formation in the brain.

Adenosine can be formed intracellularly but also extracellularly by action of a highly active cascade of ecto-nucleotidases. There is accumulating evidence that extracellular adenosine formation is a physiological process whereby adenine nucleotides released from cells are the primary source (e.g., Borst and Schrader, 1991). It is, however, still a matter of debate by which mechanisms the adenine nucleotides are released. The multidrug-transporter MDR1 was suggested to mediate nucleotide efflux (Abraham , et al., 1993). It is unknown, however, whether the P-glycoprotein is the only nucleotide transporter in the cell membrane, and which factors control nucleotide release under physiological conditions.

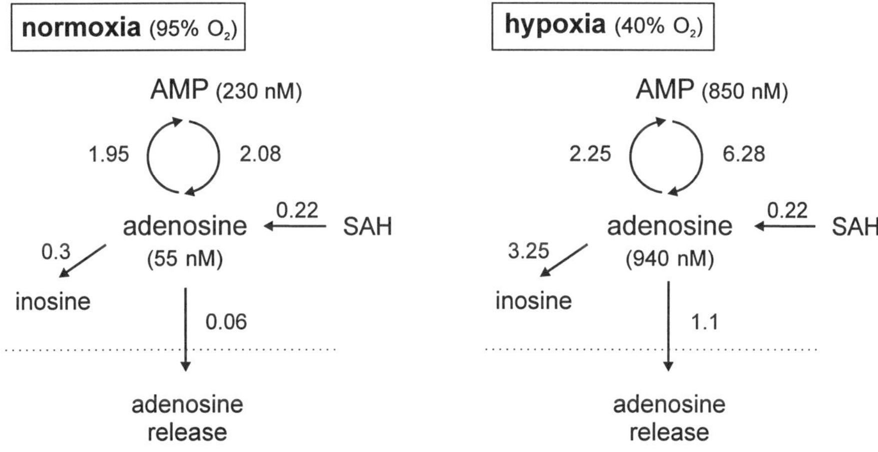

Figure 2: Fluxes and cytosolic concentrations in cardiac adenosine metabolism. Cytosolic concentrations of AMP, adenosine formation (AMP → adenosine + SAH → adenosine) and coronary venous adenosine release rates were experimentally determined as was the relative increase in free cytosolic adenosine (SAH-technique). Basal concentrations of cytosolic adenosine were taken to be 55 nmol/L based on model predictions and in agreement with previously determined data (Deussen, et al., 1988). Adenosine formation from SAH (transmethylation rate) was estimated on the basis of previously measured data (Deussen, et al., 1989). Flux through adenosine kinase and adenosine deaminase was estimated based on experimental results obtained with selective blockade of adenosine kinase (10 µmol/L iodotubercidin) and adenosine deaminase (5 µmol /L EHNA). Adapted from Decking, et al. (1997).

The principles of regulation of the adenosine metabolism elaborated in some detail in cardiac tissue appear to have a broader application. Adenosine kinase as well as cytosolic 5′-nucleotidase are found in almost every organ and cell (Arch and Newsholme, 1978; Snyder and Lukey, 1982). Pharmacological blockade of adenosine kinase was found to result in a significant rise in adenosine in liver and brain (Bontemps, et al., 1983; Pak, et al., 1994), suggesting a rapid turnover of the AMP-adenosine metabolic cycle. Provided there is a general decrease of adenosine kinase activity this should result in a substantial increase in the overall rate of net adenosine formation. Since the half live of plasma adenosine is rather short, most of this nucleoside is channeled into the catabolic pathway via inosine, hypoxanthine and finally to uric acid. This metabolic cascade is highly active in endothelial cells and assures that the extracellular biological activity of adenosine is maintained within a narrow and low concentration range. The possibility that an enhanced uric acid production such as in gouty patients may be caused by a decreased activity of adenosine kinase has not been explored so far but appears to be an attractive hypothesis which can be solved by standard enzymatic and genetic techniques.

V. REFERENCES

Abraham EH, Prat AG, Gerweck L, Seneveratne T, Arceci AJ, Kramer R, Guidotti G, Cantiello HF. The multidrug-resistance (mdr1) gene product functions as an ATP channel. *Proc Natl Acad Sci USA, 90:* 312-316, 1993.

Arai AE, Pantely GA, Anselone CG, Bristow J, Bristow JD. Active down-regulation of myocardial energy requirements during prolonged moderate ischemia in swine. *Circ Res, 69:* 1458-1469, 1991.

Arch JRS, Newsholme EA. Activities and some properties of 5'-nucleotidase, adenosine kinase and adenosine deaminase in tissues from vertebrates and invertebrates in relation to the control of the concentration and the physiological role of adenosine. *Biochem J, 174:* 965-977, 1978.

Balaban RS, Kantor HL, Katz LA, Briggs RW. Relation between work and phosphate metabolites in the in vivo paced mammalian heart. *Science, 232:* 1121-1123, 1986.

Bardenheuer H, Schrader J. Supply-to-demand ratio for oxygen determines formation of adenosine by the heart. *Am J Physiol, 250:* H173-H180, 1986.

Bardenheuer H, Whelton B, Sparks HV, Jr. Adenosine release by the isolated guinea pig heart in response to isoproterenol, acetylcholine, and acidosis: the minimal role of vascular endothelium. *Circ Res, 61:* 594-600, 1987.

Becker BF, Gerlach E. Uric acid, the major catabolite of cardiac adenine nucleotides and adenosine, originates in the coronary endothelium. In: *Topics and Perspectives in Adenosine Research*, Gerlach E, Becker BF (eds.), Berlin: Springer, 1987, pp 209-221.

Belardinelli L, Linden J, Berne RM. The cardiac effects of adenosine. *Prog Cardiovasc Dis, 32:* 73-97, 1989.

Belardinelli L, Shryock JC, Song Y, Wang D, Srinivas M. Ionic basis of the electrophysiological actions of adenosine on cardiomyocytes. *FASEB J, 9:* 359-365, 1995.

Berne RM. Cardiac nucleotides in hypoxia: possible role in regulation of coronary blood flow. *Am J Physiol, 204:* 317-322, 1963.

Berne RM. The role of adenosine in the regulation of coronary blood flow. *Circ Res, 47:* 807-813, 1980.

Bontemps F, van den Berghe G, Hers HG. Evidence for a substrate cycle between AMP and adenosine in isolated hepatocytes. *Proc Natl Acad Sci USA, 80:* 2829-2833, 1983.

Borst M, Schrader J. Adenine nucleotide release from isolated perfused guinea pig hearts and extracellular formation of adenosine. *Circ Res, 68:* 797-806, 1991.

Chagoya de Sanchez V. Circadian variations of adenosine and of its metabolism. Could adenosine be a molecular oscillator for circadian rhythms? *Can J Physiol Pharmacol, 73:* 339-355, 1995.

Chance B, Leigh JS, Jr., Kent J, McCully K, Nioka S, Clark BJ, Maris JM, Graham T. Multiple controls of oxidative metabolism in living tissues as studied by phosphorus magnetic resonance. *Proc Natl Acad Sci USA, 83:* 9458-9462, 1986.

De Jong JW. Partial purification and properties of rat-heart adenosine kinase. *Arch Int Physiol Biochim, 85:* 557-569, 1977.

Decking UKM, Arens S, Schlieper G, Schulze K, Schrader J. Dissociation between adenosine release, cardiac energy status and oxygen consumption in working guinea pig hearts. *Am J Physiol, 272:* H371-H381, 1997.

Decking UKM, Schlieper G, Kroll K, Schrader J. Hypoxia-induced inhibition of adenosine kinase potentiates cardiac adenosine release. *Circ Res, 81:* 154-164, 1997.

Deussen A, Bading B, Kelm M, Schrader J. Formation and salvage of adenosine by macrovascular endothelial cells. *Am J Physiol 264:* H692-H700, 1993.

Deussen A, Borst M, Kroll K, Schrader JKK. Formation of S-adenosylhomocysteine in the heart. II: A sensitive index for regional myocardial underperfusion. *Circ Res, 63:* 250-261, 1988.

Deussen A, Lloyd HGE, Schrader J. Contribution of S-adenosylhomocysteine to cardiac adenosine formation. *J Mol Cell Cardiol, 21:* 773-782, 1989.

Deussen A, Möser GH, Schrader J. Contribution of coronary endothelial cells to cardiac adenosine production. *Pflügers Arch, 406:* 608-614, 1986.

Drury AN, Szent-Györgi A. The physiological activity of adenine compounds with especial reference to their action upon the mammalian heart. *J Physiol, 68:* 213-237, 1929.

Gerlach E, Deuticke B, Dreisbach RH. Der Nucleotid-Abbau im Herzmuskel bei Sauerstoffmangel und seine mögliche Bedeutung für die Coronardurchblutung. *Die Naturwissenschaften, 50:* 228-229, 1963.

Gorman MW, He M-X, Hall CS, Sparks HV. Inorganic phosphate as regulator of adenosine formation in isolated guinea pig hearts. *Am J Physiol, 272:* H913-H920, 1997.

Heineman FW, Balaban RS. Control of mitochondrial respiration in the heart in vivo. *Ann Rev Physiol, 52:* 523-542, 1990.

Heusch G, Schulz R. Hibernating myocardium: a review. *J Mol Cell Cardiol, 28:* 2359-2372, 1996.

Hochachka PW. Defense strategies against hypoxia and hypothermia. *Science, 231:* 234-241, 1986.

Ko KR, Ngai AC, Winn, HR. Role of adenosine in regulation of regional cerebral blood flow in sensory cortex. *Am J Physiol, 259:* H1703-H1708, 1990.

Kroll K, Decking UKM, Dreikorn K, Schrader J. Rapid turnover of the AMP-adenosine metabolic cycle in the guinea pig heart. *Circ Res, 73:* 846-856, 1993.

Kroll K, Feigl EO. Adenosine is unimportant in controlling coronary blood flow in unstressed dog hearts. *Am J Physiol, 249:* H1176-H1187, 1985.

Kroll K, Martin GV. Steady-state catecholamine stimulation does not increase cytosolic adenosine in canine hearts. *Am J Physiol, 265:* H503-H510, 1994.

Kroll K, Schrader J, Piper HM, Henrich M. Release of adenosine and cyclic AMP from coronary endothelium in isolated guinea pig hearts: relation to coronary flow. *Circ Res, 60:* 659-665, 1987.

Kroll K, Stepp DW. Adenosine kinetics in canine coronary circulation. *Am J Physiol, 270:* H1469-H1483, 1996.

Lasley RD, Mentzer RMJ. Protective effects of adenosine in the reversibly injured heart. *Ann Thorac Surg, 60:* 843-846, 1995.

Li JM, Fenton RA, Cutler BS, Dobson JG, Jr.. Adenosine enhances nitric oxide production by vascular endothelial cells. *Am J Physiol, 269:* C519-C523, 1995.

Linden J. Cloned adenosine A_3 receptors: Pharmacological properties, species differences and receptor functions. *Trends Pharmacol Sci 15:* 298-306, 1994.

Marban E. Myocardial stunning and hibernation. *Circulation, 83:* 681-688, 1991.

Mattig S, Gruber M, Deussen A. Intracellular concentration and transmembraneous gradient of adenosine in macrovascular endothelial cells. *Pflügers Arch, 431:* R128, 1996 (Abstract).

Mistry G, Drummond, GI. Adenosine metabolism in microvessels from heart and brain. *J Mol Cell Cardiol, 18:* 13-22, 1986.

Mustafa SJ, Marala RB, Abebe W, Jeansonne N, Olanrewaju HA, Hussain T. Coronary adenosine receptors: subtypes, localization, and function. In: *Adenosine and Adenine Nucleotides: From Molecular Biology to Integrative Physiology*, Belardinelli L, Pelleg A (eds.), Boston: Kluwer, 1995, pp 229-239.

Nees S, Herzog V, Becker BF, Böck M, Des Rosiers Ch, Gerlach E. The coronary endothelium: a highly active metabolic barrier for adenosine. *Basic Res Cardiol, 80:* 515-519, 1985.

Ning X-H, He M-X, Gorman MW, Romig GD, Sparks HV. Adenosine formation and energy status in isolated guinea pig hearts perfused with erythrocytes. *Am J Physiol, 262:* H1075-H1080, 1992.

Ogawa S, Menon RS, Tank DW, Kim S-G, Merkle H, Ellermann JM, Ugurbil K. Functional brain mapping by blood oxygenation level-dependent contrast magnetic resonance imaging. *Biophys J, 64:* 803-812, 1993.

Olah ME, Stiles GL. Adenosine receptor subtypes: Characterization and therapeutic regulation. *Annu Rev Pharmacol Toxicol, 35:* 581-606, 1995.

Olsson RA. A brief history of adenosine research. *Pharm Pharmacol Lett, 2:* 3-4, 1992.

Olsson RA, Pearson JD. Cardiovascular Purinoceptors. *Physiol Rev 70:* 761-845, 1990.

Pak MA, Haas HL, Decking UKM, Schrader J. Inhibition of adenosine-kinase increases endogenous adenosine and depresses neuronal activity in hippocampal slices. *Neuropharmacology, 33:* 1049-1053, 1994.

Palella TD, Andres CM, Fox IH. Human placental adenosine kinase. Kinetic mechanism and inhibition. *J Biol Chem 255:* 5264-5269, 1980.

Raichle ME. Circulatory and metabolic correlates of brain function in normal humans. In: *Handbook of Physiology: The nervous system*, Baltimore: Williams & Wilkens, 1987, pp 643-674.

Rongen GA, Floras JS, Lenders JW, Thien T, Smits P. Cardiovascular pharmacology of purines. *Clin Sci, 92:* 13-24, 1997.

Schrader J. Adenosine: A homeostatic metabolite in cardiac energy metabolism. *Circulation, 81:* 389-391, 1990.

Schrader J, Gerlach E. Compartmentation of cardiac adenine nucleotides and formation of adenosine. *Pflügers Arch, 267:* 129-135, 1976.

Schrader J, Haddy FJ, Gerlach E. Release of adenosine, inosine and hypoxanthine from the isolated guinea pig heart during hypoxia, flow-autoregulation and reactive hyperemia. *Pflügers Arch, 369:* 1-6, 1977.

Schumacker PT, Chandel M, Agusti AGN. Oxygen conformance of cellular respiration in hepatocytes. *Am J Physiol, 265:* L395-L402, 1993.

Shen WK, Kurachi Y. Mechanisms of adenosine-mediated actions on cellular and clinical cardiac electrophysiology. *Mayo Clin Proc, 70:* 274-291, 1995.

Skladanowski AC, Smolenski RT, Tavenier M, De Jong JW, Yacoub MH, Seymour A-ML. Soluble forms of 5'-nucleotidase in rat and human heart. *Am J Physiol 270:* H1493-H1500, 1996.

Smolenski RT, Schrader J, De Groot H, Deussen A. Oxygen partial pressure and free intracellular adenosine of isolated cardiomyocytes. *Am J Physiol, 260:* C708-C714, 1991.

Snyder FF, Lukey T. Kinetic considerations for the regulation of adenosine and deoxyadenosine metabolism in mouse and human tissues based on a thymocyte model. *Biochim Biophys Acta, 696:* 299-307, 1982.

Sobrevía L , Jarvis SM, Yudilevich DL. Adenosine transport in cultured human umbilical vein endothelial cells is reduced in diabetes. *Am J Physiol, 267:* C39-C47, 1994.

Sonntag M, Deussen A, Schultz J, Loncar R, Hort W, Schrader J. Spatial heterogeneity of blood flow in the dog heart. I. Glucose uptake, free adenosine and oxidative / glycolytic enzyme activity. *Pflügers Arch, 432:* 439-450, 1996.

Sparks HV, Jr., Bardenheuer H. Regulation of adenosine formation by the heart. *Circ Res, 58:* 193-201, 1986.

Stepp DW, Van Bibber R, Kroll K. Quantitative relation between interstitial adenosine concentration and coronary blood flow. *Circ Res, 79:* 601-610, 1996.

Stumpe T, Schrader J. Phosphorylation potential, adenosine formation, and critical PO_2 in stimulated rat cardiomyocytes. *Am J Physiol, 273:* H756-H766, 1997.

Sylvén C. Mechanisms of pain in angina pectoris - A critical review of the adenosine hypothesis. *Cardiovasc Drugs Ther, 7:* 745-759, 1993.

THE ROLE OF PERIPHERAL AUTONOMIC NEURONS WITH P_1- AND/OR P_2-PURINORECEPTORS IN CARDIAC REGULATION

J. Andrew Armour, Dalhousie University, Canada

I. SUMMARY

Neurons in different intrathoracic ganglia which are involved in cardiac regulation exhibit non-coupled behavior, even when they are mutually entrained to cardiac events by cardiovascular afferent feedback. A redundancy of cardioregulatory control is exhibited by the various populations of intrathoracic neurons, each responding differently to similar cardiac interventions. This is indicative of the selective nature of the feedback mechanisms which exist at successive levels of the cardiac nervous system. Neurons in different intrathoracic ganglia receive inputs from intrathoracic afferent neurons associated with cardiac sensory neurites, many of which possess P_1- and/or P_2-purinoceptors. Such cardiac afferent neurons are sensitive to local myocardial ischemia. That local circuit and efferent neurons in different peripheral ganglia display dissimilar responses when the sensory neurites of intrathoracic cardiac afferent neurons are exposed to adenosine or ATP is indicative of the complexity of their functional interconnectivity. Given that, selective blockade of P_1- and P_2-purinoceptors associated with ischemia sensitive cardiac afferent neurons may influence not only cardiac symptoms but cardiovascular reflexes associated with myocardial ischemia. Interactions among purinergic peripheral and central neurons must be taken into account when considering the development of pharmacological strategies to modify such cardiac pathology.

II. BACKGROUND

Intrathoracic autonomic neurons, under the influence of central neurons, interact via a complex hierarchy to maintain adequate cardiac function (Armour, 1994). Information arising from afferent neurons associated with mechanosensory and/or chemosensory neurites in cardiac tissues and intrathoracic vessels influence not only central neurons, but cardiac efferent neurons in various intrathoracic ganglia via local circuit neurons. Some sensory neurites located in the carotid bulb or upper limb influence intrathoracic cardiac neurons indirectly via spinal cord neurons. The interactions which occur among intrathoracic neurons involved in cardiac regulation employ a variety of chemicals, including purinergic compounds. This review focuses on the role that purine sensitive peripheral autonomic neurons play in regulating the heart

III TEXT

This review begins with a discussion about the putative function of cardiac afferent neurons associated with purinergic sensory neurites. Then the function of intrathoracic local circuit and efferent cardiac neurons which are sensitive to purinergic agents will be discussed. The data reviewed in this chapter must be taken in the context that much remains to be known about the role of purine sensitive peripheral autonomic neurons in cardiac regulation, including how other chemicals influence purinergic cardiac neurons.

A. Cardiac afferent neurons.

Afferent neurons associated with sensory neurites in cardiac tissue are located in dorsal root, nodose, stellate, middle cervical, mediastinal and intrinsic cardiac ganglia (Armour, 1994). Although similar numbers of cardiac afferent neurons are located in dorsal root and nodose ganglia (Hopkins & Armour, 1989), each population displays unique functional characteristics. Whereas about 85% of dorsal root ganglion afferent neurons associated with epicardial sensory neurites display polysensory characteristics (Huang et al, 1996), only about 12% of nodose ganglion afferent neurons associated with such sensory neurites do so (Armour et al, 1994). Epicardial sensory neurites associated with intrathoracic afferent neurons, including those in intrinsic cardiac ganglia, are polysensory in nature as well (Armour, 1994).

The somata of isolated afferent neurons are known to be sensitive to adenosine (Macdonald et al, 1986). ATP and, to a lesser extent, adenosine influence skin sensory neurites associated with dorsal root ganglion neurons (Burnstock and Wood, 1996). The importance of adenosine in the genesis of cardiac pain became evident when Christer Sylvén and his colleagues administered adenosine into the blood stream of patients' diseased coronary arteries (Sylvén, 1989; Sylvén et al., 1986). The symptoms induced in these patients mimicked those which they experienced during effort (Crea et al, 1990; Gaspardone et al., 1992; Lagerqvist, 1991; Sylvén et al., 1987). These data are in accord with the finding that purine sensitive afferent neurons in dorsal root ganglia play an important role in the genesis of limb pain (Burnstock & Wood, 1996).

1. Nodose ganglion cardiac afferent neurons.

Cardiac sensory neurites associated with nodose ganglion afferent neurons are modified by chemical (63% of cardiac afferent neurons), mechanical (22%) or mechanical and chemical (15%) stimuli (Armour et al, 1994). The transmembrane properties of isolated nodose ganglion afferent neurons can be influenced by exogenously applied purinergic agents (Krishtal et al, 1988). In agreement with that, the activity generated by cardiac sensory neurites associated with nodose ganglion afferent neurons increases many fold when those sensory neurites are exposed to purinergic agents (from 0.1 to 35 Hz). The ventricular sensory neurites associated with about 55% of nodose ganglion cardiac afferent neurons possess P_1-purinoreceptors (Armour et al, 1994), some responding to local application of the A_1-adenosine receptor analogue N^6-Cyclopentlyadenosine (Fig. 1). About one third of the ventricular sensory neurites associated with nodose ganglion afferent neurons are sensitive to ß, g-methylene adenosine 5'-triphosphate. As ß, γ-methylene adenosine 5'-triphosphate does not rapidly degrade to adenosine, it appears that some

ventricular sensory neurites associated with nodose ganglion afferent neurons possess
P₂-purinoreceptors.

2. Dorsal root ganglion cardiac afferent neurons.

Adenosine affects the activity generated by about 70% of spontaneously
active cardiac afferent neurons in dorsal root ganglia (Huang et al, 1996).   The
spontaneous activity generated by ventricular sensory neurites associated with these
neurons increases when exposed to adenosine by, on average, about 500%   (peak
value of about 60 impulses per second from a baseline value of 1.8±0.6 impulses per
second). Ventricular sensory neurites associated with a number of dorsal root
ganglion afferent neurons (~60% of total) also respond to locally applied ATP, their
activity increasing by about 300% (peak activity of 49 impulses per second). There
remains a population (~10% of total) of dorsal root ganglion neurons associated with
ventricular sensory neurites which are quiescent in physiological states. The activity
generated by their ventricular sensory neurites increases from 0 to about 40 impulses
per second when they are exposed to adenosine and/or ATP (Huang et al, 1996).
The activity generated by ventricular sensory neurites associated with dorsal root
ganglion neurons can occur at different frequencies of their activity power spectra (0.2
- 40 Hz range), depending on their sensory characteristics as well as the type and
magnitude of the mechanical and chemical stimuli they receive at any time.

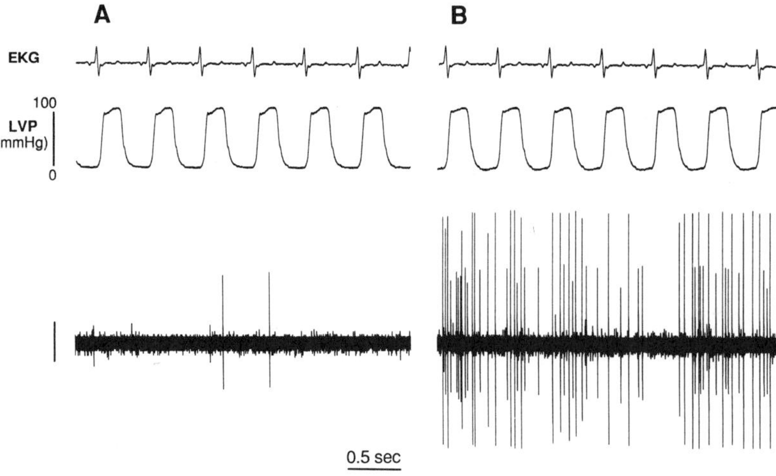

Figure 1. <u>Sensitivity of nodose ganglion cardiac afferent neurons to the A₁-adenosine
receptor analogue N⁶-Cyclopentlyadenosine (CPA).</u> When sensory neurites located
in the left ventricular epicardium were exposed to a 1 cm x 1 cm pledglet soaked
with CPA for 100 msec. (between panels A & B), the activity generated by one
associated nodose ganglion afferent neuron increased (action potentials with the
smaller signal-to-noise ratio), while a previously quiescent neuron became active
(that generating activity with the greatest signal-to-noise ratio). Monitored cardiac

328

variables were not altered. Data presented in panel B were obtained 12 minutes after exposing epicardial sensory neurites to CPA.

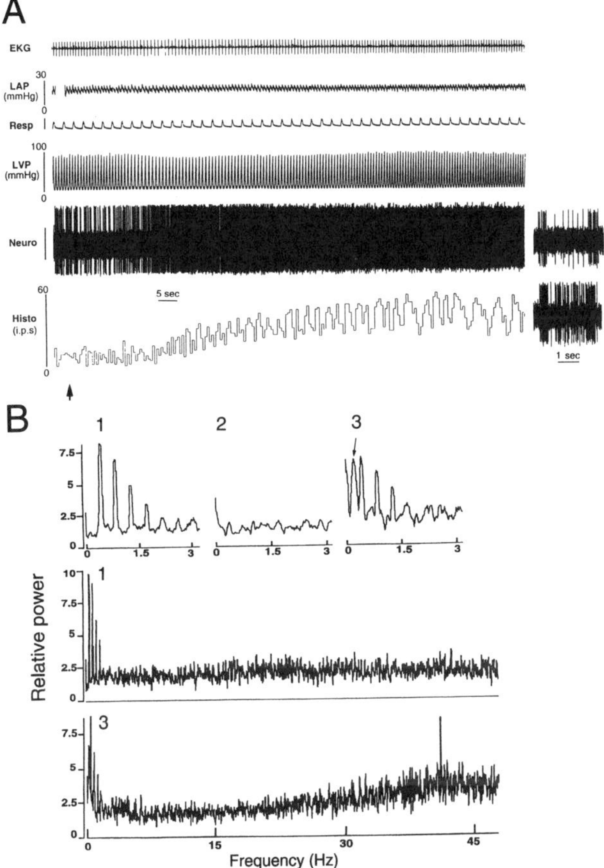

Figure 2. <u>Activity generated by sensory neurites on the ventral surface of the right ventricular conus associated with an afferent neuron in a left T$_3$ dorsal root ganglion.</u>

Panel A. This neuron generated activity with a respiratory related pattern in control states, as exemplified by the bursts of activity which coincided with inflation (fast trace in the right upper panel and histogram below). Activity generated by this neuron increased when its sensory field was exposed to adenosine (arrow at bottom of panel A). EKG = electrocardiogram; LAP = left atrial chamber pressure; Resp = tracheal pressure; LVP = left ventricular chamber pressure; Neuro = afferent neuronal activity. Vertical calibration bar to the left of the neurogram = 0.1 mV. Panel B. The two panels labeled # 1 demonstrate activity generated by this afferent neuron during control states occurring in the low frequency domains of its power spectrum. #2 During cessation of respiration this activity was eliminated. #3 Epicardial application of adenosine induced novel activity in the high (38 Hz; #3, lowest trace) and low (0.03 Hz; #3 upper panel) frequency domains of its activity power spectrum.

During control states, they often generate activity patterns in the heart (0.25 Hz)and/or respiratory (0.4 Hz) frequency domains (Figs. 2 & 3). When exposed to purinergic agents, the activity that these sensory neurites generate usually occurs at multiple frequencies of their power spectral bands. Thus although adenosine, ATP and other stimuli, including mechanical ones, may excite ventricular sensory neurites associated with dorsal root ganglion cardiac afferent neurons to similar degrees, the activity patterns that each stimulus induces in individual afferent

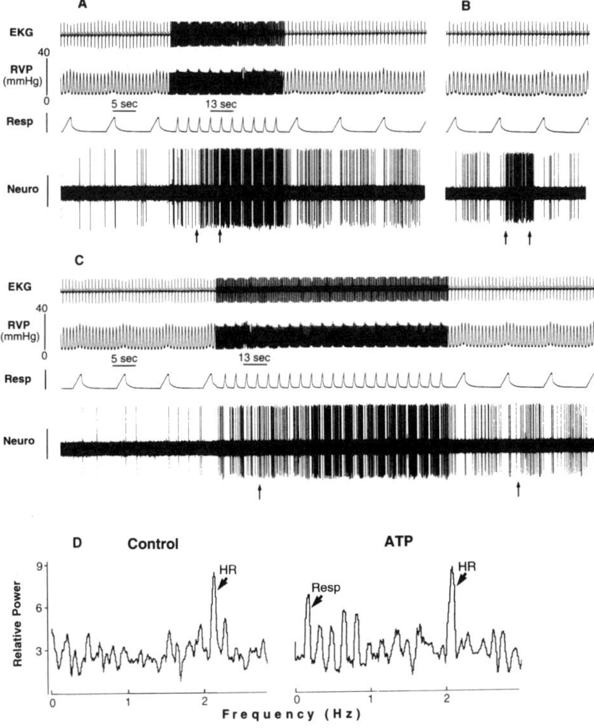

Figure 3. <u>Activity generated by sensory neurites on the ventral surface of the right ventricular conus associated with one neuron in a right T_2 dorsal root ganglion.</u>

This neuron generated respiratory related activity which was greatest during pulmonary inflation. Activity also increased when the sensory field of this neuron was exposed to substance P (A; between arrows below), touch (B) or ß, g-m ATP (C). The vertical calibration bar to the left of the neurogram = 0.1 mV. Panel D represents power spectral analysis data of activity generate by this neuron during control states (left side) and when its sensory field was exposed to ATP (right side). Note that ATP induced novel activity in low frequency power domains. EKG = electrocardiogram; LAP = left atrial chamber pressure; RVP = right ventricular intramyocardial pressure; Resp = tracheal pressure; Neuro = afferent neuronal activity. neurons usually occurs in specific frequency domains of their activity power spectra. This may provide a mechanism whereby the limited population of dorsal root ganglion cardiac afferent neurons inform spinal cord neurons of the varied nature of the stimuli they receive at any time.

That the spontaneous activity generated by many dorsal root ganglion afferent neurons associated with cardiac sensory neurites is suppressed following application of a selective A_1- or A_2-adenosine receptor antagonist to their sensory neurites indicates that such sensory neurites receive tonic inputs via purinergic receptors *in situ* (Huang et al, 1995). That most ventricular sensory neurites associated with dorsal root ganglion neurons are influenced by agonists orantagonists to either receptor subtype suggests that many of these sensory neurites possess either A_1- or A_2-adenosine receptors. Ventricular sensory neurites associated with dorsal root ganglion neurons also display mechano-transduction capabilities in the presence of a non-selective adenosine receptor antagonist. These data indicate that P_1- and P_2-purinoreceptors associated with such cardiac sensory neurites may function relatively independent of their mechanosensory transduction properties (Huang et al, 1995).

3. Intrathoracic extracardiac afferent neurons

Anatomical data indicate the presence of unipolar neurons in intrathoracic extracardiac ganglia (Armour, 1986; Horackova et al, 1996). Functional studies support this anatomical evidence in as much as the activity generated by ventricular sensory neurites associated with intrathoracic extracardiac afferent neurons increases when exposed to adenosine, ATP, substance P, bradykinin or, for that matter, local sensory field deformation (Armour, 1985; Armour et al, 1998). These data illustrate the polysensory nature of cardiac afferent neurons located in intrathoracic extracardiac ganglia.

4. Intrinsic cardiac afferent neurons.

Afferent neurons located wholly within the intrinsic cardiac nervous system: Histological evidence indicates that mammalian intrinsic cardiac ganglia contain unipolar neurons (Armour et al, 1997; Kuntz, 1934; Robb, 1965; Yuan et al, 1994). These neurons share morphological characteristics with unipolar neurons in dorsal root and nodose ganglia. The activity generated by this population of intrinsic cardiac neurons is modified when their sensory neurites are exposed to mechanical or chemical stimuli (Armour et al, 1998). This occurs even when the intrinsic cardiac nervous system is chronically disconnected from intrathoracic extracardiac and central neurons (Ardell et al, 1991). Thus not only are there afferent neurons on the heart, some of them maintain their function in the absence of inputs from extracardiac neurons. The sensory neurites associated with these intrinsic cardiac afferent neurons respond to local application of adenosine and/or ß, γ-m ATP (Armour et al 1998). Purine sensitive intrinsic cardiac afferent neurons respond to myocardial ischemia, even when disconnected from more centrally located neurons (Armour et al, 1998). These data indicate that polysensory afferent neurons located wholly within the intrinsic cardiac nervous system are capable of transducing ischemic events.

Intrinsic cardiac afferent neurons which project axons centrally: Reflexly induced cardiodynamic changes occur when a sufficient population of intrinsic cardiac afferent neurons are excited by locally applied purinergic agents in situ (Huang et al, 1993d). These cardiovascular reflexes become obtunded or eliminated when all connections between intrinsic cardiac and central neurons are severed (Huang et al, 1993d). These data indicate that a population of purinergic intrinsic cardiac afferent neurons project axons to central neurons. As most of the reflexes so induced are obtunded when the cervical vagosympathetic complexes are severed, it appears that most of their centrally projecting axons course in vagal as opposed to sympathetic nerves (Huang et al, 1993d). The concept that one population of intrinsic cardiac afferent neurons project axons to the medulla oblongata has received anatomical confirmation (Cheng et al, 1997). That ventricular sensory neurites associated with such afferent neurons respond to local application of adenosine and/or ATP indicates they possess P_1-and/or P_2-purinoreceptors.

5 Clinical correlates of purinergic cardiac afferent neurons

The data described above indicate that many ventricular sensory neurites associate with afferent neurons in intrinsic cardiac, extracardiac intrathoracic, nodose and dorsal root ganglia possess P_1-purinoreceptors and/or P_2-purinoreceptors. A certain amount of adenosine is liberated from cardiac tissues normally, even more being released during myocardial ischemia (Rubio et al, 1969). In that regard, greater than normal amounts of adenosine are found in human pericardial fluid in the presence of a transmural myocardial infarction (Kollain, 1997). Studies in man indicate that symptoms of cardiac origin depend to a great extent on the presence of P_1-purinoreceptors associated with cardiac sensory neurons (Sylvén et al, 1987), some of which are A_1-adenosine receptors (Gaspardone et al, 1995). Although these clinical findings may depend in part on adenosine induced coronary vascular steal phenomena, data derived from animal studies support the contention that locally released adenosine can activate human cardiac afferent neurons (Sylvén, 1989). Such data also are in accord with the fact that ischemia induced enhancement of activity generated by ventricular sensory neurites associated with many cardiac afferent neurons does not occur following local A_1- and/or A_2-adenosine receptor blockade (Huang et al, 1995).

ATP activates ventricular sensory neurites associated with many cardiac afferent neurons in nodose, dorsal root or intrathoracic ganglia to a degree that is similar to that induced when they are exposed to adenosine (Armour et al, 1998; Huang et al, 1996). It should be recalled that adenosine or ATP usually induce varied activity patterns from ventricular sensory neurites associated with cardiac afferent neurons even when similar levels of peak activity are achieved (c.f., section 2.3). Thus individual cardiac sensory neurites associated with each population of cardiac afferent neurons usually generate different activity patterns, depending on the chemical and mechanical stimuli they receive at any time as well as their transduction properties.

B. <u>Intrathoracic local circuit neurons and efferent cardiac neurons</u>.

Recent evidence indicates that intrathoracic extracardiac ganglia contain, in addition to sympathetic efferent postganglionic neurons, local circuit neurons which are involved in cardiac regulation (Armour, 1994; Ardell, 1994). On the other hand, intrinsic cardiac ganglia possess not only parasympathetic efferent postganglionic neurons, but sympathetic efferent postganglionic neurons, local circuit and afferent neurons (Armour, 1994; Armour & Hopkins, 1990; Butler et al, 1990; Gagliardi et al, 1988; Horackova et al, 1996). As an interneuron represents a neuron with cell processes which project only to adjacent neurons, this definition may be too limited to characterize many of the neurons found in intrathoracic ganglia, including those on the heart (Armour & Hopkins 1990; Gagliardi et al, 1988; Horackova et al, 1996). The term local circuit neuron (Hamos et al, 1985) has been used to describe neurons which project axons either to other neurons within their ganglion or in other intrathoracic ganglia (Armour, 1994; Yuan et al, 1994). In agreement with that, a sizable population of neurons in mediastinal, middle cervical and stellate ganglion are not activated after a consistent latency when electrical stimuli are delivered in a

332

repetitive fashion to the nerves connected to their respective ganglia (Armour, 1985; Armour & Janes, 1986). Thus many neurons in intrathoracic ganglia do not project axons outside their respective ganglia.

1. Intrinsic cardiac neurons

Consistent with the varied anatomy displayed by neurons in intrinsic cardiac ganglia, the transmembrane properties of intrinsic cardiac neurons differ, depending on the population of neurons studied. Three different types of intrinsic cardiac neurons have been identified based on the transmembrane properties displayed by acutely dissociated intrinsic cardiac neurons. One population generates single action potentials when relatively-long-duration currents are injected intracellularly, another generates phasic activity and the third generate repetitive action potentials throughout the period of time that they are exposed to intracellular current injection (Allen & Burnstock, 1990; Allen et al, 1994). Thus it is not surprising that local application of a chemical induces varied neuronal responses from the different populations of intrinsic cardiac neurons in situ (Armour, 1994).

Intrinsic cardiac neurons are sensitive to exogenously applied ß-adrenoceptor agonists (Armour & Butler, 1989; Huang et al, 1993b), nicotinic and muscarinic receptor agonists (Butler et al, 1990; Huang et al, 1993b), amino acids (Huang et al, 1993c), peptides (Armour et al, 1993) or nitric oxide donors (Armour et al, 1995). Intrinsic cardiac neurons are also sensitive to exogenously applied purinergic agents (Huang et al, 1993d). Freshly cultured neonatal intrinsic cardiac neurons (Allen & Burnstock, 1990; Allen et al, 1994) or adult intrinsic cardiac neurons co-cultured with adult cardiomyocytes for a few weeks (Horackova et al, 1994) respond to exogenously applied purinergic agents. The beating frequency of adult cardiomyocytes co-cultured with intrinsic cardiac neurons changes when these cells are exposed to purinergic agents; the beating frequency of cardiomyocytes cultured alone does not (Horackova et al, 1994). In accord with these findings, cardiodynamics alterations occur when purinergic agents activate a sufficient population of intrinsic cardiac neurons in situ (Huang et al, 1993d). For the most part, purine sensitive intrinsic cardiac neurons act to suppress heart rate and atrial contractile force. Intrinsic cardiac purinergic neurons can influence cardiodynamics in the absence of inputs from more centrally located neurons, but to a lesser extent than in the intact state. Thus their connectivity with more centrally located neurons affects their functional characteristics. As cardiac responses induced by adenosine as versus ATP sensitive intrinsic cardiac neurons differ, the effects that each induce will be dealt with separately.

Adenosine sensitive neurons.   Central neurons possess adenosine cell surface receptors which are negatively linked to adenylcyclase (Murphy et al, 1991). Neural elements in guinea-pig sympathetic ganglia possess purinergic A1-receptors (Borasio et al, 1995). Adenosine can potentate the sympathomimetic effects that nicotine exerts on sympathetic efferent postganglionic neurons (Evoniuk et al, 1986). Adenosine also affects the transmembrane properties of cultured neonatal mammalian intrinsic cardiac neurons (Allen & Burnstock, 1990). As cardiomyocytes possess P1-purinoceptors (Stein et al, 1994), negative inotropic effects are induced when isolated atrial preparations are exposed to adenosine (Hoyle et al, 1996). Adenosine induces negative chronotropic effects on sino-atrial tissues and negative dromotropic effects on the atrio-ventricular node (Vassort et al, 1992). Thus when adenosine is administered systemically in sufficient doses it acts to suppress these cardiac indices (Hoyle et al, 1996). On the other hand, P2 purinoceptors associated with cardiomyocytes act to increase calcium entry into the cardiomyocytes thereby inducing positive inotropic effects (Vassort et al, 1994).

One can successfully maintain the contractile properties of Guinea-pig adult cardiomyocytes for a number of weeks (beating frequency of about 120 per minute) when they are co-cultured with intrinsic cardiac neurons (Horackova et al, 1993). The morphological characteristics of cardiomyocytes co-cultured with intrathoracic autonomic neurons are similar to those displayed by freshly dissociated cardiomyocytes. The beating frequency of adult cardiomyocytes cultured alone remains relatively low (beating frequency about 40 beats per minute) for a week or so before cardiomyocytes dedifferentiate and eventually cease to contract. That adult mammalian cardiomyocytes lose their ability to beat spontaneously when cultured in the absence of cardiac neurons and yet retain this behavior for months when co-cultured with intrathoracic neurons implies that peripheral autonomic neurons tonically influence such cardiomyocytes in vitro.

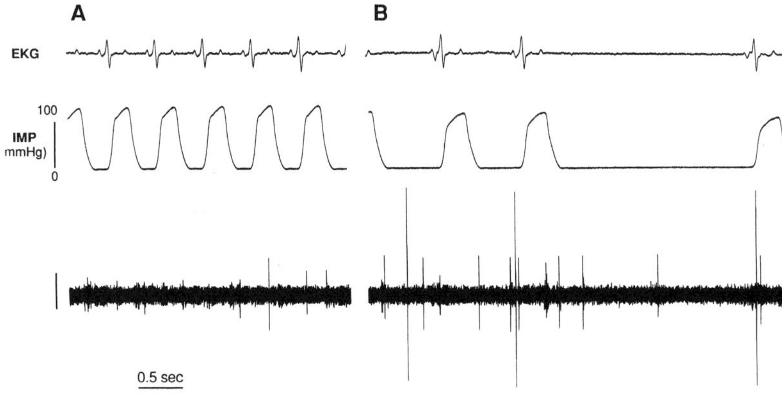

Figure 4. Administration of the A_1-adenosine receptor analogue N^6-Cyclopentlyadenosine to spontaneously active right atrial neurons via their local arterial blood supply (between panel A & B) not only enhanced their ongoing activity, but activated previously quiescent neurons. Heart rate and left ventricular intramyocardial systolic pressure (IMP) fell.

When adenosine (10 μM) is administered to adult cardiomyocytes cultured alone their beating rate is unaffected. On the other hand, when the same dose of adenosine is administered to adult cardiomyocytes co-cultured with intrinsic cardiac neurons the beating frequency of these cardiomyocytes falls (Horackova et al, 1994). In accord with that, modification of cardiac indices occurs when a sufficient population of intrinsic cardiac neurons is exposed to relatively low doses (5 - 10 μM) of adenosine in situ (Fig. 3; Huang et al, 1993). Adenosine is known to act on cardiac sympathetic efferent postganglionic nerve terminals to inhibit their release of noradrenaline during normal (Rubino et al, 1990) and ischemic (Richardt et al, 1994) states. As mentioned above, when cardiac behavior changes following excitation of intrinsic cardiac neurons by locally administered adenosine it is unlikely that the cardiac effects are due to adenosine entering the circulation in sufficient quantities to affect cardiomyocytes directly. Intrinsic cardiac neurons with P_1-purinoceptors act to either suppress or augment cardiac rate and force, depending on the population of intrinsic cardiac neurons affected (Armour, 1994). That a population of intrinsic cardiac neurons is sensitive to the adenosine A_1 receptor agonist N^6-Cyclopentlyadenosine (CPA) whereas another is sensitive to the A_2 receptor agonist 5'-(N-ethylcarboxamido)-adenosine (NECA) indicates that different populations of intrinsic cardiac neurons involved in cardiac regulation possess either A_1 or A_2 receptors (Huang et al, 1993d).

ATP sensitive neurons. ATP acts to increase calcium currents of isolated cardiomyocytes (Hoyle et al, 1996) thereby inducing positive inotropic effects (Legssyer et al, 1988). Despite that, the beating frequency of adult mammalian ventricular myocytes cultured alone is not affected significantly by ß, γ-mATP administered in 10^{-5} M doses. On the other hand the beating rate of cardiomyocytes co-cultured with intrinsic cardiac neurons increases when exposed to 10^{-5} M of ß, γ-mATP (Horackova et al, 1994). ATP is known to affect the transmembrane electrical properties of isolated neonatal guinea-pig intrinsic cardiac neurons (Allen & Burnstock, 1990; Fieber & Adams, 1991), thereby enhancing cardiomyocyte contractility (Legssyer et al, 1988). It also influences isolated sympathetic efferent neurons (Rubino et al, 1992) to release noradrenaline (Allgaier et al, 1995). That the beating frequency of adult cardiomyocytes co-cultured with intrinsic cardiac neurons increases in the presence of a relatively low dose of ATP, but not when cardiac myocytes are cultured alone, supports the contention that intrinsic cardiac neurons with P_2-purinoceptors are involved in cardiac regulation. On the other hand, ATP also affects cardiac sympathetic efferent postganglionic neurons to inhibit their release of noradrenaline (Kügelen et al, 1995). Perhaps this accounts for the fact that the beating rate of some adult cardiomyocytes co-cultured with intrinsic cardiac neurons can be suppressed when exposed to ATP (Horackova et al, 1994). Taken together, all of these data indicate that populations of intrinsic cardiac neurons possess P_1- and/or P_2-purinoreceptors.

2 Intrathoracic extracardiac neurons

Stellate and middle cervical ganglia also contain afferent and local circuit neurons in addition to sympathetic efferent postganglionic neurons (Armour, 1986 & 1994; Armour and Hopkins, 1984). Neurons in intrathoracic extracardiac ganglia are sensitive to a number of chemicals, including purinergic agents (Armour, 1994). The beating frequency of adult cardiomyocytes co-cultured with stellate ganglion neurons increases when these cells are exposed to ATP. That the beating frequency of adult cardiomyocytes cultured in the absence of autonomic neurons is not affected by ATP supports the contention that a population of intrathoracic extracardiac neurons which is involved in cardiac regulation possesses P_2-purinoreceptors (Horackova et al, 1994). As preliminary data indicate that adenosine does not modify such preparations (Horackova et al, 1994), it remains to be established whether intrathoracic extracardiac neurons involved in cardiac regulation possess P_1-purinoreceptors

C. <u>Interactions among intrathoracic neurons regulating the heart</u>.

The intrathoracic nervous system is made up of nested feedback loops, with its intrinsic cardiac component acting as the final distributive network for cardiac regulation (Ardell, 1994; Armour, 1994; Armour et al, 1998). As described above, many of the neurons in this distributive network are sensitive to purinergic agents. Some of the complex interneuronal interactions which occur among these purine sensitive peripheral cardiac neurons are summarized below.

335

1. Adenosine.

Activation of purine sensitive intrinsic cardiac neurons induces, in addition to ventricular augmentation, either bradycardia (Fig. 4) or tachycardia depending on the population of neurons studied (Huang et al, 1994). That such responses persist following decentralization of the intrathoracic nervous system, albeit in a modified form, indicates that purine sensitive intrinsic cardiac neurons are capable of modifying cardiodynamics independent of central neurons (Armour et al, 1998). Furthermore, some of these purine sensitive neurons can be involved in the genesis of ventricular arrhythmias (Huang et al, 1994).

Intrinsic cardiac afferent axons, when activated, can release adenosine stored in their sensory neurites (Rubino & Burnstock, 1996a). This locally released adenosine affects adjacent sympathetic efferent postganglionic nerve terminals to release their stores of norepinephrine, thereby increasing regional cardiomyocyte contractile behavior. This local control process, termed sensory-motor neurotransmission by Rubino & Burnstock (1996b), is another mechanism whereby purine sensitive neurons interact to regulate cardiodynamics. Such sensory-motor neurotransmission is thought to utilize A_1 adenosine receptors located on cardiac efferent sympathetic postganglionic nerve terminals (Rubino & Burnstock, 1996a).

2. ATP.

Intrinsic cardiac afferent axons can also release stores of ATP (Rubino et al, 1992) to affect adjacent sympathetic efferent postganglionic nerve terminals to release some of their stores of norepinephrine (Rubino & Burnstock, 1996b). That ATP influences intrathoracic extracardiac neurons cocultured with adult cardiomyocytes such that alterations in cardiomyocyte beating frequency occurs indicates that a population of intrathoracic extracardiac neurons which are involved in cardiac regulation possesses P_2- purinoceptors (Horackova et al, 1994). Taken together, these data indicate that some interactions among intrathoracic cardiac neurons utilize P_2-purinoreceptors.

D. <u>Interactions among intrathoracic and central cardiac neurons</u>.

As was mentioned above, many of the ventricular sensory neurites associated with nodose (Armour et al, 1994; Huang et al, 1995; Katchanov et al, 1996 & 1997; Pelleg et al, 1985 & 1996) or dorsal root (Huang et al, 1996) ganglion afferent neurons are sensitive to purinergic agents. When activated, they influence medullary and spinal cord neurons respectively. It should be recalled that ventricular ischemia can be accompanied by cardiovascular reflexes as well as symptoms (Maseri et al, 1985; Sylvén, 1989). Purine sensory neurites associated with cardiac afferent neurons, once activated, can induce cardiovascular reflexes in animal models too (Armour et al, 1994; Huang et al, 1996).

Parasympathetic efferent postganglionic neurons which innervate the heart are modified via reflexes initiated by purinergic sensitive nodose ganglion cardiac afferent neurons (Katchanov et al, 1996 & 1997; Pelleg et al, 1995 & 1996). Bradycardia occurs when cardiac parasympathetic efferent neurons are activated in a reflex fashion following ATP application to the ventricular sensory neurites associated with a sufficient population of nodose ganglion afferent neurons (Pelleg et al, 1996). On the other hand, exposure of cardiac sensory neurites associated with dorsal root ganglion neurons to adenosine results in the activation of sympathetic

336

efferent neurons which innervate blood vessels (Dibner-Dunlap et al, 1993; Malliani, 1982; Smith & Thames, 1994; Thames et al, 1993) and the heart (Huang et al, 1996). Furthermore, activation of intrinsic cardiac neurons sensitive to purinergic agents may induce cardiodynamic changes via central neuronal reflexes, some of which elicit ventricular arrhythmias (Huang et al, 1994). Coordination of autonomic outflows to cardiomyocytes depends to a large extent upon the sharing of inputs from higher centers along with the interactions among neurons in various intrathoracic ganglia. The sharing of cardiac afferent information with intrathoracic and brainstem/spinal cord feedback loops permits an overall coordination of effector control, much of which depends upon synapses utilizing P_1- and/or P_2-purinoreceptors.

IV. CONCLUSIONS AND FUTURE DIRECTIONS

The importance of purine sensitive peripheral autonomic neurons in cardiac regulation has only recently become apparent. The role that such neurons play in the maintenance of cardiac function in the presence of cardiac pathologies is just beginning to be explored. Data derived from *in situ* and *in vitro* animal models indicate that not only do various populations of peripheral cardiac neurons possess purinoreceptors, but that each population may influence cardiodynamics in a unique fashion. The challenge remains to establish the role that each population of purine sensitive cardiac neurons plays in normal and diseased cardiac states.

ACKNOWLEDGMENTS

This work was supported by a grant from the Medical Research Council of Canada (MT - 10122).

REFERENCES

Allen TGJ, Burnstock G. The actions of adenosine 5-triphosphate on guinea-pig intracardiac neurones in culture. Br J Pharmacol 1990; 100:269-76.

Allen TGJ, Hassall CJS, Burnstock G. "Mammalian intrinsic cardiac neurons in cell culture." In Neurocardiology, J. Andrew Armour and Jeffrey L. Ardell, eds. New York, NY: Oxford University Press, 1994.

Allgaier C, Wellmann H, Schobert A, Kurtz G, von Kügelen I. Cultured chick sympathetic neurons: ATP-induced noradrenaline release and its blockade by nicotinic receptor antagonists. Naunyn-Schmied Arch Pharmacol 1995; 352:25-30.

Ardell JL. "Structure and function of mammalian intrinsic cardiac neurons." In Neurocardiology, J. Andrew Armour and Jeffrey L. Ardell, eds. New York, NY: Oxford University Press, 1994.

Ardell JL, Butler CK, Smith FM, Hopkins DA, Armour JA. Activity of in vivo atrial and ventricular neurons in chronically decentralized canine hearts. Am J Physiol 1991; 260:H713-21.

Armour JA. Activity of in situ middle cervical ganglion neurons in dogs using extracellular recording techniques. Can J Physiol Pharmacol 1985; 63:704-716.

Armour JA. Neuronal activity recorded extracellularly in chronically decentralized in situ canine middle cervical ganglia. Can J Physiol Pharmacol 1986; 64:1038-1046.

Armour JA. "Peripheral autonomic neuronal interactions in cardiac regulation." In Neurocardiology, J. Andrew Armour and Jeffrey L. Ardell, eds. New York, NY: Oxford University Press, 1994.

Armour JA, Butler C. Intrathoracic ganglionic beta-adrenergic receptors involved in the efferent sympathetic regulation of the canine heart. Can J Cardiology 1989; 5:275-283.

Armour JA, Collier K, Kimber G, Ardell JL. Differential selectivity of cardiac neurons in separate intrathoracic ganglia. Amer J Physiol, in press, 1998.

Armour JA, Hopkins DA. Activity of in vivo canine ventricular neurons. Am J Physiol 1990; 258:H320-36.

Armour JA, Huang MH, Pelleg A, Sylvén C. Responsiveness of in situ canine nodose ganglion afferent neurones to epicardial mechanical or chemical stimuli. Cardiovasc Res 1994; 28:1218-1225.

Armour JA, Huang MH, Smith FM. Peptidergic modulation of in situ canine intrinsic cardiac neurons. Peptides 1993; 14:191-202.

Armour JA, Janes RD. Neuronal activity recorded extracellularly from in situ mediastinal ganglia. Can J Physiol Pharmacol 1988; 66:119-127.

Armour JA., D.A.Murphy DA, Yuan B-X, MacDonald, S, Hopkins DA. Anatomy of the human intrinsic cardiac nervous system. Anat Record 1997; 297:289-298.

Armour JA, Smith FM, Losier AM, Ellenberger HH, Hopkins DA. Modulation of intrinsic cardiac neuronal activity by nitric oxide donors induces cardiodynamic changes. Am J Physiol 1995; 268:R403-R413.

Borasio PG, Pavan B, Fabbri F, Ginanni-Corradini F, Arcelli D, Ploi A. Adenosine analogs inhibit acetylcholine release and cyclic AMP synthesis in the guinea-pig superior cervical ganglion. Neurosc Lett 1995; 184: 97-100.

Bosnjak Z, Kampine JP. Cardiac sympathetic afferent cell bodies are located in the peripheral nervous system of the cat. Circ Res 1989; 64:554-562.

Burnstock G, Wood JN. Purinergic receptors: their role in nociception and primary afferent neurotransmission. Cur Opinion Biol 1996; 6:526-532.

Butler CK, Smith FM, Cardinal R, Murphy DA, Hopkins DA, Armour JA. Cardiac responses to electrical stimulation of discrete loci in canine atrial or ventricular ganglionated plexi. Amer J Physiol 1990; 259:H1365-H1373.

Cheng Z., Powley TL, Schwaber JS, Doyle FJ. Vagal afferent innervation of the atria of the rat heart reconstructed with confocal microscopy. J Com Neurol 1997; 381:1-17.

Crea F, Pupita G, Galassi AR, Tamimi HE, Kaski JC, Davies G, Maseri A. Role of adenosine in pathogenesis of anginal pain. Circulation 1990; 81:164-172.

Dibner-Dunlap ME, Kinugawa T, Thames MD. Activation of cardiac sympathetic afferents: effects of exogenous adenosine and adenosine analogues. Am J Physiol 1993; 265 (Heart Circ. Physiol. 34):H395-H400.

Evoniuk GE, von Borstel RW, Wurtman RJ. Adenosine affects synaptic neurotransmission at multiple sites in vivo. J Pharmacol Exp Therap 1986; 236:350-355.

Fieber L, Adams D. Adenosine triphosphate-evoked currents in cultured neurones dissociated from rat parasympathetic cardiac ganglia. J Physiol (Lond) 1991; 434:239-256.

Gagliardi M, Randall WC, Beiger D, Wurster RD, Hopkins DA, Armour JA. Activity of in vivo canine cardiac plexus neurons. Am J Physiol 1988; 255:H789-800.

Gaspardone A, Crea F, Tomai F, Versaci F, Lamele M, Gioffre G, Chiariello L, Gioffré PA. Muscular and cardiac adenosine-induced pain is mediated by A_1 receptors. J Am Coll Cardiol 1995; 25:251-257.

338

Hamos JE, vanHorn SC, Raczkowski D, Uhlrich DJ, Sherman SM. Synaptic connectivity of a local circuit neurone in lateral geniculate nucleus of the cat. Nature (London) 1985; 341:197-211.

Hopkins DA, Armour JA. Ganglionic distribution of afferent neurons innervating the canine heart and physiologically identified cardiopulmonary nerves. J Autonomic Nerv Syst 1989; 26:213-222.

Horackova M, Croll RP, Hopkins DA, Losier AM, Armour JA. Morphological and immunohistochemical properties of primary long-term cultures of adult guinea-pig ventricular cardiomyocytes with peripheral cardiac neurons. Tissue and Cell 1996; 128:411-428.

Horackova M, Huang MH, Armour JA. Purinergic modulation of adult guinea pig cardiomyocytes in long term cultures and co-cultures with extracardiac or intrinsic cardiac neurones. Cardiovasc Res 1994; 28: 673-679.

Horackova M., Huang MH, Armour JA, Hopkins DA, Mapplebeck C. Co-cultures of ventricular myocytes with stellate ganglion and intrinsic cardiac neurons from adult guinea pig: spontaneous activity and pharmacological properties. Cardiovascular Res 1993; 27:1101-1108.

Hoyle CHV, Ziganshin AU, Pintor J, Burnstock G. The activation of P1- and P2-purinoceptors in the guinea-pig left atrium by diadenosine polyphosphates. Br J Pharmacol 1996; 118:1294-1300.

Huang MH, Negoescu RM, Horackova M, Wolf S, Armour JA. Polysensory response characteristics of dorsal root ganglion neurones that may serve sensory functions during myocardial ischaemia. Cardiovasc Res 1996; 32: 503-515.

Huang MH, Smith FM, Armour JA. Modulation of in situ canine intrinsic cardiac neuronal activity by nicotinic, muscarinic and ß-adrenergic agonists. Am J Physiol 1993b; 265:R659-69.

Huang MH, Smith FM, Armour JA. Amino acids modify the activity of canine intrinsic cardiac neurons involved in cardiac regulation. Am J Physiol 1993c; 264:H1275-H1282.

Huang MH, Sylvén C, Horackova M, Armour JA. Ventricular sensory neurons in canine dorsal root ganglia: effects of adenosine and substance P. Am J Physiol 1995; 269:R318-R324.

Huang MH, Sylvén C, Pelleg A, Smith FM, Armour JA. Modulation of in situ canine intrinsic cardiac neurons by locally applied adenosine, ATP or their analogs. Am J Physiol 1993d; 265:R914-22.

Huang MH, Wolf SG, Armour JA. Ventricular arrhythmias induced by chemically modified intrinsic cardiac neurons. Cardiovasc Res 1994; 28:636-42.

Katchanov G, Xu J, Clay A, Pelleg A. Electrophysiological-anatomical correlates of ATP-triggered vagal reflex in the dog. IV. Role of LV vagal afferents. Am J Physiol 1997; 272:H1898-H1903.

Katchanov G, Xu J, Hurt CM, Pelleg A. Electrophysiological-anatomical correlates of ATP-triggered vagal reflex in the dog. III. Role of cardiac afferents. Am J Physiol 1996; 270:H1785-H1790.

Kollain, M. Budapest, Hungary, personal communication, 1997.

Krishtal O., Marchenko SM, Obukhov AG, Volkova TM. Receptors for ATP in rat sensory neurons: the structure-function relationship for ligands. Br J Pharmacol 1988; 95:1057-1062.

von Kügelgen I, Stoffle D, Starke K. P$_2$-purinoreceptor-mediated inhibition of noradrenaline release in the rat atria. Br J Pharmacol 1995; 115:247-254, 1995.

Kuntz A. The autonomic nervous system. Philadelphia: Lea & Febiger, 1934.

Lagerquist B. Studies on patients with angina pectoris and normal coronary arteries with special reference to adenosine as a modulator of pain. Acta Univ. Upsaliensis, 310, Uppsala, 1991.

Legssyer A, Piggiolo J, Renard D, Vassort G. ATP and other adenine compounds increase mechanical activity and inositol triphosphate production in rat heart. J Physiol 1988; 401:185-199.

Macdonald R L, Skerrit JH, Werz MA. Adenosine agonists reduce voltage-dependent calcium conductance of mouse sensory neurones in cell culture. J Physiol Lond 1986; 370:75-90.

Malliani A. Cardiovascular sympathetic afferent fibers. Rev Physiol Biochem Pharmacol 1982; 94:11-74.

Maseri A, Chierchia S, Davies G, Glazier J. Mechanism of ischemic cardiac pain and silent myocardial ischemia. Am J Med 1985; 79(Suppl 3A): 7-11.

Murphy MG, Moak CM, Byczko Z, MacDonald WF. Adenosine-dependent regulation of cyclic AMP accumulation in primary cultures of rat astrocytes and neurons. J Neurosc Res 1991; 30:631-640.

Pan H-L, Longhurst JC. Lack of a role of adenosine in activation of ischemically sensitive cardiac sympathetic afferents. Am J Physiol 1995; 269:H106-H113.

Pelleg A. Cardiac cellular electrophysiological actions of adenosine and adenosine triphosphate. Am Heart J 1985; 110: 688-693.

Pelleg A, Katchanov G, Xu J. Purinergic modulation of neural control of cardiac function. J Auton Pharmacol 1996;16: 401-405.

Pelleg A, Mitamura H, Michelson EL. Evidence for vagal involvement in the electrophysiologic actions of exogenous adenosine and adenosine triphosphate in the canine heart. J Auton Pharmacol 1985; 5:2070-212.

Reeke FM, Burnstock G. Some effects of purines on neurones of Guinea-pig superior cervical ganglia. Gen Pharmacol 1994; 25:143-148.

Richardt G., Blessing R, Schömig A. Cardiac noradrenaline release accelerates adenosine formation in the ischemic rat heart: role of neuronal noradrenaline carrier and adrenergic receptors. J Mol Cell Cardiol 1994; 26: 1321-1328.

Robb JS. Comparative Basic Cardiology. Grune & Stratton, New York, 1965

Rubino, A., Amerini S, Ledda F, Mantelli L. ATP modulates the efferent function of capsaicin-sensitive neurones in guinea-pigs isolated atria. Br J Pharmacol 1992; 105:516-520.

Rubino A, Burnstock G. Possible role of diadenosine polyphosphates as modulators of cardiac sensory-motor neurotransmission in guinea-pigs. J Physiol 1996a; 495:515-523.

Rubino A, Burnstock G. Capsaicin-sensitive sensory-motor neurotransmission in the peripheral control of cardiovascular function. Cardiovasc Res 1996b; 31:467-479.

Rubino A, Mantelli L, Amerini S, Ledda F. Adenosine modulation of non-adrenergic non-cholinergic neurotransmission in isolated guinea-pig atria. Naunyn-Schmeid Arch Pharmacol 1990; 342:520-522.

Rubio R, Berne RM, Katori M. Release of adenosine in reactive hyperemia of the dog heart. Am J Physiol 1969; 216:56-62.

Smith ML., Thames MD. Cardiac receptors: discharge characteristics and reflex effects. In: Armour JA, Ardell JL, eds. Neurocardiology. New York: Oxford University Press, 1994: 19-52.

Stein B, Schmitz W, Scholz H, Seeland C. Pharmacological characterization of A_2-adenosine receptors in guinea-pig ventricular cardiomyocytes. J Mol Cell Cardiol 1994; 26:403-414.

Sylvén C. Angina pectoris: Clinical characteristics, neurophysiological and molecular mechanisms. Pain 1989; 36:145-167.

Sylvén C, Beermann B, Jonzon B, Brandt R. Angina pectoris-like pain provoked by intravenous adenosine in health volunteers. Br Med J 1986; 293:227-230.

Sylvén C., Jonzon B, Brandt R, Beermann B. Adenosine-provoked angina pectoris-like pain: time characteristics, influence of autonomic blockade and naloxone. Eur Heart. J 1987; 8:738-743.

Thames M D, Kinugawa T, Dibner-Dunlap ME. Reflex sympathoexcitation by cardiac sympathetic afferents during myocardial ischemia: role of adenosine. Circulation 1993; 87:1698-1704.

Vassort G, Pucéat M, Scamps F. Modulation of myocardial activity by extracellular ATP. Trends Cardiovasc Med 1994; 4:236-240.

Vassort G, Scamps F, Pucéat M. Multiple effects of extracellular ATP in cardiac tissues. N I PS 1002; 7:212-215.

Yuan B-X, Ardell JL, Hopkins DA, Armour JA. Gross and microscopic anatomy of canine intrinsic cardiac neurons. Anat Rec 1994; 239:75-87.

EXTRACELLULAR METABOLISM OF NUCLEOTIDES AND ADENOSINE IN THE CARDIOVASCULAR SYSTEM

Herbert Zimmermann, Biozentrum der Universität, Frankfurt/Main, Germany;
Jeremy D. Pearson, King's College, University of London, UK

I. SUMMARY

Extracellular nucleotides such as ATP, or diadenosine polyphosphates and also adenosine play important receptor-mediated functions in the cardiovascular system by contributing to the regulation of vascular tone, myocardial contractility, and the function of blood cells. Both the inactivation of extracellular nucleotides and the extracellular production of adenosine are key functional roles of ecto-nucleotidases. The final hydrolysis product adenosine may be salvaged by adjacent cells or become deaminated extracellularly to inosine by ecto-adenosine deaminase. The extracellular hydrolysis pathways have been identified by vascular perfusion or by analysis of isolated endothelial or smooth muscle cells or of cardiomyocytes. In principle, a considerable variety of enzymes can be involved in these catalytic functions. The diadenosine polyphosphatases have not yet been identified in molecular terms. Recent progress has led to the molecular cloning and functional expression of members of a new family of proteins that hydrolyze either nucleoside-5'-di- and/or triphosphates (ecto-ATP diphosphohydrolase, ecto-ATPase). The same compounds are substrates also of members of another family of ecto-nucleotidases, the phosphodiesterase/nucleotide pyrophosphatases. These enzymes have nucleotide pyrophosphatase as well as phosphodiesterase activity and can also cleave extracellular cAMP. The hydrolysis of extracellular AMP to adenosine is catalyzed by ecto-5'-nucleotidase. Finally, ecto-alkaline phosphatase is capable of releasing inorganic phosphate from a variety of organic compounds including the degradation of ATP, ADP, AMP. It should be noted that these enzymes have a broad substrate specificity hydrolyzing purine and pyrimidine nucleotides. The information concerning the distribution of the individual enzymes in the cardiovascular system and its specific cellular elements is still sparse. There is good support for a major role of ecto- ATP diphosphohydrolase in the degradation of ATP and ADP in the vascular system with an inhibitory effect on platelet activation. The expression of the individual enzymes may vary between cells in culture and their in situ location. Future studies need to unravel the cellular distribution of the enzymes by in situ-hybridization and immunocytological methods, analyze their functional role by intervention with specific inhibitors, reveal their regulation of expression, and evaluate the deletion of the various ecto-nucleotidase genes.

II. BACKGROUND

A. Why extracellular nucleotide metabolism in the cardiovascular system is important

As the other chapters in this book amply demonstrate, there is a wealth of evidence for the importance of extracellular nucleotides and adenosine as signaling molecules, acting via specific cell surface receptors to modulate cardiovascular function in important ways - most obviously the regulation of vascular tone and myocardial contractility. There are several sources of extracellular nucleotides. ATP is released as a cotransmitter from many sympathetic nerves to act directly on P2 receptors on cardiac myocytes or vascular smooth muscle cells. Blood platelet activation triggers the secretion of ATP and ADP from granular stores to act predominantly on endothelial cell P2 receptors leading to vasodilatation via the synthesis of nitric oxide. Following physical damage, or during periods of ischemia, it is well established that ATP (as well as adenosine) is released into the extracellular space and the bloodstream and in vitro experiments suggest that significant selective release of ATP can be evoked from cells such as endothelial cells by physical shear forces or treatment with thrombin, in the absence of general cell dysfunction (Pearson, 1985; Hellewell & Pearson, 1987).

Vascular homeostasis requires that purinoceptor-mediated responses can be regulated by rapid removal or inactivation of the agonist. In the case of adenosine, transport into cells is ubiquitous and rapid: circulating adenosine has a half life of seconds due to uptake by red cells and endothelium. In the case of ATP or ADP, inactivation is due to dephosphorylation to AMP (which has little or no action on purinoceptors) and then to adenosine by ecto-nucleotidase enzymes. There is good evidence that endothelial ecto-nucleotidases, by converting pro-aggregatory ADP to anti-aggregatory adenosine, limit the extent of intravascular platelet aggregation (Marcus et al., 1991). Smooth muscle ecto-nucleotidases similarly regulate the action of ATP secreted from nerves, generating adenosine which often acts presynaptically to inhibit further release. Extracellular catabolism of nucleotides is thus of cardinal importance for the correct regulation of vascular function. Within the bloodstream, although blood cells also possess ecto-nucleotidases, catabolism is primarily due to endothelial enzymes, with early studies in isolated perfused organs demonstrating rapid and efficient catabolism of ATP on a single passage (Hellewell & Pearson, 1987).

B. Overview of known ecto-nucleotidase pathways

In principle, the hydrolysis of extracellular nucleotides can be catalyzed by members of several enzyme families. These include the apyrase family of ecto-nucleotidases (Zimmermann, 1996a,b), the ecto-phosphodiesterase/ pyrophosphatase family (Goding et al., 1998) and the ecto-alkaline phosphatase family (Whyte 1996) (Fig. 1).

Regarding substrate specificity, the most selective enzymes are the members of the ecto-apyrase family. They hydrolyze either nucleoside-5'-triphosphates (ecto-ATPase) or both nucleoside-5'-triphosphates and nucleoside-5'-diphosphates (ecto-ATP diphosphohydrolase, ecto-apyrase) (E.C. 3.6.1.5). A much broader spectrum of substrate specificities is observed for the members of the family of phosphodiesterase/pyrophosphatases. At present, the family is also referred to as PC-1 family or as phosphodiesterase/nucleotide pyrophosphatase (PDNP) family. The complicated name reflects the surprisingly broad substrate specificity of the enzymes that includes phosphodiesterase (EC 3.1.4.1) as well as nucleotide pyrophosphatase (EC 3.6.1.9) activity. Members of the enzyme family would be capable of hydrolyzing 3',5'-cAMP to AMP, ATP to AMP and PP_i, ADP to AMP and P_i, or NAD^+ to AMP and nicotinamide mononucleotide. Alkaline phosphatase (E.C. 3.1.3.1) is another ecto-enzyme with a broad substrate specificity. It is capable of releasing inorganic phosphate from a variety of organic compounds including the degradation of ATP, ADP, AMP, and other nucleoside phosphates. One single enzyme could thus catalyze the entire hydrolysis chain from the nucleoside-5'-triphosphate to the respective nucleoside.

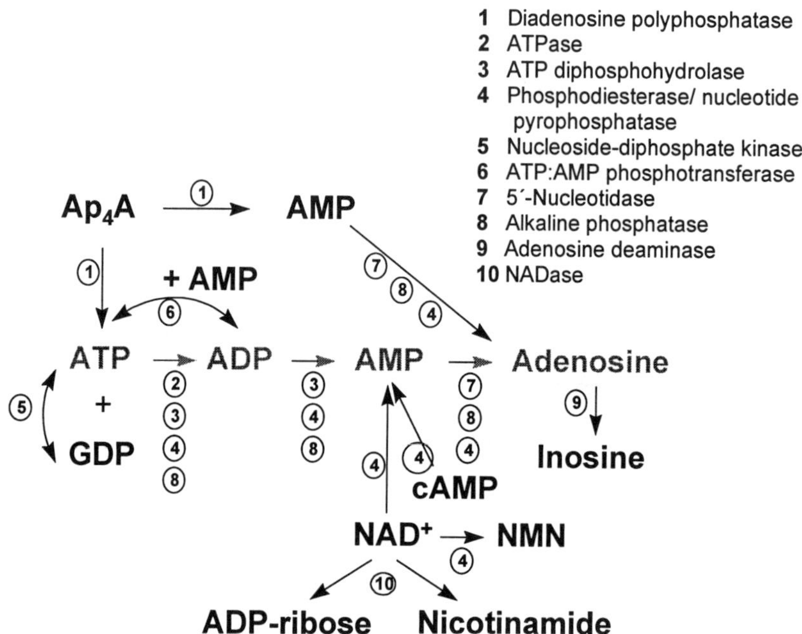

Figure 1. <u>Extracellular pathways in the degradation of nucleotides.</u> The numbers indicate the type of enzyme that can be involved in the reaction. Note that the enzymes may have a differential distribution between cell types of a tissue and also between tissues. NMN, nicotinamide mononucleotide.

In contrast, ecto-5'-nucleotidase (EC 3.1.3.5) is specific for the hydrolysis of nucleoside-5'-monophosphates such as AMP, UMP or GMP (Zimmermann, 1992). It is generally thought to play a major role in the production of adenosine from extracellular AMP. Extracellular adenosine can in turn be deaminated to inosine by ecto-adenosine deaminase (EC 3.5.4.4, Franco et al., 1997). Cell surfaces may also contain converting enzymes such as ATP:ADP phosphotransferase (myokinase, EC 2.7.4.3) or nucleoside-diphosphate kinase (EC 2.7.4.6) (Plesner, 1995). The molecular identity of the latter two enzymes in their ecto-form has not yet been determined.

The distribution of the enzymes may differ between tissues and cell types within a tissue, or also during ontogenesis. There is, however, evidence that several of these enzymes may coexist within a given tissue or even at the surface of same cell. It is a major challenge to attribute the catalytic activities measured in individual tissues or on cells to molecularly identified enzymes. This is of particular interest when nucleotide analogues are to be applied as specific inhibitors of ecto-nucleotidases or nucleotide receptors or when the release of endogenous nucleotides from a tissue is to be analyzed.

III. TEXT

<u>A. Patterns of metabolism of extracellular nucleotides by cardiovascular cells</u>

ATP, ADP and AMP can be hydrolyzed by microvascular endothelial cells, aortic endothelial cells, smooth muscle cells and cardiac myocytes (Gordon et al., 1986; 1989, Meghji et al., 1992; 1995). In each cell type the time course for catabolism is consistent with sequential dephosphorylation, though the relative levels of each activity differ (Fig. 2). Thus, in experiments with vascular cells cultured on microcarrier beads and superfused there is rapid conversion of ATP to ADP and AMP. In endothelial cell experiments, the rate of dephosphorylation of AMP was relatively slow, and actually decreased as the initial concentration of ATP or ADP was raised. These data are consistent with the presence of ecto-5'-nucleotidase, which is inhibited strongly by ADP and ATP (see III F). In smooth muscle cell experiments adenosine was produced rapidly, and kinetic analysis indicated that, unlike in the endothelial cells, AMP generated at the cell surface was hydrolyzed more efficiently than AMP added to the bulk phase, indicative of preferential local delivery. In experiments with adult rat cardiac myocytes the pattern of catabolism indicated that adenosine production was again enhanced by local delivery and was faster than adenosine uptake.

In these experiments it was assumed that sequential dephosphorylation was due to separate ecto-ATPase, ecto-ADPase and 5'-nucleotidase enzymes. Non-specific phosphatase activity was not present. AMP breakdown was blocked by the selective ecto-5'-nucleotidase inhibitor adenosine 5'-[α,ß-methylene] diphosphate or by antibodies to the enzyme. There is good evidence from earlier inhibitor studies that in pig aortic endothelial cells, ADP catabolism is inhibited by a range of poorly hydrolyzable ATP analogues that have little effect on ATP catabolism (Hellewell & Pearson; 1985). Additionally, the rate of dephosphorylation of [³H]ATP was not affected by excess ADP (Pearson et al., 1980). However, in the light of the more recent purification of ecto-apyrase and demonstration of its presence in vascular cells (see below) the identity of the enzymes needs reconsidering.

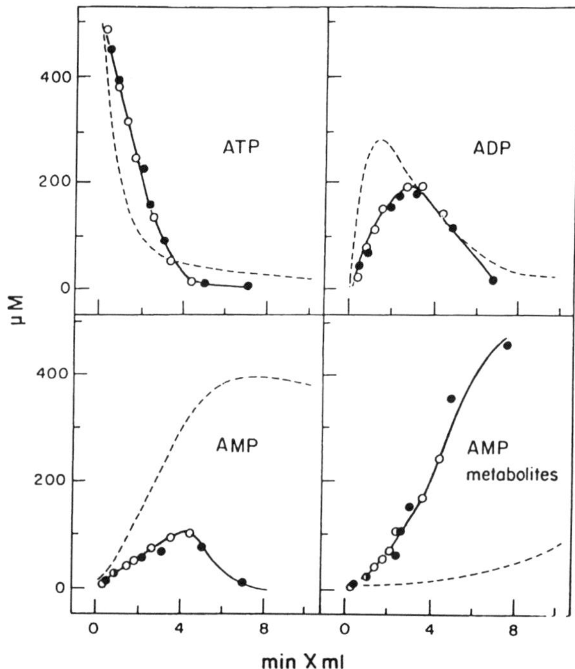

Figure 2. <u>Comparison of time course of hydrolysis of ATP on pig aortic endothelial and smooth muscle cells.</u> Cells were cultured on microcarrier beads and superfused; ATP was added at a concentration of 500 µM. Dashed lines, fitted data from endothelial cells (mean of two estimates). Solid lines, data from two experiments with smooth muscle cells. (From Gordon et al., 1989, with permission).

B. Ecto-ATPase/ADPase activities: Molecular identification of enzymes

1. Good evidence for the ecto-apyrase family.

Northern hybridization experiments reveal that both members of the ecto-apyrase family, ecto-ATPase and ecto-apyrase are expressed in rat heart (Kaczmarek et al., 1996; Kegel et al., 1997). The two enzymes recently cloned from rat brain libraries are related in molecular terms (40% amino acid identity) (Kegel et al., 1997; Wang et al., 1997). Since they can hydrolyze a broad variety of purine and pyrimidine nucleoside-5'-phosphates they represent ecto-nucleoside 5'-tri/and or di-phosphatases rather than specific ATPases or ATP diphosphohydrolases. Both are glycoproteins with apparent molecular masses in the order of 70 to 80 kDa. They represent integral membrane proteins with two putative transmembrane domains at the N- and C-terminus respectively (Fig. 3). Whereas ecto-ATPase hydrolyzes ADP only to a marginal extent, ecto-apyrase hydrolyzes ADP and ATP about equally well. Both enzymes are activated optimally by Ca^{2+} or Mg^{2+} at millimolar concentrations and at alkaline pH. Related enzymes have also been cloned from human and mouse tissues (Chadwick & Frischauf, 1997; Chadwick et al., 1998) but an ecto-ADPase has not yet been identified. These ecto-enzymes are members of a larger family of nucleotidases that occur also in plants and protozoans. Some of the enzymes are soluble. They share four major sequence motifs referred to as apyrase conserved regions (Handa & Guidotti, 1996; Kegel et al., 1997).

Ecto-apyrase isolated from human placenta has a broad and alkaline pH optimum (pH 7-8) and K_m values for ATP and ADP of 10 and 20 μM, respectively (Christoforidis et al., 1995). The apparent K_m values for ATP and ADP are very similar for an ecto-apyrase from bovine aorta (14 and 12 μM, respectively, Picher et al., 1996). In many cellular systems ecto-ATPase activity was found to be resistant to inhibitors of intracellular ATPases but sensitive to inhibitors of P_2-receptors such as the trypanocidal drug suramin and Evans blue or the pyridoxalphosphate analogue PPADS (pyridoxalphosphate-6-azophenyl-2′,4′-disulphonic acid) (Zimmermann, 1996a,b). This applies also to the apyrase on cultured bovine pulmonary artery endothelium (Chen et al., 1996). Ecto-nucleotidases from various sources, including on vascular cells, are inhibited by a number of nucleotide analogues, in particular non-hydrolyzable analogues such as methylene isoesters of ATP (α,β-methylene ATP or β,γ-methylene ATP) or the corresponding imido compounds. These properties are confirmed for the ecto-apyrase isolated from bovine aorta (Picher et al., 1996).

Ecto-apyrase is identical to CD39, a surface marker of activated B lymphocytes which can also mediate homotypic B cell adhesion (Maliszewski et al., 1994). Monoclonal antibodies to CD39 immunostain endothelial cells as well as smooth muscle cells in blood vessels in various lymphoid tissues (Kansas et al., 1991). Similarly, a polyclonal antibody directed against a peptide sequence of ecto-apyrase reveals immunohistochemical localization of the enzyme on both endothelial and smooth muscle cells of the bovine aorta media (Sévigny et al., 1997). Since the peptide sequence chosen for immunodetection is highly conserved between ecto-ATPase and ecto-apyrase this does not exclude the simultaneous presence of ecto-ATPase on the same cells, which would be consistent with the kinetic and inhibitor data described above. Coexpression of the two enzymes has been shown for a neural cell line (PC12 cells) (Kegel et al., 1997). Two isoforms of ecto-apyrase can be isolated from bovine heart. Not only coronary vessels and endothelial cells but also Purkinje fiber cells and the majority of myocardial cells were immunostained (Beaudoin et al., 1997).

The data thus suggest that much of the extracellular hydrolysis of ATP and ADP by coronary endothelial cells, by smooth muscle cells and also myocardial cells is due to the molecularly defined ecto-apyrase and possibly also ecto-ATPase. In support of this notion more than 95% of the ecto-apyrase activity can be immunoprecipitated from human umbilical vein endothelial cells by a monoclonal anti-CD39 antibody (Marcus et al., 1997).

2. Ecto-apyrase and platelet aggregation.
The antithrombotic effect of ecto-apyrase has been verified by demonstrating that COS cells transfected with human ecto-apyrase (CD39) inhibit ADP-induced aggregation in platelet-rich plasma. This ability is lost following oxidative stress (Kaczmarek et al., 1996; Marcus et al., 1997). Intravenous administration of soluble apyrase prolongs discordant xenograft survival in rats that underwent heterotopic cardiac xenografting from guinea pigs. This is presumably due to the systemic

antiaggregatory effects of the administered apyrase. Indeed, histological analyses revealed that apyrase administration abrogated local platelet aggregation and activation (Koyamada et al., 1996). Activation of endothelial cells as it accompanies inflammatory diseases results in a loss of ecto-apyrase and may thus be crucial for the progression of vascular injury (Robson et al., 1997).

C. The PDNP-family

At present, less is known about the kinetic and pharmacological properties and the distribution of the members of the newly identified PDNP family (Figs. 1, 3). Vascular smooth muscle cells metabolize cAMP to adenosine via AMP (Dubey et al., 1996) and cAMP is metabolized to adenosine in the renal vasculature (Jackson et al., 1997). This suggests the presence of ecto-phosphodiesterase. The same enzyme (PDNP) could be involved in the hydrolysis of ATP. Removal of pyrophosphate followed by hydrolysis of the AMP formed by ecto-5'-nucleotidase would be sufficient for a complete dephosphorylation chain. Three members of the family have been sequenced from human sources and may be referred to as PC-1 (PDNP1), PD-Iα/autotaxin (PDNP2) and PD-Iß (PDNP3) (Jin-Hua et al., 1997). PDNP-Iα and autotaxin appear to be splice variants of the same gene (Kawagoe et al., 1995). All have apparent molecular masses in the order of 110-120 kDa and occur in a membrane-bound as well as in a soluble, proteolytically cleaved form (Belli et al., 1993; Clair et al., 1997). Similar isoforms exist in the rat (Narita et al., 1994; Deissler et al., 1995; Scott et al., 1997). The association of these proteins with the cardiovascular system has not yet been investigated in any detail. Immuncytochemical data on mouse tissues using a monoclonal antibody against PC-1 suggest that the enzyme is associated with blood capillaries in the brain only (Harahap & Goding, 1988). In addition to their complex nucleotide hydrolyzing capacity members of the PDNP family have a number of additional common features. They contain the somatomedin B-like domain of vitronectin, the RGD-tripeptide which is recognized by several integrin receptors, and the consensus sequence of the EF-hand putative calcium-binding region (Kawagoe et al., 1995; Goding et al., 1998). The functional role of these motives is not yet understood but they imply a potential interaction with cell surface receptors and a functional Ca^{2+}-dependency. Rat PD-Iß has a very alkaline pH optimum (pH 9.7), a K_m value of 160 µM for the hydrolysis of thymidine 5'-monophosphate nitrophenylester and it can be inhibited by EDTA (Deissler et al., 1995).

D. Alkaline phosphatase

Alkaline phosphatase comprises a group of glycosylphosphatidyl inositol (GPI)-anchored glycoproteins with molecular masses in the order of 70 kDa (Fig. 3). At least four genes encode alkaline phosphatase isoenzymes in humans. Three alkaline phosphatase genes are expressed in a tissue specific manner and produce a placental, placental-like and intestinal alkaline phosphatase isoenzyme. The fourth alkaline phosphatase gene encodes a family of proteins that differ from one another by posttranslational modification only (Whyte et al., 1995). These isoenzymes are present throughout the body (tissue non-specific alkaline phosphatase) and are abundant e.g. in hepatic, skeletal and renal tissue. Alkaline phosphatase is essential for the proper mineralization of cartilage and bone (Harrison et al., 1995) and functions in the recycling of nutrients and phosphate readsorption (Fernley, 1971).

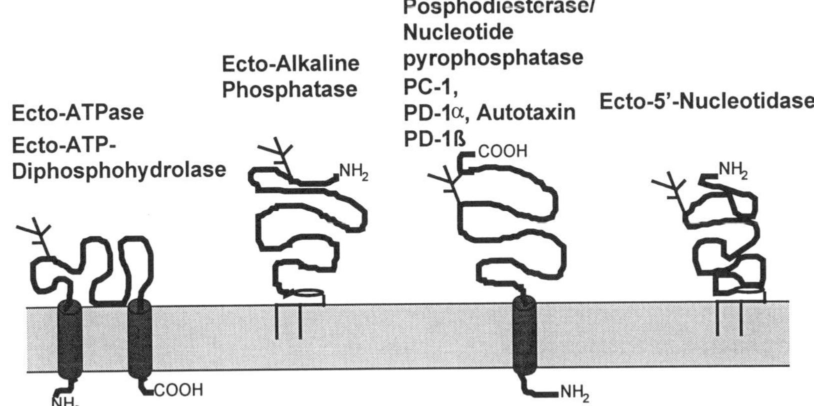

Figure 3. <u>Predicted membrane topography of key enzymes involved in the extracellular hydrolysis of nucleotides.</u> 5'-Nucleotidase and alkaline phosphatase are anchored to the plasma membrane via a glycosylphosphatidyl inositol anchor. Some of the enzymes may occur as dimers or possibly also as trimers.

Its function in the vascular system has not been investigated in any detail. As revealed by enzyme histochemistry, activity of alkaline phosphatase is specifically associated with endothelial cells of microvessels forming the blood brain barrier and absent from other vessels. It is thus regarded as a marker for the blood-brain barrier. Conditioned media of glial cell lines (interleukin-6) can induce alkaline phosphatase activity in cultured artery endothelial cells (Takemoto et al., 1994). An ultrastructural analysis of capillaries from rat epididymal fat identified alkaline phosphatase only to the capillary luminal membrane and vesicles associated with the membrane. In contrast, 5'-nucleotidase was localized on the basal or abluminal membrane (Roberts & Sandra, 1993). Enzyme histochemical investigations suggest that the enzyme is associated also with lymph endothelium and serves as a marker of lymphatic vessels (Rudert & Tillmann, 1993).

Alkaline phosphatase has a considerably higher K_m value for the hydrolysis of nucleotides (high μM to low mM range) and also a much higher pH optimum (pH 8.5 and higher) than ecto-ATP diphosphohydrolase or ecto-5'-nucleotidase. In contrast to ecto-5'-nucleotidase it is effectively inhibited by free phosphate (Fox 1978). The physiological requirements for the activation of alkaline phosphatase and the other phosphatases may thus differ.

E. Hydrolysis of diadenosine polyphosphates

Human platelets contain and release the diadenosine polyphosphates Ap₃A, Ap₄A, Ap₅A and Ap₆A together with ADP, ATP, and serotonin (Lüthje & Ogilvie, 1983, Mateo et al., 1997). The diadenosine polyphosphates may exert direct vascular actions by either lowering or increasing blood pressure (Schlüter et al., 1994, 1996) and also modulate platelet aggregation (Ogilvie, 1992). Enzymes are present in blood plasma that hydrolyze the nucleotides in an asymmetrical manner (Fig. 1). Ap₃A and Ap₄A are degraded to ADP and ATP respectively. The half-life of diadenosine polyphosphates is considerably longer than that of ATP (Lüthje & Ogilvie, 1988). Ecto-enzymes for the hydrolysis of diadenosine polyphosphates were shown to be associated with vascular endothelial cells from bovine aorta (Goldman et al., 1986, Ogilvie et al., 1989) and adrenal medulla (Mateo et al., 1997). The enzyme from adrenomedullary vascular endothelial cells hydrolyzes asymmetrically fluorogenic analogues of Ap₃A, Ap₄A and Ap₅A with K_m values of about 0.5 μM and a pH optimum around 9. It is activated by Mg^{2+} and Mn^{2+} and inhibited by Ca^{2+}, F⁻ and ATP. The ecto-nucleotidase has not yet been identified in molecular terms but its kinetic parameters and sensitivity to ions differ from the corresponding enzyme(s) on chromaffin cells or presynaptic plasma membranes (Mateo et al., 1997). Members of the PDNP family would be potential candidates for the hydrolysis of diadenosine polyphosphates.

<u>F. Ecto-5'-nucleotidase</u>

1. Identification and properties.

The GPI-anchored ecto-5'-nucleotidase (Fig. 3) occurs in essentially all tissues. The glycoprotein exists mainly as a homodimer with interchain disulfide bridges. ATP and ADP act as very effective competitive inhibitors with K_i values in the low micromolar range or even below. The enzyme is competitively inhibited by the nucleotide analogue adenosine 5'-[α,ß-methylene] diphosphate. Ecto-5'-nucleotidase is a zinc binding metalloenzyme and metal ion chelators such as EDTA inhibit enzyme activity (Zimmermann, 1992). The enzyme isolated from rat heart reveals a molecular mass of 74 kDa, a K_m value for AMP of 5.6 μM, and a pH optimum of 7.5 (Naito & Lowenstein, 1981). The primary structure of ecto-5'-nucleotidase has been determined for rat and bovine liver, human placenta, mouse kidney, and electric ray brain (Zimmermann, 1996b). The sequence is highly conserved between vertebrate species. The membrane anchor may be cleaved by GPI-specific phospholipase C with the formation of a soluble form. Soluble forms of 5'-nucleotidase occur in serum (Novo & Tutor, 1992). Ecto-5'-nucleotidase contains three sequence motifs conserved between various enzymes capable of phosphoester bond cleavage, the "phosphoesterase signature motif" (Koonin, 1994; Zhuo et al., 1994). These enzymes include a variety of phosphoesterases including Ser/Thr phosphoprotein phosphatases, sphingomyelin phosphomonoesterases, and the bacterial members of the 5'-nucleotidase family. Presumably a common catalytic strategy is employed to accomplish phosphate ester hydrolysis.

Ecto-5'-nucleotidase is also known as the lymphocyte surface protein CD73. It is a maturation marker for both T and B lymphocytes (Thompson et al., 1990). The importance of CD73 in the production of extracellular adenosine is underlined by the analysis of transgenic mice overexpressing 5'-nucleotidase on thymocytes (Resta et al., 1997). CD73 may participate in transmembrane signaling pathways (Resta & Thompson, 1997). It serves as an accessory T-lymphocyte activation molecule when cross-linked by monoclonal antibodies (Thompson et al., 1989). Like a variety of other surface-located enzymes ecto-5'-nucleotidase may function via its catalytic activity as well as via its capacity for cell adhesion or adhesion to the extracellular matrix. The enzyme can bind to the laminin/nidogen complex and to fibronectin (Olmo et al., 1992; Stochaj & Mannherz, 1992), and is identical to the lymphocyte-vascular adhesion protein-2 (Airas et al., 1995). This protein is involved in lymphocyte binding to the endothelium. Binding of peripheral blood lymphocytes can be prevented by anti-CD73 antibodies, and COS cells transfected with CD73 cDNA have a significantly increased capacity for binding of freshly isolated lymphocytes (Airas et al., 1995; Arvilomni et al., 1997). CD73 also mediates adhesion between B cells and follicular dendritic cells (Airas & Jalkanen, 1996).

2. Occurrence on cardiovascular cells.

Biochemical experiments clearly demonstrate that AMP is hydrolyzed by ecto-5'-nucleotidase on cultured endothelial and smooth muscle cells of pig aorta (Pearson et al., 1980), rat heart cultured microvascular endothelial cells (Meghji et al, 1995) and isolated rat cardiomyocytes (Meghji et al., 1992; Kitakaze et al., 1996a). However, in situ enzyme cytochemical and immunocytochemical data suggest that ecto-5'-nucleotidase is often not detectable in endothelium or on the surface of heart muscle cells. In rat heart, immunoreactivity was associated rather with interstitial cells identified as pericytes or fibroblasts (Mlodzik et al., 1995). Absence of ecto-5'-nucleotidase from endothelium has been observed also in other tissues such as kidney, liver, and brain (Ghandi et al., 1990; Schmid et al., 1994; Braun et al., 1994; Zimmermann et al., 1993). The reasons for these discrepancies are not clear, but they suggest that there may be significant regional variation in the level of ecto-5'-nucleotidase in cardiovascular cells, or that the enzyme may be upregulated by cell culturing (Roberts & Sandra, 1993). AMP released in the heart in situ by cardiomyocytes or other sources may become dephosphorylated by ecto-5'-nucleotidase situated mainly on interstitial cells.

3. Ecto-5'-nucleotidase in cardiac preconditioning

Ischemic preconditioning is accompanied by a marked increase in the activity of ecto-5'-nucleotidase in the canine heart. The enzyme is activated by protein kinase C. Whether the increased capacity for the formation of extracellular adenosine is responsible for the subsequent cardioprotection is controversially discussed (Kitakaze et al. 1996b,c; Przyklenk et al., 1996; de Jong et al., 1996; Node et al., 1997).

G. Ecto-adenosine deaminase

Extracellular adenosine may be cleared by reuptake into adjacent cells or become inactivated by deamination to inosine (Fig. 1) . Adenosine deaminase is a 41 kDa protein that catalyzes the conversion of adenosine and deoxyadenosine to inosine and deoxyinosine, respectively. Its primary structure has been resolved (Wiginton et al., 1986). The enzyme occurs in essentially all tissues and is highly expressed in lymphocytes. The hydrophilic enzyme is mainly cytosolic but it also occurs on the surface of a variety of cellular systems (Franco et al., 1997) including human mononuclear blood cells (Martín et al., 1995a,b) and capillary endothelium in the lung (Hellewell & Pearson, 1983). There it can terminate the receptor-mediated functions of adenosine. The pH optimum of the enzyme is at about pH 7.4 and the K_m value for adenosine is in the order of 50 to 150 μM.

At present it is not known how soluble adenosine deaminase can be released from cells. It may be released for example from injured erythrocytes (Jackson et al., 1996) or it may also be secreted. Surface-located adenosine deaminase is associated with specific binding proteins. One has been identified as an ecto-peptidase (dipeptidyl peptidase IV, CD26, 110 kDa), an enzyme that cleaves NH_2-terminal dipeptides from polypeptides (Kameoka et al., 1993). Binding of adenosine deaminase to CD26 produces a costimulatory response in T cell activation (Martín et al., 1995b). At present the signal transduction mechanism to the cell interior is not defined. Interestingly, adenosine deaminase can also be associated with A_1 adenosine receptors (Ciruela et al., 1996; Franco et al., 1997). Binding of adenosine deaminase to the adenosine receptor increases ligand affinity and enhances ligand-mediated second messenger production independent of its catalytic activity. Taken together these results suggest that ecto-adenosine deaminase can fulfill multiple functions. It acts as an ecto-enzyme, it may be involved in signal transduction possibly via its binding to CD26 and, in addition, it may facilitate signaling via the A_1 adenosine receptor by specific interaction with the protein.

IV. CONCLUSION AND FUTURE DIRECTIONS

Ecto-nucleotidases in the cardiovascular system are clearly important for the inactivation of P2 agonists, though direct tests of their functional significance are still needed. This is partly because the molecular identity of the enzymes present on each cell type has not yet been identified unequivocally, nor (with the exception of ecto-5'-nucleotidase) have suitable selective inhibitors been fully characterized. On a practical level, the presence of ecto-nucleotidases is sufficient to confound estimates of the relative potency and efficacy of ATP analogues acting on P2 receptors : clear differences in the potency and rank order of analogues can be obtained when results are compared between whole tissue responses (e.g. smooth muscle contraction), where agonist hydrolysis can be substantial, and single cells with local application of agonist (e.g. measuring membrane currents). A recently available ATP analogue (ARL67156), which is a selective ecto-ATPase inhibitor that poorly interacts with P2 receptors, potentiated contraction of guinea pig bladder due to electrical stimulation (i.e. neuronal release of ATP) or to added ATP demonstrating that ecto-nucleotidases modulate the response in this smooth muscle (Westfall et al., 1997).

Further tests of the importance of vascular ecto-nucleotidases will be possible if knockout mice are generated. These animals would be valuable to define directly the contribution of individual enzymes to extracellular nucleotide degradation (e.g. using isolated perfused tissues), to the control of vascular or cardiac responses induced by ATP, and to the control of platelet aggregation.

Future studies are also needed to examine the regulation of expression of cardiovascular ecto-nucleotidases, about which almost nothing is known. As noted above CD39 and CD73 are developmentally regulated in lymphocytes, and inflammatory stimulation of endothelial cells downregulates ecto-apyrase. In contrast, IL-1 upregulates 5'-nucleotidase on mesangial cells (Savic et al, 1990), while IL-6 dramatically increases alkaline phosphatase activity on non-brain endothelium (Takemoto et al, 1994; Gallo et al., 1997). Much more work is needed to determine the range of modulators of enzyme activity and to understand the pathological significance of the changes seen; and perhaps to resolve the apparent difference between high levels of 5'-nucleotidase activity on cultured endothelial cells compared to the poorly detectable levels found by immunocytochemistry in vivo.

Finally, it should be borne in mind that ecto-nucleotidases may also have separate biological functions as adhesion molecules. Nucleotidase activity, or its inhibition, may have significant effects on cell-cell or cell-matrix interactions.

V. REFERENCES

Airas L, Hellman J, Salmi M, Bono P, Puurunen T, Smith DJ, Jalkanen S. CD73 is involved in lymphocyte binding to the endothelium: Characterization of lymphocyte vascular adhesion protein 2 identifies it as CD73. J Exp Med 1995; 182:1603-1608.

Airas L, Jalkanen S. CD73 mediates adhesion of B cells to follicular dendritic cells. Blood 1996; 88:1755-1764.

Arvilommi AM, Salmi M, Airas L, Kalimo K, Jalkanen S. CD73 mediates lymphocyte binding to vascular endothelium in inflamed human skin. Eur J Immunol 1997; 27:248-254.

Beaudoin AR, Sevigny J, Grondin G, Daoud S, Levesque FP. Purification, characterization, and localization of two ATP diphosphohydrolase isoforms in bovine heart. Amer J Physiol 1997; 273:H673-H681.

Belli SJ, van Driel IR, Goding JW. Identification and characterization of soluble form of the plasma cell membrane glycoprotein PC-1 (5'-nucleotide phosphodiesterase). Eur J Biochem 1993; 217:421-428.

Braun JS, Le Hir M, Kaissling B. Morphology and distribution of ecto-5'-nucleotidase- positive cells in the rat choroid plexus. J Neurocytol 1994; 23:193-200.

Chadwick BP, Frischauf AM. Cloning and mapping of a human and mouse gene with homology to ecto-ATPase genes. Mamm Genome 1997; 8:668-672.

Chadwick BP, Williamson J, Sheer D, Frischauf AM. cDNA cloning and chromosomal mapping of a mouse gene with homology to NTPases. Mamm Genome 1998; 9:162-164.

Chen BC, Lee CM, Lin WW. Inhibition of ecto-ATPase by PPADS, suramin and reactive blue in endothelial cells, C-6 glioma cells and RAW 264.7 macrophages. Br J Pharmacol 1996; 119:1628-1634.

Christoforidis S, Papamarcaki T, Galaris D, Kellner R, Tsolas O. Purification and properties of human placental ATP diphosphohydrolase. Eur J Biochem 1995; 234:66-74.

Ciruela F, Saura C, Canela EI, Mallol J, Lluis C, Franco R. Adenosine deaminase affects ligand-induced signalling by interacting with cell surface adenosine receptors. FEBS Lett 1996; 380:219-223.

Clair T, Lee HY, Liotta LA, Stracke ML. Autotaxin is an exoenzyme possessing 5'-nucleotide phosphodiesterase/ATP pyrophosphatase and ATPase activities. J Biol Chem 1997; 272:996-1001.

Deissler H, Lottspeich F, Rajewsky MF. Affinity purification and cDNA cloning of rat neural differentiation and tumor cell surface antigen gp130[RBI3-6] reveals relationship to human and murine PC-1. J Biol Chem 1995; 270:9849-9855.

de Jong JW, de Jonge R, Marchesani A, Janssen M, Bradamante S. Controversies in preconditioning. Cardiovasc Drug Therapy 1997; 10:767-773.

Dubey RK, Mi ZC, Gillespie DG, Jackson EK. Cyclic AMP adenosine pathway inhibits vascular smooth muscle cell growth. Hypertension 1996; 28:765-771.

Fernley HN. "Mammalian alkaline phosphatases." In *The enzymes*, Boyer PD, ed. New York, London: Academic Press, 1971. p. 417-447.

Fox IH. "Degradation of purine nucleotides." In *Uric Acid*, Kelley WN, Weiner IM, eds. Berlin: Springer, 1978. p. 93-124.

Franco R, Casado V, Ciruela F, Saura C, Mallol J, Canela EI, Lluis C. Cell surface adenosine deaminase: Much more than an ecto-enzyme. Prog Neurobiol 1997; 52:283-294.

Gallo RL, Dorschner RA, Takashima S, Klagsbrun M, Eriksson E, Bernfield M. Endothelial cell surface alkaline phosphatase activity is induced by IL-6 released during wound repair. J Invest Dermatol 1997;109:597-603.

Gandhi R, Le Hir M, Kaissling B. Immunolocalization of ecto-5'-nucleotidase in the kidney by a monoclonal antibody. Histochemistry 1990; 95:165-174.

Goding JW, Terkeltaub R, Maurice M, Deterre P, Sali A, Belli SI. Ecto-phosphodiesterase/pyrophosphatase of lymphocytes and non-lymphoid cells: Structure and function of the PC-1 family. Immunol Rev 1998; 161:11-26.

Goldman SJ, Gordon EL, Slakey LL. Hydrolysis of diadenosine 5',5"-P',P"-triphosphate (Ap$_3$A) by porcine aortic endothelial cells. Circ Res 1986; 59:362-366.

Gordon EL, Pearson JD, Slakey LL. The hydrolysis of extracellular adenine nucleotides by cultured endothelial cells form pig aorta. Feed-forward inhibition of adenosine production at the cell surface. J Biol Chem 1986; 261:15496-15504.

Gordon EL, Pearson JD, Dickinson ES, Moreau D, Slakey LL. The hydrolysis of extracellular adenine nucleotides by arterial smooth muscle cells; Regulation of adenosine production at the cell surface. J Biol Chem 1989; 264:18986-18992.

Handa M, Guidotti G. Purification and cloning of a soluble ATP-diphosphohydrolase (apyrase) from potato tubers (*Solanum tuberosum*). Biochem Biophys Res Commun 1996; 218:916-923.

Harahap AR, Goding JW. Distribution of murine plasma cell antigen PC-1 in non-lymphoid tissues. J Immunol 1988; 141:2317-2320.

Harrison G, Shapiro IM, Golub EE. The phosphatidylinositol-glycolipid anchor on alkaline phosphatase facilitates mineralization initiation in vitro. J Bone Miner Res 1995; 10:568-573.

Hellewell PG, Pearson JD. Metabolism of circulating adenosine by the porcine isolated perfused lung. Circ Res 1983; 53:1-7.

Hellewell PG, Pearson JD. "Adenosine transport and ecto-nucleotidase activity in pulmonary and aortic endothelial cells". In *Carrier-mediated transport of solutes from blood to tissue* Yudilevich Dl, Mamm GE eds. Longman, 1985, p.213-222.

Hellewell PG, Pearson JD. "Adenine nucleotides and pulmonary endothelium". In *Pulmonary endothelium in health and disease* Ryan US ed, Marcel Dekker, 1987, p.327-348.

Jackson EK, Mi Z., Koehler MT, Carcillo JA, Herzer WA. Injured erythocytes release adenosine demainase into the circulation. J. Pharm Exp Ther 1996, 279: 1250-1260.

Jackson EK, Mi Z, Gillespie DE, Dubey RK. Metabolism of cAMP to adenosine in the renal vasculature. J. Pharm Exp Ther 1997: 283:177-182.

Jin-Hua P, Goding JW, Nakamura H, Sano K. Molecular cloning and chromosomal localization of PD-Iß (PDNP3), a new member of the human phosphodiesterase I genes. Genomics 1997; 45:412-415.

Kaczmarek E, Koziak K, Sevigny J, Siegel JB, Anrather J, Beaudoin AR, Bach FH, Robson SC. Identification and characterization of CD39 vascular ATP diphosphohydrolase. J Biol Chem 1996; 271:33116-33122.

Kameoka J, Tanaka T, Nojima Y, Schloman SF, Morimoto C. Direct association of adenosine demainase with a T cell activation antigen, CD26. Science 1993; 261:466-469.

Kansas GS, Wood GS, Tedder TF. Expression, distribution, and biochemistry of human CD39: Role in activation-associatied homotypic adhesion of lymphocytes. J Immunol 1991; 146:2235-2244.

Kawagoe H, Soma O, Goji J, Nishimura N, Narita M, Inazawa J, Nakamura H, Sano K. Molecular cloning and chromosomal assignment of the human brain-type phosphodiesterase I/nucleotide pyrophosphatase gene (*PDNP2*). Genomics 1995; 30:380-384.

Kegel B, Braun N, Heine P, Maliszewski CR, Zimmermann H. An ecto-ATPase and an ecto-ATP diphosphohydrolase are expressed in rat brain. Neuropharmacology 1997; 36:1189-1200.

Kitakaze M, Minamino T, Node K, Komamura K, Mori H, Inoue M, Hori M, Kamada, T. Activation of ecto-5'-nucleotidase by proteinkinase C attenuates irreversible cellular injury due to hypoxia and reoxygenation in rat cardiomyocytes. J Mol Cell Cardiol 1996a; 28:1945-1955.

Kitakaze M, Minamino T, Node K, Komamura K, Hori M. Activation of ecto-5'-nucleotidase and cardioprotection by ischemic preconditioning. Basic Res Cardiol 1996b; 91:23-26.

Kitakaze M, Node K, Minamino T, Komamura K, Funaya H, Shinozaki Y, Chujo M, Mori H, Inoue M, Hori M, Kamada, T. Role of activation of protein kinase C in the infarct size-limiting effect of ischemic preconditioning through activation of ecto-5'-nucleotidase. Circulation 1996c; 93:781-791.

Koonin EV. Conserved sequence pattern in a wide variety of phosphoesterases. Protein Sci 1994; 3:356-358.

Koyamada N, Miyatake T, Candinas D, Hechenleitner P, Siegel J, Hancock WW, Bach FH, Robson SC. Apyrase administration prolongs discordant xenograft survival. Transplantation 1996; 62:1739-1743.

Lüthje J, Ogilvie A. The presence of diadenosine 5',5'''-P1,P-triphosphate (Ap$_3$A) in human platelets. Biochem Biophys Res Commun 1983; 115:253-260.

Lüthje J, Ogilvie A. Catabolism of Ap$_4$A and Ap$_3$A in whole blood. The dinucleotides are long-lived signal molecules in the blood ending up as intracellular ATP in the erythrocytes. Eur J Biochem 1988; 173:241-245.

Maliszewski CR, DeLepesse GJT, Schoenborn MA, Armitage RJ, Fanslow WC, Nakajima T, Baker E, Sutherland GR, Poindexter K, Birks C, et al. The CD39 lymphoid cell activation antigen: Molecular cloning and structural characterization.. J Immunol 1994; 153:3574-3583.

Marcus AJ, Safier LB, Hajjar KA, Ullman HL, Islam N, Broekman MJ, Eiroa AM. Inhibition of platelet-function by an aspirin-insensitive endothelial-cell ADPase - Thromboregulation by endothelial cells. J Clin Invest 1991;88:1690-1696.

Marcus AJ, Broekman MJ, Drosopoulos JHF, Islam N, Alyonycheva TN, Safier LB, Hajjar KA, Posnett DN, Schoenborn MA, Schooley KA, et al. The endothelial cell ecto-ADPase responsible for inhibition of platelet function is CD39. J Clin Invest 1997; 99:1351-1360.

Martín M, Aran JM, Colomer D, Huguet J, Centelles JJ, Vivescorrons JL, Franco R. Surface adenosine deaminase - a novel B-cell marker in chronic lymphocytic leukemia. Hum Immunol 1995a; 42:265-273.

Martín M, Huguet J, Centelles JJ, Franco R. Expression of ecto-adenosine deaminase and CD26 in human T cells triggered by the tcr-cd3 complex - Possible role of adenosine deaminase as costimulatory molecule. J Immunol 1995b; 155:4630-4643.

Mateo J, Miras-Portugal MT, Rotllan P. Ecto-enzymatic hydrolysis of diadenosine polyphosphates by cultured adenomedullary vascular endothelial cells. Amer J Physiol Cell Physiol 1997; 42:C918-C927.

Meghji P, Pearson JD, Slakey LL. Regulation of extracellular adenosine production by ecto-nucleotidases of adult rat ventricular myocytes. Am J Physiol 1992; 263:H40-H47.

Meghji P, Pearson JD, Slakey LL. Kinetics of extracellular ATP hydrolysis by microvascular endothelial cells from rat heart. Biochem J 1995; 308:725-731.

Mlodzik K, Loffing J, Lehir M, Kaissling B. Ecto-5'-nucleotidase is expressed by pericytes and fibroblasts in the rat heart. Histochemistry Cell Biol 1995; 103:227-236.

Naito Y, Lowenstein JM. 5'-nucleotidase from rat heart. Biochemistry 1981; 20:5188-5194.

Narita M, Goji J, Nakamura H, Sano K. Molecular cloning, expression, and localization of a brain-specific phosphodiesterase I/nucleotide pyrophosphatase (PD-I alpha) from rat brain. J Biol Chem 1994; 269:28235-28242.

Node K, Kitakaze M, Minamino T, Tada M, Inoue M, Hori M, Kamada T. Activation of ecto-5'-nucleotidase by protein kinase C and its role in ischaemic tolerance in the canine heart. Br J Pharmacol 1997; 120:273-281.

Novo FJ. Tutor JC. Changes in alkaline phosphatase and 5'-nucleotidase multiple forms after surgical managment of bilary obstruction. Clin Biochem 1992; 38:1340-1342.

Ogilvie A. "Extracellular functions for Ap_nA." In *Ap_4A and other dinucleoside polyphosphates*, McLennan AG, ed.Boca Raton, Ann Arbor, London, Tokyo: CRC Press, 1992. p. 229-273.

Ogilvie A, Lüthje J, Pohl U, Busse R. Identification and partial charactertization of an adenosine(5')tetraphospho(5')adenosine hydrolase on intact bovine aortic endothelial cells. Biochem J 1989; 259:97-103.

Olmo N, Turnay J, Risse G, Deutzmann R, von der Mark K, Lizarbe A. Modulation of 5'-nucleotidase activity in plasma membranes and intact cells by the extracellular matrix proteins laminin and fibronectin. Biochem J 1992; 282:181-188.

Pearson JD, Carleton JS, Gordon JL. Metabolism of adenine nucleotides by ecto-enzymes of vasclular endothelial and smooth-muscle cells in culture. Biochem J 1980; 190:421-429.

Pearson JD. "Ecto-nucleotidases. Measurememt of activities and use of inhibitors." In *Methods in Pharmacology 6*, Paton DM, ed.New York: Plenum Press, 1985. p. 83-107.

Picher M, Sevigny J, D'Orléans-Juste P, Beaudoin AR. Hydrolysis of P_2-purinoceptor agonists by a purified ecto-nucleotidase from the bovine aorta, the ATP- diphosphohydrolase. Biochem Pharmacol 1996; 51:1453-1460.

Plesner L. Ecto-ATPases: identities and functions. Int Rev Cytol 1995;158:141-214.

Przyklenk K, Zhao L, Kloner RA, Elliott GT. Cardioprotection with ischemic preconditioning and MLA: Role of adenosine-regulating enzymes? Amer J Physiol Heart Circ Phy 1996; 40:H1004-H1014.

Resta R, Thompson LF. T cell signalling through CD73. Cell Signal 1997; 9:131-139.

Resta R, Hooker SW, Laurent AB, Rahman SMJ, Franklin M, Knudsen TB, Nadon NL, Thompson LF. Insights into thymic purine metabolism and adenosine deaminase deficiency revealed by transgenic mice overexpressing ecto-5'-nucleotidase (CD73). J Clin Invest 1997; 99:676-683.

Roberts RL, Sandra A. Apical-basal membrane polarity of membrane phosphatases in isolated capillary endothelium - Alteration in ultrastructural localisation under culture conditions. J Anat 1993; 182:339-347.

Robson SC, Kaczmarek E, Siegel JB, Candinas D, Koziak K, Millan M, Hancock WW, Bach FH. Loss of ATP diphosphohydrolase activity with endothelial cell activation. J Exp Med 1997; :153-163.

Rudert M, Tillmann B. Detection of lymph and blood vessels in the human intervertebral disc by histochemical and immunohistochemical methods. Ann Anatomy 1993; 175:237-242.

Savic V, Stefanovic V, Ardaillou N, Ardaillou R. Induction of ecto-5'-nucleotidase of rat cultured mesangial cells by interleukin-1-beta and tumor necrosis factor-alpha. Immunology, 1990;70:321-326.

Schlüter H, Offers O, Brüggemann G, van der Glet M, Tepel M, Nordhoff E, Karas M, Spieker C, Witzel H, Zidek W. Diadenosine phosphates and the physiological control of blood pressure. Nature 1994; 367:186-188.

Schlüter H, Tepel M, Zidek W. Vascular actions of diadenosine phosphates. J Auton Pharmacol 1996; 16:357-362.

Schmid TC, Loffing J, Lehir M, Kaissling B. Distribution of ecto-5'-nucleotidase in the rat liver: Effect of anaemia. Histochemistry 1994; 101:439-447.

Scott LJ, Delautier D, Meerson NR, Trugnan G, Goding JW, Maurice M. Biochemical and molecular identification of disctinct forms of alkaline phosphodiesterase I expressed on the apical and basolateral plasma membrane surfaces of rat hepatocytes. Hepatol 1997; 25:995-1002.

Sévigny J, Levesque FP, Grondin G, Beaudoin AR. Purification of the blood vessel ATP diphosphohydrolase, identification and localisation by immunological techniques. BBA Gen Subjects 1997; 1334:73-88.

Stochaj U, Mannherz HG. Chicken gizzard 5'-nucleotidase functions as a binding protein for the laminin/nidogen complex. Eur J Cell Biol 1992; 59:364-372.

Takemoto H, Kaneda K, Hosokawa M, Ide M, Fukushima H. Conditioned media of glial cell lines induce alkaline phosphatase activity in cultured artery endothelial cells -Identification of interleukin-6 as an induction factor. FEBS Lett 1994; 350:99-103.

Thompson LF, Ruedi JM, Glass A, Low MG, Lucas AH. Antibodies to 5'-nucleotidase (CD73), a glycosyl-phosphatidylinositol-anchored protein, cause human peripheral blood T cells to proliferate. J Immunol 1989; 143:1815-1821.

Thompson LF, Ruedi JM, Glass A, Moldenhauer G, Moller P, Low MG, Klemens MR, Massaia M, Lucas AH. Production and characterization of monoclonal antibodies to the glycosyl phosphatidylinositol-anchored lymphocyte differentiation antigen ecto-5'-nucleotidase (CD73). Tissue Antigens 1990; 35:9-19.

Wang TF, Rosenberg PA, Guidotti G. Characterization of brain ecto-apyrase: Evidence for only one ecto-apyrase (CD39) gene. Mol Brain Res 1997; 47:295-302.

Westfall TD, Kennedy C, Sneddon P. The ecto-ATPase inhibitor ARL 67156 enhances parasympathetic neurotransmission in the guinea-pig urinary bladder. Eur J Pharmacol 1997; 329:169-173.

Whyte MP. "Hypophosphatasia: Nature's window on alkaline phosphatase function in man". In Principles in Bone Biology Bilezkian J, Raisz, L, Rodan G, eds, San Diego: Academic Press 1996, p.951-968.

Whyte MP, Landt M, Ryan LM, Mulivor RA, Henthorn PS, Fedde KN, Mahuren JD, Coburn SP. Alkaline phosphatase: placental and tissue-nonspecific isoenzymes hydrolyze phosphoethanolamine, inorganic pyrophosphate, and pyrldoxal 5'-phosphate - Substrate accumulation in carriers of hypophosphatasia corrects during pregnancy. J Clin Invest 1995; 95:1440-1445.

Wiginton DA, Kaplan DJ, Staes JC, Akeson AL, Perne CM,Bily IJ, Vaighn AJ, Lattier DL, Hutton JJ. Complete sequence and structure of the gene for human adenosine deaminase. Biochemistry 1986; 25: 8234-8244.

Wilson DK, Rudplhp FB, Quicho FA. Atomic structure of adenosine demanisase complexed with a transition-stae analog: Understanding catalysis and immunodeficieny mutations. Science 1991; 252: 1278-1284.

Zhuo S, Clemens JC, Stone RL, Dixon JE. Mutational analysis of a Ser/Thr phosphatase - Identification of residues important in phosphoesterase substrate binding and catalysis. J Biol Chem 1994; 269:26234-26238.

Zimmermann H. 5'-Nucleotidase - Molecular structure and functional aspects. Biochem J 1992; 285:345-365.

Zimmermann H. Extracellular purine metabolism. Drug Develop Res 1996a; 39:337-352.

Zimmermann H. Biochemistry, localization and functional roles of ecto-nucleotidases in the nervous system. Prog Neurobiol 1996b; 49:589-618.

Zimmermann H, Vogel M, Laube U. Hippocampal localization of 5'-nucleotidase as revealed by immunocytochemistry. Neuroscience 1993; 55:105-112.

INDEX

Heart failure	111, 112, 115, 116
Heat shock protein	43, 103
HENECA (alkynyl-uronamide derivative)	9, 11
Hibernation	317
His-Purkinje	132, 133, 136
Histamine	40
Hydrogen peroxide	279
Hydroxydecanoic acid (5-HD)	68-70, 77
Hypertension	57, 243, 250
Hypotension	40
Hypoxia	44, 49-51, 55, 57, 144, 271, 308, 314, 317, 321
I-AB-MECA	11, 43
I-ABA	10, 11
IB-MECA	11, 13, 40, 74, 90, 91, 278
I_{CA} (Inward Calcium Current)	126-128, 131, 135, 232
ICAM-1 (Intracellular Adhesion Moledue I)	277
Icat (Cationic Current)	245
I_{cl} (Chloride Current)	134
Idiopathic dilated cardiomyopathy (IDC)	113, 114
I_F (Inward Potassium Current)	126, 127, 131, 134
IgG	273
$I_{K,ACh,Ado}$ (Potassium Current)	126-129, 131, 133
IL-1	54, 274, 276
IL-2	307, 308

PAPA-APEC	9
Papaverine	44
Parasympathetic efferent neuron	332, 336
P_{Ca}/P_{Na} ratios	159, 162
PDE inhibitor	274, 278, 280, 281
PDNP1, PDNP2, PDNP3	348
Pericyte	50
Peroxynitrite	282
Pertussis toxin	4, 13, 44
Phenylisopropyladenosine (PIA)	9, 23, 26, 32, 40, 274, 277
Phenyltheophylline	44, 51
Phorbol ester	307
Phosphocreatine (PCr)	101
Phosphodiesterase/nucleotide pyrophosphatase (PDNP)	343
Phospholipase A_2	52
Phospholipase C	1, 52-55, 86, 94, 187, 188, 195, 214, 216, 258, 292, 294, 350
Phospholipase D	86, 90, 94, 216
PIA	see phenylisopropyladenosine
Platelets	272, 282, 291
Platelet aggregation	218, 250
PLCγ	232
Polymorphonuclear leukocyte (PMN)	see neutrophil
Preconditioning	2, 64, 69-71, 76, 77, 86-88, 90, 91, 94, 95, 350

Vascular endothelial growth factor (VEGF)	44, 49, 55-57
Vascular smooth muscle	243, 244-246, 249, 271
Vasoconstriction	243, 245, 247
VCAM-1	277
WRC-0474	281
WRC-0571	12, 13
Xenopus P_{2Y} receptor	197
YAC-1 lymphoma cell	298
ZM241385	9, 12, 13, 41, 42
Zymosan	249, 279